Lecture Notes in Artificial Intelligence 5783

Edited by R. Goebel, J. Siekmann, and W. Wahlster

Subseries of Lecture Notes in Computer Science

T0216741

Francesca Rossi   Alexis Tsoukias (Eds.)

Algorithmic Decision Theory

First International Conference, ADT 2009
Venice, Italy, October 20-23, 2009
Proceedings

 Springer

Series Editors

Randy Goebel, University of Alberta, Edmonton, Canada
Jörg Siekmann, University of Saarland, Saarbrücken, Germany
Wolfgang Wahlster, DFKI and University of Saarland, Saarbrücken, Germany

Volume Editors

Francesca Rossi
University of Padova, Department of Pure and Applied Mathematics
Via Trieste 63, 35121 Padova, Italy
E-mail: frossi@math.unipd.it

Alexis Tsoukias
University of Paris Dauphine, LAMSADE-CNRS
Place du Maréchal De Lattre de Tassigny, 75775 Paris Cedex 16, France
E-mail: tsoukias@lamsade.dauphine.fr

Library of Congress Control Number: 2009934788

CR Subject Classification (1998): F.4.1, I.2.3, I.2.4, F.4.2, H.4.2

LNCS Sublibrary: SL 7 – Artificial Intelligence

ISSN 0302-9743

ISBN 978-3-642-04427-4 Springer Berlin Heidelberg New York

springer.com

© Springer-Verlag Berlin Heidelberg 2009

Typesetting: Camera-ready by author, data conversion by Scientific Publishing Services, Chennai, India
Printed on acid-free paper   SPIN: 12760657   06/3180   5 4 3 2 1 0

Preface

Algorithmic decision theory is a new interdisciplinary research area that aims to bring together researchers from different fields such as decision theory, discrete mathematics, theoretical computer science and artificial intelligence, in order to improve decision support in the presence of massive databases, combinatorial structures, partial and/or uncertain information and distributed, possibly interoperating, decision makers. Such problems arise in several real-world decision-making scenarios such as humanitarian logistics, epidemiology, risk assessment and management, e-government, electronic commerce, and recommender systems.

In 2007, the EU-funded COST Action IC0602 on Algorithmic Decision Theory was started, networking a number of researchers and research laboratories around Europe (and beyond). The COST Action IC0602 now gathers over 100 participants from more than 30 countries (including non-COST countries such as Australia, South Africa and the USA). For more details see www.algodec.org. Within the Action, and in cooperation with the EURO Working Group on Preferences, it was decided to start a new series of conferences on algorithmic decision theory, the goal being to provide a forum for researchers interested in this area.

This volume contains the papers presented at ADT 2009, the first International Conference on Algorithmic Decision Theory. The conference was held in San Servolo, a small island in the Venetian Lagoon, on October 20–23, 2009. The program of the conference included oral presentations, posters, invited talks, and tutorials (for more information see www.adt2009.org).

The conference received 65 submissions. Each submission was reviewed by at least 2 program committee members, and the program committee decided to accept 39 papers (of which 9 posters). The topics of these papers range from computational social choice to preference modeling, from uncertainty to preference learning, from multi-criteria decision making to game theory. We believe that many colleagues will find this collection of papers exciting and useful for the advancement of the state of the art in their respective disciplines.

We would like to take this opportunity to thank all authors who submitted papers to this conference, as well as all the program committee members and external reviewers for their useful work. ADT 2009 was made possible thanks to the support of the COST Action IC0602 on Algorithmic Decision Theory, the Association for Constraint Programming, the EURO (Association of European Operational Research Societies), the LAMSADE at the University of Paris Dauphine, the Department of Pure and Applied Mathematics of the University of Padova, DIMACS, Rutgers University, USA and the CNRS (Centre National de la Recherche Scientifique, France).

We would also like to express our appreciation of Easychair for its support in the creation of this volume.

October 2009   Francesca Rossi
                                                                                  Alexis Tsoukiàs

Conference Organization

Program Chair

Alexis Tsoukiàs

Program Committee

Ronen Brafman
Vince Conitzer
Jim Delgrande
Jon Doyle
Ulle Endriss
Jose Figueira
Judy Goldsmith
Salvatore Greco
Ulrich Junker
Werner Kiessling
Jerome Lang
Nicolas Maudet
Barry O'Sullivan
Sasa Pekec
Patrice Perny
Eleni Pratisini
Pearl Pu
David Rios Insua
Fred Roberts
Francesca Rossi
Ahti Salo
Roman Słowiński
Mike Trick
Esko Turunen
Toby Walsh

Local Organization

Francesca Rossi (Chair)
Mirco Gelain (Publicity and Web site)
Kristen Brent Venable

External Reviewers

Stephane Airiau
Gianluca Antonini
Denis Bouyssou
Sylvain Castagnos
Yann Chevaleyre
Eric Cope
Boi Faltings
Umberto Grandi
David Krieg
Michel Minoux
Maria Silvia Pini
Daniele Porello
Gautier Stauffer
Paul Weng

Sponsors

Table of Contents

Social Choice Theory

Multiple Criteria Decision Analysis

Decision under Uncertainty

Optimisation

Learning

A Complete Conclusion-Based Procedure for Judgment Aggregation*

Gabriella Pigozzi, Marija Slavkovik, and Leendert van der Torre

Individual and Collective Reasoning, Computer Science and Communication,
University of Luxembourg, 6, Rue Richard Coudenhove Kalergi,
L-1359 – Luxembourg

Abstract. Judgment aggregation is a formal theory reasoning about how a group of agents can aggregate individual judgments on connected propositions into a collective judgment on the same propositions. Three procedures for successfully aggregating judgments sets are: premise-based procedure, conclusion-based procedure and distance-based merging. The conclusion-based procedure has been little investigated because it provides a way to aggregate the conclusions, but not the premises, thus it outputs an incomplete judgment set. The goal of this paper is to present a conclusion-based procedure outputting complete judgment sets.

1 Introduction

Judgment aggregation [9,10,12] studies the aggregation of individual judgments of small groups such as expert panels, legal courts, boards and councils. We talk about judgment aggregation whenever a group of individuals needs to make a collective decision on a finite set of issues, and these propositions are logically connected. The propositions are of two kinds: *premises* and a *conclusion*. The first serve as supporting reasons to derive a certain judgment on the conclusion. If, for example [1], your department has to hire a new lecturer and the decision rule is such that a candidate X will be hired only if the candidate is good at teaching and good at research, we will say that "hiring X" is the conclusion while "good at teaching" and "good at research" are the premises.

How shall we derive a group decision given the individuals' opinions on premises and conclusion? It is assumed that each individual expresses yes/no opinions on the propositions while respecting the logical relations. If we now define the group opinion as the majority view on the issues, it turns out that the collectivity may have to endorse an inconsistent position. This means that your department may have to face a situation in which a majority does not deem X a good candidate. However, it will not be possible to provide reasons for this as a

---

* A previous version of this paper appeared in the proceedings of the workshop Cinquième Journée Francophone Modèles Formels de l'Interaction, Lannion, France, 2-5 June 2009.

F. Rossi and A. Tsoukis (Eds.): ADT 2009, LNAI 5783, pp. 1–13, 2009.

Table 1. Hiring committee example. The candidate X is hired if and only if X is good at teaching and X is good at research.

	$a = X$ is good at teaching	$b = X$ is good at research	$x =$ hire X
prof. A	*yes*	*no*	*no*
prof. B	*yes*	*yes*	*yes*
prof. C	*no*	*yes*	*no*
Majority	*yes*	*yes*	*no*

majority of people agrees that X is actually good at teaching and (another) majority deems X to be good at research. An example of such situation is presented in Table 1.

The problem is avoided if we decide to let the majority vote on the premises to dictate the final decision on the hiring process, or if the agents express their judgments only on the conclusion. Unlike the aggregation procedure on the premises [14,5], the aggregation on the conclusion has not been throughly investigated.

We claim that in many decision problems the conclusion is more relevant than the reasons for it. When deciding which candidate to hire in your department, you may be more concerned of which new colleague you will have in your department than of the reasons for choosing her. Considering only the individual judgments on the conclusions has also the advantage that it is a strategy-proof procedure. The same does not hold when you aggregate on the premises.

The problem this paper addresses is how a group can make decisions on the conclusion while providing reasons in support of the collective conclusion. Our procedure prioritizes the individual judgments on the conclusion and outputs sets of premises that support the collective decision.

The paper is structured as follows: in Section 2 we present the problem of judgment aggregation. Section 3 is devoted to our formal framework, and in Section 4 we prove some results about our procedure. Section 5 relates our approach with existing work and, finally, Section 6 concludes the paper and outlines directions for future work.

2   Judgment Aggregation

In judgment aggregation agents are required to express judgments (in the form of yes/no or, equivalently, 1/0) over *premises* and *conclusion*. As in [20], to represent the distinction between premise and conclusion in our language, we distinguish between premise variables $a, b, c, p, q \ldots$, and conclusion variable x.

In the hiring example, "X is good at teaching" is *premise* a, and "X is good at research" is *premise* b. The decision rule can be formally expressed by the rule $(a \wedge b) \leftrightarrow x$, where x is the conclusion about hiring X. Each member of the department expresses her judgment on the propositions a, b and x such that the rule $(a \wedge b) \leftrightarrow x$ is satisfied.

Suppose the three professors in the department make their judgments according to Table 1. Each member expresses a consistent opinion, i.e. she says yes to x if and only if she says yes to both a and b. However, propositionwise majority voting (consisting in the separate aggregation of the votes for each proposition a, b and x via majority rule) results in a majority for a and b and yet a majority for $\neg x$. This is an inconsistent collective result, in the sense that $\{a, b, \neg x, (a \wedge b) \leftrightarrow x\}$ is inconsistent in propositional logic. The paradox lies in the fact that majority voting can lead a group of rational agents to endorse an irrational collective judgment. The literature on judgment aggregation refers to such problems as the *doctrinal paradox* (or *discursive dilemma*).

The relevance of such aggregation problems applies to all situations in which individual binary evaluations need to be combined into a group decision. Furthermore, the problem of aggregating individual judgments is not restricted to majority voting, but it applies to all aggregation procedures satisfying some seemingly desirable conditions. For an overview, the reader is referred to [13].

Two ways to avoid the inconsistency are the *premise-based procedure* (PBP) and the *conclusion-based procedure* (CBP) [17,3]. According to the PBP, each agent votes on each premise. The conclusion is then inferred from the rule $(a \wedge b) \leftrightarrow x$ and from the judgment of the majority of the group on a and b. If the professors of the example followed the premise-based procedure, the lecturer would be hired.

Because in PBP the collective judgment on the conclusion is derived from the individual judgments on the premises, it can happen that PBP violates a unanimous vote on the conclusion. In [16] Nehring presents a variation on the discursive dilemma, which he calls the Paretian dilemma. In his example, a three-judges court has

> to decide whether a defendant has to pay damages to the plaintiff. Legal doctrine requires that damages are due if and only if the following three premises are established: 1) the defendant had a duty to take care, 2) the defendant behaved negligently, 3) his negligence caused damage to the plaintiff. ([16], p.1)

Suppose that the judges vote as in Table 2.

The Paretian dilemma is disturbing because, if the judges would follow PBP, they would condemn the defendant to pay damages contradicting the *unanimous* belief of the court that the defendant is *not* liable.

Table 2. Paretian dilemma. Premises: $a =$ duty, $b =$ negligence., $c =$ causation. Conclusion: $x = (a \wedge b) =$ damages.

Agenda	a b c	$x = (a \wedge b \wedge c)$
Judge A	1 1 0	0
Judge B	0 1 1	0
Judge C	1 0 1	0
Majority	1 1 1	0

A CBP would not lead to such a unanimity violation. According to CBP, the judges decide privately on a and b and only express their opinions on x publicly. The judgement of the group is then inferred from applying the majority rule to the agents' judgments on x. The defendant will be declared liable if and only if a majority of the judges actually believes that she is liable. In the example, contrary to PBP, the application of CBP would free the defendant. However, no reasons for the court decision could be supplied.

Unlike PBP [14,5], CBP did not receive much attention in the literature. Here we aim at filling this gap. We propose a procedure that attempts to overcome the major limit of CBP, that is the lack of reasons supporting the decision.

3   Framework

In this section we introduce our formal framework to represent judgment aggregation problems. A set of agents $N = \{1, 2, \ldots, n\}$ makes judgments on logically interconnected propositions. The set $\mathcal{P}$ of atomic propositions is defined as the union of two disjoint sets: $\mathcal{P}_p$ containing variables $a, b, c, \ldots, p, q$ for the premises, and $\mathcal{P}_c$ being a singleton $\{x\}$, where x is the variable for the conclusion. We assume that the conclusion is an atomic formula. $\mathcal{L}$ is a language built from $\mathcal{P}$, including complex formulas as $\neg a, (a \wedge b), (a \vee b), (p \rightarrow q), (a \leftrightarrow p)$.

The set of issues on which the judgments have to be made is called *agenda* and is denoted by $\Phi \subseteq \mathcal{L}$. The agenda is assumed to be finite and closed under negation: if $a \in \Phi$, then $\neg a \in \Phi$.[1] Each double negated proposition $\neg\neg a$ is identified with its corresponding non negated proposition a. We split the agenda in two parts: one containing the premises (Φ_p), and one containing the conclusion (Φ_c). We exclude agenda items such as $a \rightarrow x$, i.e. formulas containing premises and conclusion. Our procedure consists of two different aggregations: one on the individual judgments on Φ_p and one on the individual judgments on Φ_c.

A subset $J \subseteq \Phi$ is the *collective judgment set* and contains the set of propositions believed by the group. Similarly, we define individual i's judgment set $J_i \subseteq \Phi$. A collective judgment set is *consistent* if it is a consistent set in $\mathcal{L}$, and is *complete* if, for any $a \in \mathcal{L}$, $a \in J$ or $\neg a \in J$ (consistent and complete individual judgment sets are defined in the same way). We only consider consistent complete judgment sets.

A *decision rule* $\mathcal{R}$ is a formula of $\mathcal{L}$ that represents the logical connections between premises and conclusion. More precisely, $\mathcal{R}$ has the form $\Psi \leftrightarrow x$, where $\Psi \in \mathcal{L}/\{x\}$. The decision rule is not an item of the agenda. This means that the group members do not vote on $\mathcal{R}$, but each individual is required to give judgments that satisfy the given rule.

Like the agenda, each judgment set is split in two disjoint subsets: $J_{i,p}$ and $J_{i,c}$. The first is the individual i's judgment set on the premises, and $J_{i,c}$ is the

---

[1] To increase readability, in the tables we list only the positive issues, and assume that, for any issue in the agenda, an individual deems that issue to be true if and only if she deems its negation to be false.

individual i's judgment set on the conclusion. The collective judgment sets on premises and conclusion will be denoted respectively by J_p and J_c.

We say that a premise a (resp. a conclusion x) is *unanimously supported* if $a \in J_{i,p}$ for all $J_{i,p} \subseteq \Phi$ (resp. $x \in J_{i,c}$ for all $J_{i,c} \subseteq \Phi$).

A *profile* $\underline{J}$ is an n-tuple $(J_1, J_2, \ldots, J_n)$ of agents' judgment sets. An *aggregation rule* F assigns a set of collective judgment sets J to each profile $\underline{J}$. For our procedure we need to define two aggregation rules: one for the aggregation of the individual premises and one for the conclusion. To relate the two aggregation rules, we have a set of integrity constraints IC. IC indicates the set of admissible interpretations, i.e. the admissible collective judgment sets. Also, to allow for situations in which the aggregated judgment set is not unique, i.e. there are ties, we aggregate the profiles into *sets* of aggregated judgment sets.

A *premise profile* $\underline{J_p}$ is an n-tuple $(J_{1,p}, J_{2,p}, \ldots, J_{n,p})$ of agents' judgment sets on premises. A *premise aggregation rule* F_{IC} assigns a set of collective judgment sets J_p to each premise profile $(J_{1,p}, J_{2,p}, \ldots, J_{n,p})$ and set of integrity constraints IC. Conclusion profiles $(J_{1,c}, J_{2,c}, \ldots, J_{n,c})$ and conclusion aggregation rules F_c are defined similarly.

3.1   Complete Conclusion-Based Procedure

Each individual provides, simultaneously, the set of premises and conclusion that she believes. Our two-step procedure first performs a standard CBP, i.e. it aggregates the individual judgments on the conclusion by majority rule. This means that x (resp. $\neg x$) is the collective conclusion iff there are at least $\lfloor \frac{n}{2} \rfloor + 1$ agents voting for x (resp. $\neg x$). The second step consists in determining the set of reasons which support the collective conclusion. This is done by applying a *distance-based merging operator* to $J_{i,p}$.

Distance minimization merging procedures have been already applied to judgment aggregation problems [18]. In this section we briefly present a majority merging operator with integrity constraints following [8,7]. Unlike in [8,7], when the merging operator outputs ties, we take the disjunction of the formulas which completely characterize the tied alternatives.

An *interpretation* is a function $v : \mathcal{P} \to \{0, 1\}$ and it is represented as the list of the binary evaluations. For example, given three propositional variables a, b and c, the vector $(0,1,0)$ stands for the interpretation in which a and c are false and b is true. Let $\mathcal{W} = \{0, 1\}^{\mathcal{P}}$ be the set of all interpretations. An interpretation is a *model* of a propositional formula if and only if it makes the formula true in the usual truth functional way.

Let us suppose that $\Phi_p = \{a, \neg a, b, \neg b, c, \neg c\}$, and that agent 1 believes that $a, \neg b$ and $\neg c$, i.e. $J_{1,p} = \{a, \neg b, \neg c\}$. We represent $J_{1,p}$ as a 0-1 vector of length equal to the number of propositions in $J_{1,p}$, i.e. $(1, 0, 0)$. Suppose also that $\mathcal{R} = ((a \vee b) \wedge c) \leftrightarrow x$ and that, unlike agent 1, the majority of the individuals voted in favor of x. Hence, the first step of our procedure sets $v(x) = 1$. We now want to define an aggregation on $J_{i,p}$ such that the collective judgment set on the premises is one of the models of $((a \vee b) \wedge c) \leftrightarrow x$ where $v(x) = 1$. This means that J_p must be one of the following interpretations: $(1,1,1)$, $(0,1,1)$, $(1,0,1)$. The set of premises supporting the collective conclusion will constrain the aggregation procedure on $J_{i,p}$.

Given a premise profile J_p and IC, $F_{IC}(J_p)$ denotes a set of collective judgment sets on the premises resulting from the IC merging on J_p. The idea of a distance minimization merging operator is that $F_{IC}(J_p)$ will select those interpretations in IC, which are at minimal distance from J_p. A distance $d(\omega, J_p)$ between an interpretation ω and the premise profile J_p induces a total pre-order ($\leq$) on the interpretations.

In order to obtain the total pre-order on the interpretations, we first need to determine a pseudo-distance between each admissible interpretation and each $J_{i,p}$. Then, we need to aggregate all these values in order to obtain a pseudo-distance value between an interpretation and J_p. Let us see this in detail (we follow [8,7]).

A pseudo-distance between interpretations is a function $d : \mathcal{W} \times \mathcal{W} \to \mathbb{R}_+$ such that for all $\omega, \omega' \in \mathcal{W}$: $d(\omega, \omega') = d(\omega', \omega)$ and $d(\omega, \omega') = 0$ iff $\omega = \omega'$.

A pseudo-distance between an interpretation ω and J_p is defined with the help of an aggregation function $D: \mathbb{R}_+^n \to \mathbb{R}_+$ as $D^d(\omega, J_p) = D(d(\omega, J_{1,p}), \ldots, d(\omega, J_{n,p}))$ [7]. Any such aggregation function induces a total pre-order $\leqslant_{J_p}$ on the set $\mathcal{W}$ with respect to the pseudo-distances to a given J_p. Thus, an *IC majority merging operator* for a profile J_p can be defined as $\Delta_{IC}(J_p) = \min([IC], \leqslant_{J_p})$, i.e., the set of all models of IC (denoted by $[IC]$) with minimal pseudo-distance D^d to J_p. The minimal pseudo-distance identifies the final collective outcome on the premises, i.e. the set of premises that support the conclusion voted by the majority of the agents and with the minimal distance among all possible models satisfying IC.

A majority merging operator, often mentioned in the literature, is the operator $\Delta_{IC}^{d,\Sigma}$ defined as follows:

1. d is the Hamming distance — the number of propositional letters on which two interpretations differ, i.e., $d(\omega, \omega') = |\{\pi \in \mathcal{P} | \omega(\pi) \neq \omega'(\pi)\}|$ and
2. $D^d(\omega, J_p) = \sum_i d(\omega, J_{i,p})$ is the sum of componentwise distances d defined before.

For example, the Hamming distance between $\omega = (1,0,0)$ and $\omega' = (0,1,0)$ is $d(\omega, \omega') = 2$. In the following we use the Hamming distance because it is a well known and intuitive distance. But the Hamming distance is only one among many possible distance functions that we may use.

The premise aggregation rule F_{IC} outputs the disjunction of formulas which completely characterize the sets of judgments selected by $\Delta_{IC}^{d,\Sigma}$ as the reasons in support of the conclusion voted by the majority of the agents. Given a premise profile J_p, F_{IC} is defined as:

$$F_{IC}(J_p) = \bigvee \Delta_{IC}^{d,\Sigma}(J_p)$$

The constraint IC is defined as $IC = \mathcal{R} \wedge \hat{x}$, where $\hat{x}$ is the conclusion chosen by the majority.

The best way to illustrate our procedure is with an example.

Example 1. Consider a *collegium medicum* that wishes to eliminate the possibility of a patient suffering from condition X before administering a treatment. We take $v(x) = 0$ if the patient is free of X. The doctors consider the three relevant alternative medical conditions a, b and c the patient may suffer from. The patient is free of X if medical conditions a, b and c are present ($v(a) = 1$, $v(b) = 1$ and $v(c) = 1$), if all three medical conditions are absent ($v(a) = 0$, $v(b) = 0$ and $v(c) = 0$) or if the last condition is present while the previous two are absent ($v(a) = 0$, $v(b) = 0$ and $v(c) = 1$). In all other cases the patient is likely to suffer from X. Table 3 gives the truth table of $\mathcal{R}$.

Table 3. The truth table of $\mathcal{R}$ for the doctor example

a	0 0 0 0 1 1 1 1
b	0 0 1 1 0 0 1 1
c	0 1 0 1 0 1 0 1
x	0 0 1 1 1 1 1 0

Three equally qualified members of the *collegium medicum* give their opinions shown in Table 4. As Table 4 shows, the group is facing a dilemma. The majority of the conclusions from the doctors opinions indicates that the patient does not suffer from X though the majority on the premises supports the opposite conclusion.

Our procedure (see Table 5) selects the reasons that are most compatible with the doctors' different opinions, i.e. the judgment set (1,1,1).

Table 4. The dilemma faced by the doctors

Agenda	a	b	c	x
Dr. A	1	1	1	0
Dr. B	0	0	0	0
Dr. C	1	1	0	1
Majority	1	1	0	0

Table 5. Selection of the premise set from the doctors opinions under the constraint $v(x) = 0$

	$J_{1,p}$	$J_{2,p}$	$J_{3,p}$	$\Sigma_i \, d(\omega, J_{i,p})$
(1,1,1)	0	3	1	4
(0,0,0)	3	0	2	5
(0,0,1)	2	1	3	6

By applying the same procedure, the premises selected for the discursive dilemma in Table 2 with $v(x) = 0$ would be $(0, 1) \vee (1, 0)$, representing a tie between X is good at teaching but bad at research and X is good at research but bad at teaching.

Example 1 illustrates that, when aggregating the premises, we do not only take into account the judgment sets of the agents that support the aggregated conclusion, but also the judgment sets of agents that do not support the conclusion. Consider for example the selection of premises with $v(x) = 1$ in Table 4. We take also the judgment set of Dr. C into account, although she voted for $v(x) = 1$.

The justification for taking all individual judgments on the premises into account is two-folded. On the one hand, from the perspective of probability theory, if all judgments are independent, then more judgment sets mean a higher chance to get a better judgment. On the other hand, from the perspective of democracy, involving agents whose conclusion is not supported will give broader basis for the decision. However, we do not exclude the possibility that there are situations in which only the individuals' judgments that actually supported the aggregated conclusion should be taken into account when determining the reasons for that conclusion.

4   Results

We now show some properties which hold for the premise aggregation rule F_{IC} we had defined in the previous section. We start by noticing that, in the case of the aggregation of binary evaluations, there is an obvious correspondence between proposition-wise majority voting and distance minimization. This has been already observed in several contexts (see, e.g., [2]), and can be generalized to the following folk theorem.

Proposition 1. *Let $\underline{J} = (J_1, \ldots, J_n)$ be a profile over the agenda Φ. Let $J^{maj} \subset \Phi$ be a complete and consistent set. Let it hold that for every premise $a \in J^{maj}$, $a \in J_{i,p}$ for at least $\lfloor \frac{n}{2} \rfloor + 1$ premise sets in the profile J_p. Also, for the conclusion $x \in J^{maj}$, let it holds that $x \in J_{i,c}$ for at least $\lfloor \frac{n}{2} \rfloor + 1$ conclusion sets in the profile J_c. The sum of Hamming distances from J^{maj} to the judgment sets in $\underline{J}$ is minimal.*

This means that, in the absence of a Paretian dilemma (i.e. when J^{maj} satisfies the decision rule $\mathcal{R}$), proposition-wise majority voting, distance-based merging and our procedure coincide.

4.1   Unanimity Preservation

One of the desirable properties for a judgment aggregation procedure is the heeding of unanimity. If all the agents unanimously support an agenda item, then it is natural to expect the unanimously supported item will be adopted as the collective judgment. However, PBP does not necessarily preserve unanimity on the conclusion (as it was the case with the Paretian dilemma shown in Table 2).

PBP aggregates each premise independently from the other premises, but the aggregation on the conclusion depends on the collective judgments on the premises. Therefore the unanimity on the premises will be preserved, but the unanimity on the conclusion may be violated.

When aggregating according to the CBP, unanimity on the conclusion will always be maintained, but unanimity on the premises may be violated. However, our procedure offers the option to preserve unanimity on the premises as well, by constraining the models which do not support unanimity.

Table 6. A case in which unanimity on premises will be violated by the complete CBP

	p_1	p_2	p_3	p_4	p_5	p_6	p_7	p_8	p_9	p_{10}	p_{11}	p_{12}	p_{13}	x
A	1	0	0	1	0	0	1	0	0	1	0	0	1	1
B	0	1	0	0	1	0	0	1	0	0	1	0	1	1
C	0	0	1	0	0	1	0	0	1	0	0	1	1	1
Maj.	0	0	0	0	0	0	0	0	0	0	0	0	1	1

We begin by giving a formal definition on when a premise aggregation rule F_{IC} preserves unanimity. Whether or not the unanimity on the premises is preserved by our F_{IC} depends on the rule $\mathcal{R}$ as well as the agenda Φ. We show two decision rules for which the unanimity is preserved and then we use an example to show that in the case of an arbitrary rule and agenda, the unanimity of the premises is not guaranteed.

Definition 1. *Let $J_p = (J_{1,p}, \ldots, J_{n,p})$ be a premise profile on the agenda Φ and p a premise from the agenda. A premise aggregation rule F_{IC} preserves unanimity on the premises if and only if the following holds:*
If $p \in J_{i,p}$ for all $i = \{1, \ldots, n\}$ then $p \in F_{IC}(J_p)$.

Note that, since F_{IC} can select more than one premise judgment set, p needs to be in all of them for unanimity to be preserved.

The following theorem indicates two decision rules $\mathcal{R}$, and an agenda, in the presence of which unanimity is preserved on the premises by F_{IC}.

Theorem 1. *Let Φ be an agenda in which all the elements are atoms or negations of atoms. Let $\mathcal{R}$ be a decision rule of the form $(a_1 \wedge \ldots \wedge a_n) \leftrightarrow x$ or of the form $(a_1 \vee \ldots \vee a_n) \leftrightarrow x$. $\{a_1, \ldots, a_n\} \subseteq \Phi$ are premises and $x \subseteq \Phi$ is a conclusion. F_{IC} preserves unanimity on the premises for any profile $\underline{J}$ over Φ and $\mathcal{R}$.*

Due to strict page limit constraints the proofs are omitted[2].

Given an arbitrary agenda, a decision rule $\mathcal{R}$ corresponding to that agenda and an arbitrary profile $\underline{J}$, the merging operator does not necessary preserve unanimity. We show this through an example.

Consider the profile presented in Table 6. The rule $\mathcal{R}$ is such that the value of x is 1 if and only if the evaluations of the premises are one of the sets in the first column of Table 7. For all other evaluations of premises, x is 0.

Our procedure preserves the unanimity on the conclusion and selects $v(x) = 1$, but gives an aggregation for the premises which violates the unanimity on premise p_{13} (Table 7).

The preservation of unanimously held premises can be imposed by IC. This is done by making $IC = \mathcal{R} \wedge \hat{x} \wedge p^*$, where p^* is any unanimously voted premise. Admissible outcomes for J_p then are those supporting the conclusion voted by the majority and containing the premise(s) unanimously chosen.

---

[2] The proofs can be found in [19].

Table 7. Selection of premises for the counterexample

	$J_{1,p}$	$J_{2,p}$	$J_{3,p}$	$\Sigma_i d()$
(1,0,0,1,0,0,1,0,0,1,0,0,1)	0	8	8	16
(0,1,0,0,1,0,0,1,0,0,1,0,1)	8	0	8	16
(0,0,1,0,0,1,0,0,1,0,0,1,1)	8	8	0	16
(0,0,0,0,0,0,0,0,0,0,0,0,0)	5	5	5	15

4.2 Manipulability

Another property which is of interest when dealing with aggregation procedures is that of manipulability. A judgment aggregation procedure is called manipulable if an agent, who would not obtain a desired outcome by submitting her sincere premise set, can obtain a desired outcome by choosing to submit a set of premises different than her honest premise set. Under the context of complete-conclusion based procedures, we will distinguish between full and preferred manipulability.

Full manipulability means that we distinguish only whether the aggregated premise set entirely corresponds to an agent's judgments on premises or not.

A procedure is fully manipulable if an agent can obtain her complete honest premise set as an output from the procedure by submitting another (insincere) premise set that supports the same conclusion. Formally, let $J_p = (J_{1,p}, \ldots, J_{i,p} \ldots, J_{n,p})$ be a premise profile. Let $F_{IC}(J_p) = \{J^{\circ}_{1,p}, \ldots, \bar{J^{\circ}_{m,p}}\}$, i.e. the merging operator selects the premise sets $J^{\circ}_{1,p}, \ldots, J^{\circ}_{m,p}$. Let $J_{i,p}$ be the "honest" premise set of an agent i.

Definition 2. *Assume that a premise set $J^*_{i,p} \neq J_{i,p}$ exists, such that $J^*_{i,p}$ supports the same conclusion as the premise set $J_{i,p}$. The operator F_{IC} is fully manipulable if $J_{i,p} \in F_{IC}(J_{1,p}, \ldots, J^*_{i,p} \ldots, J_{n,p})$ but $J_{i,p} \notin F_{IC}(J_{1,p}, \ldots, J_{i,p}, \ldots, J_{n,p})$.*

Theorem 2. *F_{IC} is not fully manipulable.*

Let us now assume that an agent has a premise p which she holds most important (has a strong preference on the evaluation of this premise). We say that a procedure is *preferred manipulable* if an agent can ensure that the preferred projection $w(p)$ is included in the output by submitting another premise set that supports the same conclusion. Since we do not represent the preferred premise explicitly in our framework, any premise can be the preferred one, and preferred manipulability therefore means that the agent is able to change her premise set in a way such that one premise which is not a member of the aggregated set becomes member of it.

Definition 3. *Assume that a premise set $J^*_{i,p} \neq J_{i,p}$ exists, such that $J^*_{i,p}$ supports the same conclusion as the premise set $J_{i,p}$ and premise p_{pref} is in both* *of the premise sets. The operator F_{IC} is preferred manipulable if p_{pref} is in at least one premise set $J_{j,p}^{\circ} \in F_{IC}(J_{1,p}, \ldots, J_{i,p}^{*}, \ldots, J_{n,p})$, but $\neg p_{pref}$ is in all of the premise sets selected by $F_{IC}(J_{1,p}, \ldots, J_{i,p}, \ldots, J_{n,p})$.*

Theorem 3. *F_{IC} is preferred manipulable.*

Full manipulability is a relatively weak condition, in the sense that it is fairly easy to satisfy. This notion of preferred manipulability seems to conflict with the intuition of the distance measure used to aggregate the premises, which does take such distinctions into account. However, preferred manipulability is a very strong condition, since it means in practice that an agent should not be able to improve any premise (since this premise may happen to be the preferred one). Other notions of manipulability could be studied, such as the improvement of a preferred premise by changing the judgment on this premise only.

5 Related Work

One of the noted shortcomings of the CBP is that it is susceptible to path-dependence [15]. Path-dependent decisions are decisions whose outcome depends on the order in which propositions are considered. For any proposition, the collective judgment on it is decided by majority rule (or by any other suitable aggregation rule) unless this conflicts with the collective judgments of previously aggregated propositions. In the latter case, the collective value of that proposition is deduced by logical implication from the previously aggregated propositions. List [11] provided necessary and sufficient conditions for path-dependence. Furthermore, in [4] it has been shown that the absence of path-dependence is equivalent to strategy-proofness.

Here we propose a complete CBP without assuming any order over the premises. We aimed at a procedure that treats all premises in an even-handed way. The absence of full manipulability is coherent with the results of [4].

Non-manipulability is one of the advantages of CBP over PBP. The question of manipulability under operators used for merging of propositions has been treated extensively in [6]. There, Everaere *et al.* explore a broad spectrum of manipulability for various merging operators over complete and incomplete sets of beliefs. Our work uses results from [6] on complete sets of beliefs under model-based merging operators that use the sum of the distances between belief bases.

6 Conclusions and Future Work

The complete CBP we present keeps the desirable properties of non-manipulability and it can be modified to preserve unanimity on the premises. What can be considered a shortcoming of the procedure is that it may select more than one premise judgment set to support the collective conclusion. Such "ties" in the output from aggregation are known to be resolved with an additional approval vote [2] or by random selection. A random selection is not a desirable tie-breaking solution in cases when the decisions on premises can influence some future decision making process. The approval voting requires more information to be injected in the framework and opens the questions of what incentives an agent may have to prefer one premise judgment set over another.

In future work we plan to investigate the relevance that current group decisions can have on future decisions. This "evolutionary" impact over the decision making process has been an important issue in the work that gave rise to the interest in judgment aggregation [9,10], but it has fallen out of scope in the more formal study of judgment aggregation.

Acknowledgment

We thank Franz Dietrich and Christian List for discussions on the issues raised in this paper.

References

1. Bovens, L., Rabinowicz, W.: Democratic answers to complex questions. an epistemic perspective. Synthese 150, 131–153 (2006)
2. Brams, S., Kilgour, D., Sanver, M.: A minimax procedure for negotiating multilateral treaties. In: Wiberg, M. (ed.) Reasoned Choices: Essays in Honor of Hannu Nurmi, pp. 265–282. Finnish Political Science Association (2004)
3. Chapman, B.: Rational aggregation. Politics, Philosophy and Economics 1(3), 337–354 (2002)
4. Dietrich, F., List, C.: Judgment aggregation by quota rules: majority voting generalized. Journal of Theoretical Politics 19(4), 391–424 (2007)
5. Dietrich, F., Mongin, P.: The premiss-based approach to judgment aggregation. Research memoranda. METEOR, Maastricht Research School of Economics of Technology and Organization, Maastricht (2008)
6. Everaere, P., Konieczny, S., Marquis, P.: The strategy-proofness landscape of merging. J. Artif. Intell. Res. (JAIR) 28, 49–105 (2007)
7. Konieczny, S., Lang, J., Marquis, P.: Da^2 merging operators. Artificial Intelligence 157, 49–79 (2004)
8. Konieczny, S., Pino-Pérez, R.: Merging information under constraints: a logical framework. Journal of Logic and Computation 12(5), 773–808 (2002)
9. Kornhauser, L., Sager, L.: Unpacking the court. Yale Law Journal 96, 82–117 (1986)
10. Kornhauser, L., Sager, L.: The one and the many: Adjudication in collegial courts. California Law Review 81, 1–51 (1993)
11. List, C.: A model of path-dependence in decisions over multiple propositions. American Political Science Review 98(3), 495–513 (2004)
12. List, C.: Judgment aggregation - a bibliography on the discursive dilemma, the doctrinal paradox and decisions on multiple propositions (2007), http://personal.lse.ac.uk/LIST/doctrinalparadox.htm
13. List, C., Puppe, C.: Judgment aggregation: A survey. In: Anand, P., Puppe, C., Pattanaik, P. (eds.) Oxford Handbook of Rational and Social Choice. Oxford University Press, Oxford (forthcoming)
14. Mongin, P.: Factoring out the impossibility of logical aggregation. Journal of Economic Theory 141(1), 100–113 (2008)

15. Nash, J.R.: A context-sensitive voting protocol paradigm for multimember courts. Stanford Law Review 56, 75–159 (2003)
16. Nehring, K.: The (im)possibility of a Paretian rational. Economics working papers, Institute for Advanced Study, School of Social Science (November 2005)
17. Pettit, P.: Deliberative democracy and the discursive dilemma. Philosophical Issues 11, 268–299 (2001)
18. Pigozzi, G.: Belief merging and the discursive dilemma: an argument-based account to paradoxes of judgment aggregation. Synthese 152(2), 285–298 (2006)
19. Pigozzi, G., Slavkovik, M., van der Torre, L.: A complete conclusion-based procedure for judgment aggregation. University of Luxembourg (July 2009) ISBN number: 978-2-87971-028-0
20. Pigozzi, G., Slavkovik, M., van der Torre, L.: Conclusion-based procedure for judgment aggregation satisfying premise independence. In: van der Hoek, W., Bonanno, G., Lowe, B. (eds.) Proceedings of the Eighth International Conference on Logic and the Foundations of Game and Decision Theory, LOFT 2008, Amsterdam, The Netherlands, p. 35 (2008)

A Geometric Approach to Paradoxes of Majority Voting in Abstract Aggregation Theory

Daniel Eckert and Christian Klamler

University of Graz, Institute of Public Economics, Universitaetsstr. 15, 8010 Graz,
Austria
{daniel.eckert,christian.klamler}@uni-graz.at

Abstract. In this paper we extend Saari's geometric approach to paradoxes of preference aggregation to the analysis of paradoxes of majority voting in a more general setting like Anscombe's paradox and paradoxes of judgment aggregation. In particular we use Saari's representation cubes to provide a geometric representation of profiles and majority outcomes. Within this geometric framework, we show how profile decompositions can be used to derive restrictions on profiles that avoid the paradoxes of majority voting.

1 Introduction

In the last thirty years, there have been several attempts to generalize the Arrovian framework of preference aggregation (e.g. Rubinstein and Fishburn [8] or Wilson [11]). This literature on abstract aggregation has been considerably stimulated by the growing interest in problems of judgment aggregation. The problem of judgment aggregation consists in aggregating individual judgments on an agenda of logically interconnected propositions into a collective set of judgments on these propositions (see List and Puppe [5] for a survey).

As an example of a paradox in judgment aggregation, consider a variant of the so-called discursive dilemma, in which a committee of three recruitment officers in a firm has to decide whether a job applicant should be hired or not. There is a written test and an oral interview and each of them is advised to recommend hiring the applicant if and only if the applicant passes the written test and gives a satisfiable interview. Table 1 shows the judgments of the officers and their majority decisions.

Based on the majority of the individual decisions, the job applicant will not be hired as a majority does not find her acceptable. However, a majority finds the written test as well as the interview acceptable.

Problems of judgment aggregation are structurally similar to paradoxes and problems in social choice theory like the Condorcet paradox and Arrow's general possibility theorem, but also related to paradoxes of compound majorities like the Anscombe or Ostrogorski paradoxes, both nicely analysed by Nurmi [7] (see also Nurmi [6]). Contemporarily, Saari [9] has developed and popularized a geometric approach to Arrovian social choice theory. His approach has helped

F. Rossi and A. Tsoukis (Eds.): ADT 2009, LNAI 5783, pp. 14–25, 2009.

Table 1. Discursive Dilemma

Officer	written test	oral interview	decision
Officer 1	1	1	1
Officer 2	1	0	0
Officer 3	0	1	0
Majority Outcome	1	1	0

to understand what drives many of the impossibility results and paradoxes in social choice theory.

In this paper we develop - in Saari's style - a geometric approach to abstract aggregation theory starting from the Anscombe paradox and extending this framework to typical paradoxes in judgment aggregation.

Our approach focuses on and will be exhaustive for aggregation problems that can be represented in the three-dimensional hypercube. While this is the smallest dimension in which interesting aggregation problems can be formulated and particularly illuminating for problems that naturally fall into this framework, we have to give a warning that most of our results are not easily extendable to more than three dimensions.

A major difference of judgment aggregation to social choice theory lies in the representation of the information involved. While binary relations over a set of alternatives are a canonical representation of preferences, a natural representation of judgments are binary valuations over a set of propositions, where the logical interconnections between these propositions determine the set of admissible (i.e. logically consistent) valuations. E.g. the agenda of the famous discursive dilemma $\{p, q, p \wedge q\}$ is associated the set of admissible valuations $\{(0,0,0), (1,0,0), (0,1,0), (1,1,1)\}$.

The paper is structured as follows: In section 2 we introduce the formal framework with the main definitions and properties. Section 3 discusses paradoxes of majority voting. We will use Saari's representation cubes to provide a unified geometric representation of profiles and majority rule outcomes and introduce Saari's idea of a profile decomposition. In this framework we will provide a characterization of profiles leading to the Anscombe paradox. Section 4 applies the same tools to judgment aggregation. In particular we show what drives the logical inconsistency of majority outcomes and how this can be avoided with the help of restrictions on the distribution of individual valuations, i.e. give a kind of generalized domain restriction.

2   Abstract Aggregation Theory and Majority Voting: Formal Framework and Central Properties

In the binary framework of abstract aggregation theory individual vectors of binary valuations $v = (v^1, v^2, ..., v^{|J|})$ from a set $X \subseteq \{0,1\}^{|J|}$ of admissible valuations over the set J of issues (the agenda) are aggregated into a collective valuation. (In a slight abuse of notation we will use the term valuation both for a the binary valuation of a single issue as for vectors of binary valuations.)

Such an issue $j \in J$ might be the pairwise comparison between two alternatives in preference aggregation or a proposition on which a judgment needs to be made. Typically, the interconnections between the issues constrain the set of admissible valuations. In judgment aggregation a valuation $v = (v^1, v^2, ..., v^{|J|}) \in X \subseteq \{0,1\}^{|J|}$, represents an individuals' beliefs, where $v^j = 1$ means that proposition j is believed and X denotes the set of all admissible (logically consistent) valuations (see Dokow and Holzman [3]).

Given a set N of individuals, a profile of individual valuations is then a mapping $\pi : N \to \{0,1\}^{|J|}$ which assigns to each individual a vector of binary valuations. A desirable property of an aggregation rule, stronger than nondictatorship, is of course anonymity, which requires that the same collective valuation be assigned to any permutation of the set of individuals.

If anonymity is assumed, a profile of individual valuations can be represented by a vector $\mathbf{p} = (p_1, .., p_{|X|}) \in [0,1]^{|X|}$ with $\sum_k p_k = 1$, which associates with every admissible valuation $v_k \in X$ the share p_k of individuals with this valuation. Such an anonymous representation of profiles is particularly appropriate for the analysis of majority voting, where anonymity is typically assumed. While such an anonymous representation is unique only up to a permutation of the original profile, we however use the term profile for it in the following.

Geometrically any binary valuation is a vertex of the $|J|$-dimensional hypercube and, more interestingly, any profile $\mathbf{p} \in [0,1]^{|X|}$ can be given a lower-dimensional representation by a point $x(\mathbf{p}) \in [0,1]^{|J|}$ in the $|J|$-dimensional 0/1-polytope, i.e. the convex hull of the hypercube $\{0,1\}^{|J|}$, where for each component $j \in J$, $x^j(\mathbf{p}) = \sum_{k \in \{1,...,|X|\}} p_k v_k^j$ denotes the average support for issue j. Thus the $|J|$-dimensional 0/1-polytope will be referred to as the representation polytope of the profiles.

An anonymous aggregation rule is a mapping f that associates with every profile $\mathbf{p} = (p_1, p_2, ..., p_{|X|}) \in [0,1]^{|X|}$ a valuation $v = f(x(\mathbf{p})) \in \{0,1\}^{|J|}$.

We will write $v(\mathbf{p})$ for $f(x(\mathbf{p}))$ and identify by $v^j(\mathbf{p})$ the jth component of $v(\mathbf{p})$ under the given aggregation rule.

In this framework majority voting on issues (or majority voting for short) is defined as follows:

Definition 1. *For any issue $j \in J$ and any profile $\mathbf{p} \in [0,1]^{|X|}$, $^M v^j(\mathbf{p}) \in \{0,1\}$ is the outcome of majority voting on issue j if*

$$v^j(\mathbf{p}) = 1 \Leftrightarrow x^j(\mathbf{p}) > 0.5.$$

This representation immediately provides majority with a wellknown metric rationalization in terms of the Hamming distance between binary vectors. (For any two binary vectors $v, v' \in \{0,1\}^{|J|}$, the Hamming distance $d_H(v, v')$ is the number of components in which these two vectors differ.)

Proposition 1. *(Brams et al. [1]) For any profile* $\mathbf{p} \in [0,1]^{|X|}$*, the valuation* $^M v(\mathbf{p}) \in \{0,1\}^{|J|}$ *is the majority outcome if and only if it minimizes the sum of Hamming distances weighted by the population shares, or formally,*

$$^M v(\mathbf{p}) = \arg\min_{v \in \{0,1\}^{|J|}} \sum_{k=1}^{|X|} p_k d_H(v_k, v).$$

Thus, whenever the sum of Hamming distances can be interpreted as an appropriate measure of social disutility, majority voting can be justified by its minimization.

Observe however that nothing in this characterisation prevents the majority outcome $^M v(\mathbf{p}) \in \{0,1\}^{|J|}$ from being an inadmissible valuation, i.e. that $^M v(\mathbf{p}) \in \{0,1\}^{|J|} \backslash X$.

In the hypercube, a more natural metric representation of majority voting can be given in terms of the euclidean distance d_E.

Proposition 2. *For any profile* $\mathbf{p} \in [0,1]^{|X|}$*, the valuation* $^M v(\mathbf{p}) \in \{0,1\}^{|J|}$ *is the majority outcome if and only if it minimizes the euclidean distance between the corresponding vertex and the point* $x(\mathbf{p})$ *in the representation polytope, or formally,*

$$^M v(\mathbf{p}) = \arg\min_{v \in \{0,1\}^{|J|}} d_E(x(\mathbf{p}), v).$$

Conversely, the set of all profiles for a given majority outcome $v \in \{0,1\}^{|J|}$ defines a subcube of $[0,1]^{|J|}$, $P_v = [|v^j - 0.5|]^{|J|}$, which is the set of all profiles for which v is the majority outcome. Such a subcube will be called the majority subcube of v (or simply v-subcube) and can be seen in Figure 1 for vertex $(1,0,1)$.

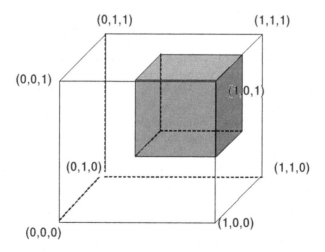

Fig. 1. Majority subcube

Table 2. Valuations in three-dimensional hypercube

valuation		valuation	
v_1	$(0,0,0)$	v_5	$(1,1,0)$
v_2	$(1,0,0)$	v_6	$(1,0,1)$
v_3	$(0,1,0)$	v_7	$(0,1,1)$
v_4	$(0,0,1)$	v_8	$(1,1,1)$

3   The Anscombe Paradox and the Irrationality of a Metric Rationalization

Because majority voting on issues has a metric rationalization in terms of distance minimization, it is quite disturbing that the majority outcome need not be the one that minimizes the distance for the majority of individuals, as the Anscombe paradox shows. In other words the Anscombe paradox states that a majority of the voters can be on the loosing side on a majority of issues. Formally, the Anscombe paradox can be defined in the following way:

Definition 2. *A profile* $\mathbf{p} = (p_1, .., p_{|X|}) \in [0,1]^{|X|}$ *exhibits the Anscombe paradox if*

$$\sum_{k \in \{1,...,|X|\}: d_H(v_k, {}^M v(\mathbf{p})) > \frac{|J|}{2}} p_k > \frac{1}{2}$$

Indeed, it is the particular distribution of individual valuations that leads to the paradox. To analyse this and further paradoxes in later sections, we will numerate the vertices of the three-dimensional hypercube as listed in Table 2.

Now, consider the profile $\mathbf{p} = (\frac{2}{5}, 0, 0, 0, \frac{1}{5}, \frac{1}{5}, \frac{1}{5}, 0)$, which specifies exactly an Anscombe paradox situation. It is easily observed, that $x(\mathbf{p}) = (\frac{2}{5}, \frac{2}{5}, \frac{2}{5})$ and hence the majority outcome is ${}^M v(\mathbf{p}) = (0,0,0)$.

Are we able to specify profiles that lead to an Anscombe type or other paradoxical majority outcome? Saari [10] identifies what he calls "Condorcet portions" as the driving part of paradoxes of preference aggregation.[1] In our three-dimensional setting for abstract aggregation problems, we can consider such portions as triples of valuations that have a common neighbor, i.e. a valuation that differs from each of the three valuations in exactly one issue.[2] Given that, we can now easily specify for every vertex in the hypercube its triple of neighbors. E.g. for v_5 the corresponding triple of neigbors is (v_2, v_3, v_8). Table 3 indicates the triples for all eight vertices, the set of all such triples will be denoted by $\mathcal{P}$.

To analyse the paradoxical outcomes and suggest restrictions to overcome them, we will use a profile decomposition technique developed by Saari [9]. From

---

[1] A "Condorcet portion" is a multiple of the set of individuals that has the following preferences over 3 alternatives a, b, c: $a \succ_1 b \succ_1 c$, $c \succ_2 a \succ_2 b$, $b \succ_3 c \succ_3 a$ leading to the the majority cycle $a \succ b \succ c \succ a$.

[2] Equivalently we could say that they are each of Hamming distance 1 from their joint neighbor.

Table 3. Triples of neighbors

valuation	triple	valuation	triple
v_1	(v_2, v_3, v_4)	v_5	(v_2, v_3, v_8)
v_2	(v_1, v_5, v_6)	v_6	(v_2, v_4, v_8)
v_3	(v_1, v_5, v_7)	v_7	(v_3, v_4, v_8)
v_4	(v_1, v_6, v_7)	v_8	(v_5, v_6, v_7)

a majority point of view it is clear that two opposite valuations about an issue cancel out, i.e. have no impact on the majority outcome. This can, however, be extended to any number of valuations by decomposing a profile into subprofiles:

Definition 3. *For any profile* $\mathbf{p} = (p_1, .., p_{|X|}) \in [0,1]^{|X|}$ *with* $\sum_k p_k = 1$ *a subprofile is a vector* $\underline{\mathbf{p}} = (\underline{p}_1, ..., \underline{p}_{|X|}) \in [0,1]^{|X|}$ *such that* $\underline{p}_k \leq p_k$ *for all* $k \in \{1, ..., |X|\}$

It is obvious that the above decomposition argument for two opposite valuations does hold for any subprofile $\underline{\mathbf{p}}$ of $\mathbf{p}$ for which $x(\underline{\mathbf{p}}) = (\frac{1}{2}, \frac{1}{2}, \frac{1}{2})$. Such a subprofile does not influence the majority outcome based on $\mathbf{p}$ at all.

As an example consider two individuals with the respective valuations v_2 and v_7. They are exact opposites, so from a majority point of view those two valuations cancel out. Hence this implies that in any profile $\mathbf{p}$, for all opposite valuations we can cancel the share of the valuation held by the smaller number of individuals (and correct for the other shares accordingly) and still have the majority outcome unchanged.

Lemma 1. *Let* $\mathbf{p}$ *and* $\mathbf{p}'$ *be two profiles such that, for each* $i \in \{1, ..., 8\}$,

$$p_i' = \frac{max\{p_i - p_{9-i}, 0\}}{\sum_{k=1}^4 |p_k - p_{9-k}|}.$$

Then $x^j(\mathbf{p}) \geq \frac{1}{2} \Leftrightarrow x^j(\mathbf{p}') \geq \frac{1}{2}$.

Proof. The average support for each of the three issues can be stated as follows:

$x^1(\mathbf{p}) = p_2 + p_5 + p_6 + p_8$
$x^2(\mathbf{p}) = p_3 + p_5 + p_7 + p_8$
$x^3(\mathbf{p}) = p_4 + p_6 + p_7 + p_8$

Now, let $x^j(\mathbf{p}) = a$ and for some i, $|p_i - p_{9-i}| = t$, and assume w.l.o.g. that $1 > a \geq \frac{1}{2}$ and $0 < t < a$. For $\frac{a}{1} \geq \frac{1}{2}$ we also get $\frac{a-t}{1-2t} \geq \frac{1}{2}$. To see this suppose this is not the case, i.e. $\frac{a-t}{1-2t} < \frac{1}{2}$. It follows that $2a - 2t < 1 - 2t$. For $a \geq \frac{1}{2}$ this is false and therefore $\frac{a-t}{1-2t} \geq \frac{1}{2}$ is true. Repeat this for all $i \in \{1, ..., 4\}$. For necessity just reverse the above arguments.

The lemma shows that in a subprofile $\mathbf{p}'$ at most 4 entries can be positive. As already previously mentioned, we can reduce a profile by any subprofile that does not change the majority outcome. Consider a subprofile $\mathbf{p}$ with positive shares only for the valuations $(0,0,0)$, $(1,1,0)$, $(1,0,1)$ and $(0,1,1)$, namely $\mathbf{p} = (\frac{1}{8}, 0, 0, 0, \frac{1}{8}, \frac{1}{8}, \frac{1}{8}, 0)$. On each issue there is the same number of individuals in favor of it and against it, i.e. $x(\mathbf{p}) = (\frac{1}{2}, \frac{1}{2}, \frac{1}{2})$. Hence, the elimination of such a subprofile does not change the majority outcome of the original profile and eventually increases the number of zero entries in the profile. The only two sets of valuations useable for such a reduction are $\{v_1, v_5, v_6, v_7\}$ and $\{v_2, v_3, v_4, v_8\}$.

Both, the pairwise reduction as well as the reduction using 4 valuations, lead to a reduced profile, the majority outcome of which is identical to the majority outcome of the original profile.

Now we can use the above concepts for a result in a three-dimensional framework, namely that the Anscombe paradox manifests itself in a particularly strong form:

Proposition 3. *For $|J| = 3$ the Anscombe paradox will always show up in its strong form, i.e. a majority of the voters has a lower Hamming distance to the valuation which is the exact opposite of the majority outcome than to the majority outcome itself.*

Proof. Assume, w.l.o.g., that we want the majority outcome to be $^M v(\mathbf{p}) = (0,0,0)$. As $|J| = 3$, each voter k among a majority of the voters needs to have $d_H(v_k, {}^M v(\mathbf{p})) \geq 2$. Starting with $^M v(\mathbf{p}) = v_1$ this leads to the following conditions needed to be satisfied for the Anscombe paradox to occur, where the first three conditions guarantee the majority outcome to be v_1 and the fourth condition ensures that a majority of voters is of a Hamming distance of at least 2 from the majority outcome:

1. $x^1(\mathbf{p}) = p_2 + p_5 + p_6 + p_8 < \frac{1}{2}$
2. $x^2(\mathbf{p}) = p_3 + p_5 + p_7 + p_8 < \frac{1}{2}$
3. $x^3(\mathbf{p}) = p_4 + p_6 + p_7 + p_8 < \frac{1}{2}$
4. $p_5 + p_6 + p_7 + p_8 > \frac{1}{2}$

Based on our previous decomposition argument (especially Lemma 1), no identical change in p_8 and p_1 would change the truth of any of the above inequalities. But this is also true for any other pair of opposite valuations. Hence we can directly look at the reduced profile $\mathbf{p}'$ with at most 4 entries. For $^M v^j(\mathbf{p}') = 0$ it is not possible that more than half of the shares are located on one plane of the cube, i.e. $p_8' + p_r' + p_s' < \frac{1}{2}$ for all $r, s \in \{5, 6, 7\}$. But this implies that $p_r' > 0$ for all $r \in \{5, 6, 7\}$ and hence $p_1' > 0$ (and therefore $p_8' = 0$ to enable $^M v(\mathbf{p}') = v_1$). Now for any $k \in \{5, 6, 7\}$, v_k is not closer to a majority of the voters' valuation than to $^M v(\mathbf{p})$, as $p_k < \frac{1}{2}$, and this would be the only voters with smaller distance. For any $k \in \{2, 3, 4\}$, v_k is not closer to a majority of the voters' valuation than to $^M v(\mathbf{p})$, as $p_r' + p_s' < \frac{1}{2}$ for all $r, s \in \{5, 6, 7\}$ and only two valuations out of $\{v_5, v_6, v_7\}$ are closer to v_k than to $^M v(\mathbf{p})$.

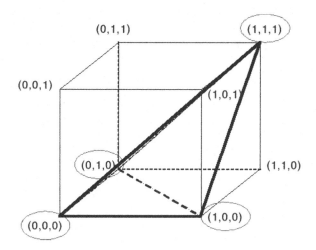

Fig. 2. Representation polytope

4 Judgment Aggregation and the Logical Inconsistency of the Majority Outcome

In judgment aggregation, the issues in the agenda are logically interconnected propositions and thus not all valuations are admissible, i.e. logically consistent. Given the binary structure of the problem, we see that the tools of the geometric approach can be used to analyse paradoxes of judgment aggregation.[3] The discursive dilemma with the agenda $\{p, q, p \wedge q\}$ and the associated set of admissible valuations $X = \{(0,0,0), (1,0,0), (0,1,0), (1,1,1)\}$ can again be analysed in our three-dimensional hypercube, in which the four admissible vertices determine the representation polytope as seen in Figure 2.

Given the set of admissible valuations X, consider the profile $\mathbf{p} = (0, \frac{1}{3}, \frac{1}{3}, 0, 0, 0, 0, \frac{1}{3})$, i.e. no voter has valuation $(0,0,0)$, one third of the voters has valuation $(1,0,0)$, and so on. As this maps into the point $x(\mathbf{p}) = (\frac{2}{3}, \frac{2}{3}, \frac{1}{3})$ - a point whose closest vertex is $(1,1,0)$ - the representation polytope obviously passes through the majority subcube of an inadmissible valuation, i.e. the set of admissible valuations X is not closed under majority voting. In this case X is called majority inconsistent.

That this type of paradox can easily occur with majority voting is seen from the following lemma:

Lemma 2. *Given any vertex $v \in \{0,1\}^{|J|}$, there exist 3 vertices v_a, v_b, v_c with respective shares p_a, p_b, p_c such that for some profile $\mathbf{p}$ with $p_k = 0$ for all $k \notin \{a, b, c\}$, $x(\mathbf{p})$ lies in the v-subcube.*

---

[3] See e.g. Saari [10] for a very brief discussion of the link of his geometric approach to judgment aggregation.

For $|J| = 3$, these 3 vertices necessarily need to have v as their common neighbor. Given that, we can now provide a simple result for the majority inconsistency of a set of valuations X, i.e. a necessary condition for X not to be closed under majority voting.

Proposition 4. *For $|J| = 3$, the set of admissible valuations X is majority inconsistent only if for some triple of vertices in the domain with a common neighbor, this common neighbor is not contained in the domain.*

In our 3-dimensional setting, we can easily specify all possible triples that could lead to inadmissible majority outcomes. The reduced profile does have an interesting feature in exactly those situations when inadmissible majority outcomes could occur:

Proposition 5. *For $|J| = 3$, if for some $v_i \in \{0,1\}^3$ with $p_i \leq p_{9-i}$, each valuation in the triple of neighbors has a larger share in $\mathbf{p}$ than its opposite valuation, then the reduced profile $\bar{\mathbf{p}}$ has at most 3 positive entries.*

Proof. Let $\mathbf{p} = (p_1, p_2, ..., p_8)$ s.t. $\sum_k p_k = 1$. From Lemma 1 we know that

$$p'_i = \frac{max\{p_i - p_{9-i}, 0\}}{\sum_{k=1}^4 |p_k - p_{9-k}|}.$$

Now in $\mathbf{p}'$ there are at most 4 positive entries. Given that it is not possible that $[p'_i > 0 \land p'_{9-i} > 0]$ for any $i = 1, ..., 8$, and that for some v_k each valuation in the triple of neigbors has a larger share than its opposite valuation, this only leaves two possibilities, namely that we have positive shares at most either for all of (p_1, p_5, p_6, p_7) or for all of (p_2, p_3, p_4, p_8). However, in both cases - as was discussed before - further reductions are possible by looking for particular subprofiles. Let - for the above two combinations - $\mathcal{A} = \{i : p'_i > 0\}$ be all valuations for which there is a positive share. Then we can reduce the profile further to profile $\bar{\mathbf{p}}$ such that

$$\bar{p}_i = \frac{max_{j \in \mathcal{A}/\{i\}}\{p'_i - p'_j, 0\}}{\sum_{i \in \mathcal{A}} p'_i - min_{j \in \mathcal{A}} p'_j}.$$

Obviously $\bar{\mathbf{p}}$ has at most 3 positive entries.

Example 1. Let us consider the following set of admissible valuations $X = \{v_1, v_2, v_3, v_8\}$, i.e. any profile $\mathbf{p} = (p_1, p_2, p_3, 0, 0, 0, 0, p_8)$, where $p_k \geq 0$ for all $k = 1, 2, 3, 8$ and $\sum_k p_k = 1$. As $v_1 = (0,0,0)$ and $v_8 = (1,1,1)$ are exact opposites, the reduced profile will have a share of 0 for the valuation held by the smaller number of individuals. In the case of $p_1 > p_4$ such a reduced profile will be $\bar{\mathbf{p}} = (\frac{p_1-p_4}{p_1+p_2+p_3-p_4}, \frac{p_2}{p_1+p_2+p_3-p_4}, \frac{p_3}{p_1+p_2+p_3-p_4}, 0, 0, 0, 0, 0)$, in the case of $p_1 \leq p_4$ we can create the reduced profile accordingly. Hence the reduced profile maps into one of the following two planes shown in Figure 3, namely either into the one determined by the vertices v_1, v_2 and v_3 or the one determined by the vertices v_2, v_3 and v_8.

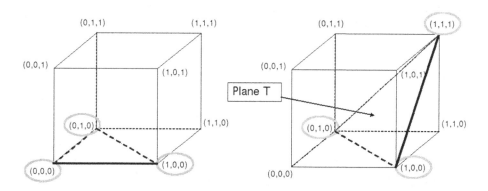

Fig. 3. Planes

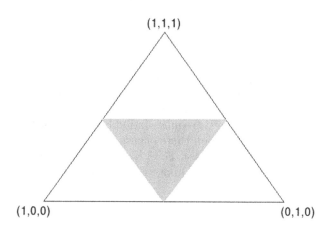

Fig. 4. Plane T

Problems may arise if the reduced profile has positive shares only for 3 valuations that constitute a triple in $\mathcal{P}$. For the above example this would be the triple (v_2, v_3, v_8) on the right side of Figure 3. Now, in Figure 4 the intersection of this plane with the $(1, 1, 0)$ majority subcube is indicated by the shaded triangle. Only if the reduced profile maps into this triangle do inadmissible majority outcomes arise.

Now, for the 3-dimensional framework we can state the following result:

Proposition 6. *A set of admissible valuations $X \subseteq \{0, 1\}^3$ is majority inconsistent if and only if for some reduced profile $\bar{\mathbf{p}}$ the following conditions are met:*

1. *$\bar{\mathbf{p}}$ has 3 positive entries*
2. *the 3 valuations with positive shares form a triple $(v_a, v_b, v_c) \in \mathcal{P}$ whose joint neighbor is not in X*

3. *the following condition holds for all $v_k \in \{v_a, v_b, v_c\}$ with corresponding shares $\bar{p}_k \in \{\bar{p}_a, \bar{p}_b, \bar{p}_c\}$:*

$$\frac{\bar{p}_k}{\bar{p}_a + \bar{p}_b + \bar{p}_c} \leq \frac{1}{2}$$

Proof. The sufficiency part is obvious from Figure 4. For necessity, it is clear that with less than 3 positive entries in $\bar{\mathbf{p}}$ no inadmissible outcome can occur. Moreover, any triple not in $\mathcal{P}$ is closed under majority rule. In the case of 4 positive entries in $\bar{\mathbf{p}}$ problems only arise in case a triple in $\mathcal{P}$ has a positive share where the joint neigbor is not contained in X. But then the fourth positive entry must be one such that the resulting profile can still be further reduced and hence this contradicts the assumption that $\bar{\mathbf{p}}$ was the reduced profile already. Now, the only further option is a triple in $\mathcal{P}$ with a joint neigbor not in X. In this situation inconsistency occurs exactly in the cut with the respective majority subcube (see Figure 4) whose points are specified by the condition above.

One interesting feature of this result is that the complementary set of profiles actually determines the domain that is closed under majority rule. As those restrictions are based on the space of profiles, this approach is more general than restrictions on the space of valuations which is usually used in the classical literature on domain restrictions. E.g. List [4] introduces the unidimensional alignment domain which has a certain resemblence to Black's single peakedness condition in social choice theory. It requires individuals to be ordered from left to right such that on each proposition there occurs only one switch from believing it to not believing it (or vice versa). For $|J| = 3$ a unidimensional alignment domain would not satisfy one of the above conditions for inadmissible majority outcomes.[4]

5   Conclusion

In this paper we have shown how geometry can be used to analyse paradoxes occuring under majoritarian aggregation and (impossibility) results in judgment aggregation, such as inadmissible majority outcomes and distance based aggregation rules. In addition we gave generalized domain conditions characterizing these paradoxes and determined the likelihood of such inadmissible majority outcomes.

Most of the stated results do not easily extend to more than three issues because of problems of dimensionality. E.g. an agenda with three propositions and their conjunction, like $\{p, q, r, p \wedge q \wedge r\}$, leads to eight admissible valuations, i.e. eight vertices out of the 16 vertices in the four-dimensional hypercube. The extensions of our (domain) restrictions and calculations of the likelihood of the occurrence of paradoxes to those higher dimensions are not obvious and need further work.

---

[4] For a more elaborated discussion on majority voting on restricted domains see also Dietrich and List [2].

References

1. Brams, S.J., Kilgour, M.D., Sanver, M.R.: A minimax procedure for negotiating multilateral treaties. In: Wiberg, M. (ed.) Reasoned choices: Essays in Honor of Hannu Nurmi. Finnish Political Science Association, Turku (2004)
2. Dietrich, F., List, C.: Majority voting on restricted domains. Mimeo (2007)
3. Dokow, E., Holzman, R.: Aggregation of binary evaluations. Journal of Economic Theory (2008), doi:10.1016/j.jet.2007.10.004
4. List, C.: The probability of inconsistencies in complex collective decisions. Social Choice and Welfare 24, 3–32 (2005)
5. List, C., Puppe, C.: Judgment aggregation. In: Anand, P., Pattanaik, P.K., Puppe, C. (eds.) The Handbook of Rational and Social Choice: An Overview of New Foundations and Applications. Oxford University Press, Oxford (2009)
6. Nurmi, H.: Comparing voting systems. D. Reidel, Dordrecht (1987)
7. Nurmi, H.: Voting paradoxes and how to deal with them. Springer, Berlin (1999)
8. Rubinstein, A., Fishburn, P.C.: Algebraic aggregation theory. Journal of Economic Theory 38, 63–77 (1986)
9. Saari, D.G.: Basic geometry of voting. Springer, Berlin (1995)
10. Saari, D.G.: Disposing dictators, demystifying voting paradoxes: Social choice analysis. Cambridge University Press, New York (2008)
11. Wilson, R.: On the theory of aggregation. Journal of Economic Theory 10, 89–99 (1975)

Manipulating Tournaments in Cup and Round Robin Competitions*

Tyrel Russell[1] and Toby Walsh[2]

[1] Cheriton School of Computer Science, University of Waterloo, Waterloo, Canada
tcrussel@cs.uwaterloo.ca
[2] NICTA and UNSW, Sydney, Australia
toby.walsh@nicta.com.au

Abstract. In sports competitions, teams can manipulate the result by, for instance, throwing games. We show that we can decide how to manipulate round robin and cup competitions, two of the most popular types of sporting competitions in polynomial time. In addition, we show that finding the minimal number of games that need to be thrown to manipulate the result can also be determined in polynomial time. Finally, we show that there are several different variations of standard cup competitions where manipulation remains polynomial.

1 Introduction

The Gibbard-Satterthwaite theorem proves that, under some modest assumptions, voting systems are always manipulable. One possible escape proposed by Bartholdi, Tovey and Trick is that the manipulation may be computationally too difficult to find [2] (but see [13] for discussion about whether manipulation is hard not just in the worst case). Like elections, sporting competitions can also be manipulated. For example a coalition of teams might throw games strategically to ensure that a desired team wins or a certain team loses. We consider here the computational complexity of computing such manipulations. We show that, for several common types of competitions, determining when a coalition can manipulate the result is polynomial. Our results adapt manipulation procedures for elections where voters can misrepresent their preferences. We consider two of the most common methods used for deciding sporting competitions, cups and round robins. These correspond to elections run using sequential majority voting (also known as the cup rule) and Copeland scoring, respectively.

Manipulating a sporting competition is slightly different to manipulating an election as, in a sporting competition, the voters are also the candidates. A tournament graph describes the outcome of all fair games between opponents. Manipulating a competition therefore modifies not votes but the tournament graph directly. Since it is hard without bribery or similar mechanisms for a team

---

* This work is funded by an NSERC Postgraduate Scholarship[1], the Department of Broadband, Communications and Digital Economy[2] and the Australian Research Council[2].

F. Rossi and A. Tsoukis (Eds.): ADT 2009, LNAI 5783, pp. 26–37, 2009.

to play better than it can, we consider manipulations where teams in the coalition are only able to throw games. By comparison, in an election, voters in the manipulating coalition can mis-report their preferences in any way they choose. Tang, Shoham and Lin [11] addressed this type of tournament manipulation in team competitions by providing conditions for truthful reporting of player strengths. Their method tries to encourage teams to rank their players honestly so that, when the teams compete in bouts, the best player on one team plays the best on the other, the second best plays the opposing second and so forth. An example of this type of competition is Davis Cup Tennis.

Conitzer, Sandholm and Lang [3] give an algorithm to determine if a coalition can manipulate the cup rule. We modify this algorithm to manipulate directly the tournament graph instead of the votes. Bartholdi, Tovey and Trick [2] discuss direct manipulations of the tournament under second order Copeland, a round robin like rule with secondary tie breaking. Using the work of Kern and Paulusma [7], we show that the manipulation of round robin competitions is directly tied to the problem of winner determination in sports problems.

Altman, Procaccia and Tenneholtz [1] construct a social choice rule that is monotonic, pairwise non-manipulable and non-imposing. Round robin and cup competitions are monotonic as a single team losing a game does no better. Pairwise non-manipulability means that no two teams are better off by manipulating the tournament. Our results show that round robin and cup competitions are pairwise manipulable and that manipulations can be calculated in polynomial time.

We modify our algorithms to calculate the smallest number of manipulations needed. For cup competitions, we add dynamic programming to Conitzer, Sandholm and Lang's algorithm. For round robin competitions, we modify the flow network used to solve winner determination to include weights on manipulations and calculate a minimum cost feasible flow. Vu, Altman and Shoham [12] used a similar method to calculate the probability that a team wins the competition. Vu et al. [12] provide several results on determining probabilities of teams winning given a seeding of the tournament. Hazon et al. [6] showed that it is NP-Complete to determine if a team wins a cup with a given probability. This is similar to determining a possible winner given random reseeding except edges in the tournament are labelled with probabilities. We look at the complexity of manipulation under reseeding in the deterministic case. Finally, we look at the complexity of double elimination cups.

2   Background

In many sporting competitions, the final winner of a competition is decided by a tree-like structure, called a *cup*. The most common type is a *single elimination cup*, a tree structure where the root and internal nodes represent games and leaves represent the teams in the tournament. A cup can include a *bye game*, a game where a team skips a game to re-balance the schedule. Usually, the top teams are given a bye game while the lower teams do not so that the number of teams in the next round is strict power of two. Cups need to be *seeded* to determine which teams play against each other in each round. One method for seeding is by rank. The most common method for ranked seeding or reseeding is to have the top team play the worst team, the second place team play the second worst team and so forth. An example of ranked seeding using this method is the National Basketball Association in the US. Another method for determining seeding is randomly, also known as a draw. An example of this is the UEFA Champions League where teams reaching the quarter finals are randomly paired for the remainder of the tournament. Seeding may also be more complex (for instance, it may be based on the group from which teams qualify or some other criteria). Another way that cups are modified is between fixed and unfixed cups. A *fixed cup* is a cup where there is a single seeding at the start of the cup. Examples of this are the National Basketball Association and the World Cup of Football. An *unfixed cup* is one where seeding may occur not only before the start but between any round. Examples with an unfixed cup are the National Hockey League and the UEFA Champions League.

Cups are not necessarily single elimination. A double elimination cup is designed so that a team can lose two games instead of one game. If a team loses, they play other teams that have also lost until they lose a second time or they win the final game of the tournament. These tournaments are organized as two cups where losers enter the second cup at various stages depending on when they lose their first game.

Finally, a *round robin competition* is a competition where each team plays every other team a given number of times. In a *single round robin competition*, each team plays every other team exactly once. Another common variant of this is for teams to play a *double round robin competition* where each team plays every other team twice, often at home and away.

3   Manipulating the Tournament

A *tournament* is a directed graph $G = (V, E)$ where the underlying undirected graph is a complete graph. We assume that the tournament is available for the remainder of the paper. Every directed edge $(v_i, v_j) \in E$ represents a victory by v_i over v_j. The number of the teams in the competition is $|V| = m$. We define a *manipulation* of the tournament as any replacement of an edge (v_i, v_j) in the graph with the edge (v_j, v_i). This is equivalent to a manipulation of votes but here we are changing the winner directly instead of just changing the vote. Note that, as in election manipulation where the electoral vote is assumed to be known, we assume that we know, via an oracle, the relative strengths of teams and can represent the winner of the contests in the tournament graph. We restrict manipulations by only allowing the manipulation of an edge (v_i, v_j) if candidate v_i is a member of the coalition. This restricts the behaviour of the manipulators to throwing games where they could have won. This restriction is due to the fact that it is simple to perform worse but more difficult to play better. We consider two different types of manipulations. A *constructive manipulation* is one that ensures a specific team wins the competition. A *destructive manipulation* is one that ensures a specific team loses the competition. For round robin competitions, we generalize the concept of the tournament beyond the simple win-loss scoring model to a complete graph where the edge (v_i, v_j) has a non-negative weight w_{ij} which represents the number of points that would be earned by v_i when playing v_j in a fair game. We define a manipulation in this case as an outcome where the points earned in the match are different to those given by the tournament. However, manipulations are restricted so that the manipulator achieves no more points and the team being manipulated achieves no less points.

In this section, we restrict ourselves to fixed cups with a known seeding. We also look just at single round robin tournaments though the results generalize to multi-round robin tournaments.

3.1   Cup Competitions

For cup competitions, finding a constructive or destructive manipulation of the tournament is polynomial. Our results make use of results in [3] which shows that a manipulation of an election using the cup rule can be found in $O(m^3 n)$ time where m is the number of candidates and n is the number of voters.

Theorem 1. *Determining if a cup competition can be constructively manipulated using manipulations of the tournament takes polynomial time.*

Proof. This proof is a bottom up version the proof of Theorem 2 from Conitzer, Sandholm and Lang (CSL)[3] but substitutes tournament manipulations for voting manipulations. The basic CSL algorithm is a recursive method that treats each node in the tree (which is not a leaf) as a sub-election (see Algorithm CSL). Conitzer et al. [3] note that a team wins a sub-election if and only if they must win one of its children and they can defeat one of the potential winners on the other side. It is perhaps simpler to understand this algorithm from a bottom up perspective. Observe that if we have two leaf nodes v_i and v_j and there exists an arc in the tournament (v_i, v_j) then v_i wins the match and is a potential winner of the sub election between v_i and v_j. Now suppose that v_i is in the member of the coalition so it is possible for them to replace (v_i, v_j) with (v_j, v_i) in the tournament and therefore v_j is also a potential winner of the sub-election via manipulation. Assume we have some sub-election in the middle of the tournament with two sets of potential winners A and B. Any team from A is a potential winner of the sub-election if there exists a team in B that they can defeat or if a coalition member in B throws a game. The same is true for teams in B. Therefore, there is a constructive manipulation if the desired winner is a member of the potential winners at the top node in the cup tree.

The original algorithm looked at $O(m^2)$ pairs of opponents as no two teams were compared more than once. Note that the original analysis provided a looser $O(m^3)$ bound on the number of comparisons, but this can be tightened by an observation of Vu et al. [12]. The difference between direct manipulation of the tournament and the method by Conitzer, Sandholm and Lang is that determining if a team could defeat another team meant summing all values of the n

Algorithm:CSL(v_w,c,T,C)

input : A team v_w, a cup tree c, a tournament graph T, and a coalition of
teams C

output: Returns true if v_w can win via manipulation and false otherwise

winners $\leftarrow$ PossibleWinners(c,T,C);

if $v_w \in$ winners then
| return true;
else
| return false;

Procedure: PossibleWinners(c,T,C)

input : A cup tree c, a tournament graph T and a coalition of teams C

output: Returns the set of possible winners of the cup tree via manipulation of
the tournament by the coalition

if leaf(c) then
| return $\{c\}$;
else
| winners $\leftarrow \{\}$;
| LeftWinners $\leftarrow$ PossibleWinners(left(c) ,T,C);
| RightWinners $\leftarrow$ PossibleWinners(right(c) ,T,C);
| forall $v_i \in$ LeftWinners do
| | if $\exists v_j \in$ RightWinners $\;$ *such that* $\;$ $(v_i, v_j) \in E \vee v_j \in C$ then
| | | add(winners,v_i);
|
| forall $v_j \in$ RightWinners do
| | if $\exists v_i \in$ LeftWinners $\;$ *such that* $\;$ $(v_j, v_i) \in E \vee v_i \in C$ then
| | | add(winners,v_j);
|
| return winners;

voters requiring $O(n)$ time whilst in the direct manipulation of the tournament
this can be done in constant time. Therefore, constructive manipulation of the
tournament under the cup rule takes just $O(m^2)$ time.   □

We observe that destructive manipulation of a competition using tournament
manipulations is similar since this simply requires determining if there is at
least one other possible winner of the tournament via manipulations.

Theorem 2. *Determining if a cup tournament can be destructively manipulated
using tournament manipulations takes polynomial time.*

Proof. We just determine if we can constructively manipulate the tournament
for each other team in turn than the one we wish to lose.   □

3.2   Round Robin Competition

For round robin competitions, manipulations of the tournament can be computed in polynomial time for a restricted class of scoring models. We define a

scoring model to be the set of tuples giving the possible outcomes of a game. Copeland scoring has a simple win-loss ($\{(0,1),(1,0)\}$) scoring model where the wining team earns one point and the losing team earns none. Bartholdi, Tovey and Trick[2] showed that constructive manipulation can be determined in polynomial time for a chess scoring model ($\{(0,1),(\frac{1}{2},\frac{1}{2}),(1,0)\}$). Faliszewski et al. [4] showed that for a range of scoring models manipulating Copeland voting is NP-Complete.

First, we discuss the problem of determining which games need to be manipulated to ensure that a given team v_w wins the competition. Clearly, there are some games that cannot be affected by the coalition and are fixed. All other games are manipulable. Games between coalition members can earn any of the possible scores allowed by the scoring model. We restrict games against non-coalition members by only allowing the manipulator to earn less points and the non-member earns more. Determining if a given team can be made a winner is analogous to determining if a team wins a round robin tournament when the fixed games have been played and the manipulable games have not been played. The restriction of the outcomes on games between coalition and non-coalition members requires that the games have outcomes within only a subset of the scoring model. Using this observation, we obtain the following theorem.

Theorem 3. *Determining if there exists a constructive manipulation of a round robin competition is polynomial if the normalized scoring model is of the form $S = \{(i, n - i) \mid 0 \le i \le n\}$ and NP-complete, otherwise.*

Proof. This proof uses the equivalence of determining whether a team can win a tournament and determining if a constructive manipulation exists with a set of fixed and manipulable games. Note that a game between a non-coalition member v_i and a coalition member v_j is unfixed but the scores that can be assigned are restricted. When the scoring model is of the form $S = \{(i, n - i) \mid 0 \le i \le n\}$ and the initial result of the game is (c_i, c_j), then the remaining valid scores that can be assigned are those from (c_i, c_j) to $(n, 0)$. By normalizing this new model, we obtain one in which the non-coalition member earns c_i points by default and the result of the game is scored from the model $\{(0, c_j), \ldots, (n - c_i, 0)\}$ which is of the form $S = \{(i, n - i) \mid 0 \le i \le n\}$. Kern and Paulusma [7] showed that determining if a team can win a tournament (i.e. is not eliminated from competition) takes polynomial time if the normalized scoring model is of the form $S = \{(i, n - i) \mid 0 \le i \le n\}$ and is NP-complete otherwise.   □

By comparison, it is always polynomial to determine if a destructive manipulation exists.

Theorem 4. *Determining if there is a destructive manipulation of a round robin competition takes polynomial time.*

Proof. Assume that v_l is the team that the coalition desires to lose. It is sufficient to check whether the maximum points of another team via manipulation is greater than the points of v_l. If v_l is a member of the coalition and therefore a manipulator, for each team i that we check for points, we apply only manipulations that increase the relative points between i and v_l. For all other teams, we apply the manipulation which decreases the points of v_l the most. If v_l is not a member of the coalition, no games involving v_l may be manipulated since we restrict manipulations to allow only those manipulations that increase the points of v_l and increase the relative gap between v_l and the manipulator. Therefore, no other team is better off when games involving v_l are manipulated. In both cases, we apply the manipulation that increase the points of the team under consideration against all other teams. If the total number of points of any other team is greater than the points of v_l under these manipulations, then there is a destructive manipulation of v_l. This algorithm can be run in $O(n^2)$ time. □

A further complication is when the goal of manipulation is just to earn a berth in the next round of the playoffs. It is NP-hard to decide these questions under most playoff systems for all scoring models [9,5].

4   Minimizing Manipulations

The number of manipulations required is an important factor. It may be advantageous for the coalition to manipulate as few games as possible to avoid detection or to minimize the cost of bribing players. We show that there is a polynomial algorithm to calculate manipulations which throw a minimal number of games. This highlights the vulnerability of the two most common types of competitions in sports to manipulation.

4.1   Minimal Number of Manipulations for Cup Competitions

Computing the minimal number of manipulations simply requires keeping a count within our algorithm for computing a manipulation. We give some notation to identify a specific sub-election in the cup. We let $s_\ell^{v_i}$ be the sub-election at level ℓ where v_i is a leaf node of a sub tree below $s_\ell^{v_i}$. We denote the level as the height from the bottom of the cup tree, which is assumed to be a perfect binary tree. We also define level 0 to be the level belonging to the leaves. We have m^2 constants c_{ij} that are 1 if $(v_j, v_i) \in M$ and 0 otherwise, where $M \subseteq E$ is the set of edges which can be manipulated by the coalition. This corresponds to $c_{ij} = 1$ when a manipulation must occur for v_i to win and 0 otherwise. Finally, we define the minimal number of manipulations needed to win a sub-election $s_\ell^{v_i}$, $m(v_i, s_\ell^{v_i})$, to be sum of the minimal number of the manipulations for v_i to win one of the children of $s_\ell^{v_i}$, and the minimum number of manipulations plus c_{ij} over all possible winners of the other child which v_i can defeat. We denote the set of teams that v_i can defeat either as described in the tournament or by manipulation as D_i. More formally, the minimal number of manipulations for v_i at $s_\ell^{v_i}$ ($\ell \geq 0$) is given by:

$$m(v_i, s_\ell^{v_i}) = \begin{cases} 0 & \text{if } \ell = 0 \\ m(v_i, s_{\ell-1}^{v_i}) + \min_{v_j \in D_i}(m(v_j, s_{\ell-1}^{v_j}) + c_{ij}) & \text{if } \ell > 0 \end{cases}.$$

Lemma 1. *The minimal number of manipulations needed to make a team v_i a winner at level n in the tree is equal to $m(v_i, s_n^{v_i})$.*

Proof. By induction. First, observe that the minimal number of manipulations at a leaf is 0. Hence, $m(v_i, s_0^{v_i}) = 0$ for all leaves v_i. Next note that at level 1 there are only 2 nodes in the possible winner sets of the leaves. Therefore if v_i can defeat v_j, $m(v_i, s_1^{v_i}) = m(v_i, s_0^{v_i}) + m(v_j, s_0^{v_j}) + c_{ij} = c_{ij}$ which is the exact number of manipulations that have occurred to make v_i a possible winner so far. We assume the premise for $1 < n \leq k$. Now, $m(v_i, s_{k+1}^{v_i}) = m(v_i, s_k^{v_i}) + \min_{v_j \in D_i}(m(v_j, s_k^{v_j}) + c_{ij})$. We know that $m(v_i, s_k^{v_i})$ is the minimal number of manipulations for v_i up to level k by the assumption and, for every $v_j \in D_i$, we know that $m(v_j, s_k^{v_j})$ is also the minimal number of manipulations for each v_j up to level k. By definition, c_{ij} is the number of manipulations for v_i to defeat v_j. Since v_i can defeat any v_j in D_i, the one with the fewest previous manipulations to reach k plus c_{ij} leads to the fewest manipulations in total to make v_i win the sub election $s_{k+1}^{v_i}$. This equals the minimum over the set D_i. Therefore the lemma holds for $k+1$ and, by induction, all n levels of the tree.   □

Theorem 5. *A modified CSL algorithm, where the team which minimizes the value of $m(v_i, s_n^{v_i})$ is selected to lose to team v_i at every node $s_n^{v_i}$, calculates the minimal number of manipulations needed to constructively or destructively manipulate a cup competition in polynomial time.*

Proof. By Lemma 1, the value of $m(v_w, s_n^{v_w})$ at the root node is the minimal number of manipulations which ensures v_w is the winner. Hence, we just need to show that the algorithm remains polynomial. The modified CSL algorithm still makes $O(m^2)$ comparisons. The only difference is that we have to calculate the minimum which can be done by storing the minimum as each team is checked. Therefore, the time complexity remains $O(m^2)$ and calculating the minimum is polynomial. Constructive manipulation requires calculating $m(v_w, s_n^{v_w})$ whilst destructive manipulation requires the minimum over all other teams.   □

4.2   Minimal Number of Manipulations for Round Robin Competitions

We consider here just Copeland scoring. We conjecture that similar methods could be developed for other scoring schemes.

Definition 1. *Given a tournament $T = (V, E)$ where $V = \{v_1, \ldots, v_n\}$, a set of manipulable edges $M \subseteq E$, and a distinguished node v_w, the Minimal Number of Manipulations under Copeland Scoring is the problem of determining the minimal number of edges in M that can be reversed such that $\forall_{v_k \in V, v_w \neq v_k} outDegree(v_w) \geq outDegree(v_k)$.*

Note that Copeland Scoring is the simple win-loss method of scoring where the winning team earns 1 point and the losing team earns 0 points. Before we show how to calculate the minimal number of manipulations, we show that we can determine the out degree, i.e. the Copeland Score, of the distinguished node using a minimal number of manipulations in isolation with a greedy algorithm. The intuition behind this is that we select manipulations to increase the out degree of v_w.

Lemma 2. *The value of $outDegree(v_w)$ can be determined in isolation by greedily using, in sequence, a minimal number of manipulations of edges $(v_i, v_w) \in M$ where $\forall_{(v_j, v_w) \in M, v_j \neq v_i} outDegree(v_i) \geq outDegree(v_j)$ until $\forall_{v_k \in V, v_w \neq v_k} outDegree(v_w) \geq outDegree(v_k)$.*

Proof. First, we prove that it always uses the least number of manipulations to increase the out degree of v_w. To reduce the out degree of two or more nodes that have an out degree larger than v_w, it takes at least two manipulations but to increase the out degree of v_w by the same amount takes just one. For a single node, it is preferred to use the manipulation involving v_w since the other node may increase the out degree of another node requiring more manipulations. Therefore, using manipulations involving v_w is most efficient.

Now we show that we never overshoot the stopping criteria and use more than a minimal number of manipulations. Assume that we use more than the minimal number of manipulations. This means that we selected an edge that did not decrease the maximum out degree when there existed an edge that would have decreased the maximum out degree of all nodes that we did not select. However, since we always selected the edge where the source node had the maximal out degree within M, we always decreased the maximum out degree whenever possible. This is a contradiction and the greedy algorithm only uses a minimal number of manipulations when reaching the stopping condition.   □

Theorem 6. *Determining the minimal number of tournament manipulations required under Copeland Scoring takes polynomial time.*

Proof. We define c to be the out degree of the distinguished node, v_w, calculated using the greedy algorithm. This corresponds to the number of wins earned by v_w. If the stopping condition has not been reached, we must use c to determine how many more manipulations are necessary. We construct a winner determination flow graph as described by Kern and Paulusma [7] and Gusfield and Martel[5](See Fig. 1, for example). We add a weight of 1 to each edge (v_i, v_j) where $(v_i, v_j) \notin V$ and therefore represents a manipulation. All other edges have the weight 0. The feasible flow which uses the fewest of the non-zero edges is the minimal number of tournament manipulations to achieve a constructive manipulation. Since the value of c can be determined in a linear number of steps, we only need to do a single min cost flow computation, which is polynomial, to determine the remainder of the minimum number of manipulations necessary to make v_w the team with the highest Copeland score.   □

Example 1. An example tournament can be seen in Fig. 1a. There are 5 teams in this tournament: v_0 to v_4. Suppose teams v_0 and v_3 form a coalition to manipulate the tournament so that v_0 wins. We want to determine the minimum

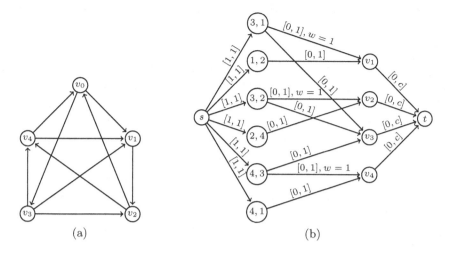

Fig. 1. (a) The tournament graph for five teams. The distinguished node in the example is v_0 which has formed a coalition with v_3. The manipulable edges are (v_3, v_1), (v_3, v_2), (v_3, v_4), (v_0, v_1) and (v_0, v_3). Edges (v_1, v_2) (v_2, v_4) and (v_4, v_1) cannot be manipulated by the coalition. (b) The min cost flow graph used to calculate the minimum number of manipulations for a given value constructed from the tournament in Fig. 1a. The distinguished team is v_0, $c = 2$ and all weights not shown are 0.

number of manipulations needed to ensure that v_0 is the winner. This requires switching any of the arcs where team v_3 wins. We know that the value of $c = 2$ since none of v_0's edges are manipulable in v_0's favour. We construct the graph seen in Fig. 1b to determine for $c = 2$ if there is a feasible solution. The solution returned has a minimum cost which is equal to the minimum number of manipulations needed to get a feasible flow with the value c plus any used in the greedy algorithm.

5   Reseeding

If we add multiple seeding rounds then computing a manipulation appears difficult. Recall that ranked reseeding matches the best remaining teams against the worst remaining teams in each round. The CSL algorithm cannot therefore be applied and a general solution is not known. However, if the size of the coalition is a constant c, then we can determine a manipulation in polynomial time.

Theorem 7. *For a ranked reseeding cup competition, if the manipulating coalition is of bounded size c, then determining a set of manipulations that makes a team win takes polynomial time.*

Proof. The key observation is that with a constant sized coalition there are only a polynomial number of ways to manipulate the games by rearranging the tournament graph. It suffices to check the winner of each of the polynomial number of fixed tournament graphs. For each fixed tournament graph, the winner can be determined in linear time as there are only $O(m)$ matches to check.

We show that there are only a polynomial number of different arrangements of manipulations. First note that at most c of the $\frac{m}{2}$ matches in the first round have more than one team as a possible winner. This means that there is at most 2^c possibilities to examine after each round. As there are $\log(m)$ rounds, we consider at most $(2^c)^{\log m}$ $(=m^c)$ possibilities. Hence there are at most $O(m^c)$ arrangements of manipulations for an unfixed cup with ranked reseeding and a constant sized coalition. It is sufficient to check each arrangement, which can be done in linear time. This gives a polynomial algorithm for bounded c.   □

With random reseeding the problem can be separated into two issues: determining whether manipulation is possible to make a team a winner under every possible seeding and determining if there exists any seeding such that the coalition can manipulate the games to make a given team the winner. It is unknown whether either of these problems have polynomial algorithms. Vu et al. [12] and Hazon et al. [6] tackle some probabilistic variants of possible winners without manipulation of games. However, the complexity of determining possible winners with a win-loss tournament graph in balanced cup trees remains open [8,6,10].

6   Double Elimination Competitions

In a double elimination competitions, a manipulation of the tournament does not automatically bounce the manipulator out of the tournament as in the single elimination case. However, it does guarantee that the manipulator will be bounced to the secondary bracket from the primary bracket on the first manipulation and out of the tournament on the second manipulation. As in the case of ranked reseeding, a general solution is not known but there is a polynomial algorithm for double elimination tournaments if the coalition is of constant size.

Theorem 8. *For double elimination tournaments, if the coalition is of a constant size c, determining whether there is a constructive manipulation takes polynomial time.*

Proof. This proof follows similar lines as the proof for ranked reseeding. We will show that there is a polynomial number of manipulation scenarios which can be checked in linear time. If there is a coalition of size c then a team can manipulate the cup only once if they wish to win the tournament and twice if they desire another team to win. At each step in the tree, a team must decided whether they wish to manipulate or not. Before and after they have manipulated once, there remains c teams which can manipulate. Only after they have manipulated a second time are they removed from the competition. This means there are at most 2^c manipulations at each of the $logm$ levels. This gives us $O(2^{\log m c})(= O(m^c))$ possibilities that can be checked in linear time, which gives a polynomial algorithm for determining if there is a constructive manipulation.   □

7   Conclusions and Open Problems

In sporting tournaments, teams can directly manipulate the tournament graph. We showed that algorithms used to compute manipulations of votes in elections can be modified to determine the manipulations needed of the tournament graph. We proved that such direct manipulation of the fixed cup and round robin competitions can be computed in polynomial time. In a similar way, we can determine the minimal number of manipulations needed. For ranked reseeding of cup competitions, we showed that it is easy to calculate the number of manipulations if the size of the manipulating coalition is bounded by a constant. We also gave a polynomial time algorithm for double elimination tournaments for a constant sized coalition. A number of open question remain. The manipulation of various variations of the cup competition have unknown complexity including the ranked and random cup competitions. For random cup competitions, the complexity of manipulation is also unknown if the size of the coalition is bounded. Similarly, the complexity of manipulating double elimination competitions is still undetermined when the size of the coalition is unbounded.

References

1. Altman, A., Procaccia, A.D., Tennenholtz, M.: Nonmanipulable Selections from a Tournament. In: Proc. of 21st Intl. Joint Conf. on AI (2009)
2. Bartholdi, J.J., Tovey, C.A., Trick, M.A.: The computational difficulty of manipulating an election. Social Choice and Welfare 6, 227–241 (1989)
3. Conitzer, V., Sandholm, T., Lang, J.: When are elections with few candidates hard to manipulate? J. ACM 54, 1–33 (2007)
4. Faliszewski, P., Hemaspaandra, E., Schnoor, H.: Copeland voting: Ties matter. In: Proc. of the 7th Int. Conf. on Autonomous Agents and Multiagent Systems (2008)
5. Gusfield, D., Martel, C.E.: The structure and complexity of sports elimination numbers. Algorithmica 32, 73–86 (2002)
6. Hazon, N., Dunne, P.E., Kraus, S., Wooldridge, M.: How to Rig Elections and Competitions. In: Proc. of 2nd Int. Workshop on Computational Social Choice (2008)
7. Kern, W., Paulusma, D.: The computational complexity of the elimination problem in generalized sports competitions. Discrete Optimization 1, 205–214 (2004)
8. Lang, J., Pini, M.S., Rossi, F., Venable, K.B., Walsh, T.: Winner Determination in Sequential Majority Voting. In: Proc. of the 20th Int. Joint Conf. on AI (2007)
9. McCormick, S.T.: Fast algorithms for parametric scheduling come from extensions to parametric maximum flow. Operations Research 47, 744–756 (1999)
10. Pini, M.S., Rossi, F., Venable, K.B., Walsh, T.: Dealing with incomplete agents' preferences and an uncertain agenda in group decision making via sequential majority voting. In: Proc. of the 11th Int. Conf. on Principles of Knowledge Representation and Reasoning (2008)
11. Tang, P., Shoham, Y., Lin, F.: Team Competition. In: Proc. of 8th Intl. Conf. on Autonomous Agents and Multiagent Systems (2009)
12. Vu, T., Altman, A., Shoham, Y.: On the Complexity of Schedule Control Problems for Knockout Tournaments. In: Proc. of 8th Intl. Conf. on Autonomous Agents and Multiagent Systems (2009)
13. Walsh, T.: Where are the really hard manipulation problems? The phase transition in manipulating the veto rule. In: Proc. of 21st Intl. Joint Conf. on AI (2009)

Iterated Majority Voting

Stéphane Airiau and Ulle Endriss

Institute for Logic, Language and Computation
University of Amsterdam
{s.airiau,u.endriss}@uva.nl

Abstract. We study a model in which a group of agents make a sequence of collective decisions on whether to remain in the current state of the system or switch to an alternative state, as proposed by one of them. Examples for instantiations of this model include the step-wise refinement of a bill of law by means of amendments to be voted on, as well as resource allocation problems, where agents successively alter the current allocation by means of a sequence of deals. We specifically focus on cases where the majority rule is used to make each of the collective decisions, as well as variations of the majority rule where different quotas need to be met to get a proposal accepted. In addition, we allow for cases in which the same proposal may be made more than once. As this can lead to infinite sequences, we investigate the effects of introducing a deadline bounding the number of proposals that can be made. We use both analytical and experimental means to characterise situations in which we can expect to see a convergence effect, in the sense that the expected payoff of each agent will become independent from the initial state of the system, as long as the deadline is chosen large enough.

1 Introduction

We consider the very general problem where a finite set of *agents* must choose one *alternative* among many, and we are interested in *decentralised* solutions. The alternatives may represent different policies, world states, or allocations of resources, etc. One simple idea is to start with an initial current alternative, let a *random* agent propose a different alternative, and organize a vote between these two alternatives. The process *iterates* using the winner of the election as current alternative. Unfortunately, with no restriction on the agents' preferences, this simple process may iterate forever. In this paper, we investigate the possibility of using a *bound* on the number of iterations. In particular, we investigate the problem of determining whether the choice of the bound can guarantee that no agent benefits unduly from the choice of the initial alternative.

An example from political science that fits our generic model is the step-wise refinement of a bill of law by means of amendments to be voted on (although in this case, repeating the same proposal twice may not be allowed). Another example concerns multiagent resource allocation problems: some work in multiagent systems has focussed on negotiation scenarios where agents approach a solution in small steps rather than computing the best solution in one go [1,2].

F. Rossi and A. Tsoukis (Eds.): ADT 2009, LNAI 5783, pp. 38–49, 2009.

In this case, every possible allocation of resources constitutes an alternative and proposing a new alternative means proposing a deal regarding the reallocation of some items. Here, the voting rule used would typically require each agent affected by the proposal to give their consent. Finally, recent work in computational social choice has shown that decomposing combinatorial voting problems into a sequence of smaller elections has a number of advantages [3].

In general, there are many possible choices of a voting rule. In this paper, we focus on the *majority rule*, which specifies that a proposal is accepted if at least half of the concerned agents (those that are not indifferent) vote in favour. This is clearly a very natural choice, and it is the only rule that is anonymous, neutral, and monotone (May's Theorem [4]). We also consider generalisations of the majority rule, where a quota different from 50% may be needed to accept a proposal. For example, for some important elections, a higher proportion of votes in favour of an alternative is needed, e.g., a two-thirds majority in parliament is needed to change the constitution. As we allow agents to make the same proposal over and over, there is the possibility for cycles. This phenomenon is linked to the fact that the preference relation we obtain when several individual preferences are aggregated by means of the majority rule need not be transitive. In social choice theory, one approach to address this problem has been to restrict the range of allowed preferences, e.g., to single-peaked preferences or preferences meeting Sen's triple-wise value restriction [5]. When we have no control over the agents' preferences, we need to modify the protocol to induce the agents to choose a good social alternative. A simple solution to the problem of cycles is to introduce a deadline that limits the number of iterations.

We assume that the agents' preferences are common knowledge and that agents are strategic: they will make proposals and vote in elections so as to maximise their expected payoff. What would be a good choice of deadline under these circumstances? If it is small, the choice of initial state will play an important role for the final outcome. If it is large, as we shall see, it is sometimes the case that the expected payoff of an agent becomes independent of the initial state, which provides some level of fairness. Our aim in this paper is to get a clearer understanding of such convergence effects.

The remainder of this paper is as follows. Section 2 further motivates and defines our model. In particular, we detail how strategic agents can compute their best moves using backward induction. Section 3 formally defines the notion of convergence as used here and establishes sufficient conditions on a game for being convergent. Then, Section 4 takes this analysis further by mapping out the convergence behaviour for a wider class of games by means of an experimental study. We conclude with a discussion about related work and present future axes of research.

2   The Model

We study games consisting of a finite set N of n *agents* and a finite set X of m *alternatives* or *states*. Each agent $i \in N$ has a *utility* function $u_i : X \to \mathbb{R}$, which is assumed to be common knowledge. Each agent is rational, i.e., it maximizes its expected utility. The utility functions are represented together as an $m \times n$ matrix U_0, with $U_0(x, i) = u_i(x)$. We do *not* assume that utility is transferable: the utility of two agents may not be comparable (e.g., agents can use different currencies). However, we do assume that agents will use their knowledge of the utility of other agents for predicting their behaviour.

A game proceeds in successive iterations. At each iteration t, there is a current alternative $x(t)$ (a given allocation, a current bill). One agent is randomly selected, with equal probability, to propose a new alternative $x^\star$ to be considered (e.g., a new allocation, an amendment to the current bill). An agent may propose an alternative that was already proposed in the past; and it may propose to maintain the *status quo* by proposing $x^\star = x(t)$. The agents vote between $x^\star$ and $x(t)$. If the proposed alternative wins the election, and we will present the criterion to win an election next, it replaces the current alternative for the next iteration. Else, the current alternative remains in place for the next iteration.

Elections are decided using a *quota system* for some fixed quota q: a proposal will be declared the winner iff it receives at least q percent of the votes. More precisely, if $n_\oplus$ agents are voting in favour and $n_\ominus$ against a proposal (and some agents may abstain), then the proposal is accepted if $n_\oplus > q \cdot (n_\oplus + n_\ominus)$. The standard majority rule is the quota system with $q = 50\%$. When the majority rule is used, cycles may occur. The same is true for quota systems with $q \neq 50\%$. In the presence of a cycle, the sequence of elections could be infinite. In order to force the eventual choice of an alternative, we propose the use of a *deadline* limiting the number of iterations to be played. The following definition summarises the components that make up a game:

Definition 1 (Game). *A game is a quadruple $\langle N, X, U_0, q \rangle$, where N with $n = |N|$ is a finite set of agents, X with $m = |X|$ is a finite set of states (or alternatives), U_0 is an $m \times n$ matrix defining the utility each agent assigns to each state, and $q \in [0, 1]$ is a quota (typically expressed in percent).*

Playing a game requires us to also specify a *deadline*, i.e., the number of iterations to be played, and an *initial state* from X.

2.1 Backward Induction

Agents are assumed to be expected-utility maximisers, i.e., the goal of a proposal or a vote is to maximise expected utility in the final state. We now discuss how to perform this strategic vote and strategic choice of proposals.

Let the matrix W_t of size $m \times m$ specify the transition between the alternatives at iteration t. An entry $W_t(x, y)$ of W_t is the probability to have alternative y become the current alternative at the next iteration when alternative x was current at iteration t. A row of W_t is a probability distribution over the alternatives that can become current at the next step. Hence, the sum of the entries of a row is equal to 1, i.e., W_t is a stochastic matrix. Furthermore, let the matrix U_t of size $m \times n$ contain the expected utility of all agents for all alternatives, i.e., $U_t(x, i)$ is the expected payoff of agent i for alternative x at iteration t. We have $U_{t+1} = W_{t+1} \cdot U_t$, and therefore:

$$U_{t+1} = \left(\prod_{\tau=t+1}^{1} W_\tau \right) \cdot U_0 = W_{t+1} \cdot W_t \cdot W_{t-1} \cdot \ldots \cdot W_1 \cdot U_0$$

Next, we discuss how to compute W_{t+1} from U_t. Let us assume that the agents know what to propose and how to vote during the tth iteration for any alternative $x \in X$. Because of the common knowledge assumption on the utility functions, the possible proposals and votes of any agents are also known, hence the matrices $W_t, W_{t-1}, \ldots, W_1$ are known. What should an agent do during iteration $t + 1$?

How to vote? The decision depends on the comparison of the expected utility of the current alternative $x(t)$ with the one of the proposed alternative $x^\star$, i.e., agent i will vote in favour of the proposal when $U_t(x^\star, i) > U_t(x(t), i)$ and against when $U_t(x^\star, i) < U_t(x(t), i)$. Note that the agent does not vote in case it is indifferent between the two alternatives.

What to propose? First, the agent needs to compute the outcome of the vote between the current state $x(t)$ and each possible alternative $x' \in X$. Let $X^w \subseteq X$ denote the set of winning alternatives against $x(t)$. For agent i, the set of best proposals is $P_i = \text{argmax}_{x' \in X^w} U_t(x', i)$. If the expected utility of the alternatives in P_i is greater than $U_t(x(t), i)$ (the expected utility of the current alternative), then agent i proposes with equi-probability one of the alternatives in P_i. Else, agent i is content with the current alternative and proposes maintaining the *status quo* (there is no decision to make since $x^\star = x(t)$). Since each agent has an equal probability to be selected to make a proposal, we can compute the probability of any alternative to be proposed. And since any alternative that is proposed is winning against the current alternative, we can compute the probability of an alternative to become current at the next iteration.

To summarise, a game $\langle N, X, U_0, q \rangle$ induces a sequence of $m \times m$ transition matrices $W_1, W_2, \ldots$, as described above, as well as a sequence of $m \times n$ matrices $U_1, U_2, \ldots$, fixing the expected payoffs for each agent at each iteration, with $U_{t+1} = W_{t+1} \cdot U_t$. Here, "iteration 1" is the final iteration/step in a play of the game, "iteration 2" is the penultimate iteration, and so forth. These matrices allow us to study the game for all possible choices of initial current alternative.

2.2   Example: A Cycle with Majority Voting

Consider the following problem with 3 agents and 3 states. The utility vectors are: $\langle 4, 1, 2 \rangle$ for state a, $\langle 2, 4, 1 \rangle$ for state b and $\langle 1, 3, 4 \rangle$ for state c. The corresponding matrix U_0 is shown in Table 1. With no deadline, the agents would be stuck in a cycle, as a majority of agents would prefer to move to a different state $(a \rightarrow b \rightarrow c \rightarrow a \rightarrow \ldots)$. Using a deadline and backward induction, the agents can break the cycle. If the agents were voting sincerely instead of strategically in this example, the final state would be entirely determined by the initial state

Table 1. Breaking a cycle with a deadline

$$U_0 = \begin{bmatrix} 4 & 1 & 2 \\ 2 & 4 & 1 \\ 1 & 3 & 4 \end{bmatrix} \quad W_1 = \begin{bmatrix} 1/3 & 0 & 2/3 \\ 2/3 & 1/3 & 0 \\ 0 & 2/3 & 1/3 \end{bmatrix}$$

$$U_1 = \frac{1}{3}\begin{bmatrix} 6 & 7 & 10 \\ 10 & 6 & 5 \\ 5 & 11 & 6 \end{bmatrix} \quad W_2 = \begin{bmatrix} 1 & 0 & 0 \\ 1/3 & 1/3 & 1/3 \\ 2/3 & 0 & 1/3 \end{bmatrix} \quad U_\infty = \begin{bmatrix} 2.0795 & 2.6549 & 2.7560 \\ 2.0795 & 2.6549 & 2.7560 \\ 2.0795 & 2.6549 & 2.7560 \end{bmatrix}$$

$$U_2 = \frac{1}{9}\begin{bmatrix} 18 & 21 & 30 \\ 21 & 24 & 21 \\ 17 & 25 & 26 \end{bmatrix} \quad W_3 = \begin{bmatrix} 1/3 & 2/3 & 0 \\ 0 & 1/3 & 2/3 \\ 2/3 & 0 & 1/3 \end{bmatrix}$$

(there is a unique cycle). First, let us explain the computation of the expected payoffs; then we will describe the properties of the outcome.

State a would lose an election against state c, and win an election against state b. If the current alternative is state a one step before the deadline, the second and third agents should propose state c, the first agent should propose the *status quo*. As the agents are chosen to make a proposal with equi-probability, the probability to stay in a is $\frac{1}{3}$, the probability to move to c is $\frac{2}{3}$, and the probability to move to b is zero. This provides the first row of the matrix W_1 in Table 1. We carry this reasoning to complete W_1. The expected utility of the agents one step before the deadline is $U_1 = W_1 U_0$. We iterate the reasoning to obtain the matrices W_2, U_2, etc. We note that, in this example, W_1, W_2 and W_3 are different. We have implemented the iterative algorithm for computing the matrices W_t and U_t. For large values of t, U_t converges to a particular matrix (see U_∞ in Table 1). That is, for t large enough, the expected utility of each agent is independent of the initial state (e.g., agent 1's expected utility approaches 2.0795). Hence, if the deadline is far enough, no agent can take advantage of the initial choice of alternative. Finally, note that, in the limit, the expected utility of an agent is not simply the average utility over the 3 alternatives.

3   Convergence

The example given in the previous section shows that there are instances of games where we will observe some kind of *convergence*. For the example in question, we have seen that the expected payoff of any given agent for any given initial state will converge to a certain value as we increase the deadline (we call this *intra-state convergence*). We have also observed that the expected payoff will become less and less dependent on the initial state as we increase the deadline (we call this *inter-state convergence*). In this section, we will define these notions of convergence formally and identify some classes of games for which convergence can be guaranteed.

3.1   Types of Convergence

We define a game as being intra-state convergent if, for any given agent and any initial state, the difference in expected payoff for small changes in the deadline will become arbitrarily small as deadlines become larger:

Definition 2 (Intra-state convergence). *A game* $\langle N, X, U_0, q \rangle$ *is said to be intra-state convergent if, for any agent* $i \in N$ *and any state* $x \in X$, $\lim_{t \to \infty} [U_t(x, i) - U_{t+1}(x, i)] = 0$.

Next, we define a game as being inter-state convergent if, for any agent and any two states, the difference in expected payoff for making either one of these states the initial state can be made arbitrarily small when we increase the deadline:

Definition 3 (Inter-state convergence). *A game* $\langle N, X, U_0, q \rangle$ *is said to be inter-state convergent if, for any agent* $i \in N$ *and any two states* $x, x' \in X$, $\lim_{t \to \infty} [U_t(x, i) - U_t(x', i)] = 0$.

Inter-state convergence provides a level of *fairness*: as long as we choose a sufficiently large deadline, the choice of initial state will not affect the expected payoff of the individual agents. This does not mean that all agents can be expected to do equally well, but it does mean that one important parameter that determines how a game is played (the initial state) does not influence the (expected) outcome. Intra-state convergence does not directly affect fairness, but offers some level of *robustness* of the mechanism: expected payoffs will not depend on the exact deadline chosen (for instance, we would avoid situations whereby an agent's expected payoff could depend on whether the chosen deadline is odd or even).

Our example did exhibit both types of convergence. Both are entailed by a third notion of convergence, expressed in terms of the transition matrices W_t induced by a game.

Definition 4 (Fundamental convergence). *A game* $\langle N, X, U_0, q \rangle$ *is said to be fundamentally convergent if the limit of the product of its transition matrices*

$$W = \lim_{t \to \infty} \prod_{\tau = t}^{1} W_\tau \text{ is a matrix in which all row vectors are identical.}$$

Definition 4 is reminiscent of the Fundamental Limit Theorem for regular Markov chains, which says that if P is the transition matrix of a regular Markov chain, then $\lim_{n \to \infty} P^n = W$ for some matrix W with identical row vectors [6]. Recall that a stochastic matrix P is a matrix defining a regular Markov chain if there exists a k such that P^k has only non-zero elements. In particular, the theorem applies when P itself only has non-zero elements. Inspection of the standard proof of the Fundamental Limit Theorem shows that the same is true for the product of several *different* matrices, provided that an infinite number of them are zero-free. While there are similarities to our scenario, we stress that the Fundamental Limit Theorem does *not* apply here, because the transition matrices generated by our games need neither be zero-free nor regular.

Before linking the three definitions of convergence, we state a simple property of stochastic matrices. The proof is standard and omitted for lack of space.

Lemma 1. *Let A and B be $m \times m$ stochastic matrices. If B has all row vectors identical, then $A \cdot B = B$.*

Proposition 1. *If a game is fundamentally convergent, then it is also inter-state convergent and intra-state convergent.*

Proof. Let $W = \lim_{t \to \infty} \prod_{\tau=t}^{1} W_\tau$ for the game under consideration. Suppose the game is fundamentally convergent, i.e., W is a matrix of identical row vectors. Then the game is also inter-state convergent: if we multiply W with the payoffs for agent i, then we get the same expected payoff for any initial state.

Next, we show that intra-convergence also follows. By Lemma 1, as $\prod_{\tau=t}^{1} W_\tau$ converges to a matrix with identical rows, $W_{t+1} \cdot \prod_{\tau=t}^{1} W_\tau$ converges to that very same matrix. The former determines $U_t(x, i)$, while the latter determines $U_{t+1}(x, i)$. Hence, their difference must converge to 0. $\square$

Furthermore, it is not hard to verify that inter-state convergence implies intra-state convergence. Intra-state convergence is weaker than the other two forms of convergence. There are games that are intra-state convergent but not inter-state convergent. A simple example would be a game with 2 agents, 2 states, and $q = 50\%$, where agent 1 prefers state a and agent 2 prefers state b. Then, if the initial state is a, this will remain the *status quo*, independently of the deadline (and analogously for the case where b is the initial state). The transition matrices for this game are all equal to the identity matrix. Hence, inter-state convergence is not satisfied (your expected payoff is equal to the utility you assign to the initial state), while intra-state convergence is (your expected payoff remains constant when we vary the deadline).

3.2   Sufficient Conditions

We will now identify sufficient conditions for a game to be convergent. There are some clear-cut cases, when the quota q takes extreme values. First, if $q = 0\%$, i.e., when a single agent in favour is sufficient for a proposal to be accepted, then all forms of convergence are satisfied. In such a game, whichever agent is chosen to make a proposal in the last iteration will propose their favourite state, and that motion will carry—independently from the current state. Hence, W_1 (the transition matrix for the last iteration) will be a matrix with all rows equal. Therefore, by Lemma 1, also the product W of all transition matrices will be such a matrix, which means that the game will satisfy the convergence condition of Definition 4, and by Proposition 1 also all other notions of convergence.

Second, if $q = 100\% - \epsilon$, i.e., when only proposals opposed by no agent are accepted, then a proposal is accepted iff it represents a Pareto improvement. Hence, the final state will be Pareto optimal—provided the deadline is chosen large enough. But if there is more than one Pareto optimal state, then the expected payoff can be different for an agent if either one of these Pareto optimal

Table 2. Example with a unique Condorcet winner: the final outcome may not be the Condorcet winner

$$
U_0 = \begin{bmatrix} 4 & 4 & 4 \\ 6 & 2 & 3 \\ 2 & 6 & 2 \\ 3 & 0 & 6 \end{bmatrix} \quad W_1 = \begin{bmatrix} 1 & 0 & 0 & 0 \\ 2/3 & 1/3 & 0 & 0 \\ 0 & 1/3 & 1/3 & 1/3 \\ 1/3 & 1/3 & 0 & 1/3 \end{bmatrix} \quad U_\infty = \begin{bmatrix} 4.3676 & 3.0288 & 3.9973 \\ 4.3676 & 3.0288 & 3.9973 \\ 4.3676 & 3.0288 & 3.9973 \\ 4.3676 & 3.0288 & 3.9973 \end{bmatrix}
$$

states is selected as the initial state. Therefore, inter-state convergence is not generally satisfied (and neither is fundamental convergence). Only in very special cases, such as when all agents are indifferent between all states, would be obtain inter-state convergence.

In addition, note that the presence of a Condorcet winner is not sufficient to guarantee convergence to that state. For example, it is possible that no agent proposes the Condorcet winner. In the example in Table 2 we observe inter-state convergence and there is a unique Condorcet winner (the state with payoffs $\langle 4, 4, 4 \rangle$), but the expected payoffs in the limit are not the payoffs of the Condorcet winner. This is because from the state with payoff $\langle 2, 6, 2 \rangle$, no agent has an incentive to propose the Condorcet winner.

We now analyse the case of games with two states in detail.

Proposition 2. *Any game with two states is intra-state convergent.*

Proof. In games $\langle N, X, U_0, q \rangle$ with $|X| = 2$, each transition matrix is of the following form:

$$
W_t = \begin{pmatrix} p & 1-p \\ 1-q & q \end{pmatrix}
$$

Two special cases are $p = q = 1$ (the identity matrix) and $p = q = 0$ (the "switch matrix"). We first claim that no transition matrix can be such a switch matrix: if all agents are indifferent between both states in iteration t, then W_t is the identity matrix. Otherwise, w.l.o.g., assume agent 1 prefers state 1 in iteration t. Then agent 1 will propose the *status quo* when in state 1, so $p \geq \frac{1}{n}$. This proves our claim.

So each transition matrix will either be the identity matrix or different from both the identity matrix and the switch matrix. We now distinguish two cases:

(i) There exists a transition matrix W_t that is the identity matrix. Then $U_{t+1} = U_t$, and W_{t+1} and all subsequent transition matrices are also the identity matrix. Hence, from point t on the expected payoff for any given state will not change anymore and we have intra-state convergence as required.

(ii) There does not exist a transition matrix that is the identity matrix. We shall prove inter-state convergence in this case (which entails intra-state convergence). W.l.o.g. we analyse the expected payoff of agent 1. Suppose x is its expected payoff for state 1 and y for state 2, at some step of the process. The expected payoffs for the next step are computed as follows:

$$\begin{pmatrix} p & 1-p \\ 1-q & q \end{pmatrix} \cdot \begin{pmatrix} x \\ y \end{pmatrix} = \begin{pmatrix} y + p \cdot (x - y) \\ x + q \cdot (y - x) \end{pmatrix}$$

Consider the difference $|x - y|$. In the next step, this difference becomes $|(y + p \cdot (x - y)) - (x + q \cdot (y - x))| = |p + q - 1| \cdot |x - y|$. The factor of change $|p + q - 1|$ is of course ≤ 1. But we can give a better bound. We know that neither $p = q = 0$ nor $p = q = 1$. We also know that p and q must be multiples of $\frac{1}{n}$, where n is the number of agents. (This follows from the rules of the game: each of the n agents has the same likelihood of being the proposer, and each agent will either propose the *status quo* or the other state with certainty—for the special case of two states no agent will ever randomise between several top proposals). Hence, $|p+q-1| \leq 1-\frac{1}{n}$. Now, if x and y are the actual utilities for the two states, t steps before the deadline the difference in expected payoffs can be at most $(1 - \frac{1}{n})^t \cdot |x - y|$. Therefore, as t goes to infinity, this difference must go to 0. This is the case for both agents, meaning that inter-state convergence holds as claimed.   □

Proposition 3. *Any game with two states and a quota of $q < 50\%$ is inter-state convergent.*

Proof. The proof of Proposition 2 shows that, first, no transition matrix can be the "switch matrix"; and second, *if* no transition matrix is the identity matrix (or the switch matrix), then inter-state convergence is satisfied. The only remaining possibility is when there exists a iteration t such that W_t is the identity matrix. We distinguish two cases:

(i) First, assume all agents are indifferent between states 1 and 2 in iteration t. Then we clearly have convergence.

(ii) Otherwise, w.l.o.g., assume agent 1 strictly prefers state 1 in iteration t. Then the only explanation for agent 1 not proposing state 1 when in state 2 is that such a proposal would not make the quota. Hence, as $q < 50\%$ by assumption, at least 50% $(1 - q)$ of the concerned agents (those not indifferent between state 1 and state 2 in iteration t) must prefer state 2. But then, when in state 1, each of these agents would propose to move to state 2 and any such proposal would get accepted, so the probability of moving from state 1 to 2 in iteration t must be > 0. Hence, W_t cannot be the identity, and we have a contradiction.□

The bound on the quota in Proposition 3 is tight: the example sketched after Proposition 1 demonstrates that inter-state convergence cannot be guaranteed anymore for quotas $q \geq 50\%$. In the next section, we will analyse our games experimentally, to see to what extent the trends reflected by Propositions 2 and 3 extend to larger games.

4 Experimental Analysis

The aim of the experiments is to investigate what parameters affect convergence. In all the experiments we performed, we have always observed intra-state convergence. In the following, we will therefore only report on inter-state convergence (and hence, we write simply "convergence" to denote inter-state convergence).

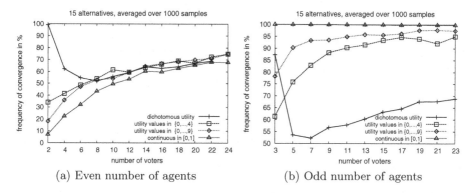

(a) Even number of agents   (b) Odd number of agents

Fig. 1. Frequency of convergence for $q = 50\%$

4.1   Varying the Utility Range

The three obvious parameters of the experimental study are the number of agents, the number of alternatives, and the quota. Another important parameter concerns the generation of utility functions. Each utility value could be drawn from a continuous or a discrete distribution (over either the interval $[0, 1]$ or a set of integers $\{0, 1, \ldots, u_{max}\}$. Here we focus on uniform distributions (other cases are also interesting and left for future work). The goal of our the first experiment is to answer the question: how does the set of possible utility values affect the rate of convergence?

. When a continuous uniform distribution is used to generate the utility values, the preference order on the set of alternatives is strict with probability one. When a discrete distribution is used, however, the agent may have a weak preference order as some alternatives may receive the same values. For a fixed number of alternatives, the smaller u_{max}, the more likely the agents have a weak order. In the extreme case of $u_{max} = 1$, agents have dichotomous preferences (i.e., either they approve or disapprove the alternative).

We fixed the number of states to 15 and varied the number of agents in $\{2 \ldots 24\}$. We generated 1000 utility matrices U_0 for each number of agents and checked for convergence for a quota of 50%. The results are provided in Figure 1.

First, we do not always observe convergence. Convergence is more frequently observed when the number of agents is large. Then, the frequency of convergence is higher when the number of agents is odd than when it is even. For an even number (Figure 1a), there is no significant difference between the different generators, except for very small numbers of agents.[1] For an odd number of agents (Figure 1b), the convergence occurred for all the tested cases with a continuous distribution. Finally, convergence is less frequent when u_{max} is small.

---

[1] For small numbers of agents and dichotomous utility functions, it is very likely (99% for 2 agents and 15 states) that at least one alternative y is preferred by all agents, which guarantees convergence.

These observations suggest that convergence is more frequent when there are fewer ties between alternatives. When ties are extremely unlikely, we observe high frequency of convergence (i.e., when the number of agents is odd and utilities are drawn from a continuous distribution; or when the number of agents is large, as exact ties would be required). When ties are likely, however, we observe a lower frequency of convergence (i.e., when the number of agents is even and small and the utility values are drawn from a continuous distribution; or when utilities are drawn from a discrete distribution and u_{max} is small). Experiments with different numbers of alternatives did not alter our conclusions.

4.2   Varying the Quota

For two-state games, we have seen that inter-state convergence is guaranteed for quotas $< 50\%$ and can fail for greater quotas. Our second experiment is aimed at checking whether the same trend can be observed for larger games, and at getting an understanding of the likelihood of convergence, when it cannot be guaranteed. We have used 15 alternatives and a population of 100 agents and we have randomly generated 1000 matrices U_0. The results are shown in Figure 2.

For $q < 50\%$, we always observe convergence, which leads us to conjecture that Proposition 3 generalises to games with any number of states. For higher quotas, we do not always observe convergence (about 80% of the time for $q = 50\%$); and for quotas $q \geq 60\%$, we have never observed convergence in our experiments. Still, even for $q = 100\%$, it is clear that cases satisfying convergence *do* exist (e.g., if all agents are indifferent between all states); such cases are just very unlikely to occur, certainly for our method of data generation.

Maybe the most striking effect we can observe in Figure 2 is the sudden and sharp decrease in the frequency of convergence at the 50% mark. This clearly singles out the majority rule as having a special status within the set of all quota rules, thus suggesting an interesting characterisation of this rule above and beyond the characterisations given by May's Theorem [4] and the Condorcet Jury Theorem.

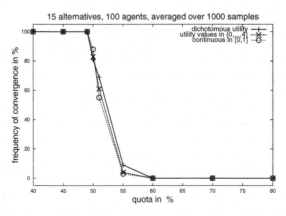

Fig. 2. Frequency of convergence with different quotas

5   Conclusion

We have proposed a model for iterated voting, where a group of agents make a social choice by implementing a sequence of binary decisions between the *status quo* and an alternative proposal. Each decision is made using a quota rule.

Our model is related to the study of *tournaments* [7] as we can think about our generic approach as a walk in the majority graph. The Markov solution is related to our work as it considers a random walk in the majority graph, and an outcome is in the solution set when it has a positive probability of being the current outcome in the limit. Our model differs by allowing strategic behaviour (*i*) in the choice of the proposed alternative and (*ii*) in the vote (an agent may vote in favor of a less preferred outcome in the short run if this promises a better outcome in the long run). Models closer to ours have been studied in political science, e.g., by Baron [8], although in that model each voter receives a payoff at every time step, while we only ascribe utility to the final state. In the work by Penn [9] another difference is that the challengers are drawn from a given probability density rather than proposed by the agents.

For the case of games with just two states, we have shown that the expected payoff of each agent converges as the deadline increases, when the initial state is fixed. For games with two states and a quota of less than 50%, we have furthermore shown that the expected payoff is independent of the initial state, which offers a level of fairness. Our experimental study shows that this trend generalises also to larger games. For the majority rule, corresponding to a quota of exactly 50%, we have seen that convergence is frequent, but cannot be guaranteed. We have also illustrated how the range of possible utility values an agent may assign to an alternative, and thereby the likelihood of ties, affect convergence.

Future work should be directed towards formulating further conditions under which convergence can be guaranteed, and prediction of a bound guaranteeing that no agent benefits from the choice of the initial state.

References

1. Rosenschein, J.S., Zlotkin, G.: Rules of Encounter. MIT Press, Cambridge (1994)
2. Endriss, U., Maudet, N., Sadri, F., Toni, F.: Negotiating socially optimal allocations of resources. Journal of Artif. Intell. Res. 25, 315–348 (2006)
3. Lang, J., Xia, L.: Sequential composition of voting rules in multi-issue domains. Mathematical Social Sciences 57(3), 304–324 (2009)
4. May, K.O.: A set of independent necessary and sufficient conditions for simple majority decisions. Econometrica 20, 680–684 (1952)
5. Sen, A.K.: Collective Choice and Social Welfare. Holden-Day (1970)
6. Grinstead, C.M., Snell, J.L.: Introduction to Probability, 2nd edn. American Mathematical Society (1997)
7. Laslier, J.F.: Tournament Solutions and Majority Voting. Springer, Heidelberg (1997)
8. Baron, D.P.: A dynamic theory of collective goods programs. The American Political Science Review 90, 316–330 (1996)
9. Penn, E.M.: A model of farsighted voting. American Journal of Political Science 53, 36–54 (2009)

Committee Selection with a Weight Constraint Based on Lexicographic Rankings of Individuals

Christian Klamler[1], Ulrich Pferschy[2], and Stefan Ruzika[3],[*]

[1] University of Graz, Institute of Public Economics, Universitaetsstr. 15, 8010 Graz, Austria
christian.klamler@uni-graz.at
[2] University of Graz, Department of Statistics and Operations Research Universitaetsstr. 15, 8010 Graz, Austria
pferschy@uni-graz.at
[3] University of Kaiserslautern, Department of Mathematics, P.O. Box 3049, 67653 Kaiserslautern, Germany
ruzika@mathematik.uni-kl.de

Abstract. We investigate the problem of selecting a committee consisting of k members from a list of m candidates. The selection of each candidate consumes a certain weight (or cost). Hence, the choice of the k-committee has to satisfy a weight (or budget) constraint: The sum of the weights of all selected committee members must not exceed a given value W. While the former part of the problem is a typical question in Social Choice Theory, the latter stems from Discrete Optimization. The purpose of our contribution is to link these two research fields: We first define reasonable ways of ranking sets of objects, i.e. candidates, and then develop efficient algorithms for the actual computation of optimal committees. We focus in particular on the running time complexity of the developed algorithms.

Keywords: knapsack constraint, committee selection.

1 Introduction

Social Choice Theory deals with the aggregation of individual preferences on sets of objects (candidates, alternatives, etc.) into a group outcome, which could either be a ranking or a set of chosen objects. Usual results focus on situations in which no other than logical restrictions are put on the possible group outcomes, an exception being Barbera et al. [2]. Such restrictions could be of various forms, obvious examples being weight, cost or size constraints on the set of chosen alternatives.

A possible example for such a situation could be a sports club and a team of managers that have to decide on who to recruit for the next season. The

---

[*] Stefan Ruzika was partially supported by Deutsche Forschungsgemeinschaft (DFG) grant 1737/7 "Algorithmik großer und komplexer Netze" and by the Rheinland-Pfalz cluster of excellence "Dependable adaptive systems and mathematical modeling".

F. Rossi and A. Tsoukis (Eds.): ADT 2009, LNAI 5783, pp. 50–61, 2009.

managers' preferences might differ based on their different foci such as technical qualities or advertising value. Moreover, the club might have a restricted budget to buy new players who all have their market values and the club probably needs exactly a certain number of players, i.e. the set of possible new players is restricted to certain subsets.

In this article we want to select a committee of a predefined size from a set of candidates given certain restrictions. The selection will depend on a social preference over the set of candidates derived from one of the voting rules provided in the social choice literature (see Brams and Fishburn [3]). Based on such a preference relation we consider the ranking of subsets of candidates. Some of those rankings of subsets, e.g. based on best and/or worst candidates in the corresponding ranking of candidates, have been thoroughly analysed in Barbera et al. [1]. We will add two new ranking options recently characterized by Klamler et al. [9].

In this contribution we want to check whether there exist applicable algorithms that compute such optimal committees with respect to the underlying ranking criterion. Given a social preference on the set of candidates and the desired conditions for ranking sets of candidates we will provide the correct algorithm to compute the committee.

This article tries to solve questions raised in Social Choice Theory by solution methods discussed in Operations Research and thus provides an interesting interdisciplinary link between these areas. Efforts along these lines were also made in Klamler and Pferschy [8], Darmann et al. [6] and Perny and Spanjaard [11] who introduce preference relations on different graph structures.

In Section 2 we will introduce the formal framework. Sections 3 and 4 present and discuss algorithms used to determine the optimal committees. Section 5 concludes the paper.

2   Formal Framework

Let X be a finite set of m alternatives (or candidates). We define a preference on X by a reflexive, complete and transitive binary relation $R \subseteq X \times X$, where $(x, y) \in R$ means that x is at least as good as y, also written as xRy.[1] The asymmetric and symmetric parts of R are written as P and I respectively. The set of all reflexive, complete and transitive binary relations is written as $\mathcal{R}$.

Let $\mathcal{X}$ denote the set of all subsets of X. A preference on $\mathcal{X}$ is a relation $\succsim \subseteq \mathcal{X} \times \mathcal{X}$. The strict preference relation and the indifference relation are denoted by $\succ$ and $\sim$ respectively.

For any $A \in \mathcal{X}$ we say that the set of best elements in A according to R is $max(A, R) = \{x \in A : \forall y \in A, xRy\}$. Conversely, the set of worst elements in A is $min(A, R) = \{x \in A : \forall y \in A, yRx\}$.[2] Whenever not mentioned otherwise we

---

[1] A binary relation R is reflexive if and only if for all $x \in X$, xRx, it is complete if and only if for all $x, y \in X$, xRy or yRx, and it is transitive if and only if for all $x, y, z \in X$, xRy and yRz imply xRz.

[2] Whenever clear from the context, $max(A, R)$ ($min(A, R)$) will be written as $max(A)$ ($min(A)$).

will also assume that the alternatives in $A = \{a_1, a_2, ..., a_k\}$ are ordered, i.e., it is the case that $a_i R a_j$ for $i < j$.

In our model every alternative $j \in X$ is assigned a weight (or cost) $w_j \in \mathbb{R}_+$, where $w = (w_1, w_2, ..., w_m)$. The total weight bound is $W > 0$. For convenience, in the remainder of the paper we will denote $w(S) = \sum_{i \in S} w_i$ for any subset S of alternatives. In addition, throughout this article we will exclusively focus on situations where we select and compare sets of a particular cardinality. Hence, let $\mathcal{X}_k$ be the set of subsets containing exactly $k \leq m$ alternatives and $\mathcal{X}_k^W$ the set of subsets S containing exactly $k \leq m$ alternatives such that $\sum_{i \in S} w_i \leq W$.

This inequality is known in the Operations Research literature as knapsack constraint and was widely studied as an optimization problem in its own right (see Kellerer et al. [7]) and as a side constraint for other problems. In particular, knapsack problems with a cardinality constraint were considered by Caprara et al. [4]. However, in the latter each candidate is assigned a certain profit value and the quality of a committee is given by the sum of profits over all candidates. This generalization of the standard knapsack problem is $\mathcal{NP}$-hard, while the preference based valuations of committees in this paper permit polynomial time algorithms. Moreover, the algorithms from [4] do not apply to our problems.

A *k-committee selection function with constraint* is a function $C : \mathcal{R} \times \mathbb{R}_+^m \rightarrow \mathcal{X}_k^W$ that assigns to every preference $R \in \mathcal{R}$ and weight profile $w \in \mathbb{R}_+^m$ the k-committee $C(R, w) \in \mathcal{X}_k^W$.

Throughout this article the quality of a committee is based on the quality (i.e. ranking position) of its candidates with respect to the social preference on the set of candidates. Hence, let $k' \leq k$ denote the k'-best position of a candidate in a committee with respect to the social preference on X. One possibility of ranking sets $A, B \in \mathcal{X}$ is to compare the quality of the candidates in the k'_1-position in A and B, respectively. Some of those rules are also lexicographic by nature, i.e., in case society is indifferent between the candidates in k'_1-position, the sets are compared according to some other k'_2-position. Further extensions up to the usual lexicographic rule, by starting with the best candidates and going down to the worst, are feasible.

Although there exist various ways of ranking sets in $\mathcal{X}$ that fall in particular into the class of (k'_1, k'_2)-rules which compare the candidate in k'_1-position first and - in case of indifference - move on to comparing candidates in k'_2-position, certain simple k'-rules have not been extensively studied yet. The special - and most intuitive - cases here are the $k' = 1$ case which compares sets according to their best alternatives and which will be called *Maximize Max Ordering*. Analogously we could have $k' = k$, i.e. sets are ranked according to their worst candidates contained. This will be called *Maximize Min Ordering*.

Naturally, choosing committees according to set orderings raises questions about the computational complexity. The computation of a committee based on the *Maximize Max Ordering* can be done in linear time. This is an improvement to the general case of k'-MAX and, surprisingly, this improvement does not carry over to the *Maximize Min Ordering*.[3]

---

[3] Beware that k'-Max for $k' > 1$ also contains the interesting "Median" ordering based on the median element(s) of the ranking of candidates.

Table 1. Overview of complexity results

Algorithm / Ordering	Computational Complexity
k'-Max	$\mathcal{O}(m \log k)$
MaximizeMax	$\mathcal{O}(m)$
(k'_1, k'_2)-Max	$\mathcal{O}(m \log k)$
$(k'_1, \ldots k'_k)$-Max	$\mathcal{O}(km \log k)$
LexiMin	$\mathcal{O}(m \cdot \min\{k, \log m\})$
LexiMax	$\mathcal{O}(m + k \log k)$

As previously mentioned, we will also discuss usual lexicographic rules.[4] Let $A, B \in \mathcal{X}_k$. The *leximin ordering* $\succsim^L_{min}$ is defined by letting for all $A, B \in \mathcal{X}_k$

$$A \succsim^L_{min} B \Leftrightarrow a_i I b_i \ \forall i \text{ or } \exists j \in \{1, \ldots k\} \text{ such that } a_i I b_i \ \forall i > j \text{ and } a_j P b_j.$$

Analogously, we can define the *leximax ordering* $\succsim^L_{max}$ by letting for all $A, B \in \mathcal{X}_k$

$$A \succsim^L_{max} B \Leftrightarrow a_i I b_i \ \forall i \text{ or } \exists j \in \{1, \ldots k\} \text{ such that } a_i I b_i \ \forall i < j \text{ and } a_j P b_j.$$

To give an overview, the computational complexities of the algorithms discussed in this paper are given in Table 1.

3 Generic Algorithms

In this section we address the generic preference relations k'-Max, (k'_1, k'_2)-Max and $(k'_1, \ldots, k'_k)$-Max and consider general algorithms for committee selection based on these relations. Efficient algorithms (in terms of asymptotic complexity) can be formulated when using adequate data structures.

Priority queues allow to assign a key value to its elements. This key defines a sorting on the elements of the queue. In our case, this key can either be the weight value of an element or a number being consistent with the preference relation of the elements in X. Using either of these values, the priority queue can be minimum or maximum ordered. In a minimum ordered priority queue the top-most element has a minimal key among all elements left in the priority queue. We use Fibonacci heaps (see [5, Sec. 20]) for implementing priority queues. Table 2 surveys the most relevant priority queue operations together with their amortized complexities.

These four operations will be applied to select committees efficiently for the general cases of the three orderings k'-Max, (k'_1, k'_2)-Max and $(k'_1, \ldots, k'_k)$-Max. Priority queues will represent different orderings depending on what suits our needs best.

Consider the k'-Max ordering for $k' \leq k$. Let Q_A, Q_B be *maximum weight ordered* priority queues. Thus, the top-most element has largest weight among all elements in the queue, i.e., Top(Q) accesses the *heaviest* element in the queue.

---

[4] Characterizations of lexicographic rules can be found in Pattanaik and Peleg [10].

Table 2. Priority queue operations - brief description and amortized complexities

MAKE()	creates empty priority queue Q	$\mathcal{O}(1)$
INSERT(Q, x)	allows insertion of element x in Q	$\mathcal{O}(1)$
TOP(Q)	accesses the top-most element of Q	$\mathcal{O}(1)$
EXTRACT-TOP(Q)	deletes the top-most element of Q	$\mathcal{O}(\log n)$

We introduce two auxiliary arrays $A[j]$ and $B[j]$ of length m with the following definition: $A[j]$ and $B[j]$ contain the total weight of the subset consisting of the $k' - 1$ and $k - k'$ items of smallest weight among all items ranked higher and lower than j, respectively, as illustrated in the following example.

Example:

element	1	2	3	4	5	6	7	8	9
weight	8	6	9	3	6	2	5	7	1

The elements are numbered in decreasing order of their ranking. Let $k = 6$, $k' = 4$ and $j = 5$. Then we have $A[5] = 8 + 6 + 3 = 17$ and $B[5] = 2 + 1 = 3$.

Algorithm 1 first computes the entries of these arrays by keeping the current subset of these items in a priority queue. Then, the best ranked k'th largest item in a feasible k-committee can be determined in a single run over these two arrays A and B by looking for the best ranked item such that an appropriate number of smallest weight elements ranked higher and lower fulfill the weight constraint.

We assume that a feasible solution exists (otherwise the while-loop would not terminate). To reconstruct the subset yielding the optimal solution (i.e. the feasible committee) we repeat the two for-loops but stop as soon as the index reaches the k'th element found before.

Proposition 1. MAX(L, W, k, k') *is correct.*

Proof. The correctness of this algorithm relies on two facts:

(i) The entry $A[j]$ of array A contains the total weight of the $k' - 1$ lowest weight items which are ranked better then j.
(ii) The entry $B[j]$ contains the total weight of the $k - k'$ lowest weight items which are ranked worse than j.

Once we have established this observation, we simply test for the best ranked element which can be completed to a feasible committee (lines 23 - 26). Let us focus on $A[j]$ now. The for-loop in lines 1 to 3 determines the weights of the $k' - 1$ top-ranked elements which are also the lowest-weight elements being ranked better than k'. Thus, with line 4, statement (i) holds for $j = k'$. Note that the entries $A[1], \ldots, A[k' - 1]$ are meaningless. In lines 6 - 9 it is tested if a subsequent element should replace the heaviest among the $k' - 1$ lowest-weight items. In line 10 the loop invariant follows for $A[j], j = k'+1, \ldots, m-(k-k'+1)$. All other entries are not important. Thus, statement (i) is shown. Analogously, it can be verified that (ii) holds.   □

Algorithm 1. A new algorithm solving the k'-Max problem

MAX(L, W, k, k')

Input: A list $L := \{e_1, \ldots, e_m\}$ representing the ranking of m elements; a weight bound $W \in \mathbb{R}$, $k \in \mathbb{N}$ and $k' \in \mathbb{N}$, $k' \leq k$.

Output: The maximal k'-element for which a feasible k-committee exists.

```
 1: for all i ← 1, . . . , k′ − 1 do
 2:     INSERT(Q_A, e_i)
 3: end for
 4: A[k′] ← w(Q_A)
 5: for all j ← k′, . . . , m − (k − k′ + 1) do
 6:     if w(e_j) < w(TOP(Q_A)) then
 7:         EXTRACT-TOP(Q_A)
 8:         INSERT(Q_A, e_j))
 9:     end if
10:     A[j + 1] ← w(Q_A)
11: end for
12: for all i ← m − (k − k′) + 1, . . . , m do
13:     INSERT(Q_B, e_i)
14: end for
15: B[m − k + k′] ← w(Q_B)
16: for all j ← m − (k − k′), . . . , k′ + 1 do
17:     if w_j < w(TOP(Q_B)) then
18:         EXTRACT-TOP(Q_B)
19:         INSERT(Q_B, e_j))
20:     end if
21:     B[j − 1] ← w(Q_B)
22: end for
23: j ← k′
24: while A[j] + w_j + B[j] > W do
25:     j ← j + 1
26: end while
27: return e_j
```

Proposition 2. MAX(L, W, k, k') *can be performed in $\mathcal{O}(m \log k)$ time.*

Proof. Note that the sizes of the priority queues Q_A and Q_B are bounded by k. The dominating operation is EXTRACT-TOP which is in $\mathcal{O}(\log k)$. The number of EXTRACT-TOP-operations is in $\mathcal{O}(m)$. Thus, we get an overall asymptotic complexity of $\mathcal{O}(m \log k)$ time.   □

Let us now consider the ordering (k'_1, k'_2)-MAX as an extension of the k'-MAX ordering. In this case, we first fix the k'_1-best ranked element using the previous algorithm. Then, we use the same algorithm for finding the k'_2-best ranked element. This leads to a $\mathcal{O}(m \log k)$ algorithm the pseudocode of which is given in Algorithm 2.

This idea can now be iteratively extended to as many as k alternatives. The corresponding algorithm is then a $\mathcal{O}(km \log k)$ algorithm.

Algorithm 2. An algorithm for (k_1', k_2')-MAX Ordering

(k_1', k_2')-MAX
Input: A list $L := \{e_1, \ldots, e_m\}$ representing the ranking of m elements; a weight bound $W \in \mathbb{R}$, $k \in \mathbb{N}$, $k_1' \in \mathbb{N}$, $k_2' \in \mathbb{N}$, $k_1' \leq k_2' \leq k$.
Output: The best possible pair (k_1', k_2') for which a feasible k-committee exists.
1: Find the best possible k_1' element by algorithm MAX(L, W, k, k_1')
2: Delete the first k_1' elements from L
3: $W \leftarrow W - \sum_{i=1}^{k_1'} w(e_i)$
4: Find the best possible k_2' element by calling algorithm MAX($L, W, k - k_1', k_2'$)
5: **return** (k_1', k_2')

In the following section, we improve the running times originating from this generic approach for the special cases of $k' = 1$, i.e., the MAXIMIZE-MAX problem and for the leximax and leximin orders, respectively, which can be written as $(1, 2, \ldots, k)$-MAX and $(k, k-1, \ldots, 1)$-MIN, respectively.

4   Improvements

4.1   Maximize-Max or $k' = 1$-Max Order

An optimal k-committee with respect to the maximize-max order can be computed in linear time thus improving upon the $\mathcal{O}(m \log k)$ bound given in Proposition 2. The main idea is quite simple: Starting from a solution S containing the k lightest elements we look for the highest ranked element e_j which can be added to S and preserves feasibility. To maintain feasibility, the best choice of removing an element from S is given by the heaviest element e^h.

Proposition 3. *Algorithm 3 computes an optimal k-committee with respect to the maximize-max order in $\mathcal{O}(m)$.*

Proof. Observe that S always includes the $k-1$ lightest elements. At termination of the algorithm, let $S := \{e_{i_1}, \ldots, e_{i_k}\}$. Without loss of generality, we may assume that $e_{i_l} Re_{i_{l+1}}$ for all $l = 1, \ldots, k-1$. Let $\bar{S} := \{e_{j_1}, \ldots, e_{j_k}\}$ be an optimal solution for the maximize-max order with $e_{j_l} Re_{j_{l+1}}$ for all $l = 1, \ldots, k-1$, $e_{j_1} Re_{i_1}$ and $\sum_{l=1}^{k} w(e_{j_l}) \leq W$.

Case 1: e_{j_1} is among the $k-1$ lightest elements. Then $e_{j_1} \in S$. Thus, $e_{i_1} Re_{j_1}$ and therefore $e_{i_1} Ie_{j_1}$.

Case 2: e_{j_1} is not among the $k-1$ lightest elements. Then, consider $\hat{S}$ consisting of e_{j_1} together with the $k-1$ lightest elements. $\hat{S}$ is weight feasible since $w(\hat{S}) \leq w(\bar{S})$. Without loss of generality, we may assume that e_{j_1} is the best ranked element in $\hat{S}$. Due to the while-loop of the algorithm, e_{i_1} is the overall best ranked element which — together with the $k-1$ lightest elements — forms a feasible committee. Thus, $e_{i_1} Re_{j_1}$ and therefore $e_{i_1} Ie_{j_1}$.

The straightforward details of the running time are omitted.   □

Algorithm 3. An algorithm for finding an optimal solution to the maximize max order

MAXIMIZEMAX(L, W, k)

Input: A list $L := \{e_1, \ldots, e_m\}$ representing the ranking of m elements; a weight bound $W \in \mathbb{R}$, $k \in \mathbb{N}$.
Output: A feasible set S containing k elements of L.

1: Find the kth lightest element of L
2: Denote S the set containing the k lightest elements
3: Denote e^{br} the best ranked element in S
4: Denote e^h the heaviest element in S
5: stop ← false, $j \leftarrow 1$.
6: **while** stop = false and $e_j \neq e^{br}$ **do**
7:   **if** $w(S) - w(e^h) + w(e_j) \leq W$ **then**
8:     $S \leftarrow S \setminus \{e^h\} \cup \{e_j\}$
9:     stop ← true
10:   **end if**
11:   $j \leftarrow j + 1$
12: **end while**
13: **return** S

4.2   Leximax-Order

Obviously, an optimal k-committee with respect to the leximax order can be computed by executing k iterations of the generic Algorithm 1 in $\mathcal{O}(km \log k)$ time. However, we can do much better.

In principle, the underlying idea of algorithm MAXIMIZEMAX can be extended to solve also the leximax ordering by continuing the exchange operation as long as possible. However, the details of the resulting algorithm require considerable attention. Throughout the algorithm we will keep a priority queue Q containing a set of the lightest elements sorted in decreasing order of weights. The solution set S is constructed iteratively by trying to insert an item ranked better than the currently best item in Q denoted by e^{br}. If no such item is found then e^{br} is inserted in S and the procedure is continued. Since we cannot explicitly delete an arbitrary element from Q, we postpone the removal of e^{br} from Q to the point when it is accessed as the heaviest element of Q. The necessary update of e^{br} can be done easily by linking the elements of Q when they are first generated and keeping labels to check whether an element was inserted into S.

Proposition 4. LEXIMAX(L, W, k) *is correct.*

Proof. omitted.   □

Proposition 5. LEXIMAX(L, W, k) *runs in* $\mathcal{O}(m + k \log k)$.

Proof. As pointed out in the proof of Proposition 3 the items in Q can be found in $\mathcal{O}(m)$ time and inserted into a priority queue in $\mathcal{O}(k)$ time. The loop in lines 6-22 is performed at most $\mathcal{O}(m)$ times, while the EXTRACT-TOP(Q) command

---

Algorithm 4. An algorithm for finding an optimal solution to the leximax order

LexiMax(L, W, k)

Input: A list $L := \{e_1, \ldots, e_m\}$ representing the ranking of m elements; a weight bound $W \in \mathbb{R}$, $k \in \mathbb{N}$.
Output: A feasible set S containing k elements of L.

1: $S \leftarrow \emptyset$
2: Find the kth lightest element of L
3: Insert the k lightest elements of L into Q
4: Denote e^{br} the best ranked element in $Q \setminus S$
5: $j \leftarrow 1$
6: **repeat**
7:   **if** $e_j = e^{br}$ **then**
8:     $S \leftarrow S \cup \{e_j\}$
9:     **if** $e^{br} = \text{Top}(Q)$ **then**
10:       Extract-Top(Q)
11:     **end if**
12:     Update e^{br}
13:   **else if** $w(S) + w(Q) - w(\text{Top}(Q)) + w(e_j) \leq W$ **then**
14:     $S \leftarrow S \cup \{e_j\}$
15:     Extract-Top(Q)
16:     **while** Top(Q) $\in S$ **do**
17:       Extract-Top(Q)
18:     **end while**
19:     Update e^{br}
20:   **end if**
21:   $j \leftarrow j + 1$
22: **until** $Q = \emptyset$
23: **return** S

---

requiring $\mathcal{O}(\log k)$ time can be executed at most k many times before $Q = \emptyset$. Note that we never insert elements into Q.

Updating e^{br} can be done in constant time if the items in Q are linked in decreasing order of rank when they are identified in line 3. Whenever an element is deleted from Q this linked structure can be easily updated in constant time (e.g. by using a double-linked list). This proves the statement.   □

4.3   Leximin-Order

An optimal k-committee with respect to the maximize-min order can be computed in a straightforward way by performing Max(L, W, k, k) in $\mathcal{O}(m \log k)$ time. For the leximin-order it remains to adjust the remaining $k - 1$ elements. A trivial approach would apply Max(L', W', j, j) iteratively for $j = k - 1, k - 2, \ldots, 1$ with adapted sets of elements L' and weight bounds W'. Along these lines an $\mathcal{O}(k\, m \log k)$ algorithm for the leximin-order is derived.

The remaining $k - 1$ elements can be computed more efficiently. Algorithm LexiMin(L, W, k) starts with a feasible solution consisting of $S = \{e_\ell\}$ determined by Max(L, W, k, k) and the $k - 1$ elements of smallest weight ranked

---

Algorithm 5. An algorithm for finding an optimal solution to the leximin order

LexiMin(L, W, k)

Input: A list $L := \{e_1, \ldots, e_m\}$ representing the ranking of m elements; a weight bound $W \in \mathbb{R}$, $k \in \mathbb{N}$.
Output: A feasible set S containing k elements.
1: Run Max(L, W, k, k) to compute a committee with a worst ranked element e_ℓ
2: $S \leftarrow \{e_\ell\}$
3: Insert the $k - 1$ lightest elements from $\{e_1, \ldots, e_{\ell-1}\}$ into Q_S
4: Insert the remaining elements $\{e_1, \ldots, e_{\ell-1}\} \setminus Q_S$ into Q_R
5: **while** $Q_S \neq \emptyset$ **and** $Q_R \neq \emptyset$ **do**
6:     **while** rank(Top(Q_R)) worse than rank(Top(Q_S)) **do**
7:         Extract-Top(Q_R)
8:     **end while**
9:     **if** $w(S) + w(Q_S) - w(\text{Top}(Q_S)) + w(\text{Top}(Q_R)) \leq W$ **then**
10:         Extract-Top(Q_S)
11:         Insert(Q_S, Top(Q_R))
12:         Extract-Top(Q_R)
13:     **else**
14:         $S \leftarrow S \cup \text{Top}(Q_S)$
15:         Extract-Top(Q_S)
16:     **end if**
17: **end while**
18: **return** S

---

better than e_ℓ which are stored in Q_S. Throughout the algorithm, elements of the optimal solution are computed iteratively and added to S while Q_S keeps items complementing S to a feasible solution. All items ranked better than the best element currently in S but not contained in Q_S are candidates for an improvement of the current feasible solution and stored in Q_R.

Both, Q_S and Q_R, are organized as Fibonacci heaps. Q_S is maximum-ordered with respect to the ranking, i.e., the top element is the worst-ranked one, since this is the first element which we try to replace in the current feasible solution. The second heap Q_R is minimum-ordered with respect to weight, since the lightest element is most likely to fit into a new feasible solution.

After initializing this structure in lines 3-4 we enter the main loop of the algorithm. There, we first remove all items in Q_R which are ranked worse than the worst element in Q_S (line 6). Then we try to improve the current feasible solution (line 9) by removing its worst element (among those which are still not fixed). The best chance to construct a new feasible solution is given by the element with smallest weight (among those with better ranking), i.e. Top(Q_R). If the improvement step was successful we update Q_S and Q_R (lines 10-12), otherwise we have to include the lowest ranked element of Q_S permanently in the solution and thus move it into S (lines 14-15).

Proposition 6. LexiMin(L, W, k) *is correct.*

Proof. omitted. □

Proposition 7. LexiMin(L, W, k) *runs in* $\mathcal{O}(m \log m)$.

Proof. Running Max(L, W, k, k) takes $\mathcal{O}(m \log k)$ time (see Proposition 2) while the generation of the two heaps requires $\mathcal{O}(m)$ time. In every execution of the main while-loop the number of operations is dominated by the number of Extract-Top operations. Q_R contains at most $m - k$ elements at the beginning and no elements are ever added to Q_R. Q_S starts with $k - 1$ items and may receive at most $|Q_R|$ additional elements from Q_R during the execution of the algorithm. Therefore, the total number of Extract-Top operations is bounded by $m - k + (k - 1 + m - k) < 2m$ which proves the statement. □

For small values of k we can formulate a more efficient algorithm which is however not discussed here for lack of space.

Proposition 8. *An optimal k-committee for the leximin order can be computed in $\mathcal{O}(mk)$ time.*

Proof. See our technical report [9]. □

Comparing LexiMin(L, W, k) and Proposition 8 it turns out that for $k > \log m$ LexiMin(L, W, k) dominates its competitor. On the other hand, for $k \leq \log m$ the algorithm quoted in Proposition 8 is superior. Hence, a combination of the two algorithms yields a $\mathcal{O}(m \cdot \min\{k, \log m\})$ running time bound.

5   Conclusion

In this paper we try to link social choice theory with methods from discrete optimization. Based on a social ordering of the candidates and a weight or budget constraint on that set, we derive and analyze efficient algorithms for the computation of socially optimal committees. Particular attention is devoted to the running time complexities of the algorithms.

Further work on this subject might include analyzing different set orderings, additional constraints or the relationship to conceptual approaches that might provide certain insight, such as "prudent orders" in group decision problems.

References

1. Barbera, S., Bossert, W., Pattanaik, P.K.: Ranking Sets of Objects. In: Barbera, S., Hammond, P.J., Seidl, C. (eds.) Handbook of Utility Theory, vol. 2. Kluwer Academic Publishers, Boston (2004)
2. Barbera, S., Masso, J., Neme, A.: Voting by committees under constraints. Journal of Economic Theory 122, 185–205 (2005)
3. Brams, S.J., Fishburn, P.C.: Voting Procedures. In: Arrow, K.J., Sen, A.K., Suzumura, K. (eds.) Handbook of Social Choice and Welfare, vol. 1, pp. 173–236. Elsevier, Amsterdam (2002)

4. Caprara, A., Kellerer, H., Pferschy, U., Pisinger, D.: Approximation algorithms for knapsack problems with cardinality constraints. European Journal of Operational Research 123, 333–345 (2000)
5. Cormen, T.H., Leiserson, C.E., Rivest, R.L., Stein, C.: Introduction to Algorithms, 2nd edn. MIT Press, Cambridge (2001)
6. Darmann, A., Klamler, C., Pferschy, U.: Finding socially best spanning trees. Social Science Research Network (2008), http://ssrn.com/abstract=1328916
7. Kellerer, H., Pferschy, U., Pisinger, D.: Knapsack Problems. Springer, Heidelberg (2004)
8. Klamler, C., Pferschy, U.: The traveling group problem. Social Choice and Welfare 29, 429–452 (2007)
9. Klamler, C., Pferschy, U., Ruzika, S.: Committee selection under weight constraints. Technical Report, available in: Social Science Research Network (2009), http://ssrn.com/abstract=1361770
10. Pattanaik, P.K., Peleg, B.: An axiomatic characterization of the lexicographic maximin extension of an ordering over a set to the power set. Social Choice and Welfare 1, 113–122 (1984)
11. Perny, P., Spanjaard, O.: A preference-based approach to spanning trees and shortest paths problems. European Journal of Operational Research 162, 584–601 (2005)

The Effects of Noise and Manipulation on the Accuracy of Collective Decision Rules

Peter Bodo

Department of Economics and Finance, Southern Connecticut State University,
501 Crescent Street, New Haven, CT 06515
bodop1@southernct.edu

Abstract. How do noise and manipulation affect the accuracy of collective decision rules? This paper presents simulation results that measure the accuracy of ten well known collective decision rules under noise and manipulation. When noise is low these rules can be divided into accurate ("good") and inaccurate ("bad") groups. The bad rules' accuracy improves, sometimes significantly, when noise increases while the good rules' performance steadily worsens with noise. Also, when noise increases the accuracy of the good rules deteriorates at different rates. Manipulation delays the effects of noise: accuracy improvement and deterioration due to noise emerge only at higher noise levels with manipulation than without it. In some cases at high noise levels there is only a negligible difference between the accuracy of good and bad collective decision rules.

Keywords: Collective decision rules, noise sensitivity, manipulability of collective decisions.

1 The Problem

Let us assume that a committee has to rank candidates. Suppose that there exists an objective ranking of the candidates but because of human error (later: system noise or noise) committee members' perceptions about the *objective* ranking are inevitably imperfect and different.

In order to generate a collective ranking first committee members individually rank candidates, then the committee aggregates the individual rankings according to a method earlier agreed upon. Several collective decision rules are available to the committee to come up with a collective preference. Suppose that the committee is aware of the fact that the method chosen for preference aggregation may affect the accuracy of the collective rankings and that candidates may try to manipulate committee members to get a better ranking. What is the most accurate and least manipulable method of preference aggregation in a noisy setting? This paper compares how ten well known collective decision rules perform under noise and in the presence of manipulation.

1.1 An Example

The following example illustrates the importance of understanding the effects of noise and manipulation on the accuracy of collective decision rules. Consider the task of

F. Rossi and A. Tsoukis (Eds.): ADT 2009, LNAI 5783, pp. 62–73, 2009.

distributing organs among patients waiting for organ transplant. A fair distribution should be based on, among other things, medical urgency. There must exist an objective ranking of patients considering their medical condition. Sometimes this ranking can be reproduced using objective criteria e.g. blood test results. In other cases the ranking of patients is a result of committee decisions that include judgment calls. When human error cannot be excluded committee rankings will not perfectly reproduce the objective ranking of patients, i.e. committee members' rankings will be noisy. In addition, if it is known that the patients' final ranking is affected by judgment calls, patients may try to influence the committee members' judgment in the hope of a better ranking. The decision rule chosen by the committee will be used repeatedly. The ranking of patients may determine life or death. What kind of collective decision rule should be selected by the committee in order to produce a ranking of patients that is the most consistent with their medical conditions and the most resistant to manipulation?

1.2 The Ranking Rules Examined

The ten collective decision rules I will examine are as follows: anti-plurality, Borda, Coombs, Copeland, Jech, Kemeny-Young, median rank, minimax, plurality, and single transferable vote[1]. Although all of these rules can be used to rank alternatives some rules are better for choosing a winner[2] (plurality, anti-plurality, minmax, single transferable vote) while others are better for ranking candidates (Coombs, Jech, Borda Copeland, Kemeny-Young, median rank)[3].

1.3 Control versus Misrepresentation of Preferences

Manipulation of collective decision making often involves attempts to control the process. Individuals may try to achieve a favorable outcome by adding or removing or partitioning committee members and/or candidates. In other cases manipulation means that committee members and/or candidates try to change some committee members' preferences by misrepresenting their true type or their true preferences. In the following I will not consider control issues. I will use the term *manipulation* as a synonym for misrepresentation of one's preferences or type and investigate how manipulation, coupled with noise, affects the accuracy of collective decisions.

---

[1] The basic tenets of collective decision making, including the description and analysis of the collective decision rules I investigate can be found in most public choice textbooks, see for example Mueller [9]. Levin and Nalebuff [7] also discuss all the rules considered here except the median rank rule, which is analyzed in detail by Basset and Persky [2].

[2] See [7] about this partitioning of collective decision rules.

[3] There are three reasons why I examine the accuracy of rankings produced by decision rules that are known to be ill-suited for rankings: 1) some rules considered to be inaccurate without noise may become relatively accurate when noise is present. 2) it could be interesting to *measure* the accuracy of the most popular collective decision rules under noise; and 3) I investigate the effects of noise in conjunction with the effects of manipulation. Measuring the manipulability of these rules in a noisy environment when they are used for rankings may produce some insights about the manipulability of these rules under noise when they are used for picking a winner.

1.4 Short Literature Review

Tovey and Trick published several papers in which they treated the manipulation of collective decisions as a computational problem and produced hardness results for cases with a large number of voters and candidates. For example, in Bartholdi, Tovey, and Trick [1], they showed that it is not very difficult to manipulate popular collective decision rules, and only a few NP hard rules exist. Conitzer and Sandholm [3] and Conitzer, Sandholm, and Lang [4] produced hardness results for collective decision rules in cases with limited number of players. Faliszewski, Hemaspaandra, Hemaspaandra, and Rothe [5] provided hardness results for different versions of Copeland elections.

Other authors capture the manipulability of collective decision methods by the proportion of voting profiles for which micro manipulation could be successful. Saari [11] measures and compares manipulability of positional voting schemes with a small number of candidates. Smith [12] discusses cases with a small number of candidates and voters and uses computational methods to measure the manipulability of some collective decision rules.

Kalai [6] analyzes the noise sensitivity of some social welfare functions. He investigates the chaotic consequences of noise when the number of voters tends to infinity and finds that majority rules are the most stable. Procaccia, Rosenstein, and Kaminka [10] measure the k-robustness of certain voting rules, which is the worst case probability that k independent errors in the preferences of voters will change the results of the elections. These authors rank some voting rules according to their k-robustness.

Truchon and Gordon [13] use simulation models to analyze the noise sensitivity of a few collective choice rules. They assume that the probability of error in recognizing the true rank of alternatives is a known function of the distance between alternatives and there exists a loss function that measures social loss due to erroneous rankings. They use Monte Carlo simulations to compare the performance of five ranking rules and show that the maximum likelihood rule is the most accurate.

Mitlöhner, Eckert and Klamler [8] discuss the combined effects of noise and manipulation. They investigate how effective manipulation of rank aggregation is under different ranking rules if voters receive noisy information about other voters' preferences. By using simulation models they show how the number of manipulable voting profiles change with noise and how manipulation attempts may backfire in a noisy environment.

2 Simulation Models to Measure the Accuracy and Manipulability of Collective Decision Rules

I use Monte Carlo simulations to measure the accuracy and the manipulability of the earlier mentioned ten collective decision rules in a noisy environment. The following models work with weaker assumptions than similar simulation models in the literature. Also, they are especially well suited for the investigation of cases with not too small but limited number of candidates and committee members. The next simulation models make it possible to treat uniformly mathematically very different collective decision rules, and produce comparable and generalizable results. On the negative side simulation models by default are less robust than mathematical models. It is also true that simulation models are not well suited for the investigation of one of the above rules, the Kemeny-Young rule, because it is a computationally difficult, NP hard rule[4].

3 Description of the Simulations

3.1 Case N: Noise Only

I distinguished among the following three scenarios in which a committee has to rank candidates: 1) noisy environment without manipulation (N); 2) noisy environment with manipulation (NM), and 3) noisy environment with noisy manipulation (NNM).

In case N, in each simulation, random number of committee members and candidates are generated. Committee members and candidates are placed on a 100 unit line; their locations are also random. Committee members can perfectly observe the candidates' locations. At the beginning of the simulation all candidates are given a random initial score (objective strength). Committee members observe the candidates' objective strength imperfectly.

A simple discounting formula, the *parametric reward function,* converts the objective strength of a candidate at a given position into the strength observed by a committee member at another location. The parametric reward function is as follows:

$$f(s,d;z) = (1-d^z) \cdot s, \tag{1}$$

where s is the candidate's objective strength, d is the normalized distance (distance divided by the length of the field) between a member and a candidate, and z is a fixed reward parameter (later: RewPar). I assume that all committee members use the same reward parameter. I considered the following ten z (reward parameter) values: 0.05, 0.3, 0.6, 0.9, 1, 1.2, 1.6, 2, 4, 6. How do these values change the type of strength-discounting? The overall pattern is that as noise increases i.e. the reward parameter decreases, the accuracy of the perceived strengths of closer candidates deteriorates. When $z = 6$ the system is almost noise free. There is almost no discounting, committee members observe clearly the objective strength of even very distant candidates. If $z = 0.05$ noise is extremely strong. In this case there is hardly any clarity regarding the objective strengths of individual candidates. Only the closest candidates' strengths are observable to committee members, the other candidates' objective strengths are heavily and almost equally discounted. The other eight z parameter values represent eight different noise levels more or less equally spaced between the two extremes.

In the simulations a reward parameter is selected from the above list at random. Every committee member processes each candidate's strength score according to the parametric reward function with the selected reward parameter and produces a unique, individual ranking based on the perceived (processed) scores. Then the committee uses

---

[4] To be able to handle computation problems I used a modified version of the Kemeny-Young rule in the simulations. The *KYStartWithBorda* rule approximates the Kemeny-Young ranking. It starts with the Borda ranking of candidates, then in 2000 iterations finds the best improvement of the Borda ranking according to the Kemeny-Young criterion.

one of the ten ranking rules to generate the aggregate ranking of candidates. Next, the Spearman rank correlation coefficient between the actual collective ranking and the objective ranking is computed in conjunction with a critical value for the coefficient. The results (correlation coefficient and the critical value) are stored with the reward parameter, which represents the noise level, and the name of the aggregation rule. This procedure is performed for all ten ranking rules with the same parameter settings (same number of members, candidates, same locations, and the same reward parameter), and it is repeated 5000 times.

3.2   Case NM: Noisy Environment with Manipulation

All the conditions and procedures for case N apply except the candidates' locations are not fixed during the simulation: candidates are allowed to move. After generating a random number of committee members and candidates all players receive their initial random locations. Candidates can perfectly observe the committee members' locations. Players learn the fixed reward parameter committee members will use to process strength. In the light of his own initial location, the committee's location, and the reward parameter each candidate determines his optimal location and moves to it[5]. After candidates stopped moving the simulation goes according to case N.

3.3   Case NNM: Noisy Environment with Noisy Manipulation

All the conditions and procedures for case NM apply except candidates have imperfect information about the reward parameter: they are not certain how committee members process strength based on distance. It is assumed that candidates learn the *expected value* of the reward parameter, and this information is accurate, but the actual reward parameter they use to determine their optimal locations will be different for each candidate. It is a random number drawn from a uniform distribution with the known expected value and fixed standard deviation. Once these individual reward parameter values are determined the simulation goes according to case NM. Also, as in the NM case, candidates are given ample time to move to their optimal locations.

4   The Measurement of the Accuracy of the Ranking Rules

The simulations produced large data sets. Based on these data sets I computed two accuracy measures. One was the mean accuracy of rules at different noise (reward parameter) levels. It was computed for all three cases by finding the *mean* of the Spearman rank correlations for a rule at all of the different noise levels. The other measure was the *mean proportion* of significant Spearman rank correlation coefficients at different noise levels. This was also computed for all the rules at all the different reward parameter values.

---

[5] A candidate has an incentive to move away from his location if by this move his perceived total strength improves. In the simulations of the NM cases candidates are allowed to make a maximum of 90 consecutive, either one-step-to-the-left or one-step-to-the-right moves on a 100 unit long line, which implies that they practically always end up at their optimal locations regardless of their original location.

5 Summary of the Simulation Results

No matter which accuracy measure one uses or which case one considers at minimum noise level (z = 6), just like the literature predicts, accuracy-wise there are two types of collective decision rules. The accurate, "good" rules are the Coombs, Jech, Borda Copeland, Kemeny-Young, and median rank rules, while the inaccurate, "bad" rules are the plurality, anti-plurality, minmax, and single transferable vote rules. The anti-plurality rule, however, is an outlier, it is unusually inaccurate in a noisy environment.

Noise affects the accuracy of the above mentioned two types of rules differently. As noise increases, i.e. the value of the reward parameter declines, the bad rules' performance improves, in some cases dramatically, until noise becomes very intense (z = 0.3). When noise becomes extremely high (z = 0.05) the accuracy of bad rules slightly declines. On the other hand, the good rules' accuracy constantly deteriorates when noise increases. However, in this group accuracy declines with noise at different rates. While at low noise levels there is hardly any difference among the accuracy of the rules in the good group, at higher noise levels one can distinguish two subgroups: group 1, which includes the Jech, median rank, and Borda rules, always outperforms group 2 with the Copeland, Coombs, and Kemeny-Young rules.

Manipulation, to some extent, shields most of the rules from the effects of noise. With noisy manipulation or pure manipulation the bad rules improve slower while the good rules deteriorate slower as noise increases. Also, with manipulation at high noise levels the accuracy of the rules is slightly higher across the board. If one uses the mean proportion of significant rank correlations to measure accuracy, at high noise levels with manipulation there is only a minimal difference between the accuracy of good and bad rules. With manipulation there seems to be a sweet spot, z = 0.3: at this noise level all the rules, good and bad, are almost equally accurate.

6 Simulation Results in More Detail

Next, six graphs will be presented summarizing the simulation results in more detail. The simulations revealed that the anti-plurality rule is an outlier; in most situations it is much less accurate than the other nine rules. For this reason the graphs present this rule's performance separately from other rules. The remaining nine rules are partitioned into three homogeneous groups. The following graphs capture the mean accuracy of the rules in the three groups under noise and/or manipulation. Groups 1 and 2 contain the good, accurate rules: in group 1 there are the Jech, median rank and Borda rules, and group 2 includes the Copeland, Coombs and KYStartWithBorda methods. The two groups are separated because their response to noise and manipulation follows somewhat different patterns. Group 3 includes the bad rules, except the outlier anti-plurality rule. In group 3 there are the minmax, the plurality and the single transferable vote methods. In some cases the KYStartWithBorda rule or the minmax rule shows atypical patterns. In these cases these rules will be presented individually, separately from their groups.

6.1 Mean Rank Correlations Compared, Case N

Fig. 1 shows that as z decreases, i.e. noise increases, mean accuracy in the first two groups declines and the mean values of these groups diverge. Group 3 rules react to higher noise differently. Mean accuracy in group 3 constantly increases until z = 0.3, then declines as noise becomes extreme (z = 0.05). The mean accuracy of minmax is always higher than the mean accuracy of the other two rules in this group. At high noise levels minmax gets quite close to group 2.

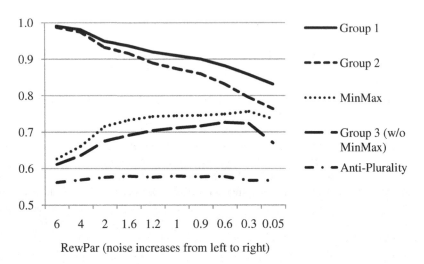

RewPar (noise increases from left to right)

Fig. 1. Mean Rank Correlations with Noise

6.2 Mean Rank Correlations Compared, Case NNM

The accuracy of anti-plurality is still bad, but not constant anymore: it mildly increases with noise, which is a group 3 pattern. Also, at very low noise anti-plurality's accuracy is very similar to the accuracy of rules in group 3. In this case minmax is in group 3 from the outset and stays with this group for all reward parameter values. In group 3 mean accuracy increases sharply with noise between z = 0.9 - 0.3, then it declines. At extremely high noise (z = 0.05) the presence of noisy manipulation slightly improves mean accuracy for all rules. In the NNM case the trends regarding the good rules are similar to the trends in case N. Accuracy is near perfect until z = 1.6. Then, as noise increases the mean accuracy in groups 1 and 2 declines but until z = 0.9 at a slower rate than in the previous case. At lower reward parameters the decline of the mean values in groups 1 and 2 is faster. Also, the divergence of means in groups 1 and 2 also becomes faster. The KYStartWithBorda rule shows a slightly different pattern than the other two rules in group 2. In sum, compared with case N, noisy manipulation preserves trends but somewhat delays the effects of noise on the accuracy of collective decision rules.

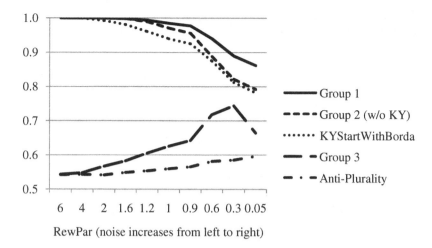

Fig. 2. Mean Rank Correlation with Noise and Noisy Manipulation

6.3 Mean Rank Correlations Compared, Case NM

The delaying effects of manipulation are even more evident in the pure manipulation case than under noisy manipulation. As Fig. 3 shows the good rules are almost perfectly accurate until $z = 0.9$, except KYStartWithBorda. Then a sudden, steep decline and divergence of accuracy figures follows. Between $z = 6 - 1.2$ group 3 rules' accuracy is indistinguishable from the worst performing anti-plurality rule's performance: it is practically constant, close to 0.55. From $z = 1.2$ these rules' performance

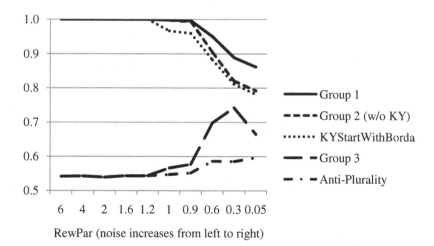

Fig. 3. Mean Rank Correlation with Noise and Manipulation

improves. Between z = 0.9 - 0.3 it increases very fast, then accuracy significantly declines at z = 0.05. At the highest noise level all the rules have the same mean correlation coefficients as in the NNM case.

6.4 Proportion of Significant Correlation Coefficients Compared, Case N

Next let us investigate how the mean proportion of significant correlation coefficients vary with noise. Fig. 4 reveals how some patterns change when one uses this measure to assess the accuracy of the aggregation rules in question.

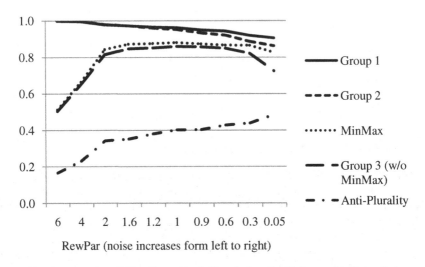

Fig. 4. Proportion of Significant Rank Correlation Coefficients with Noise Only

The anti-plurality rule's mean proportions constantly increase with noise but remain significantly below the mean proportions of the other nine rules. Accuracy in groups 1 and 2 decline and diverge slower than in cases where accuracy was measured by the mean of rank correlation coefficients. Group 1 and 2 rules' performance diverge only for noise levels higher than z = 0.9. Even at extreme noise overall accuracy is quite high and accuracy differences between rules in groups 1 and 2 are small. Between z = 6 - 2 the accuracy of group 3 rules improves fast, then it practically stagnates until z = 0.6. Interestingly, group 3 catches up with group 2 at z = 0.6. There is a minimal difference among the mean proportions in the three groups at this noise level. Minmax remains quite accurate even at extremely high noise level and for z < 0.6 it practically joins to rules in group 2. When noise becomes extreme the accuracy of the other two rules in group 3 declines.

6.5 Proportion of Significant Correlation Coefficients Compared, Case NNM

As Fig. 5 shows when we add noisy manipulation to the simulations, the first two groups' accuracy starts do decline and diverge from z = 0.9. However, even at z = 0.05 the

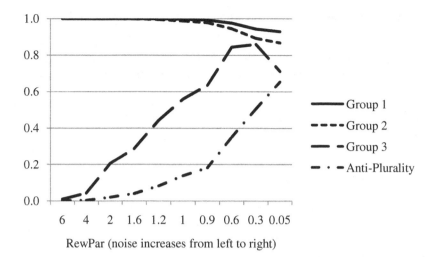

Fig. 5. Proportion of Significant Rank Correlation Coefficients with Noise and Noisy Manipulation

accuracy of rules in the first two groups is higher than in the previous case without manipulation. Group 3 (including minmax) starts at approximately zero mean significant proportion level. In this group, between z = 4 - 0.6, accuracy improves fast, almost in a linear fashion, and at z = 0.3 it gets close to the accuracy level in group 2. Then, at z = 0.05 accuracy sharply declines. It is interesting how the accuracy of the anti-plurality rule varies. It starts at 0 but increases steadily with noise. At z = 0.05 it almost reaches the accuracy of group 3 rules, and it surpasses its own accuracy without manipulation.

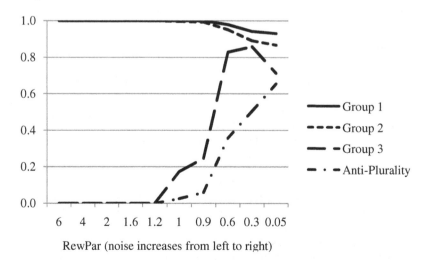

Fig. 6. Proportion of Significant Rank Correlation Coefficients with Noise and Manipulation

6.6 Proportion of Significant Correlation Coefficients Compared, Case NM

The performance of group 1 and 2 rules in the pure manipulation case is almost identical to their performance in the previous noisy manipulation case. Group 3 rules' performance, however, changes significantly. Just like anti-plurality, in the NM case group 3 rules are completely inaccurate between $z = 6 - 1.2$. Then until $z = 0.6$ the accuracy in this group (and the accuracy of anti-plurality at a different level) increases very fast. Between $z = 0.6 - 0.05$ accuracy figures in the NM case coincide with the appropriate values from the NNM case.

7 Conclusion

How do noise and manipulation affect the accuracy of collective ranking rules? The paper presented simulation results that *measured* the performance of ten frequently used collective decision rules under noise and manipulation. Among the ten ranking rules there was an outlier, the anti-plurality rule, which in most cases performed significantly weaker than the other nine rules. Not counting the outlier, at low noise the rules can be divided into two groups: good and bad. Noise improves the bad rules' accuracy and worsens the good rules' performance. At low noise the rules in the good group perform almost equally well. When noise increases the accuracy of the good rules deteriorates at different rates. At higher noise levels two different subgroups emerge: group 1 with the Jech, median rank and Borda rules, and group 2 with the Copeland, Coombs, and (a version of the) Kemeny-Young rule. At high noise group 1 rules always outperform group 2 rules. Manipulation delays the effects of noise for some time: performance improvement (bad rules) and performance deterioration and divergence (good rules) emerge only at higher noise levels with manipulation than without it. It was also shown that in some cases at higher noise levels the difference between the accuracy of good and bad rules is negligible.

References

1. Bartholdi, J.J., Tovey, C.A., Trick, M.A.: Voting schemes for which it can be difficult to tell who won the election. Social Choice and Welfare 6, 157–165 (1989)
2. Basset, G.W., Persky, J.: Robust Voting. Public Choice 99, 299–310 (1999)
3. Conitzer, V., Sandholm, T.: Universal Voting Protocol Tweaks to Make Manipulation Hard. In: Proceedings of the 18th International Joint Conference on Artificial Intelligence, pp. 781–788. Morgan Kaufmann, San Francisco (2003)
4. Conitzer, V., Sandholm, T., Lang, J.: When Are Elections with Few Candidates Hard to Manipulate. Journal of the ACM 54, Article 14 (2007)
5. Faliszewski, P., Hemaspaandra, E., Hemaspaandra, L.A., Rothe, J.: Llull and Copeland Voting Computationally Resist Bribery and Control. Univ. of Rochester Comp. Sci. Dept. Technical Report TR-2008-933 (2008)
6. Kalai, G.: Noise Sensitivity and Chaos in Social Choice Theory. Discussion Paper Series dp399. Center for Rationality and Interactive Decision Theory, Hebrew University, Jerusalem (2005)

7. Levin, J., Nalebuff, B.: An Introduction to Vote-Counting Schemes. Journal of Economic Perspectives. Winter 9, 3–26 (1995)
8. Mitlöhner, J., Eckert, D., Klamler, C.: Simulating the Effects of Misperception on the Manipulability of Voting Rules. In: Endriss, U., Lang, J. (eds.) Proceedings of the 1st International Workshop on Computational Social Choice (COMSOC 2006), pp. 345–355. Institute for Logic, Language and Computation, Universiteit van Amsterdam (2006)
9. Mueller, D.C.: Public Choice II. Cambridge University Press, Cambridge (1989)
10. Procaccia, A.D., Rosenschein, J.S., Kaminka, G.A.: On the Robustness of Preference Aggregation in Noisy Environments. In: Proceedings of the 6th International Joint Conference on Autonomous Agents and Multiagent Systems, pp. 416–422. IFAAMAS (2007)
11. Saari, D.G.: Susceptibility to Manipulation. Public Choice 64, 21–41 (1990)
12. Smith, D.A.: Manipulability Measures of Common Social Choice Functions. Social Choice and Welfare 16, 639–661 (1999)
13. Truchon, M., Gordon, S.: Statistical Comparison of Aggregation Rules for Votes. Mathematical Social Sciences (2008), doi:10.1016/j.mathsocsci.2008.11.001

Subset Weight Maximization with Two Competing Agents

Gaia Nicosia[1], Andrea Pacifici[2], and Ulrich Pferschy[3]

[1] Dipartimento di Informatica e Automazione, Università degli studi "Roma Tre", Italy
nicosia@dia.uniroma3.it
[2] Dipartimento di Ingegneria dell'Impresa, Università degli Studi di Roma "Tor Vergata", Italy
pacifici@disp.uniroma2.it
[3] Department of Statistics and Operations Research, University of Graz, Austria
pferschy@uni-graz.at

Abstract. We consider a game of two agents competing to add items into a solution set. Each agent owns a set of weighted items and seeks to maximize the sum of their weights in the solution set. In each round each agent submits one item for inclusion in the solution. We study two natural rules to decide the winner of each round: Rule 1 picks among the two submitted items the item with larger weight, Rule 2 the item with smaller weight. The winning item is put into the solution set, the losing item is discarded.

For both rules we study the structure and the number of efficient solutions, i.e. Pareto optimal solutions. For Rule 1 they can be characterized easily, while the corresponding decision problem is NP-complete under Rule 2. We also show that there exist no Nash equilibria. Furthermore, we study the best-worst ratio, i.e. the ratio between the efficient solution with largest and smallest total weight, and show that it is bounded by two for Rule 1 but can be arbitrarily high for Rule 2. Finally, we consider *preventive* or *maximin* strategies, which maximize the objective function of one agent in the worst case, and *best response* strategies for one agent, if the items submitted by the other agent are known before either in each round (on-line) or for the whole game (off-line).

Keywords: multi-agent optimization, games.

1 Introduction

We consider a multi-agent problem where agents compete to fill a joint solution set with their items. We focus on the following situation: There are two agents, each of them owning one of two disjoint sets of weighted items. The agents have to select items from their set for putting them in a common solution set. This process proceeds in a fixed number of rounds. In every round each of the two agents selects exactly one of its items and submits the item for possible inclusion in the solution set. A central decision mechanism chooses one of the

F. Rossi and A. Tsoukis (Eds.): ADT 2009, LNAI 5783, pp. 74–85, 2009.
© Springer-Verlag Berlin Heidelberg 2009

items as "winner" of this round. The winning item is permanently included in the solution set while the losing item is permanently discarded. Each agent wants to maximize its total solution value which is given by its own items' weights in the solution set. The sets if items are known to both agents while the submission of items in each round occurs simultaneously. Now the problem is how to compute solutions which take into account each agent's solution function, and that can be used to support the negotiation among the agents. Although the addressed problem does not come directly from a real world application, it is such a natural and simple game (and yet not trivial) to make its investigation interesting and stimulating.

This problem can also be regarded as a single-suit card game in which each of two players chooses a card from its hand. The highest (or lowest) value card wins and each card can be used only once. In [5] the authors study a zero-sum game in which the cards are submitted simultaneously and the players want to maximize the total value of the won cards. In [3,6] the so called *whistette* game is addressed. There is a totally ordered suit of $2n$ cards, distributed between the two players. The player who has the lead plays first on each trick. The player with the higher card wins a trick and obtains the lead. Players want to maximize the number of won tricks.

The problem here addressed is also related to seeds assignment in team sport tournaments, e.g. chess leagues, where players of each team are ordered and each player faces the opponent on the same ordered position of the other teams' list. However, the ordering of players may be restricted to obey an established ranking (e.g. ELO points) to some extend. There the objective is to maximize the number of wins while our problem aims at the maximization of the total weight of all winning rounds.

1.1 Formal Problem Setting

In the following, A resp. B indicate the agents' names each of them owning a set of n items, where item i has weight a_i resp. b_i. Sometimes we will identify items by their weight. All information about the input list of items is public.

The game is performed over c rounds. In each round both of the two agents simultaneously submit one of their items, say a_i and b_j, in secret. We consider the two most natural rules for deciding which of the two submitted items wins and is added to the solution set:

Rule 1: if $a_i \geq b_j$ then A wins;

Rule 2: if $a_i \leq b_j$ then A wins.[1]

Recall that in both cases the losing item is discarded and can not be submitted a second time.

Throughout this paper we assume the items to be sorted in decreasing order of weights, i.e. $a_1 \geq a_2 \geq \ldots \geq a_n$ resp. $b_1 \geq b_2 \geq \ldots \geq b_n$.

---

[1] In case of a tie we assume that A always wins. Two analogous rules where B wins all ties can be represented by exchanging the roles of A and B.

This problem can be represented by a graph model. Each agent's item is associated to a node of a weighted complete bipartite graph $G = (V^A \cup V^B, E^A \cup E^B)$. An arc (i, j) belongs to E^A or to E^B depending on the winner of a comparison of a_i and b_j, which of course depends on the applied selection rule.

Rule 1: arc (i, j) with weight $w_{ij} = \max\{a_i, b_j\}$ belongs to E^A if $a_i \geq b_j$, i.e. if A wins, otherwise it belongs to E^B;

Rule 2: arc (i, j) with weight $w_{ij} = \min\{a_i, b_j\}$ belongs to E^A if $a_i \leq b_j$, i.e. if A wins, otherwise it belongs to E^B;

Every pair of items (a_i, b_j) submitted simultaneously in one round can be represented by an arc (i, j) between the two corresponding nodes. Hence, after completing c rounds any solution may be represented as a c-matching M on G. Recall that a c-matching assigns c nodes in V^A to c nodes in V^B. The total weight of items in the solution for agent A is given by $w^A(M) = \sum_{ij \in M \cap E^A} w_{ij}$ and that of B by $w^B(M) = \sum_{ij \in M \cap E^B} w_{ij}$. Thus, determining a global optimum maximizing the sum of the two agents' weights can be done in polynomial time by solving a weighted cardinality assignment problem [2]. Indeed, in the next section, we show that the global optimum under both rules can be found in an straightforward way, without recurring to matching techniques.

Note that matching problems on graphs with edges partitioned into two sets are addressed in [4].

2   Structure of Efficient Solutions

In this section we investigate the solution structure in an offline perspective, i.e. we consider the problem from a centralized and static point of view and look for Pareto efficient solutions. Then, we show that under both rules, no Nash equilibria exist, in general.

2.1   Pareto Efficient Solutions

Using the same terminology as in multicriteria optimization a solution M is called *Pareto efficient* or simply *efficient* or *nondominated* if there exist no solution M' such that $w^A(M) \leq w^A(M')$, $w^B(M) \leq w^B(M')$ and $w^A(M) + w^B(M) < w^A(M') + w^B(M')$.

We start our investigation with Rule 1 and notice that in this case each agent always submits its c largest items. So, in the remainder of this section, without loss of generality, under Rule 1 we assume $c = n$.

Theorem 1. *Under Rule 1, there are at most $c + 1$ efficient solutions. These can be computed in polynomial time.*

Proof. Suppose that a feasible solution exists such that agent A wins k rounds and therefore B wins the remaining $c - k$ rounds. W.l.o.g. we assume $a_1 > b_1$ (otherwise we can exchange the roles of A and B). Since a_1 may win in any case, if A is rational there is $k \geq 1$, otherwise, if A "voluntarily" decides to lose all rounds by submitting at each round an item smaller than the corresponding B item, k may be equal to 0.

By a simple pair interchange argument it is easy to see that in this case the solution where A (B resp.) wins with its heaviest k ($c - k$ resp.) items is also feasible. Clearly, it is also nondominated. If $k \geq 1$, the structure of the resulting solution is such that item a_i is matched to b_{c-k+i} for $i = 1, \ldots, k$, and to b_{i-k} for $i = k+1, \ldots, c$. Otherwise, we have an additional efficient solution with $w^A = 0$ and $w^B = \sum_{i=1}^{c} b_i$.

Computing the solution values for all $c + 1$ efficient solutions can be done in linear time by one scan through the list of items after sorting the items. Giving also the corresponding pairs of items as output would require $O(n^2)$ time.

A straightforward consequence of the above result is that, finding the global optimum, i.e. the solution M^* for which $w^A(M^*) + w^B(M^*)$ is maximum, can be done in a simple way in polynomial time also without solving a matching problem, since M^* is a Pareto efficient solution.

Turning to Rule 2, we observe that it is not a trivial task to select the "best" c items to submit among the n available. Picking only large items may results in too many losses, a restriction to the smallest items increases the chances to win many rounds but the gain from these victories may be quite small. Hence, we first restrict our attention to the case $c = n$. In this case we can show the following results.

Theorem 2. *Under Rule 2, there can be an exponential number of efficient solutions for $c = n$.*

Consider the following example for some small $\varepsilon > 0$:

Example 1.

agent A		agent B		
a_1	0	b_1	ε	
a_i	2^{i-2}	b_i	$a_i + \varepsilon$	$i = 2, \ldots, c$

For any set $S \subseteq \{2, \ldots, c - 1\}$ consider the solution M where an item a_i wins and, at the same time, b_i loses if and only if $i \in S \cup \{1\}$. It is always possible to construct such an M: clearly, a_1 always wins; then, if $i \in S$, a_i is matched to b_i, else $(i \notin S)$ a_i is matched to some b_j, with $j < i$, $j \notin S$. In M, a_1 is matched to b_1 or b_c and a_c is always losing.

Neglecting the ε values, the sum of the weights $w^A(M) + w^B(M)$ is constant and equals $2^{c-1} - 1$. Therefore, for any possible choice of S, it is possible to obtain an efficient solution since the resulting values of $w^A(M)$ are all different. Thus, there are exponentially many efficient solutions.

We now show that, under Rule 2, the problem of deciding whether a certain combination of weights for agents A and B can be reached is $\mathcal{NP}$-complete by reduction from PARTITION. This implies that finding efficient solutions under Rule 2 is in general an $\mathcal{NP}$-hard task.

Let the decision problem be defined as follows.

SUBSET WEIGHT PROBLEM (SWP):
Instance: positive integers a_i and b_i, $i = 1, \ldots, n$; two positive values W^A and W^B.
Question: Is there a solution (matching) M such that $w^A(M) \geq W^A$ and $w^B(M) \geq W^B$?

Theorem 3. SUBSET WEIGHT PROBLEM *is (binary)* $\mathcal{NP}$*-complete.*

Proof. Consider an instance I of PARTITION with integer valued items $\{v_1, v_2, \ldots, v_n\}$ and $\sum_{i=1}^{n} v_i = V$. Assume $v_1 < v_2 < \ldots < v_n$.

For $T > v_n$ and $\varepsilon < 1/(n+1)$, build an instance I_{SWP} of SUBSET WEIGHT PROBLEM as follows:

$$
(I_{SWP}) \quad
\begin{aligned}
a_0 &= T, & b_0 &= \varepsilon, \\
a_i &= v_i, & b_i &= v_i + \varepsilon, \quad i = 1, \ldots, n, \\
W^A &= W^B = \tfrac{V}{2}.
\end{aligned}
$$

Let I be a YES-instance of PARTITION. Then it is easy to build a solution M of I_{SWP} such that $w^A(M) = V/2$ and $w^B(M) \geq V/2$. Let $S \subseteq \{1, 2, \ldots, n\}$ be such that $\sum_{i \in S} v_i = V/2$ and $\bar{S} = \{0, 1, \ldots, n\} \setminus S$. The required solution M is given as follows: Each item i of A with $i \in S$ is matched to the corresponding item i of B, while each item j of B with $j \in \bar{S}$ is matched to

$$\arg \min\{a_h \mid h \in \bar{S}, a_h > b_j\}.$$

Clearly, the two agents total weights are: $w^A(M) = \sum_{i \in S} a_i = V/2$ and $w^B(M) = \sum_{i \in \bar{S}} b_i \geq V/2$.

Now, let I_{SWP} be a YES-instance of SUBSET WEIGHT PROBLEM and M be a solution of I_{SWP} such that $\bar{S} \subseteq \{0, 1, 2, \ldots, n\}$ is the set of "winning" items of agent B. Note that $0 \in \bar{S}$, since item 0 of B always wins. So,

$$w^B(M) = \sum_{i \in \bar{S}} b_i = \sum_{i \in \bar{S}} a_i + \varepsilon \geq V/2$$

and $w^A(M) \geq V/2$. Note that, since each winning item i of A is matched to a losing item of B with larger weight, the total weight L^B of the "losing" items of agent B is such that $L^B = \sum_{i \in S} b_i \geq w^A(M) \geq V/2$. Since, $L^B + w^B(M) = V + (n+1)\varepsilon$, the v_i are all integers and $\varepsilon < 1/(n+1)$, then $\sum_{i \in S} v_i = \sum_{i \in \bar{S}} v_i = V/2$.

Observe that the above proof shows that even if $c = n$, i.e. the items to be submitted must not be selected, the problem is $\mathcal{NP}$-hard.

Although finding efficient solutions under Rule 2 is $\mathcal{NP}$-hard, the global optimum M^* can be found in polynomial time using matching techniques [2] as already mentioned in Section 1.1. However, it is easy to see that for any value of $c \leq n$, we may compute a global optimal solution in a straightforward way by setting $M^* = \{(a_1, b_1), (a_2, b_2), \ldots, (a_c, b_c)\}$.

Table 1. Payoffs of agents A and B with data of Example 2 under Rule 1 and 2

	$\langle b_1, b_2 \rangle$	$\langle b_2, b_1 \rangle$
$\langle a_1, a_2 \rangle$	0,18	10,12
$\langle a_2, a_1 \rangle$	10,12	0,18

	$\langle b_1, b_2 \rangle$	$\langle b_2, b_1 \rangle$
$\langle a_1, a_2 \rangle$	15,0	5,6
$\langle a_2, a_1 \rangle$	5,6	15,0

2.2 Nash Equilibrium

It is easy to show that, under both rules, no Nash equilibria exist, except for trivial instances where only one Pareto optimum exists. The following example presents an instance in which there are no Nash equilibria.

Example 2.

agent A		agent B	
a_1	10	b_1	12
a_2	5	b_2	6

There are $c = n = 2$ rounds. Table 1 gives the normal-form representations of the game under the two rules. Assume, under Rule 1, that A submits a_1 and a_2 in the first and second round, respectively. Then the best sequence for B is $\langle b_1, b_2 \rangle$. Clearly, when B submits the latter sequence, A changes its own, by swapping a_1 with a_2. So, A's strategy $\langle a_1, a_2 \rangle$ is not in a Nash equilibrium. An analogous reasoning applies for the only other possible sequence for A, $\langle a_2, a_1 \rangle$. So, under Rule 1, no Nash equilibrium exists. The same argument applies for Rule 2.

3 Best-Worst Ratio

The *price of anarchy* is usually defined as the worst possible ratio between the value of the global (social) optimum and the value of a solution derived by a selfish optimization. In this context, we adopt a slightly different definition, considering the ratio between the best and worst values of two efficient solutions, and we denote it as the *best-worst ratio* (BWR). Formally,

$$\text{BWR} = \max_{M \in \mathcal{E}} \left\{ \frac{w^*}{w^A(M) + w^B(M)} \right\},$$

where $\mathcal{E}$ is the set of efficient solutions and w^* is the global optimum value.

It will be shown that this ratio is equal to 2 in case of Rule 1, while it can be arbitrarily high in case of Rule 2.

Theorem 4. *The best-worst ratio under Rule 1 is 2.*

Proof. Let $X = \{a_1, a_2, \ldots, a_c\} \cup \{b_1, b_2, \ldots, b_c\}$ and number the items of X in non-increasing order of weights, $X = \{x_1, x_2, \ldots, x_{2c}\}$. Obviously, the global optimum value is bounded from above by the sum of the c largest items, i.e.

$w^* \leq \sum_{i=1}^c x_i$. On the other hand, even the value of the efficient solution $\bar{M}$ with the *worst* global value can be bounded from below. In fact, in $\bar{M}$, the set X can be partitioned into the two sets X^V and X^L of winning and losing items. Clearly $|X^V| = |X^L| = c$. Since, under Rule 1, each winning item is matched to an item with smaller weight, then $w(X^V) \geq w(X^L)$. Therefore $w^A(\bar{M}) + w^B(\bar{M}) = w(X^V) \geq w(X)/2$. Since $w(X) \geq w^*$, the statement follows.

To show that the bound of 2 can be reached consider the following example for a large constant T:

Example 3.

agent A		agent B	
a_1	T	b_1	$T-1$
a_2	2	b_2	1

Let $c = 2$. Agent A submits items $\langle a_1, a_2 \rangle$. There are only two efficient solutions: If B submits b_1 first and b_2 second thus losing both rounds, the global solution value is $T + 2$. If B exchanges the order of its submissions then A wins the first round and B the second yielding a global solution value of $2T - 1$. Therefore, for T tending to infinity, the best-worst ratio tends to 2.

Theorem 5. *The best-worst ratio under Rule 2 can be arbitrarily high.*

Consider the following trivial example:

Example 4.

agent A		agent B	
a_1	T	b_1	$T+1$
a_2	2	b_2	1

There is only one round. If agent B submits b_1 it will lose in any case. Therefore, any selfish strategy of B will submit b_2 gaining in any case $w^B = 1$, while agent A will lose and have $w^A = 0$. A global optimum would submit a_1 and b_1 yielding a total weight of T.

4   Strategies of Agents

In this section we address the problem of devising a *strategy*, that is an algorithm, that suggests an agent which item to submit at each round, in order to maximize the agent's weight under different information scenarios.

In particular, in the *off-line* case, where an agent, say B, knows all c submissions of A before deciding on its c submissions, the problem of selecting the best c items for B is easy under both rules. Take the bipartite graph G defined in Section 1.1 and restrict V^A to the c nodes representing the items submitted by A. Now compute a maximum weight V^A-perfect matching, where the weight of every arc in E^A is set to 0 while arcs in E^B keep the weight of the corresponding winning item in B. Actually, such c-matchings can be computed through simple greedy algorithms depending on the rule in force.

The above remark implies that, in the *on-line* case in which an agent does not know in advance the sequence of items submitted by the other, no strategy of B can guarantee a limited *competitive ratio* [1], since A can always minimize B's weight by maximizing its own weight. For instance, consider the problem given in Example 2, whatever the strategy of B is, under Rule 2 in the worst case, $w^B = 0$, which makes the competitive ratio unbounded. (Under Rule 1, the same example applies by exchanging the role of the two agents).

In what follows we consider the scenario under Rule 1 where agent B, gets to know the item submitted by agent A in every round *before* making its own move. We will try to answer the question how B should select its items under such an advantageous asymmetry of information. In this particular context, the strategy adopted by agent B is referred to as *best response* strategy.

Somewhat surprisingly, it will turn out that for Rule 2 an optimal best response strategy can be found while for Rule 1 the existence of such a strategy can be ruled out.

4.1 Best Response Strategies for Rule 1

We now consider the scenario where B only knows the submission of A in the current round and has to react immediately before both agents move on to the next round. If B knows in advance which items A would submit (for instance when $n = c$, or A behaves in a rational way and only uses its largest items $a_1, \ldots, a_c$) the problem is easy. By solving the matching problem described above, B knows the optimal answer to every item submitted by A. The ordering of these submissions does not matter, since B simply uses the answer indicated by the optimal matching to every item of A.

The problem becomes more difficult if we cannot restrict the choices of A but have to assume that any item $i \in \{1, \ldots, n\}$ could be submitted by A. In this case, there is no optimal strategy for B. Indeed we can show that no strategy of B can have a competitive ratio (compared to the best off-line strategy) better than $1.618\ldots$. On the other hand, we will show that a simple greedy-type algorithm has a worst-case competitive ratio of 2.

Theorem 6. *No on-line strategy of agent B against an arbitrary strategy of agent A can have a competitive ratio smaller than $\frac{\sqrt{5}+1}{2} = 1.618034\ldots$ under Rule 1.*

Consider the following example:

Example 5.

agent A		agent B	
a_1	2	b_1	$1 + \varepsilon$
a_2	1	b_2	y
a_3	ε	b_3	ε

There are $c = 2$ rounds and parameter y is chosen such that $2\varepsilon < y < 1$. We will consider two (suboptimal) strategies of agent A: In both cases A starts by submitting a_3. In strategy S_1 it is followed by a_1, in strategy S_2 by a_2.

Given the first item a_3 there are only two ways for agent B to react:

In Case 1, B submits b_2 thus winning this round. Against strategy S_1, B loses the second round and the worst-case total weight for Case 1 is $W_1^B = y$.

In Case 2, B submits b_1 and wins again. But against strategy S_2, B loses again the second round and in the worst-case stops with a total weight $W_2^B = 1 + \varepsilon$.

However, an optimal off-line strategy of B against S_1 would be $\langle b_1, b_2 \rangle$ gaining a weight of $W_1^* = 1 + \varepsilon$, while against S_2 the submission of $\langle b_2, b_1 \rangle$ would yield $W_2^* = y + 1 + \varepsilon$.

Altogether we get a lower bound for the worst-case competitive ratio of

$$\max_{2\varepsilon < y < 1} \min_{i=1,2} \left\{ \frac{W_i^*}{W_i^B} \right\} = \max_{2\varepsilon < y < 1} \min \left\{ \frac{1 + \varepsilon}{y}, \frac{y + 1 + \varepsilon}{1 + \varepsilon} \right\}.$$

Clearly, this maximum is attained if the expressions of the minimum are equal. An elementary calculation yields $y = (1 + \varepsilon) \frac{\sqrt{5}-1}{2}$. Plugging in this value of y in the above equation yields $\frac{\sqrt{5}+1}{2}$ as a lower bound for the competitive ratio and proves Theorem 6.

In the above analysis B submits its $c = 2$ largest items; however it is not hard to see that the same result is attained when B may decide to submit b_3.

The following greedy-type algorithm tries to win against every item submitted by A with the largest item still available. If this fails and a loss cannot be avoided in the current round the smallest remaining item is submitted.

Algorithm 1. Algorithm for agent B responding to the submissions of agent A under Rule 1

GREEDY RESPONSE RULE 1

1: $f \leftarrow 1$ {index of largest available item}
2: $\ell \leftarrow c$ {index of smallest available item}
3: **repeat**
4:      A submits item a'
5:      **if** $b_f > a'$ **then**
6:          B submits b_f and wins
7:          $f \leftarrow f + 1$
8:      **else**
9:          B submits b_ℓ and loses
10:         $\ell \leftarrow \ell - 1$
11:     **end if**
12: **until** $f < \ell$ {all c rounds finished}

Theorem 7. *Algorithm* GREEDY RESPONSE RULE 1 *has a tight competitive ratio of* 2.

Proof. Let S^A be the complete set of items submitted by A. At the end of the execution of GREEDY RESPONSE RULE 1 agent B has added its k largest items into the solution and reached a total weight of W^B ($k = f - 1$ at the end of the algorithm).

It is easy to see that an optimal off-line solution against the same set S^A will always consist of the largest k^* items with $k^* \geq k$. In this optimal strategy the items in set $D^B = \{k + 1, \ldots, k^*\}$ of B all win against some items in S^A. This means that at least the smallest items in S^A must be potential losers against D^B. Let j be the index of the $c - (k^* - k - 1)$ largest item in S^A. Then we have $b_{k+1} > a_j$.

Considering the item $b_f = b_\ell$ submitted by B in the last round of the game there are two cases to distinguish.

Case 1: b_f loses. Since $b_f = b_{k+1}$ would win against any of the $k^* - k$ smallest items in S^A, these must have been submitted by A before. Moreover, at the rounds where these items were submitted, item b_{k+1} was still available and would yield a win for B. By definition of GREEDY RESPONSE RULE 1 some other item of B must have won these rounds. Therefore, the total number of wins for B must be at least $k^* - k$, i.e. $k \geq k^* - k$, which proves the statement of the theorem.

Case 2: b_f wins. In all rounds, in which A submitted one of the $k^* - k$ smallest items in S^A, item b_f was still available and would have won against these items. As above, by the definition of GREEDY RESPONSE RULE 1 some other item of B must have won these rounds and again $k \geq k^* - k$.

To show that the bound of 2 is tight consider the simple Example 6.

Example 6.

agent A		agent B	
a_1	$1 + \varepsilon$	b_1	$1 + 2\varepsilon$
a_2	$1 - \varepsilon$	b_2	1

There are $c = 2$ rounds. Agent A submits a_2 in the first round. GREEDY RESPONSE RULE 1 reacts with b_1 and loses the second round (a_1 against b_2) gaining a total weight of $1 + 2\varepsilon$.

An optimal off-line strategy of B would react with b_2 in the first round thus winning both rounds and gaining $2 + 2\varepsilon$.

4.2   Best Response Strategies for Rule 2

Even under Rule 2, if B knows in advance which items A would submit (e.g., when $n = c$), it is easy to devise an optimal response algorithm using the same matching argument holding for Rule 1.

As in Rule 1, the following greedy-type algorithm for Rule 2 tries to win against every item submitted by A with the largest possible winning item. If no such item exists, a losing item is determined which can not worsen the remainder of the solution for B. To avoid a tedious special treatment of ties we will assume in this subsection that all $2n$ items have different weights.

Algorithm 2. Algorithm for agent B responding to the submissions of agent A under Rule 2

GREEDY RESPONSE RULE 2

1: $R^A \leftarrow \{a_1, \ldots, a_n\}$ {set of remaining items for A}
2: $R^B \leftarrow \{b_1, \ldots, b_n\}$ {set of remaining items for B}
3: **repeat**
4:     A submits item a', remove a' from R^A
5:     $b_{\min} \leftarrow \min\{b_i \mid b_i \in R^B\}$
6:     **if** $a' > b_{\min}$ **then**
7:         $b_{win} \leftarrow \max\{b_i \mid b_i < a', b_i \in R^B\}$
8:         B submits b_{win} and wins, remove b_{win} from R^B
9:     **else**
10:         $A_j \leftarrow \{a_i \mid a_i \geq b_j, a_i \in R^A\}$, $j = 1, \ldots, |R^B|$
11:         $B_j \leftarrow \{b_i \mid b_i \geq b_j, b_i \in R^B\}$, $j = 1, \ldots, |R^B|$
12:         determine the minimum index $\ell \geq 1$ such that $|A_\ell| < |B_\ell|$
13:         B submits b_ℓ and loses, remove b_ℓ from R^B
14:     **end if**
15: **until** all c rounds finished

Theorem 8. *Algorithm* GREEDY RESPONSE RULE 2 *yields the optimal off-line solution under Rule 2.*

Proof. We will consider round 1, where agent A submits a', and show that also an optimal algorithm OPT can not do better than GREEDY RESPONSE RULE 2 and that the later does not diminish the range of options for an optimal strategy in the subsequent rounds. Repeating this argument over all rounds yields the statement of the theorem.

Case 1: B wins. What are the alternatives for OPT? If OPT submits an item $b_j < b_{win}$ it earns a smaller weight in round 1. In contrary to GREEDY RESPONSE RULE 2 it can use b_{win} to win in one of the subsequent rounds. However, in this case GREEDY RESPONSE RULE 2 could submit b_j in this future round and win as well since $b_j < b_{win}$. Obviously, the total weight of these two rounds would be the same for both algorithms.

If OPT chooses to submit an item $b_k \geq a'$ and loose "voluntarily", it can again use b_{win} to win in a later round. GREEDY RESPONSE RULE 2 might be forced to loose in that round, but again the two algorithms end up with the same total weight for these two rounds.

Case 2: B looses. First, let us note that in the special case $b_1 > a_1$ (considering only items in R^A and R^B) GREEDY RESPONSE RULE 2 will submit $b_\ell = b_1$. Since b_1 cannot win in any round, it clearly cannot violate optimality to submit b_1 in a round which B will lose in any case.

Also for $b_1 < a_1$ index ℓ is well defined. Note that $b_j \in B_j$, $\forall j$. For $j = 1$ the defining inequality in line 12 is clearly violated since $A_1 \supseteq \{a_1\}$ and $B_1 = \{b_1\}$. Since $a' < b_{\min}$ and $|R^A| = |R^B|$, an index ℓ satisfying $|A_\ell| < |B_\ell|$ will be determined for $b_{\min}$ at the latest.

Note that in the ordering of the items in $R^A \cup R^B$ there must occur an item $b_{\ell-1} \in R^B$ directly before b_ℓ. Otherwise, since there was $|A_{\ell-1}| \geq |B_{\ell-1}|$, one (or more) items of R^A plugged in between $b_{\ell-1}$ and b_ℓ would not lead to the required change in the balance of cardinalities. This means that $A_{\ell-1} = A_\ell$.

Now consider that $B_{\ell-1}$ can be matched to win against $A_{\ell-1}$. This can be seen by a simple scan through the items in $R^A \cup R^B$ in decreasing order, starting with a_1 being matched to b_1. A failure of such a scan would immediately yield a new smaller value of ℓ. In fact, the existence of such a matching follows from Hall's marriage theorem. Clearly, also every subset of items of $A_{\ell-1}$ ($= A_\ell$) can be matched to be beaten by the corresponding number of items of $B_{\ell-1}$ (e.g. by taking the smallest items in $B_{\ell-1}$).

On the other hand, it follows from the definition of $|A_\ell|$ and $|B_\ell|$ that there are more items in R^B with weight larger or equal b_ℓ than in R^A. Therefore, it is impossible for any strategy that *all* items in B_ℓ will win against some items in R^A, the items in A_ℓ being the only candidates for losers.

Since it is impossible for agent B to beat all items in A_ℓ, but every strict subset of A_ℓ can be beaten also by items in $B_{\ell-1}$, the removal of b_ℓ does not change the options of OPT in any subsequent round.

5   Conclusions

This work puts several directions forward for future research. One is the design of algorithms to efficiently enumerate all Pareto-optimal solutions under Rule 2. Another is to extend our results to further relevant rules controlling how an agent's item is added to the solution set. For instance, the case when losing items are not discarded but are *reusable* (meaning that an agent can submit them in successive rounds), seems particularly significant. Furthermore, it would be interesting to investigate the problem where both agents ignore the weights of the items of the other agent.

References

1. Borodin, A., El-Yaniv, R.: Online Computation and Competitive Analysis. Cambridge University Press, Cambridge (1998)
2. Dell'Amico, M., Martello, S.: The k-cardinality assignment problem. Discrete Applied Mathematics 76, 103–121 (1997)
3. Kahn, J., Lagarias, J.C., Witsenhausen, H.S.: Single-Suit Two-Person Card Play. International Journal of Game Theory 16(4), 291–320 (1987)
4. Nomikos, C., Pagourtzis, A., Zachos, S.: Randomized and Approximation Algorithms for Blue-Red Matching. In: Kučera, L., Kučera, A. (eds.) MFCS 2007. LNCS, vol. 4708, pp. 715–725. Springer, Heidelberg (2007)
5. Schlag, K.H., Sela, A.: You play (an action) only once. Economic Letters 59, 299–303 (1998)
6. Wästlund, J.: A solution of two-person single-suit whist. The Electronic Journal of Combinatorics 12 (2005)

The Complexity of Probabilistic Lobbying[*]

Gábor Erdélyi[1], Henning Fernau[2], Judy Goldsmith[3],
Nicholas Mattei[3], Daniel Raible[2], and Jörg Rothe[1]

[1] Universität Düsseldorf, Institut für Informatik, 40225 Düsseldorf, Germany
[2] Universität Trier, FB 4—Abteilung Informatik, 54286 Trier, Germany
[3] University of Kentucky, Department of Computer Science, Lexington, KY 40506, USA

Abstract. We propose various models for lobbying in a probabilistic environment, in which an actor (called "The Lobby") seeks to influence the voters' preferences of voting for or against multiple issues when the voters' preferences are represented in terms of probabilities. In particular, we provide two evaluation criteria and three bribery methods to formally describe these models, and we consider the resulting forms of lobbying with and without issue weighting. We provide a formal analysis for these problems of lobbying in a stochastic environment, and determine their classical and parameterized complexity depending on the given bribery/evaluation criteria. Specifically, we show that some of these problems can be solved in polynomial time, some are NP-complete but fixed-parameter tractable, and some are W[2]-complete. Finally, we provide (in)approximability results.

1 Introduction

In the American political system, laws are passed by elected officials who are supposed to represent their constituency. Many factors can affect a representative's vote on a particular issue: a representative's personal beliefs about the issue, campaign contributions, communications from constituents, communications from potential donors, and the representative's own expectations of further contributions and political support.

It is a complicated process to reason about. Earlier work considered the problem of meting out contributions to representatives in order to pass a set of laws or influence a set of votes. However, the earlier computational complexity work on this problem made the assumption that a politician who accepts a contribution will in fact—if the contribution meets a given threshold—vote according to the wishes of the donor.

It is said that "An honest politician is one who stays bought," but that does not take into account the ongoing pressures from personal convictions and opposing lobbyists and donors. We consider the problem of influencing a set of votes under the assumption that we can influence only the *probability* that the politician votes as we desire. The methods for exerting influence on the voters is discussed in the section on bribery criteria while the notion of sufficient influence for a voter is discussed in the section on evaluation criteria.

---

[*] Supported in part by DFG grants RO 1202/11-1 and RO 1202/12-1, the European Science Foundation's EUROCORES program LogICCC, the Alexander von Humboldt Foundation's TransCoop program, and NSF grant ITR-0325063.

F. Rossi and A. Tsoukis (Eds.): ADT 2009, LNAI 5783, pp. 86–97, 2009.

Lobbying has been studied formally by economists, computer scientists, and special interest groups since at least 1983 [13] and as an extension to formal game theory since 1944 [15]. Each discipline has considered mostly disjoint aspects of the process while seeking to accomplish distinct goals with their respective formal models. Economists have formalized models and studied them as "economic games," as defined by von Neumann and Morgenstern [15]. This analysis is focused on learning how these complex systems work and deducing optimal strategies for winning the competitions [13,1, 2]. This work has also focused on how to "rig" a vote and how to optimally dispense the funds among the various individuals [1]. Economists are interested in finding effective and efficient bribery schemes [1] as well as determining strategies for instances of two or more players [1,13,2]. Generally, they reduce the problem of finding an effective lobbying strategy to one of finding a winning strategy for the specific type of game. Economists have also formalized this problem for bribery systems in both the United States [13] and the European Union [6].

In the emerging field of computational social choice, voting and preference aggregation are studied from a computational perspective, with a particular focus on the complexity of winner determination, manipulation, procedural control, and bribery in elections (see, e.g., the survey [9] and the references cited therein), and also with respect to lobbying in the context of direct democracy where voters vote on multiple referenda. In particular, Christian et al. [5] show that "Optimal Lobbying" (OL) is complete for the (parameterized) complexity class W[2]. The OL problem is a deterministic and non-weighted version of the problems that we present in this paper. Sandholm noted that the "Optimal Weighted Lobbying" (OWL) problem, which allows different voters to have different prices, can be expressed as and solved via the "binary multi-unit combinatorial reverse auction winner-determination problem" (see [14]).

We extend the models of lobbying, and provide algorithms and analysis for these extended models in terms of classical and parameterized complexity. Our problems are still related to the reverse auction winner-determination problem—in particular, our extensions of the optimal lobbying problem allow the seller to express desire over the objects, thus crucially changing the original problem in both the economic and complexity-theoretic senses. This change is a result of the probabilistic modeling of the seller's reaction to the bribery. We also show novel computational and algorithmic approaches to these new problems. In this way we add breadth and depth to not only the models but also the understanding of lobbying behavior.

2 Models for Probabilistic Lobbying

2.1 Initial Model

We begin with a simplistic version of the PROBABILISTIC LOBBYING PROBLEM (PLP, for short), in which voters start with initial probabilities of voting for an issue and are assigned known costs for increasing their probabilities of voting according to "The Lobby's" agenda by each of a finite set of increments.

The question, for this class of problems, is: Given the above information, along with an agenda and a fixed budget B, can The Lobby target its bribes in order to achieve its agenda? The complexity of the problem seems to hinge on the evaluation criterion for what it means to "win a vote" or "achieve an agenda." We discuss the possible interpretations of evaluation and bribery later in this section. First, however, we will formalize the problem by defining data objects needed to represent the problem instances.

Let $\mathbb{Q}_{[0,1]}^{m \times n}$ denote the set of $m \times n$ matrices over $\mathbb{Q}_{[0,1]}$ (the rational numbers in the interval $[0,1]$). We say $P \in \mathbb{Q}_{[0,1]}^{m \times n}$ is a probability matrix (of size $m \times n$), where each entry $p_{i,j}$ of P gives the probability that voter v_i will vote "yes" for referendum (synonymously, for issue) r_j. The result of a vote can be either a "yes" (represented by 1) or a "no" (represented by 0). Thus, we can represent the result of any vote on all issues as a $0/1$ vector $X = (x_1, x_2, \ldots, x_n)$, which is sometimes also denoted as a string in $\{0,1\}^n$.

We now associate with each pair (v_i, r_j) of voter/issue, a discrete price function $c_{i,j}$ for changing v_i's probability of voting "yes" for issue r_j. Intuitively, $c_{i,j}$ gives the cost for The Lobby of raising or lowering (in discrete steps) the ith voter's probability of voting "yes" on the jth issue. A formal description is as follows.

Given the entries $p_{i,j} = a_{i,j}/b_{i,j}$ of a probability matrix $P \in \mathbb{Q}_{[0,1]}^{m \times n}$, choose some $k \in \mathbb{N}$ such that $k+1$ is a common multiple of all $b_{i,j}$, where $1 \leq i \leq m$ and $1 \leq j \leq n$, and partition the probability interval $[0,1]$ into $k+1$ steps of size $1/(k+1)$ each. For each $i \in \{1,2,\ldots,m\}$ and $j \in \{1,2,\ldots,n\}$, $c_{i,j} : \{0, 1/(k+1), 2/(k+1), \ldots, k/(k+1), 1\} \to \mathbb{N}$ is the (discrete) price function for $p_{i,j}$, i.e., $c_{i,j}(\ell/(k+1))$ is the price for changing the probability of the ith voter voting "yes" on the jth issue from $p_{i,j}$ to $\ell/(k+1)$, where $0 \leq \ell \leq k+1$. Note that the domain of $c_{i,j}$ consists of $k+2$ elements of $\mathbb{Q}_{[0,1]}$ including 0, $p_{i,j}$, and 1. In particular, we require $c_{i,j}(p_{i,j}) = 0$, i.e., a cost of zero is associated with leaving the initial probability of voter v_i voting on issue r_j unchanged. Note that $k = 0$ means $p_{i,j} \in \{0,1\}$, i.e., in this case each voter either accepts or rejects each issue with certainty and The Lobby can only flip these results.[1] The image of $c_{i,j}$ consists of $k+2$ nonnegative integers including 0, and we require that, for any two elements a, b in the domain of $c_{i,j}$, if $p_{i,j} \leq a \leq b$ or $p_{i,j} \geq a \geq b$, then $c_{i,j}(a) \leq c_{i,j}(b)$. This guarantees monotonicity on the prices.

We represent the list of price functions associated with a probability matrix P as a table C_P whose $m \cdot n$ rows give the price functions $c_{i,j}$ and whose $k+2$ columns give the costs $c_{i,j}(\ell/(k+1))$, where $0 \leq \ell \leq k+1$. Note that we choose the same k for each $c_{i,j}$, so we have the same number of columns in each row of C_P. The entries of C_P can be thought of as "price tags" that The Lobby must pay in order to change the probabilities of voting.

The Lobby also has an integer-valued budget B and an "agenda," which we will denote as a vector $Z \in \{0,1\}^n$, where n is the number of issues, containing the outcomes The Lobby would like to see on the corresponding issues. For simplicity, we may assume that The Lobby's agenda is all "yes" votes, so the target vector is $Z = 1^n$. This assumption can be made without loss of generality, since if there is a zero in Z at position j, we can flip this zero to one and also change the corresponding probabilities $p_{1,j}, p_{2,j}, \ldots, p_{m,j}$ in the jth column of P to $1 - p_{1,j}, 1 - p_{2,j}, \ldots, 1 - p_{m,j}$ (see the evaluation criteria in Section 2.3 for how to determine the result of voting on a referendum).

Example 1. We create a problem instance with $k = 9$, $m = 2$ (number of voters), and $n = 3$ (number of issues). We will use this as a running example for the rest of this

---

[1] This is the special case of Optimal Lobbying.

paper. In addition to the above definitions for k, m, and n, we also give the following matrix for P. (Note that this example is normalized for an agenda of $Z = 1^3$, which is why The Lobby has no incentive for lowering the acceptance probabilities, so those costs are omitted below.)

Our example consists of a probability matrix P:

	r_1	r_2	r_3
v_1	0.8	0.3	0.5
v_2	0.4	0.7	0.4

and the corresponding cost matrix C_P:

$c_{i,j}$	0.0	0.1	0.2	0.3	0.4	0.5	0.6	0.7	0.8	0.9	1.0
$c_{1,1}$	--	--	--	--	--	--	--	--	0	100	140
$c_{1,2}$	--	--	--	0	10	70	100	140	310	520	600
$c_{1,3}$	--	--	--	--	--	0	15	25	70	90	150
$c_{2,1}$	--	--	--	--	0	30	40	70	120	200	270
$c_{2,2}$	--	--	--	--	--	--	--	0	10	40	90
$c_{2,3}$	--	--	--	--	0	70	90	100	180	300	450

In Section 2.2, we describe three bribery methods, i.e., three specific ways in which The Lobby can influence the voters. These will be referred to as B_i, $i \in \{1, 2, 3\}$. In addition to the three bribery methods described in Section 2.2, we also define two ways in which The Lobby can win a set of votes. These evaluation criteria are defined in Section 2.3 and will be referred to as C_j, $j \in \{1, 2\}$. They are important because votes counted in different ways can result in different outcomes depending on voting and evaluation systems (cf. Myerson and Weber [11]).

We now introduce the six basic problems that we will study. For $i \in \{1, 2, 3\}$ and $j \in \{1, 2\}$, we define:

Name: B_i-C_j PROBABILISTIC LOBBYING PROBLEM (B_i-C_j-PLP, for short).
Given: A probability matrix $P \in \mathbb{Q}_{[0,1]}^{m \times n}$ with table C_P of price functions, a target vector $Z \in \{0, 1\}^n$, and a budget B.
Question: Is there a way for The Lobby to influence P (using bribery method B_i and evaluation criterion C_j, without exceeding budget B) such that the result of the votes on all issues equals Z?

2.2 Bribery Methods

We begin by first formalizing the bribery methods by which The Lobby can influence votes on issues. We will define three methods for donating this money.

Microbribery (B_1). The first method at the disposal of The Lobby is what we will call *microbribery*. We define microbribery to be the editing of individual elements of the P matrix according to the costs in the C_P matrix. Thus The Lobby picks not only which voter to influence but also which issue to influence for that voter. This bribery method allows the most flexible version of bribery, and models private donations made to candidates in support of specific issues.

Issue Bribery (B$_2$). The second method at the disposal of The Lobby is *issue bribery*. We can see from the P matrix that each column represents how the voters think about a particular issue. In this method of bribery, The Lobby can pick a column of the matrix and edit it according to some budget. The money will be equally distributed among all the voters and the voter probabilities will move accordingly. So, for d dollars each voter receives a fraction of d/m and their probability of voting "yes" changes accordingly. This can be thought of as special-interest group donations. Special-interest groups such as PETA focus on issues and dispense their funds across an issue rather than by voter. The bribery could be funneled through such groups.

Voter Bribery (B$_3$). The third and final method at the disposal of The Lobby is *voter bribery*. We can see from the P matrix that each row represents what an individual voter thinks about all the issues on the docket. In this method of bribery, The Lobby picks a voter and then pays to edit the entire row at once with the funds being equally distributed over all the issues. So, for d dollars a fraction of d/n is spent on each issue, which moves accordingly. The cost of moving the voter is generated using the C_P matrix as before. This method of bribery is analogous to "buying" or pushing a single politician or voter. The Lobby seeks to donate so much money to an individual voter that he or she has no choice but to move his or her votes toward The Lobby's agenda.

2.3   Evaluation Criteria

Defining criteria for how an issue is won is the next important step in formalizing our models. Here we define two methods that one could use to evaluate the eventual outcome of a vote. Since we are focusing on problems that are probabilistic in nature, it is important to note that no evaluation criteria will guarantee a win. The criteria below yield different outcomes depending on the model and problem instance.

Strict Majority (C$_1$). For each issue, a strict majority of the individual voters have probability at least some threshold, t, of voting according to the agenda. In our running example (see Example 1), with $t = 50\%$, the result of the votes would be $X = (0, 0, 0)$, because none of the issues has a strict majority of voters with above 50% likelihood of voting according to the agenda.

Average Majority (C$_2$). For each issue, r_j, of a given probability matrix P, we define: $\overline{p_j} = \left(\Sigma_{i=1}^{m} p_{i,j}\right)/m$. We can now evaluate the vote to say that r_j is accepted if and only if $\overline{p_j} > t$ where t is some threshold. This would, in our running example, with $t = 50\%$, give us a result vector of $X = (1, 0, 0)$.

2.4   Issue Weighting

Our modification to the model will bring in the concept of issue weighting. It is reasonable to surmise that certain issues will be of more importance to The Lobby than others. For this reason we will allow The Lobby to specify higher weights to the issues that they deem more important. These weights will be defined for each issue.

   We will specify these weights as a vector $W \in \mathbb{Z}^n$ with size n equal to the total number of issues in our problem instance. The higher the weight, the more important that particular issue is to The Lobby. Along with the weights for each issue we are also given an objective value $O \in \mathbb{Z}^+$ which is the minimum weight The Lobby wants to see passed. Since this is a partial ordering, it is possible for The Lobby to have an ordering such as: $w_1 = w_2 = \cdots = w_n$. If this is the case, we see that we are left with an instance of B_i-C_j-PLP.

We now introduce the six probabilistic lobbying problems with issue weighting. For $i \in \{1,2,3\}$ and $j \in \{1,2\}$, we define:

Name: B_i-C_j-PROBABILISTIC LOBBYING PROBLEM WITH ISSUE WEIGHTING (B_i-C_j-PLP-WIW, for short).

Given: A probability matrix $P \in \mathbb{Q}_{[0,1]}^{m \times n}$ with table C_P of price functions and a lobby target vector $Z \in \{0,1\}^n$, a lobby weight vector $W \in \mathbb{Z}^n$, an objective value $O \in \mathbb{Z}^+$, and a budget B.

Question: Is there a way for The Lobby to influence P (using bribery method B_i and evaluation criterion C_j, without exceeding budget B) such that the total weight of all issues for which the result coincides with The Lobby's target vector Z is at least O?

3   Complexity-Theoretic Notions

We assume the reader is familiar with standard notions of (classical) complexity theory, such as P, NP, and NP-completeness. Since we analyze the problems stated in Section 2 not only in terms of their classical complexity, but also with regard to their *parameterized* complexity, we provide some basic notions here (see, e.g., Downey and Fellows [7] for more background). As we derive our results in a rather specific fashion, we will employ the "Turing way" as proposed by Cesati [4].

A *parameterized problem* $\mathscr{P}$ is a subset of $\Sigma^* \times \mathbb{N}$, where Σ is a fixed alphabet and $\mathbb{N}$ is the set of nonnegative integers. Each instance of the parameterized problem $\mathscr{P}$ is a pair (I,k), where the second component k is called the *parameter*. The language $L(\mathscr{P})$ is the set of all YES instances of $\mathscr{P}$. The parameterized problem $\mathscr{P}$ is *fixed-parameter tractable* if there is an algorithm (realizable by a deterministic Turing machine) that decides whether an input (I,k) is a member of $L(\mathscr{P})$ in time $f(k)|I|^c$, where c is a fixed constant and f is a function whose argument k is independent of the overall input length, $|I|$. The class of all fixed-parameter tractable problems is denoted by FPT.

There is also a theory of parameterized hardness (see, e.g., [7]), most notably the W[t] hierarchy, which complements fixed-parameter tractability: $FPT = W[0] \subseteq W[1] \subseteq W[2] \subseteq \cdots$. It is commonly believed that this hierarchy is strict. Only the second level, $W[2]$, will be of interest to us in this paper (see, e.g., [7] for the definition).

The complexity of a classical problem depends on the chosen parameterization. For problems that involve a budget $B \in \mathbb{N}$ (and hence can be viewed as minimization problems), the most obvious parameterization would be the given budget bound B. In this sense, we state parameterized results in this paper. (For other applications of fixed-parameter tractability and parameterized complexity to problems from computational social choice, see, e.g., [10].)

Table 1. Complexity results for B_i-C_j-PLP

Bribery Criterion	Evaluation Criterion	
	C_1	C_2
B_1	P	P
B_2	P	P
B_3	W[2]-complete	W[2]-complete

4   Classical Complexity Results

We now provide a formal complexity analysis of the probabilistic lobbying problems for all combinations of evaluation criteria and bribery methods.

Table 1 summarizes our results for B_i-C_j-PLP, $i \in \{1, 2, 3\}$ and $j \in \{1, 2\}$. Some of these results are known from previous work by Christian et al. [5], as will be mentioned below. In this sense, our results generalize the results of [5] by extending the model to probabilistic settings.

4.1   Microbribery

The following result can be easily seen.

Theorem 1. B_1-C_1-PLP *is in* P.

The complexity of microbribery with evaluation criterion C_2 is somewhat harder to determine. We use the following auxiliary problem. Here, a *schedule S* of q jobs (on a single machine) is a sequence $J_{i(1)}, \ldots, J_{i(q)}$ such that $J_{i(r)} = J_{i(s)}$ implies $r = s$. The *cost of schedule S* is $c(S) = \sum_{k=1}^{q} c(J_{i(k)})$. S is said to *respect the precedence constraints* of graph G if for every (path)-component $P_i = J_{i,1}, \ldots, J_{i,p(i)}$ and for each k with $2 \leq k \leq p(i)$, we have: If $J_{i,k}$ occurs in the schedule S then $J_{i,k-1}$ occurs in S before $J_{i,k}$.

Name: PATH SCHEDULE
Given: A set $V = \{J_1, \ldots, J_n\}$ of jobs, a directed graph $G = (V, A)$ consisting of pairwise disjoint paths $P_1, \ldots, P_z$, two numbers $C, q \in \mathbb{N}$, and a cost function $c : V \to \mathbb{N}$.
Question: Can we find a schedule $J_{i(1)}, \ldots, J_{i(q)}$ of q jobs of cost at most C respecting the precedence constraints of G?

PATH SCHEDULE is in P by dynamic programming. Then we show how to reduce B_1-C_2-PLP to PATH SCHEDULE, which implies that B_1-C_2-PLP is in P as well.

Lemma 1. PATH SCHEDULE *is in* P.

Theorem 2. B_1-C_2-PLP *is in* P.

Proof. Let (P, C_P, Z, B) be a given B_1-C_2-PLP instance, where $P \in \mathbb{Q}_{[0,1]}^{m \times n}$, C_P is a table of price functions, $Z \in \{0, 1\}^n$ is The Lobby's target vector, and B is its budget. For $j \in \{1, 2, \ldots, n\}$, let d_j be the minimum cost for The Lobby to bring referendum r_j into line with the jth entry of its target vector Z. If $\sum_{j=1}^{n} d_j \leq B$ then The Lobby can

Table 2. Complexity results for B_i-C_j-PLP-WIW

Bribery Criterion	Evaluation Criterion	
	C_1	C_2
B_1	NP-compl., FPT	NP-compl., FPT
B_2	NP-compl., FPT	NP-compl., FPT
B_3	W[2]-complete	W[2]-complete

achieve its goal that the votes on all issues equal Z. We now focus on the first task. For every r_j, create an equivalent PATH SCHEDULING instance. First, compute for r_j the minimum number b_j of bribery steps needed to achieve The Lobby's goal on r_j. That is, choose the smallest $b_j \in \mathbb{N}$ such that $\overline{p_j} + b_j/(k+1)m > t$. Now, for every voter v_i, derive a path P_i from the price function $c_{i,j}$. Let s, $0 \le s \le k+1$, be minimum with the property $c_{i,j}(s) \in \mathbb{N}_{>0}$. Then create a path $P_i = p_s, \dots, p_{k+1}$, where p_h represents the hth entry of $c_{i,j}$ (viewed as a vector). Assign the cost $\hat{c}(p_h) = c_{i,j}(h) - c_{i,j}(h-1)$ to p_h. Observe that $\hat{c}(p_h)$ represents the cost of raising the probability of voting "yes" from $(h-1)/(k+1)$ to $h/(k+1)$. In order to do so, we must have reached an acceptance probability of $(h-1)/(k+1)$ first. Now, let the number of jobs to be scheduled be b_j. Note that one can take b_j bribery steps at the cost of d_j dollars if and only if one can schedule b_j jobs with a cost of d_j. Hence, we can decide whether or not (P, C_P, Z, B) is in B_1-C_2-PLP by using Lemma 1. $\qquad\Box$

4.2   Issue Bribery

A greedy strategy succeeds for proving:

Theorem 3. B_2-C_1-PLP *and* B_2-C_2-PLP *are in* P.

4.3   Probabilistic Lobbying with Issue Weighting

Table 2 summarizes our results for B_i-C_j-PLP-WIW, $i \in \{1,2,3\}$ and $j \in \{1,2\}$. The most interesting observation is that introducing issue weights raises the complexity from P to NP-completeness for all cases of microbribery and issue bribery by using KNAPSACK in the reduction (though it remains the same for voter bribery). Nonetheless, we show later as Theorem 6 that these NP-complete problems are fixed-parameter tractable.

Theorem 4. *For* $i, j \in \{1,2\}$, B_i-C_j-PLP-WIW *is* NP-*complete.*

5   Parameterized Complexity Results

5.1   Voter Bribery

Christian et al. [5] proved that the following problem is W[2]-complete. We state this problem here as is common in parameterized complexity:

Name: OPTIMAL LOBBYING (OL, for short).

Given: An $m \times n$ matrix E and a $0/1$ vector Z of length n. Each row of E represents a voter. Each column represents an issue in the election. The vector Z represents The Lobby's target outcome.

Parameter: A positive integer k (representing the number of voters to be influenced).

Question: Is there a choice of k rows of the matrix (i.e., of k voters) that can be changed such that in each column of the resulting matrix (i.e., for each issue) a majority vote yields the outcome targeted by The Lobby?

Christian et al. [5] proved this problem to be W[2]-complete by a reduction from k-DOMINATING SET to OL (showing the lower bound) and from OL to INDEPENDENT-k-DOMINATING SET (showing the upper bound). To employ the W[2]-hardness result of Christian et al. [5], we show that OL is a special case of B_3-C_1-PLP and thus (parameterized) polynomial-time reduces to B_3-C_1-PLP. The "Turing" approach suggested by Cesati [4] shows membership in W[2]. Analogous arguments apply to B_3-C_2-PLP.

Theorem 5. *For $j \in \{1,2\}$, B_3-C_j-PLP (parameterized by the budget) is W[2]-complete.*

5.2   Probabilistic Lobbying with Issue Weighting

Recall from Theorem 4 that B_i-C_j-PLP-WIW, where $i,j \in \{1,2\}$, is NP-hard. Theorem 6 says that each of these problems is fixed-parameter tractable when parameterized by the budget, using KNAPSACK again.

Theorem 6. *For $i,j \in \{1,2\}$, B_i-C_j-PLP-WIW (parameterized by the budget) is in FPT.*

Voter bribery with issue weighting remains W[2]-complete for both evaluation criteria; the membership proof is somewhat more involved than the one in the unweighted case.

Theorem 7. *For $j \in \{1,2\}$, B_3-C_j-PLP-WIW (parameterized by the budget) is W[2]-complete.*

6   Approximability

As seen in Tables 1 and 2, many problem variants of probabilistic lobbying are NP-complete. Hence, it is interesting to study them not only from the viewpoint of parameterized complexity, but also from the viewpoint of approximability.

    The budget constraint on the bribery problems studied so far gives rise to natural minimization problems: Try to minimize the amount spent on bribing. For clarity, let us denote these minimization problems by prefixing the problem name with MIN, leading to, e.g., MIN-OL.

    The already mentioned reduction of Christian et al. [5] (that proved that OL is W[2]-hard) is parameter-preserving (regarding the budget). It further has the property that a possible solution found in the OL instance can be re-interpreted as a solution to the DOMININATING SET instance the reduction started with, and the OL solution and the

DOMININATING SET solution are of the same size. This in particular means that inapproximability results for DOMININATING SET transfer to inapproximability results for OL. Similar observations are true for the interrelation of SET COVER and DOMINATING SET, as well as for OL and B_3-C_1-PLP-WIW (or B_3-C_2-PLP-WIW).

The known inapproximability results [3,12] for SET COVER hence give the following result (see also Footnote 4 in [14]).

Theorem 8. *There is a constant $c > 0$ such that MIN-OL is not approximable within factor $c \cdot \log(n)$ unless $\mathrm{NP} \subset \mathrm{DTIME}(n^{\log\log(n)})$, where n denotes the number of issues.*

Since OL can be viewed as a special case of both B_3-C_i-PLP and B_3-C_i-PLP-WIW for $i \in \{1,2\}$, we have the following corollary.

Corollary 1. *For $i \in \{1,2\}$, there is a constant $c_i > 0$ such that both MIN-B_3-C_i-PLP and MIN-B_3-C_i-PLP-WIW are not approximable within factor $c_i \cdot \log(n)$ unless $\mathrm{NP} \subset \mathrm{DTIME}(n^{\log\log(n)})$, where n denotes the number of issues.*

A *cover number* $c(r_j)$ is associated with each issue r_j, indicating by how many levels voters must raise their acceptance probabilities in order to arrive at average majority for r_j. The cover numbers can be computed beforehand for a given instance. Then, we can also associate cover numbers to sets of issues (by summation), which finally leads to the cover number $N = \sum_{j=1}^{n} c(r_j)$ of the whole instance.

When we interpret an OL instance as a B_3-C_2-PLP instance, the cover number of that resulting instance equals the number of issues, assuming that the votes for all issues need amendment. Thus we have the following corollary:

Corollary 2. *There is a constant $c > 0$ such that MIN-B_3-C_2-PLP is not approximable within factor $c \cdot \log(N)$ unless $\mathrm{NP} \subset \mathrm{DTIME}(N^{\log\log(N)})$, where N is the cover number of the given instance. A fortiori, the same statement holds for MIN-B_3-C_2-PLP-WIW.*

Let H denote the harmonic sum function, i.e., $H(r) = \sum_{i=1}^{r} 1/i$. It is well known that $H(r) = O(\log(r))$. More precisely, it is known that

$$\lfloor \ln r \rfloor \leq H(r) \leq \lfloor \ln r \rfloor + 1.$$

We show the following theorem by providing and analyzing a greedy approximation algorithm.

Theorem 9. *MIN-B_3-C_2-PLP can be approximated within a factor of $\ln(N) + 1$, where N is the cover number of the given instance.*

In the strict-majority scenario, cover numbers would have a different meaning—we thus call them *strict cover numbers*: For each referendum, the corresponding strict cover number tells in advance how many voters have to change their opinions (bringing them individually over the given threshold t) to accept this referendum. The strict cover number of a problem instance is the sum of the strict cover numbers of all given issues.

Theorem 10. *MIN-B_3-C_1-PLP can be approximated within a factor of $\ln(N) + 1$, where N is the strict cover number of the given instance.*

Note that this result is in some sense stronger than Theorem 9 (which refers to the average-majority scenario), since the cover number of an instance could be larger than the strict cover number.

This approximation result is complemented by a corresponding hardness result.

Corollary 3. *There is a constant $c > 0$ such that* MIN-B_3-C_1-PLP *is not approximable within factor $c \cdot \log(N)$ unless* NP $\subset$ DTIME$(N^{\log \log(N)})$, *where N is the strict cover number of the given instance. A fortiori, the same statement holds for* MIN-B_3-C_1-PLP-WIW.

Unfortunately, those greedy algorithms do not (immediately) transfer to the case when issue weights are allowed.

7 Conclusions

We have studied six lobbying scenarios in a probabilistic setting, both with and without issue weights. Among the twelve problems studied, we identified those that can be solved in polynomial time, those that are NP-complete yet fixed-parameter tractable, and those that are hard (namely, W[2]-complete) in terms of their parameterized complexity with suitable parameters. It would be interesting to study these problems in different parameterizations. Finally, we investigated the approximability of hard probabilistic lobbying problems (without issue weights) and obtained both approximation and inapproximability results. A number of related results can be found in the full version [8]. An interesting open question is whether one can find logarithmic-factor approximations for voter bribery with issue weights.

References

1. Baye, M., Kovenock, D., de Vries, C.: Rigging the lobbying process: An application of the all-pay auction. The American Economic Review 83(1), 289–294 (1993)
2. Baye, M., Kovenock, D., de Vries, C.: The all-pay auction with complete information. Economic Theory 8(2), 291–305 (1996)
3. Bellare, M., Goldwasser, S., Lund, C., Russell, A.: Efficient probabilistically checkable proofs and applications to approximations. In: Proceedings of the 25th ACM Symposium on Theory of Computing, pp. 294–304. ACM Press, New York (1993)
4. Cesati, M.: The Turing way to parameterized complexity. Journal of Computer and System Sciences 67, 654–685 (2003)
5. Christian, R., Fellows, M., Rosamond, F., Slinko, A.: On complexity of lobbying in multiple referenda. Review of Economic Design 11(3), 217–224 (2007)
6. Crombez, C.: Information, lobbying and the legislative process in the European Union. European Union Politics 3(1), 7–32 (2002)
7. Downey, R., Fellows, M.: Parameterized Complexity. Springer, Heidelberg (1999)
8. Erdélyi, G., Fernau, H., Goldsmith, J., Mattei, N., Raible, D., Rothe, J.: The complexity of probabilistic lobbying. Technical Report arXiv:0906.4431 [cs.CC]. ACM Computing Research Repository (CoRR) (June 2009)

9. Faliszewski, P., Hemaspaandra, E., Hemaspaandra, L., Rothe, J.: A richer understanding of the complexity of election systems. In: Ravi, S., Shukla, S. (eds.) Fundamental Problems in Computing: Essays in Honor of Professor Daniel J. Rosenkrantz, ch. 14, pp. 375–406. Springer, Heidelberg (2009)

10. Lindner, C., Rothe, J.: Fixed-parameter tractability and parameterized complexity, applied to problems from computational social choice. In: Holder, A. (ed.) Mathematical Programming Glossary. INFORMS Computing Society (October 2008)

11. Myerson, R., Weber, R.: A theory of voting equilibria. The American Political Science Review 87(1), 102–114 (1993)

12. Raz, R., Safra, S.: A sub-constant error-probability low-degree test, and a sub-constant error-probability PCP characterization of NP. In: Proceedings of the 29th ACM Symposium on Theory of Computing, pp. 475–484. ACM Press, New York (1997)

13. Reinganum, J.: A formal theory of lobbying behaviour. Optimal Control Applications and Methods 4(4), 71–84 (1983)

14. Sandholm, T., Suri, S., Gilpin, A., Levine, D.: Winner determination in combinatorial auction generalizations. In: Proceedings of the 1st International Joint Conference on Autonomous Agents and Multiagent Systems, pp. 69–76. ACM Press, New York (2002)

15. von Neumann, J., Morgenstern, O.: Theory of Games and Economic Behavior. Princeton University Press, Princeton (1944)

On the Complexity of Efficiency and Envy-Freeness in Fair Division of Indivisible Goods with Additive Preferences

Bart de Keijzer[1], Sylvain Bouveret[2], Tomas Klos[1], and Yingqian Zhang[1]

[1] Delft University of Technology
B.deKeijzer@student.tudelft.nl, {T.B.Klos,Yingqian.Zhang}@tudelft.nl
[2] Onera-DTIM, Toulouse
Sylvain.Bouveret@onera.fr

Abstract. We study the problem of allocating a set of indivisible goods to a set of agents having additive preferences. We introduce two new important complexity results concerning efficiency and fairness in resource allocation problems: we prove that the problem of deciding whether a given allocation is Pareto-optimal is coNP-complete, and that the problem of deciding whether there is a Pareto-efficient and envy-free allocation is Σ_2^p-complete.

1 Introduction

The problem of allocating a set of indivisible goods to a set of agents arises in a wide range of applications including, among others, auctions, divorce settlements, frequency allocation, airport traffic management, fair and efficient exploitation of Earth Observation Satellites [1]. In many such real-world problems, one needs to find *efficient* and *fair* solutions, where an efficient solution can be seen informally as ensuring the greatest possible satisfaction to the agents, and where fairness refers to the need for compromises between the agents' (often antagonistic) objectives.

In this paper, we study the resource allocation problem from the point of view of computational complexity. We restrict our setting to *additive* preferences. In other words, the preferences of each agent are represented by a set of weights $w(o)$, standing for the utility (or satisfaction) she enjoys for each single object o. The utility of an agent for a subset of objects S is then given by the sum of the weights of all the objects o in S.

Moreover, we restrict our study to two particular definitions of efficiency and fairness: *Pareto-efficiency* (or Pareto-optimality) and *envy-freeness*. Pareto-efficient allocations are such that we cannot increase the satisfaction of an agent without strictly decreasing the satisfaction of another agent. An allocation is envy-free if and only if each agent likes her share at least as much as the share of any other agent.

In this paper, we introduce two new complexity results concerning the resource allocation problem with additive preferences. Even if the setting seems

F. Rossi and A. Tsoukis (Eds.): ADT 2009, LNAI 5783, pp. 98–110, 2009.

restrictive, we advocate that the particular problems we address are important enough to justify an extensive study for the following reasons. Firstly, one of the most natural ways of (compactly) modeling cardinal preferences over sets of objects (or more generally over combinatorial domains) is to suppose that they are additive. Notice that this goes far beyond resource allocation: matching problems, weighted path in a graph, valued constraints satisfaction problems, etc. Secondly, Pareto-efficiency is one the most prominent notion of efficiency used in collective decision making problems. Thirdly, envy-freeness is a key concept in the literature about resource allocation (see *e.g.* [2]), as it provides an elegant way of encoding the notion of fairness and does not require, contrary to Rawlsian egalitarianism, the interpersonal comparison of utilities.

This paper contributes to fill a gap. On the one hand resource allocation with additive preferences have been extensively studied in economics[1] (see *e.g.* [2,3]), but computational issues (and *a fortiori* complexity) have rarely been considered. On the other hand, computational issues in resource allocation with additive preferences have been studied extensively in computer science (see e.g. [4,5,6]). However, these works mainly concern the optimization of the system's performance as a whole. The properties of Pareto-efficiency and fairness are rarely addressed. Two notable exceptions are the work from Lipton *et al.* [7] that studies envy-freeness in fair resource allocation problems mainly from an algorithmic point of view, and the work from Bouveret and Lang [8] that introduces complexity results for fair resource allocation problems under different hypotheses, including additive preferences. In the latter paper, one result of importance misses, though being conjectured: the complexity of the problem of deciding whether there is a Pareto-efficient and envy-free allocation in a resource allocation problem with additive preferences. This is one of the two main complexity results introduced in our paper and is studied in section 4. The other main complexity result is about the related problem of deciding whether a given allocation is Pareto-efficient when agents have additive preferences. This result is more easily obtained, and is explained in section 3.

2   Background and Notations

In what follows, we will write vectors using arrowed letters (*e.g.* $\overrightarrow{v}$), or brackets for their explicit representations (*e.g.* $\langle v_1, \ldots, v_n \rangle$). v_i will denote the ith component of a vector. Moreover, for any finite set X, $|X|$ will denote the cardinal of X.

In a resource allocation problem, a set of resources must be divided among a set of agents. Since we will focus on additive utility functions only, it suffices to use the following definition of a resource allocation instance:

Definition 1 (Resource allocation instance). *A resource allocation problem is a triple* $\mathcal{P} = \langle A, O, w \rangle$, *where A is a set of* agents, *O is a set of* indivisible items, *and* $w : A \times O \to \mathbb{R}$ *is a* weight function.

---

[1] In most social choice studies, utilities stand for amounts of money. Thus additivity is a very natural assumption in this framework.

We define an allocation as follows:

Definition 2 (Allocation). *An allocation for* $\mathcal{P} = \langle A, O, w \rangle$ *is a vector* $\overrightarrow{\pi} = \langle \pi_1, \ldots, \pi_n \rangle \in (2^O)^n$ *such that for all* $i, j \in A$, $i \neq j \Rightarrow \pi_i \cap \pi_j = \emptyset$. *If for every* $o \in O$ *there exists an* i *such that* $o \in \pi_i$ *then* $\overrightarrow{\pi}$ *is a* complete *allocation.*

Thus, in the problems that we will focus on, the items are non-sharable.

Definition 3 (Individual utility, utility profile). *Let* $\mathcal{P} = \langle A, O, w \rangle$ *be a resource allocation instance. For all* $i \in A$ *and* $\pi_i \subseteq O$, $u_i(\pi_i) = \sum_{o \in \pi_i} w(i, o)$ *is agent* i's *individual utility regarding* π_i. *Given an allocation* $\overrightarrow{\pi}$, *the vector* $\langle u_1(\pi_1), \ldots, u_n(\pi_n) \rangle$ *is the* utility profile *associated to* $\overrightarrow{\pi}$.

Two properties that we will focus on, are *Pareto-efficiency* and *envy-freeness*.

Definition 4 (Pareto-efficiency). *Let* $\overrightarrow{\pi}, \overrightarrow{\pi}'$ *be two allocations.* $\overrightarrow{\pi}$ *Pareto-dominates* $\overrightarrow{\pi}'$ *if and only if (a) for all* i, $u_i(\pi_i) \geq u_i(\pi_i')$, *and (b) there exists an* i *such that* $u_i(\pi_i) > u_i(\pi_i')$. $\overrightarrow{\pi}$ *is* (Pareto-)efficient *(or* Pareto-optimal*) if and only if there is no* $\overrightarrow{\pi}'$ *such that* $\overrightarrow{\pi}'$ *Pareto-dominates* $\overrightarrow{\pi}$.

Definition 5 (Envy & envy-freeness). *We say that an agent* $i \in A$ *envies another agent* $j \in A$ *iff* $u_i(\pi_j) > u_i(\pi_i)$. *An allocation* $\overrightarrow{\pi}$ *is* envy-free *if and only if* $u_i(\pi_i) \geq u_i(\pi_j)$ *holds for all* i *and* $j \neq i$.

In this paper, we will refer to some complexity classes located in the polynomial hierarchy. We assume that the reader is familiar with the complexity class NP and its complementary class coNP. $\Sigma_2^p = \text{NP}^{\text{NP}}$ is the class of all languages recognizable by a nondeterministic Turing machine working in polynomial time using NP oracles. Its complementary class is denoted by Π_2^p.

3 Complexity of Deciding Pareto-optimal Allocations for Agents with Additive Utility

In this section we prove that it is coNP-complete to decide whether an allocation of resources is Pareto-optimal if the agents have additive utility functions. coNP-completeness has already been proved for a generalized case where agents express their utilities explicitly for each bundle of items [9]. coNP-completeness is also known for the case where the agents have k-additive utility functions and $k \geq 2$ [10].[2] This is not explicitly stated in [10], but it follows directly from their proof that it is NP-complete to decide whether it is possible to increase the utilitarian collective utility (*i.e.* sum of individual utilities) of a given allocation, when the agents have 2-additive utility functions. In addition to [10], the problem of maximizing utilitarian collective utility is also explored in [11].

---

[2] Informally, an agent has k-additive utility if she has a coefficient associated for every set of k items, and her individual utility is the sum of all coefficients associated to the sets of k items that she gets.

The problem we deal with is the following.

Problem 1. Pareto-optimality with additive utility functions (PO-ADD)

INSTANCE: A resource allocation instance $\mathcal{P} = \langle A, O, w \rangle$, an allocation $\overrightarrow{\pi}$.
QUESTION: Is $\overrightarrow{\pi}$ Pareto-optimal?

Theorem 1. PO-ADD *is coNP-complete.*

Membership of coNP is easy to establish: a nondeterministic Turing machine could guess an allocation $\overrightarrow{\pi}'$, and check whether $\overrightarrow{\pi}'$ Pareto-dominates $\overrightarrow{\pi}$.

To prove coNP-hardness, we give a Karp reduction (*i.e.* polynomial time many-one reduction) from the coNP-complete language 3UNSAT.

Problem 2. Unsatisfiability of propositional 3CNF formulas (3UNSAT)

INSTANCE: A set of clauses C denoting a propositional formula in 3CNF.
QUESTION: Is C unsatisfiable?

Let C be a set of propositional clauses of size 3 (we will suppose w.l.o.g. that the same literal does not appear more than once in each clause), $L(C)$ be the set of literals in C, and $V(C)$ be the set of variables in C. We will write $\mathcal{P}(C)$ to denote the following resource allocation instance:

Agents: $2|V(C)|+|C|+2$ agents: $\bigcup_{v \in V(C)} \{a_v, \overline{a_v}\} \cup \bigcup_{c \in C} \{a_c\} \cup \{a_{un}, a_{sat}\}$,

Objects: $4|C| + |V(C)| + 1$ objects: $\bigcup_{c \in C} \{o_{c,l} \mid l \in c\} \cup \bigcup_{v \in V(C)} \{o_v\} \cup$
$\bigcup_{c \in C} \{o_c\} \cup \{o_{sat}\}$,

Preferences: $w(i, o) = 0$ for all i and all o, except:
- $w(a_v, o_v) = |\{c \mid v \in c \in C\}|$, and $w(\overline{a_v}, o_v) = |\{c \mid \neg v \in c \in C\}|$ for all $v \in V(C)$;
- $w(a_v, o_{c,v}) = 1$ if $v \in c$, and $w(\overline{a_v}, o_{c,\neg v}) = 1$ if $\neg v \in c$ for each $v \in V(C)$ and each $c \in C$;
- $w(a_c, o_{c,l}) = 1$ for each $c \in C$ and each $l \in c$;
- $w(a_c, o_c) = 1$ for all $c \in C$;
- $w(a_{un}, o_v) = 1$ for all $v \in V(C)$;
- $w(a_{sat}, o_c) = 1$ for all $c \in C$;
- $w(a_{un}, o_{sat}) = |V(C)| + 1$;
- $w(a_{sat}, o_{sat}) = |C|$.

Let I be a partial truth assignment of the variables in C. We will define its corresponding allocation $\overrightarrow{\pi}(I)$ as follows:

- $\pi(I)_{a_v} = \{o_v\}$ if $I(v) = \textbf{true}$ and $\pi(I)_{a_v} = \{o_{c,v} \mid v \in c \in C\}$ otherwise, for each $v \in V(C)$;
- $\pi(I)_{\overline{a_v}} = \{o_v\}$ if $I(v) = \textbf{false}$ and $\pi(I)_{\overline{a_v}} = \{o_{c,\neg v} \mid \neg v \in c \in C\}$ otherwise, for each $v \in V(C)$;
- for each $c \in C$: $\pi(I)_{a_c} = \{o_c\}$ if $I \nvdash c$, and $\pi(I)_{a_c} = \{o_{c,l} \mid l \in C \land I \vDash l\}$ otherwise;

- $\pi(I)_{a_{un}} = \{o_{sat}\}$ if I is complete, and $\bigcup_{v \in V(C), I(v) \notin \{\text{true}, \text{false}\}} \{o_v\}$ otherwise;
- $\pi(I)_{a_{sat}} = \{o_{sat}\}$ if I is partial, and $\{o_c \mid I \vDash C\}$ otherwise.

It should be clear that for each assignment I, $\overrightarrow{\pi}(I)$ is well-defined.

Let $I_\emptyset$ be the empty assignment (*i.e.* the partial truth-assignment that leaves all variables unassigned). Our reduction transforms a 3UNSAT instance C into the PO-ADD-instance $\langle \mathcal{P}(C), I_\emptyset \rangle$.

We will now give an example of this reduction. Consider the 3UNSAT instance given by the set of clauses $\{c_1 = \{v_1, v_2, \neg v_3\}, c_2 = \{\neg v_1, \neg v_2, \neg v_3\}\}$.

If we run the reduction process on this instance, we get the PO-ADD-instance that is displayed in the table below. The columns of the table represent the agents and the rows of the table represent the items. The entries in the table are the weights. An entry is displayed in **boldface italic** and between brackets if the item of the corresponding row is allocated to the agent of the corresponding column. Empty cells in the table should be regarded as containing zero-weights.

	a_{c_1}	a_{c_2}	a_{v_1}	$\overline{a_{v_1}}$	a_{v_2}	$\overline{a_{v_2}}$	a_{v_3}	$\overline{a_{v_3}}$	a_{un}	a_{sat}
o_{v_1}			1	1					*[1]*	
o_{v_2}					1	1			*[1]*	
o_{v_3}								2	*[1]*	
o_{c_1}	*[1]*									1
o_{c_2}		*[1]*								1
o_{c_1,v_1}	1		*[1]*							
o_{c_1,v_2}	1				*[1]*					
$o_{c_1,\neg v_3}$	1							*[1]*		
$o_{c_2,\neg v_1}$		1		*[1]*						
$o_{c_2,\neg v_2}$		1				*[1]*				
$o_{c_2,\neg v_3}$		1						*[1]*		
o_{sat}									4	*[2]*

Lemma 1. *For each model M for C, $\overrightarrow{\pi}(M)$ Pareto-dominates $\overrightarrow{\pi}(I_\emptyset)$.*

Proof. In $\overrightarrow{\pi}(M)$, agent a_{un} has strictly higher utility: $u_{a_{un}}(\pi(M)_{a_{un}}) = |C| + 1$, while $u_{a_{un}}(\pi(I_\emptyset)_{a_{un}}) = |C|$. By definition of $\overrightarrow{\pi}(M)$, the utility of all other agents is in $\overrightarrow{\pi}(M)$ at least as high as in $\overrightarrow{\pi}(I_\emptyset)$.   □

Lemma 2. *If C is unsatisfiable, then $\overrightarrow{\pi}(I_\emptyset)$ is Pareto-optimal.*

Proof. Suppose for contradiction that there is an allocation $\overrightarrow{\pi}'$ that Pareto-dominates $\overrightarrow{\pi}(I_\emptyset)$.

There is at least one agent a such that $u_a(\pi_a') > u_a(\pi(I_\emptyset)_a)$. It can be easily proved that, starting from $I_\emptyset$, strictly increasing the utility of any agent in $A \setminus \{a_{un}, a_{sat}\}$ implies reallocating at least one item o_v from agent a_{un}'s share to another agent. Then, the only solution for not decreasing a_{un}'s utility is to give her o_{sat}.

So then $\{o_c \mid c \in C\} \in \pi_{a_{sat}}'$. Consequently, for all a_c with $c \in C$ we must have $|\{o_{c,l} \mid l \in c \wedge o_{c,l} \in \pi_{a_c}'\}| \geq 1$. Let $o \in \{o_{c,l} \mid l \in c \wedge o_{c,l} \in \pi_{a_c}'\}$. Let a' be the agent for which it holds that $o \in \pi(I_\emptyset)_{a'}$ (so $a' \in \{a_v, \overline{a_v}\}$). Let $\overline{a'}$ be the agent in $\{a_v, \overline{a_v}\}$ that does not equal a'. It must now be that $o_v \in \pi_{a'}'$, and as a consequence we now know that $\pi(I_\emptyset)_{\overline{a}} = \pi_{\overline{a}}'$.

So, for each $v \in V(C)$ there are no two objects $o_{c,v}$ and $o_{c',\neg v}$ with $v \in c$, $\neg v \in c'$, and $c, c' \in C$ such that $o_{c,v} \in \pi_{a_c}$ and $o_{c',\neg v} \in \pi'_{a_{c'}}$. It then follows immediately from the construction of the reduction that there is a complete interpretation I such that $\overrightarrow{\pi}' = \overrightarrow{\pi}(I)$. Moreover, one can check that $I \vDash c$ for each $c \in C$. Therefore, I is a model of C, and we have a contradiction. $\square$

Proof (Theorem 1). Let C be a set of clauses of size 3. By Lemma 1 and 2, we have that C is unsatisfiable if and only if $\overrightarrow{\pi}(I_\emptyset)$ is a Pareto-optimal allocation for $\mathcal{P}(C)$. The reduction from C to $\langle \mathcal{P}(C), \overrightarrow{\pi}(I_\emptyset) \rangle$ can clearly be done in polynomial time, hence coNP-hardness is proved. $\square$

4   Complexity of Deciding Existence of Efficient and Envy-Free Allocations for Agents with Additive Utility

For this section, we are interested in finding allocations that are both Pareto-efficient and envy-free.

We will now state the problem and prove that this problem is Σ_2^p-complete.

Problem 3. Efficient & envy-free allocation existence with additive utility functions ($\exists$-EEF-ADD)

INSTANCE: A resource allocation instance $\mathcal{P} = \langle A, O, w \rangle$.

QUESTION: Does there exist an allocation that is both Pareto-efficient and envy-free?

Theorem 2. $\exists$-EEF-ADD *is Σ_2^p-complete.*

We will prove Σ_2^p-completeness by a Karp reduction from the complement of the Π_2^p-complete language $\forall \exists 3 \text{CNF}$ [12].

Problem 4. Doubly quantified 3CNF satisfiability ($\forall \exists 3 \text{CNF}$)

INSTANCE: A set $V_\forall$ of propositional variables, a set $V_\exists$ of propositional variables, a set C of clauses of three literals over the variables $V_\forall \cup V_\exists$.

QUESTION: Does there exist for each assignment to the variables in $V_\forall$ an assignment to the variables in $V_\exists$ that satisfies C?

Let $F = \langle V_\forall, V_\exists, C \rangle$ be an instance of $\forall \exists 3 \text{CNF}$. We will assume w.l.o.g. that every possible literal occurs at least once in C, and that a literal does not appear more than once in each clause. Let $\#\text{occ}\forall$ be the number of literal occurrences in C of variables in $V_\forall$. We will write $L_\forall$ and $L_\exists$ for the sets of literals of variables in $V_\forall$ and $V_\exists$ respectively. We will write $\mathcal{P}(F)$ to denote the following resource allocation instance (also see the example that follows after Definition 6):

Agents: $4|V_\forall| + 2|V_\exists| + |C| + \#\text{occ}\forall + 3$ agents: $\bigcup_{v \in V_\forall} \{a_v^+, a_v^{+cl}, a_v, \overline{a_v}\} \cup \bigcup_{v \in V_\exists} \{a_v, \overline{a_v}\} \bigcup_{c \in C} \{a_c\} \cup \bigcup_{c \in C} \{a_{c,l}^{en} \mid l \in c \cap L_\forall\} \cup \{a_{un}, a_{sat}, a_{en}\}$

Objects: $4|V_\forall| + |V_\exists| + 5|C| + L_\forall + 3$ objects: $\bigcup_{v \in V_\forall} \{o_v^{en}, o_v^{cmp}, o_v^h, \overline{o_v^h}\} \cup$
$\bigcup_{v \in V_\exists} \{o_v\} \cup \bigcup_{c \in C, l \in c} \{o_c, o_c^{cmp} o_{c,l}\} \cup \bigcup_{c \in C} \{o_{c,l}^{en} \mid l \in c \cap L_\forall\} \cup$
$\{o_{sat}, o_{en1}, o_{en2}\}$

Preferences: $w(i, o) = 0$ for all i and all o, except:

- For all $v \in V_\forall$ and $o \in \{o_v^{en}, o_v^c, o_v^h, \overline{o_v^h}\}$: $w(a_v^+, o) = 1$ and $w(a_v^{+cl}, o) = 1$;[3]

- For all $v \in V_\forall$: $w(a_v, o_v^h) = |\{c \mid v \in c\}|$, $w(\overline{a_v}, \overline{o_v^h}) = |\{c \mid \neg v \in c\}|$, $w(a_v, o_{c,v}) = 1$ for all $c \in C$ where $v \in c$, and $w(\overline{a_v}, o_{c,\neg v}) = 1$ for all $c \in C$ where $\neg v \in c$;

- For all $v \in V_\exists$: $w(a_v, o_v) = |\{c \mid v \in c\}|$, $w(\overline{a_v}, o_v) = |\{c \mid \neg v \in c\}|$, $w(a_v, o_{c,v}) = 1$ for all $c \in C$ where $v \in c$, and $w(\overline{a_v}, o_{c,\neg v}) = 1$ for all $c \in C$ where $\neg v \in c$;

- For all $c \in C$: $w(a_c, o_c) = M$, $w(a_c, o_c^{cmp}) = M - 1$, and $w(a_c, o_{c,l}) = 1$ for all $l \in c$;

- For all (c, l) where $c \in C$, $l \in c \cap L_\forall$: $w(a_{c,l}^{en}, o_c) = M$, $w(a_{c,l}^{en}, o_{c,l}) = 1$, $w(a_{c,l}^{en}, o_{c,l}^{en}) = M$;

- For all $c \in C$: $w(a_{sat}, o_c) = 1$, $w(a_{sat}, o_{sat}) = |C|$, $w(a_{sat}, o_{en1}) = \frac{1}{2}$;

- For all $c \in C$: $w(a_{un}, o_c^{cmp}) = 1$; For all $v \in V_\forall$: $w(a_{un}, o_v^{cmp}) = 1$; For all $v \in V_\exists$: $w(a_{un}, o_v) = 1$;

- $w(a_{un}, o_c^{sat}) = |V_\exists| + |V_\forall| + |C| + 1$; $w(a_{un}, o_{en1}) = 2(|V_\exists| + |V_\forall| + |C| + 1)$; $w(a_{un}, o_{en2}) = 3(|V_\exists| + |V_\forall| + |C| + 1) - 1$; $w(a_{en}, o_{en2}) = M$;

where M is a large number. It suffices to take for M the sum of all weights that are not defined in terms of M.

For our proof that $\exists$-EEF-ADD is Σ_2^p-complete, we need the notion of a special type of allocation for $\mathcal{P}(F)$. An example of an $X_\forall$-allocation is given in the example-instance that follows after this proof.

Definition 6 ($V_\forall$-assignments and $V_\forall$-allocations). *For F, we define a $V_\forall$-assignment as any complete assignment to the variables in $V_\forall$ only. Given a $V_\forall$-assignment I, we define a corresponding allocation $\overrightarrow{\pi}(I)$ for $\mathcal{P}(F)$ in the following way:*

- $\pi(I)_{a_v^+} = \{o_v^{en}\}$ *for each $v \in V_\forall$;*
- $\pi(I)_{a_v^{+cl}} = \{o_v^h\}$ *if $I(v) = $ **true** and $\{\overline{o_v^h}\}$ otherwise, for each $v \in V_\forall$;*
- $\pi(I)_{a_v} = \{o_{c,v} \mid v \in c\} \cup \{o_v^h\}$ *if $I(v) = $ **false**, and $\{o_{c,v} \mid v \in c\}$ otherwise, for each $v \in V_\forall$;*
- $\pi(I)_{\overline{a_v}} = \{o_{c,\neg v} \mid \neg v \in c\} \cup \{\overline{o_v^h}\}$ *if $I(v) = $ **true**, and $\{o_{c,\neg v} \mid \neg v \in c\}$ otherwise, for each $v \in V_\forall$;*
- $\pi(I)_{a_v} = \{o_{c,l} \mid l \in c\}$ *for each $l \in L_\exists$;*
- $\pi(I)_{a_c} = \{o_c\}$ *for each $c \in C$;*
- $\pi(I)_{a_{c,l}^{en}} = \{o_{c,l}^{en}\}$ *for each c, l where $c \in C$ and $l \in c \cap L_\forall$;*
- $\pi(I)_{a_{un}} = \{o_c^{cmp} \mid c \in C\} \cup \{o_v \mid v \in V_\exists\} \cup \{o_v^{cmp} \mid v \in V_\forall\} \cup \{o_{en1}\}$;

---

[3] So, a_v^{+cl} is a *clone* of a_v^+.

- $\pi(I)_{a_{sat}} = \{o_{sat}\}$;
- $\pi(I)_{a_{en}} = \{o_{en2}\}$;

Given a $V_\forall$-assignment I, we define the set of $V_\forall$-allocations corresponding to I as follows: any allocation that can be obtained from $\pi(I)$ by a sequence of swaps of the bundles of a_v^+ and a_v^{+cl} for any $v \in V_\forall$, followed by a sequence of reallocations of $o_{c,l}$ to $a_{c,l}^{en}$ for any l,c with $c \in C$, $l \in c \cap L_\forall$, and $I \not\vdash l$.[4]

Let us give an example of this reduction, together with a $V_\forall$-allocation. Let the $\forall\exists3\text{CNF}$-instance be $F = \langle V_\forall = \{v_1\}, V_\exists = \{v_2\}, C = \{c_1 = \{v_1, \neg v_1, v_2\}, c_2 = \{v_2, \neg v_2, v_1\}\}\rangle$. Then $\mathcal{P}(F)$ looks as follows.

	a_{c_1}	a_{c_2}	$a_{v_1}^+$	$a_{v_1}^{+cl}$	a_{v_1}	$\overline{a_{v_1}}$	a_{v_2}	$\overline{a_{v_2}}$	a_{c_1,v_1}^{en}	$a_{c_1,\neg v_1}^{en}$	a_{c_2,v_1}^{en}	a_{un}	a_{sat}	a_{en}
o_{c_1}	[M]								M	M			1	
o_{c_2}		[M]									M		1	
$o_{c_1}^{cmp}$	M-1											[1]		
$o_{c_2}^{cmp}$		M-1										[1]		
o_{c_1,v_1}	1				1				[1]					
$o_{c_1,\neg v_1}$	1					[1]				1				
o_{c_1,v_2}	1						[1]							
o_{c_2,v_2}		1					[1]							
$o_{c_2,\neg v_2}$		1						[1]						
o_{c_2,v_1}		1			1							[1]		
$o_{v_1}^h$			1	1	[2]									
$o_{v_1}^h$			[1]	1		1								
$o_{v_1}^{en}$			1	[1]										
$o_{v_1}^{cmp}$			1	1								[1]		
o_{v_2}							2	1				[1]		
o_{c_1,v_1}^{en}									[M]					
$o_{c_1,\neg v_1}^{en}$										[M]				
o_{c_2,v_1}^{en}											[M]			
o_{sat}												5	[2]	
o_{en1}												[10]	$\frac{1}{2}$	
o_{en2}												14		[M]

A $V_\forall$-allocation corresponding to a $V_\forall$-assignment I with $I(v_1) = \textbf{false}$ is displayed in **boldface italic** and between brackets. This allocation has been obtained from $\pi(I)$ by swapping the bundles of $a_{v_1}^+$ and $a_{v_1}^{+cl}$, and reallocating item o_{c_1,v_1} to a_{c_1,v_1}^{en}, and item o_{c_2,v_1} to a_{c_2,v_1}^{en}.

In the following proofs, we will restrict attention to *non-wasting* allocations, that is, allocations π such that for all $(o,a) \in O \times A$, $o \in \pi(a) \Rightarrow w(a,o) > 0$. It is obvious that every Pareto-efficient allocation is a non-wasting one.

Lemma 3. *Let $\overrightarrow{\pi}$ be an allocation. $\overrightarrow{\pi}$ is envy-free if and only if $\overrightarrow{\pi}$ is a $V_\forall$-allocation.*

Proof. ($\Leftarrow$) For any arbitrary $V_\forall$-allocation, it is easy (although a bit tedious) to check for each type of agent that she does not envy any other agent.

($\Rightarrow$) We show this by reasoning about how resources should be allocated in order to prevent envy. We start by noticing that o_{en2} must necessarily be allocated to a_{en}. As a consequence o_{en1} must be allocated to a_{un}, after which

---

[4] To remove any confusion, see the example allocation that follows, together with the explanation.

o_{sat} should go to a_{sat}. In order prevent a_{un} from envying a_{en}, we should give a_{un} all remaining items for which his weight is positive. Now that we know we cannot allocate o_c^{cmp} to a_c (for any $c \in C$), we must give o_c to a_c. For the same reason, for all $v \in V_\exists$ we cannot give o_v to a_v or $\overline{a_v}$, thus these agents should receive all remaining resources for which they have a positive weight. Next, we notice that we should give $o_{c,l}^{en}$ to $a_{c,l}^{en}$ because $a_{c,l}^{en}$ is the only agent with positive weight for this item. Now we see that for all c, l with $l \in c \cap L_\forall$ and $c \in C$, we cannot give $o_{c,l}$ to a_c, because $a_{c,l}^{en}$ would then envy a_c. Next, for all $v \in V_\forall$ we must allocate o_v^{en} to either a_v^+ or a_v^{+cl}. Because the weights of both agents are exactly the same, suppose w.l.o.g. that we allocate o_v^{en} to a_v^+. In order to prevent envy between the two agents, we need to allocate either o_v^h or $\overline{o}_v^h$ (but not both) to a_v^{+cl} (we denote by o this item and $\overline{o}$ the other one of the pair). Since $\overline{o}$ cannot be allocated to a_v^+ nor to a_v^{+cl}, $\overline{o}$ must go to a_v if $\overline{o} = o_v^h$ and $\overline{a_v}$ otherwise. There is only one agent left that we can allocate o to. Lastly, let $a \in \{a_v, \overline{a_v}\}$ be the agent that does not get $\overline{o}$. All items that a has positive weight for should now be allocated to a, in order to prevent a from envying the agent that gets $\overline{o}$.

The restrictions that we just deduced, restrict the set of possibly efficient and envy-free allocations to the set of $V_\forall$-allocations. □

Lemma 4. *No two $V_\forall$-allocations dominate each other.*

Proof. Let $\overrightarrow{\pi}$ and $\overrightarrow{\pi}'$ be two $V_\forall$-allocations. If $\overrightarrow{\pi}$ and $\overrightarrow{\pi}'$ correspond to the same $V_\forall$-assignment, then $\overrightarrow{\pi}$ does not dominate $\overrightarrow{\pi}'$ because swapping the bundles of o_v^+ and o_v^{+cl} for any $v \in V_\forall$ does not increase nor decrease the utility of both agents. Reallocating $o_{c,l}$ between the agents $a_{c,l}^{en}$ and a_v (or $\overline{a_v}$) for any l, c with $l \in c \cap L_\forall$ and $c \in C$ can never result in a dominating allocation either, because both agents have exactly the same weights for all of these items.

For the case that $\overrightarrow{\pi}$ and $\overrightarrow{\pi}'$ correspond to different $V_\forall$-assignments, let I and I' be the two $V_\forall$-assignments respectively, and let v be a variable such that $I(v) \neq I'(v)$. We will show that $\overrightarrow{\pi}$ does not Pareto-dominate $\overrightarrow{\pi}'$. Assume w.l.o.g. that $I(v) = \mathbf{true}$ and $I'(v) = \mathbf{false}$. In $\overrightarrow{\pi}$ we have that $\overline{o}_v^h$ is allocated to $\overline{a_v}$; in $\overrightarrow{\pi}'$ this is not the case. Because of the weights that $\overline{a_v}$ has, we now know that $u_{\overline{a_v}}(\pi_{\overline{a_v}}) \leq u_{\overline{a_v}}(\pi'_{\overline{a_v}})$. We can divide this up in two cases: in the case that $u_{\overline{a_v}}(\pi_{\overline{a_v}}) < u_{\overline{a_v}}(\pi'_{\overline{a_v}})$, we have immediately that $\overrightarrow{\pi}$ does not Pareto-dominate $\overrightarrow{\pi}'$. In the other case that $u_{\overline{a_v}}(\pi_{\overline{a_v}}) = u_{\overline{a_v}}(\pi'_{\overline{a_v}})$, any item $o_{c,\neg v}$ in the set $\{o_{c,\neg v} \mid v \in c \in C\}$ is allocated to $a_{c,\neg v}^{en}$ under allocation $\overrightarrow{\pi}$, but not under allocation $\overrightarrow{\pi}'$, so in this case we have $u_{\overline{a_{c,\neg v}}}(\pi_{\overline{a_{c,\neg v}}}) < u_{\overline{a_{c,\neg v}}}(\pi'_{\overline{a_{c,\neg v}}})$, hence $\overrightarrow{\pi}$ does not Pareto-dominate $\overrightarrow{\pi}'$. □

Lemma 5. *Given a $V_\forall$-assignment I for F, and a $V_\forall$-allocation $\overrightarrow{\pi}$ for $\mathcal{P}(F)$ that corresponds to I; if C is satisfiable on I (i.e. I can be extended such that C is satisfied), then there is an allocation $\overrightarrow{\pi}'$ that Pareto-dominates $\overrightarrow{\pi}$.*

Proof. Let I' be a complete assignment that satisfies C such that $I \subseteq I'$. The following allocation $\overrightarrow{\pi}'$ Pareto-dominates $\overrightarrow{\pi}$.

- For all $c \in C$: $\pi'_c = \{o_{c,l} \mid l \in c \wedge I' \vDash l\} \cup \{o_c^{cmp}\}$.
- For all $v \in V_\forall$: Let $a = a_v$, $\bar{a} = \overline{a_v}$, $o = o_v^h$ if $I(v) = \textbf{true}$, and let $a = \overline{a_v}$, $\bar{a} = a_v$, $o = o_v^h$ otherwise. $\pi'_a = \{o\}$ and $\pi'_{\bar{a}} = \pi_{\bar{a}}$. Moreover, if $o \in \pi_{a_v^+}$ then $\pi'_{a_v^+} = \pi_{a_v^+} \setminus \{o\} \cup \{o_v^{cmp}\}$ and $\pi'_{a_v^{+cl}} = \pi_{a_v^{+cl}}$; otherwise $\pi'_{a_v^{+cl}} = \pi_{a_v^{+cl}} \setminus \{o\} \cup \{o_v^{cmp}\}$ and $\pi'_{a_v^+} = \pi_{a_v^+}$.
- For all $v \in V_\exists$: If $I'(v) = \textbf{true}$, then $\pi'_{a_v} = \{o_v\}$ and $\pi'_{\overline{a_v}} = \pi_{\overline{a_v}}$; otherwise $\pi'_{\overline{a_v}} = \{o_v\}$ and $\pi'_{a_v} = \pi_{a_v}$.
- For all c, l such that $c \in C$ and $l \in c \cap L_\forall$: $\pi'_{a_{c,l}^{en}} = \pi_{a_{c,l}^{en}}$.
- $\pi'_{a_{un}} = \{o_{sat}, o_{en1}\}$; $\pi'_{a_{sat}} = \{o_c \mid c \in C\}$; $\pi'_{a_{en}} = \{o_{en2}\}$.

In $\overrightarrow{\pi'}$, the utility of a_{un} is strictly higher than in $\overrightarrow{\pi}$. Moreover, one can easily check that in $\overrightarrow{\pi'}$ the utilities of all other agents are at least as high as in $\overrightarrow{\pi}$. $\square$

Lemma 6. *Given a $V_\forall$-assignment I for F, and a $V_\forall$-allocation $\overrightarrow{\pi}$ in $\mathcal{P}(F)$ that corresponds to I; if C is unsatisfiable on I, then $\overrightarrow{\pi}$ is Pareto-efficient.*

Proof. We will first show that in any $\overrightarrow{\pi'}$ that Pareto-dominates $\overrightarrow{\pi}$ we necessarily have $\{o_c \mid c \in C\} \subseteq \pi'_{a_{sat}}$. We do this by exhaustion on the type of agent. Let $\overrightarrow{\pi'}$ Pareto-dominate $\overrightarrow{\pi}$. Let a be an agent such that $u_a(\pi'_a) > u_a(\pi_a)$. We show for each of the following cases of a that necessarily $\{o_c \mid c \in C\} \subseteq \pi'_{a_{sat}}$.

Case 1: $a = a_{un}$: In this case we clearly have $o_{sat} \in \pi'_a$, and because $w(a_{sat}, o_{sat}) = |C|$ we have as a consequence that $\{o_c \mid c \in C\} \subseteq \pi'_{a_{sat}}$.

Case 2: $a \in \{a_v, \overline{a_v} \mid v \in V_\exists\}$: Let $a = a_v$ for an arbitrary $v \in V_\exists$ (the case that $a = \overline{a_v}$ is analogous). It must be that $o_v \in \pi'_a$, so then, since o_v has been removed from a_{un}'s share, we need to give her o_{sat} as a compensation. From the argument in the previous case we get $\{o_c \mid c \in C\} \subseteq \pi'_{a_{sat}}$.

Case 3: $a \in \{a_v^+ \, a_v^{+cl} \mid v \in V_\forall\}$: Let $a = a_v^+$ for an arbitrary $v \in V_\forall$ (the case that $a = a_v^{+cl}$ is analogous). Assume w.l.o.g. that $I(v) = \textbf{true}$. Because a_v^+ and a_v^{+cl} have identical weights, we may also w.l.o.g. assume that $\pi'_{a_v^+} \cap \pi_{a_v^{+cl}} = \emptyset$. If $o_v^{cmp} \in \pi'_a$, then from the argument in case 1 it follows that $\{o_c \mid c \in C\} \subseteq \pi'_{a_{sat}}$. If $o_v^h \in \pi'_a$, then it must be that $\{o_{c,\neg v} \mid \neg v \in c \in C\} \cap \pi_{\overline{a_v}} = \emptyset$ and $\{o_{c,\neg v} \mid \neg v \in c \in C\} \subseteq \pi'_{\overline{a_v}}$. Thus, for all $c \in C$ with $\neg v \in c$, we have $o_c \in \pi'_{a_{c,\neg v}^{en}}$. Consequently, we get $o_c^{cmp} \in \pi'_{a_c}$; and therefore by our argument that we gave in Case 1, $\{o_c \mid c \in C\} \subseteq \pi'_{a_{sat}}$.

Case 4: $a \in \{a_v, \overline{a_v} \mid v \in V_\forall\}$: Assume w.l.o.g. that $I(v) = \textbf{true}$. For any arbitrary $v \in V_\forall$, let $a = \overline{a_v}$. From the last part of the argument that we gave for the previous case, it follows directly that $\{o_c \mid c \in C\} \subseteq \pi'_{a_{sat}}$. Now let $a = a_v$. Necessarily we have $o_v^h \in \pi'_{a_v}$, and from our reasoning in Case 3 it follows that $\{o_c \mid c \in C\} \subseteq \pi'_{a_{sat}}$.

Case 5: $a \in \{a_c \mid c \in C\}$: Let $a = a_c$ for an arbitrary $c \in C$. If $o_c^{cmp} \in \pi'_{a_c}$, it follows from Case 1 that $\{o_c \mid c \in C\} \subseteq \pi'_{a_{sat}}$. If $o_{c,l} \in \pi'_{a_c}$ then the same conclusion follows, but this time from the last part of the proof of Case 3.

Case 6: $a \in a_{c,l}^{en} \mid c \in C \wedge l \in c \cap L_\forall\}$: Let $a = a_{c,l}^{en}$ for an arbitrary c, l with $c \in C$ and $l \in c \cap L_\forall$. We must have that $o_c \in \pi'_{a_{c,l}^{en}}$ or $o_{c,l} \in \pi'_{a_{c,l}^{en}}$ (or both), in both cases it follows from the last part of the proof of Case 3 that $\{o_c \mid c \in C\} \subseteq \pi'_{a_{sat}}$.

Case 7: $a = a_{sat}$**:** Let C' be any strict subset of C. If $\{o_c \mid c \in C'\} \subseteq \pi'_a$, then $\forall c \in C : o_c^{cmp} \in a_c$, so by the arguments in Case 1, $\{o_c \mid c \in C\} \subseteq \pi'_a$. If $o_{en1} \in \pi'_a$, then the same follows, also from the proof in Case 1.

Case 8: $a = a_{en}$**:** This case is obviously impossible.

Now we will finish the proof by obtaining the contradiction that an extension I' of I can be made to the variables in $V_\exists$, such that I' satisfies C.

Recall that we assume that $\overrightarrow{\pi}'$ Pareto-dominates $\overrightarrow{\pi}$, and as we have just shown, $\{o_c \mid c \in C\} \subseteq \pi'_{a_{sat}}$. For all $c \in C : o_c^{cmp} \cup L_c \subseteq \pi'_{a_c}$, where L_c is any subset of $\{o_{c,l} \mid l \in c\}$. Let $o_{c,l} \in L_c$ and let $v \in l$. There are two cases: either $v \in V_\forall$ or $v \in V_\exists$.

Suppose $v \in V_\forall$. Let $a = a_v$, $\overline{a} = \overline{a_v}$ if $I(v) = \mathbf{true}$, and let $a = \overline{a_v}$, $\overline{a} = a_v$ otherwise. It is easy to see that $\nexists o_{c,l} \in L_c : o_{c,l} \in \pi_{a_{c,l}^{en}}$ and $\nexists o_{c,l} \in L_c : o_{c,l} \in \pi_{\overline{a}}$, so $o_{c,l} \in \pi_a$. As a consequence, we know that l is satisfied by I. Hence, it must hold that if $o_{c,l} \in L_c$, then c is satisfied by I.

Suppose $v \in V_\exists$. Let $a = \overline{a_v}$ if $l = \neg v$ and let $a = a_v$ otherwise. Then it must be that $o_{c,l} \in \pi_a$ and $o_v \in \pi'_a$.

From the construction of the reduction, it follows that there must exist an assignment to the variables in $V_\exists$ that satisfies all clauses not satisfied by I, *i.e.*, we obtain the contradiction that C is satisfiable on I. □

Proof (Theorem 2). Membership is easily established: A nondeterministic $\mathsf{NP^{NP}}$ Turing machine that decides this problem could work as follows. On input $\langle A, \mathcal{O}, w \rangle$:

1. Guess an allocation $\overrightarrow{\pi}$.
2. Check whether $\overrightarrow{\pi}$ is envy-free. If not, then REJECT.
3. Check whether $\overrightarrow{\pi}$ is Pareto-optimal by querying the oracle. If it is, then ACCEPT. Otherwise, REJECT.

The difficult part is proving Σ_2^p-hardness.

Given a $\forall\exists$3CNF-instance $F = \langle V_\forall, V_\exists, C \rangle$, we can clearly construct $\mathcal{P}(F)$ in polynomial time.

If F is a NO-instance of $\forall\exists$3CNF, then there is a $V_\forall$-assignment I that cannot be extended to an assignment that satisfies C. Let $\overrightarrow{\pi}$ be a $V_\forall$-allocation for $\mathcal{P}(F)$ that corresponds to I. By Lemma 3, $\overrightarrow{\pi}$ is envy-free and by Lemma 6, $\overrightarrow{\pi}$ is Pareto-efficient. Hence, $\mathcal{P}(F)$ is a YES-instance of $\exists$-EEF-ADD.

If F is a YES-instance of $\forall\exists$3CNF, then for any $V_\forall$-assignment I that we pick, C is satisfiable on I. Let $\overrightarrow{\pi}$ be any $V_\forall$-allocation for $\mathcal{P}(F)$ that corresponds to I. By Lemma 5, there is an allocation $\overrightarrow{\pi}'$ that Pareto-dominates $\overrightarrow{\pi}$. By Lemma 4, $\overrightarrow{\pi}$ is not a $V_\forall$-allocation. Finally, because $\overrightarrow{\pi}$ is not a $V_\forall$-allocation, it follows from Lemma 3 that $\overrightarrow{\pi}'$ is not envy free. Because we had taken $\overrightarrow{\pi}$ to be an arbitrary $V_\forall$-allocation for an arbitrary $V_\forall$-assignment I, it follows that $\mathcal{P}(F)$ is a NO-instance of $\exists$-EEF-ADD.

Therefore we conclude that $\exists$-EEF-ADD is Σ_2^p-hard. □

5   Discussion

We have introduced in this paper two new complexity results for the resource allocation problem with additive preferences, thus filling an important gap in the previous complexity studies of this problem, mainly in [8]. Our main result shows that, even with very simple preferences (additive), deciding whether there is a Pareto-efficient and envy-free allocation is computationally very hard. This goes slightly beyond the results in [8], as it shows that the high complexity of the problem is not only related to the presence of preferential dependencies (complementarity or substitutability) between objects, since the hardness holds under the assumption of additive independence.

There are several natural ways of overcoming this high complexity. The first one could be to impose some restrictions on the setting to decrease the complexity. However, as stated in [8], the natural restrictions of the problem imply a huge loss of generality, and thus are of limited practical interest. Another solution is to relax envy-freeness or Pareto-efficiency, such as in [7], where envy-freeness is replaced by a measure of envy, and where allocations are only required to be *complete* (that is, all objects must be allocated) instead of being Pareto-efficient. An idea could be to mix collective utility maximization (*e.g.* classical utilitarian or egalitarian) with envy-minimization.[5] And lastly, designing efficient approximation algorithms could be a way of getting around the high complexity of the problem.

References

1. Chevaleyre, Y., Dunne, P.E., Endriss, U., Lang, J., Lemaître, M., Maudet, N., Padget, J., Phelps, S., Rodríguez-Aguilar, J.A., Sousa, P.: Issues in multiagent resource allocation. Informatica 30, 3–31 (2006); Survey paper
2. Brams, S.J., Taylor, A.: Fair Division: From Cake-Cutting to Dispute Resolution. Cambridge Univ. Press, Cambridge (1996)
3. Demko, S., Hill, T.P.: Equitable distribution of indivisible items. Mathematical Social Sciences 16, 145–158 (1998)
4. Shehory, O., Kraus, S.: Methods for Task Allocation via Agent Coalition Formation. Artificial Intelligence 101(1-2), 165–200 (1998)
5. Modi, P.J., Jung, H., Tambe, M., Shen, W.M., Kulkarni, S.: A dynamic distributed constraint satisfaction approach to resource allocation. In: Walsh, T. (ed.) CP 2001. LNCS, vol. 2239, pp. 685–700. Springer, Heidelberg (2001)
6. de Weerdt, M.M., Zhang, Y., Klos, T.B.: Distributed task allocation in social networks. In: Huhns, M., Shehory, O. (eds.) Proceedings of the 6th International Conference on Autonomous Agents and Multiagent Systems, Bradford, UK, pp. 488–495. IFAAMAS, Research Publishing Services (2007)
7. Lipton, R.J., Markakis, E., Mossel, E., Saberi, A.: On approximately fair allocations of indivisible goods. In: EC 2004: Proceedings of the 5th ACM conference on Electronic commerce, pp. 125–131. ACM Press, New York (2004)

---

[5] A similar idea is proposed in [13].

8. Bouveret, S., Lang, J.: Efficiency and envy-freeness in fair division of indivisible goods: Logical representation and complexity. Journal of Artificial Intelligence Research (JAIR) 32, 525–564 (2008)

9. Dunne, P.E., Wooldridge, M., Laurence, M.: The complexity of contract negotiation. Artificial Intelligence 164(1-2), 23–46 (2005)

10. Chevaleyre, Y., Endriss, U., Estivie, S., Maudet, N.: Multiagent resource allocation with k-additive utility functions. In: Proceedings of the First International Workshop on Computer Science and Decision Theory, Paris, France, pp. 83–100 (2004)

11. Conitzer, V., Sandholm, T., Santi, P.: Combinatorial auctions with k-wise dependent valuations. In: Proceedings of the 20th National Conference on Artificial Intelligence (AAAI 2005), pp. 248–254. AAAI Press, Menlo Park (2005)

12. Schaefer, M., Umans, C.: Completeness in the polynomial-time hierarchy: a compendium. SIGACT News (September 2002)

13. Brams, S.J., King, D.L.: Efficient fair division: Help the worst off or avoid envy? Rationality and Society 17(4), 387–421 (2005)

On Low-Envy Truthful Allocations*

Ioannis Caragiannis, Christos Kaklamanis, Panagiotis Kanellopoulos,
and Maria Kyropoulou

Research Academic Computer Technology Institute and
Department of Computer Engineering and Informatics
University of Patras, 26500 Rio, Greece

Abstract. We study the problem of allocating a set of indivisible items to
players having additive utility functions over the items. We consider allocations in which no player envies the bundle of items allocated to the other
players too much. We present a simple proof that deterministic truthful allocations do not minimize envy by characterizing the truthful mechanisms
for two players and two items. Also, we present an analysis for uniformly
random allocations which are naturally truthful in expectation. These
results simplify or improve previous results of Lipton et al.

1   Introduction

Resource allocation [9] has been an important problem in several areas such as
Computer Science, Artificial Intelligence, and Economics since their early days.
In the era of the Internet with a vast amount of computational, communication,
and storage resources available worldwide, the problem is still of paramount
importance. Besides efficiency, fairness is another important aspect that resource
allocation must satisfy. Additional constraints such as the selfish behavior of
resource owners and users make the variations of the problem very challenging.

A simple but foundational resource allocation problem is the well-known *cake-
cutting* problem [6,20]. In cake-cutting, we are given n players with different utilities for different parts of a cake. The objective is to allocate pieces of the cake
to the players in such a way that they are satisfied. Traditionally, satisfaction of
players has been measured by two different notions: *envy-freeness* and *proportionality*. Envy-freeness means that each player prefers her allocated pieces to
the pieces allocated to any other player. Proportionality means that the utility
of each player for the pieces allocated to her is at least $1/n$ times her utility for
the whole cake. Due to the continuity of the cake and the utilities of the players,
both objectives are always feasible.

A similar problem concerns the fair allocation of *indivisible items*; indivisibility
implies that an item cannot be broken into parts and must be allocated to a single
player. Here, we again have a set $\mathcal{N}$ of n players and a set $\mathcal{M}$ of m indivisible

* This work is partially supported by the European Union under IST FET Integrated
  Project FP6-015964 AEOLUS and Cost Action IC0602 "Algorithmic Decision Theory", and by a "Caratheodory" basic research grant from the University of Patras.

F. Rossi and A. Tsoukis (Eds.): ADT 2009, LNAI 5783, pp. 111–119, 2009.

items. Each player p has a non-negative utility function $u_p : 2^{\mathcal{M}} \to \mathbb{R}_0^+$. The objective is to assign to each player p a bundle of items $\mathcal{M}_p \subseteq \mathcal{M}$, so that $\bigcup_p \mathcal{M}_p = \mathcal{M}$ and some criterion concerning fairness is maintained. An important special case is that of *additive utilities*. In this case, each player p has a utility $u_{p,i}$ for each item $i \in \mathcal{M}$ and her utility for a bundle of items is simply the sum of her utilities on these items. In contrast to the cake-cutting problem, envy-freeness and proportionality are not always feasible goals in this setting even in the case of additive utilities. Here, *envy minimization* is among the most prominent measures of fairness. Given an allocation A in which players p and q are assigned bundles $\mathcal{M}_p$ and $\mathcal{M}_q$, the envy $e_{pq}(A)$ of player p for player q is $e_{pq}(A) = u_p(\mathcal{M}_q) - u_p(\mathcal{M}_p)$. Then, envy of A is defined as $e(A) = \max_{p,q \in \mathcal{N}} e_{pq}$. Clearly, A is envy-free if $e(A) = 0$.

An implicit assumption in the above definitions is that the players express their true utilities which are used by the algorithm (i.e., the allocation function) in order to compute an allocation. In practice, players are usually *selfish* in the sense that they aim to increase their benefit, i.e., their total utility on the bundle of items the algorithm allocates to them. In order to do so, they may report false *valuations* of items to the algorithm (i.e., different than their true utilities). *Truthful allocation functions* guarantee that the allocation is based on the true utilities of the players. A deterministic allocation function is *truthful* if the benefit obtained by a player when reporting false valuations on the items is not greater than the benefit she would have obtained by telling the truth. Similarly, a randomized allocation function is *truthful in expectation* if the expected benefit of a player is maximized when revealing her true utilities.

Related work. Research concerning fair allocations originated in the 1940s with a focus on cake-cutting [21]. Since then, the problem of achieving a proportional allocation with the minimum number of operations has received much attention and is now well-understood [12,13,6,20,24]. The problem of achieving envy-freeness has been proven to be much more challenging [8,5,22]; in fact, under the most common computational model of cut and evaluation queries [20], no algorithm with bounded running time is known for more than 3 players. Very recently, envy-freeness was proved to be a harder property to achieve than proportionality [19,23]. Better solutions exist for different computational models (e.g., moving knife algorithms [7]).

Lipton et al. [16] studied envy minimization with indivisible items. Among other results, they proved that allocations with envy bounded by the *marginal utility* always exist and can be computed in polynomial time. In the case of additive utilities, marginal utility translates to the maximum per item utility over all players. They also present algorithms that compute allocations that approximate the minimum *envy-ratio*; the envy ratio of a player p for a player q is the utility of player p for the items allocated to player q over p's utility for the items allocated to her. Complexity considerations about envy-freeness for indivisible items and non-additive utilities are presented in [4]. The papers [10,11] study the problem of achieving envy-free and efficient allocations in distributed settings and when the allocation of items is accompanied by monetary side payments (in this case, envy-freeness is always a feasible goal). Lipton et al. [16] also consider truthful allocations; they show that any deterministic allocation function that returns an allocation with minimum possible envy cannot be truthful; their proof uses an instance with two players and many items. Finally, they present an analysis of the randomized allocation function that assigns each item to one of the players uniformly at random and independently of the allocations of the other items. This allocation function is truthful in expectation. For the case where the sum of utilities of each player over the items is 1, they prove that, with high probability, the envy of the resulting allocation is $O(\sqrt{\alpha}n^{1/2+\epsilon})$, where α is the maximum utility per item over all players and ϵ is an arbitrarily small positive number. We remark that the study of truthful allocation functions belongs to the recent line of research on *algorithmic mechanism design* [18]. In particular, Mu'alem and Schapira [17] prove lower bounds on the envy of truthful allocation functions. However, unlike the model of [16] which we also follow in the current paper, [17] and most of the studies in algorithmic mechanism design allow monetary transfers between the players.

For indivisible items, a fairness objective that has been extensively considered recently is *max-min fairness*. Here, the objective is to compute an allocation in which the benefit of the least happy player is maximized. The problem was studied by Bezáková and Dani [3] and Golovin [14] who obtained approximation algorithms that provably return a solution that is always a factor of $O(n)$ within the optimal value. The problem was popularized by Bansal and Sviridenko [2] as the *Santa Claus problem*, where Santa Claus aims to distribute presents to the kids so as to maximize the happiness of the least happy kid. Subsequently, Asadpour and Saberi [1] presented an $O(\sqrt{n}\log^3 n)$-approximation algorithm for this problem.

Our results. In this paper, we consider allocation of indivisible items to players having additive utility functions over the items. We present an alternative proof that no deterministic truthful allocation function minimizes envy by characterizing the deterministic truthful allocation functions for two players and two items. Our proof actually shows that for any truthful allocation function, there are instances in which the envy is almost maximized. Our proof simplifies the proof of Lipton et al. [16] that uses a large number of items. Our impossibility result trivially extends to the case of many players and many items and also to the more general case of non-additive utility functions. We also present an improved analysis of uniformly random allocations of m items over n players. We show that the envy is at most $O(\alpha\sqrt{m \ln n})$ with high probability, where α is the maximum utility per item over all players and items. For the case where the sum of utilities of each player is 1, we prove a bound of $O(\sqrt{\alpha \ln n})$. This improves the previous bound of $O(\sqrt{\alpha}n^{1/2+\epsilon})$ for any $\epsilon > 0$ [16]. Our proof follows similar lines to the proof of [16] but we exploit the fact that the allocation of each item is independent and use the Hoeffding bound instead of the Chebychev inequality in order to bound the envy.

Roadmap. Our characterization of the deterministic truthful allocations for two players and two items is presented in Section 2. The analysis of random allocations is presented in Section 3.

2   Truthful Allocations for Two Players and Two Items

In this section we present a characterization of deterministic truthful allocations with two players and two items.

In general, the first player will have utilities $u_1 x$ for the first item and $u_1(1-x)$ for the second one while the utilities of the second player are $u_2 y$ and $u_2(1-y)$, respectively. Here, $x, y \in [0,1]$ and u_1, u_2 are the sums of utilities of the two players for both items. So, an allocation function gets as input u_1, u_2, x, and y and computes an allocation of the items to the players. We denote each of the four possible allocations as a 2×2 matrix with entries 1 and 0. The columns correspond to the players and the rows to the items. An 1 in an entry of such a matrix indicates that the item corresponding to the row is allocated to the player corresponding to the column.

We use the term non-boundary values to denote real numbers in $[0,1]$ different than 0, 1/2, and 1. We consider only non-boundary values for x and y since they suffice for proving our main result on the envy. Our characterization can be easily extended to boundary values of x and y as well.

We begin with an observation that simplifies the allocation functions that have to be considered.

Lemma 1. *For non-boundary values of x and y, no truthful allocation function f depends on u_1 and u_2.*

Proof. Assume that this is not the case and that f computes different allocations on inputs (u_1, u_2, x, y) and (u'_1, u_2, x, y) where x, y have non-boundary values and $u_1 \neq u'_1$.

When x has a non-boundary value, the four different possible allocations $\begin{pmatrix} 1 & 0 \\ 1 & 0 \end{pmatrix}$, $\begin{pmatrix} 1 & 0 \\ 0 & 1 \end{pmatrix}$, $\begin{pmatrix} 0 & 1 \\ 1 & 0 \end{pmatrix}$, and $\begin{pmatrix} 0 & 1 \\ 0 & 1 \end{pmatrix}$ yield different benefit to player 1 when her utilities on the items are $u_1 x$ and $u_1(1-x)$, namely u_1, $u_1 x$, $u_1(1-x)$, and 0.

Now assume that the function f returns an allocation of higher benefit to player 1 when she reports u'_1 instead of u_1. Then, player 1 has an incentive to lie. If this is not the case and f returns an allocation of lower benefit when player 1 reports u'_1, then consider the case when player 1 has true utilities $u'_1 x$ and $u'_1(1-x)$ on the two items. In this case, player 1 would have an incentive to lie and report $u_1 x$ and $u_1(1-x)$ as her valuation.   □

By Lemma 1, we may assume that f depends only on x and y when they have non-boundary values. Without loss of generality, we also assume that $u_1 = u_2 = 1$ in the following.

Lemma 2. *A truthful allocation function f has the following properties:*

(a) *If $f(x^*, y^*)$ assigns both items to the same player for some non-boundary values x^*, y^*, then $f(x, y)$ assigns both items to that player for any non-boundary values x, y.*

(b) *If $f(x^*, y^*)$ assigns to player 1 the item which she prefers the least for some non-boundary values x^*, y^*, then $f(x, y^*)$ assigns that item to player 1 for any non-boundary value x.*

(c) *If $f(x^*, y^*)$ assigns to player 2 the item which she prefers the least for some non-boundary values x^*, y^*, then $f(x^*, y)$ assigns that item to player 2 for any non-boundary value y.*

Proof. (a) Assume that the allocation function assigns both items to player 1 for some non-boundary values and at most one of the items for some other non-boundary values. Then, one of the following must hold:

- There exist non-boundary values x^*, y^*, x' such that $f(x^*, y^*)$ assigns both items to player 1 and $f(x', y^*)$ assigns at most one of the items to player 1. In this case, if the true utility of player 1 for item 1 is x', she has an incentive to lie and report x^* in order to get both items.

- There exist non-boundary values x^*, y^*, y' such that $f(x^*, y^*)$ assigns both items to player 1 and $f(x^*, y')$ assigns at most one of the items to player 1. In this case, if the true utility of player 2 for item 1 is y^*, she has an incentive to lie and report y' in order to get at least one item.

The case in which the allocation function assigns both items to player 2 is symmetric.

(b) Consider the case with $x^* < 1/2$ (the case $x^* > 1/2$ is symmetric) so that $f(x^*, y^*)$ assigns item 1 to player 1. Assume otherwise that there exists a non-boundary value x' such that $f(x', y^*)$ assigns item 2 to player 1. Then, if the true utility of player 1 for item 1 is x^*, player 1 has an incentive to lie and report x' in order to get item 2 which she prefers the most.

(c) The proof of this case is very similar to (b).   □

The properties of Lemma 2 yield the following.

Lemma 3. *The only truthful allocations with respect to non-boundary item valuations are those depicted in Figure 1.*

Proof. Figure 1 contains the eight possible allocation functions that satisfy the properties of Lemma 2. Truthfulness follows since for each player, given the valuation of the other player, these allocation functions either assign her the most preferred item or the allocation does not depend on her valuations.   □

We are now ready to prove the main statement of this section.

Theorem 1. *No truthful allocation minimizes envy.*

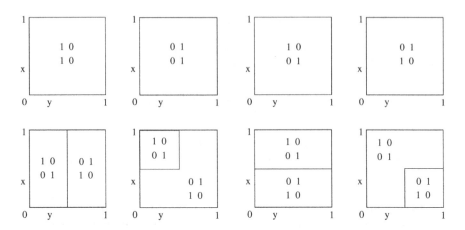

Fig. 1. The eight truthful allocation functions for two players

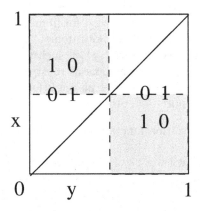

Fig. 2. The minimum envy allocation function when the sum of utilities of each player on the two items is 1

Proof. Clearly, for the two first fixed truthful allocation functions of Figure 1, both items are assigned to one player and hence the other player always has envy 1. Let $\epsilon \in (0, 1/4)$. Consider the two valuation pairs $(1 - \epsilon, 1/2 + \epsilon)$ and $(1/2 + \epsilon, 1 - \epsilon)$. The allocations $\begin{pmatrix} 1 & 0 \\ 0 & 1 \end{pmatrix}$ and $\begin{pmatrix} 0 & 1 \\ 1 & 0 \end{pmatrix}$ yield envy 2ϵ, respectively. For any of the six last truthful allocation functions of Figure 1, in one of these valuations pairs, the allocation yields benefit ϵ for one player and $1/2 + \epsilon$ to the other. Hence, one player has envy $1 - 2\epsilon$. By setting ϵ very close to 0, we have that the envy is actually maximized. □

We remark that the four non-fixed allocation functions in Figure 1 produce an envy-free allocation if one exists. Figure 2 presents the allocations that minimize envy when the sum of the utilities of each player on the items is 1. The grey areas indicate the cases of envy-free allocations. The two pairs of valuations considered in the proof of Theorem 1 have been selected to be outside but very close to these areas.

3   Improved Analysis of Random Allocations

In this section we consider the randomized allocation function that allocates each item to a player selected uniformly at random among the n players and in such a way that the allocation of an item is independent of the other allocations. Note that the allocation function does not depend on the valuations of the players. Hence, it is truthful in expectation since no player has an incentive to report a false valuation in order to increase her expected benefit. We present an upper bound on the envy of the resulting allocations using Hoeffding inequality [15].

Theorem 2 (Hoeffding [15]). *Let $X_1, \ldots, X_k$ be independent random variables with $\Pr(X_i \in [a_i, b_i]) = 1$ for $1 \leq i \leq k$. Then, for the sum of these variables $S = \sum_{i=1}^{k} X_i$, we have*

$$\Pr(S - \mathbb{E}[S] \geq t) \leq \exp\left(-\frac{2\,t^2}{\sum_{i=1}^{k}(b_i - a_i)^2}\right).$$

So, the particular version of Hoeffding inequality upperbounds the probability that a random variable which can be expressed as the sum of independent random variables exceeds its expectation by a certain amount. Our statement is the following; besides the number of players and items, it is also expressed in terms of the maximum utility per item over all players. We assume that the number n of players is large and the term high probability denotes a probability of $1 - 1/n$. We also denote by $v_{p,i}$ the utility of player p for item i.

Theorem 3. *Consider an instance with n players and m items and let $\alpha = \max_{p,i} v_{p,i}$.*

(a) With high probability, the random allocation yields an envy of at most $O(\alpha\sqrt{m \ln n})$.

(b) If the sum of utilities of each player is 1, then with high probability, the random allocation yields an envy of at most $O(\sqrt{\alpha \ln n})$.

Proof. Consider two players p and q. We define the random variable Y_i^{pq} indicating the contribution of item i to the envy of player p for player q. Then, the envy S^{pq} of player p for player q is $S^{pq} = \sum_{i=1}^{m} Y_i^{pq}$. Observe that

- $Y_i^{pq} = v_{p,i}$ if item i is allocated to q (and this happens with probability $1/n$),
- $Y_i^{pq} = -v_{p,i}$ if item i is allocated to p (and this happens with probability $1/n$), and
- $Y_i^{pq} = 0$ if item i is not allocated to p or q (this happens with probability $1 - 2/n$).

Clearly, the random variables Y_i^{pq} are independent, $\Pr(Y_i^{pq} \in [-v_{p,i}, v_{p,i}]) = 1$ and $\mathbb{E}[S^{pq}] = 0$. By applying the Hoeffding bound for any $t \geq 0$, we have

$$\Pr(S^{pq} \geq t) \leq \exp\left(-\frac{t^2}{2\sum_{i=1}^m v_{p,i}^2}\right). \tag{1}$$

In order to prove (a), we use inequality (1) by setting $t = \alpha\sqrt{6m\ln n}$ and the fact that $\sum_{i=1}^m v_{p,i}^2 \leq m\alpha^2$ to obtain

$$\Pr(S^{pq} \geq \alpha\sqrt{6m\ln n}) \leq 1/n^3.$$

Since there are at most n^2 pairs of players p, q, by applying the union bound we have that the probability that the maximum envy between any two players exceeds $\alpha\sqrt{6m\ln n}$ is at most $1/n$.

In order to prove (b), we set $t = \sqrt{6\alpha\ln n}$ and use the fact that $\sum_{i=1}^m v_{p,i}^2 \leq \alpha$ when $\sum_{i=1}^m v_{p,i} = 1$ and $v_{p,i} \geq 0$. By (1), we obtain that

$$\Pr(S^{pq} \geq \sqrt{6\alpha\ln n}) \leq 1/n^3.$$

Again, by applying the union bound we have that the probability that the maximum envy between any two players exceeds $\sqrt{6\alpha\ln n}$ is at most $1/n$.   □

The first upper bound should be compared to the lower bound of α [16] on the envy of allocations in which the maximum utility per item among all players is α. This bound is shown to be tight in [16] but the upper bound is not obtained by a truthful allocation function. Our second upper bound significantly improves the upper bound of $O(\sqrt{\alpha}n^{1/2+\epsilon})$ for the case where the sum of utilities of each player is 1. Whether there exist better allocation functions (i.e., that yield allocations with smaller envy) that are truthful in expectation is an interesting open problem.

Acknowledgments. We thank Ariel Procaccia for helpful discussions.

References

1. Asadpour, A., Saberi, A.: Max-min fair allocation of indivisible goods. In: Proceedings of the 39th ACM Symposium on Theory of Computing (STOC 2007), pp. 114–121 (2007)
2. Bansal, N., Sviridenko, M.: The Santa Claus problem. In: Proceedings of the 38th ACM Symposium on Theory of Computing (STOC 2006), pp. 31–40 (2006)
3. Bezáková, I., Dani, V.: Allocating indivisible goods. SIGecom Exchanges 5(3), 11–18 (2005)
4. Bouveret, S., Lang, J.: Efficiency and envy-freeness in fair division of indivisible goods: Logical representation and complexity. Journal of Artificial Intelligence Research 32, 525–564 (2008)
5. Brams, S.J., Taylor, A.D.: An envy-free cake division protocol. The American Mathematical Monthly 102(1), 9–18 (1995)

6. Brams, S.J., Taylor, A.D.: Fair division: From cake-cutting to dispute resolution. Cambridge University Press, Cambridge (1996)
7. Brams, S.J., Taylor, A.D., Zwicker, W.S.: A moving-knife solution to the four-person envy free cake division problem. Proceedings of the American Mathematical Society 125(2), 547–554 (1997)
8. Busch, C., Krishnamoorthy, M.S., Magdon-Ismail, M.: Hardness results for cake-cutting. Bulletin of the EATCS 86, 85–106 (2005)
9. Chevaleyre, Y., Dunne, P.E., Endriss, U., Lang, J., Lemaître, M., Maudet, N., Padget, J., Phelps, S., Rodríguez-Aguilar, J.A., Sousa, P.: Issues in multiagent resource allocation. Informatica 30, 3–31 (2006)
10. Chevaleyre, Y., Endriss, U., Estivie, S., Maudet, N.: Reaching envy-free states in distributed negotiation settings. In: Proceedings of the 19th International Joint Conference on Artificial Intelligence (IJCAI 2007), pp. 1239–1244 (2007)
11. Chevaleyre, Y., Endriss, U., Maudet, N.: Allocating goods on a graph to eliminate envy. In: Proceedings of the 22nd AAAI Conference on Artificial Intelligence (AAAI 2007), pp. 700–705 (2007)
12. Edmonds, J., Pruhs, K.: Cake-cutting really is not a piece of cake. In: Proceedings of the 17th Annual ACM-SIAM Symposium on Discrete Algorithms (SODA 2006), pp. 271–278 (2006)
13. Even, S., Paz, A.: A note on cake-cutting. Discrete Applied Mathematics 7, 285–296 (1984)
14. Golovin, D.: Max-min fair allocation of indivisible goods. Technical Report, Carnegie Mellon University, CMU-CS-05-144 (2005)
15. Hoeffding, W.: Probability inequalities for sums of bounded random variables. Journal of the American Statistical Association 58(301), 13–30 (1963)
16. Lipton, R., Markakis, E., Mossel, E., Saberi, A.: On approximately fair allocations of indivisible goods. In: Proceedings of the 5th ACM Conference on Electronic Commerce (EC 2004), pp. 125–131 (2004)
17. Mu'alem, A., Schapira, M.: Setting lower bounds on truthfulness. In: Proceedings of the 18th Annual ACM-SIAM Symposium on Discrete Algorithms (SODA 2007), pp. 1143–1152 (2007)
18. Nisan, N.: Introduction to mechanism design (for computer scientists). In: Nisan, N., Roughgarden, T., Tardos, E., Vazirani, V.V. (eds.) Algorithmic game theory. Cambridge University Press, Cambridge (2007)
19. Proccacia, A.: Thou shalt covet thy neighbor's cake. In: Proceedings of the 21st International Joint Conference on Artificial Intelligence, IJCAI 2009 (to appear, 2009)
20. Robertson, J.M., Webb, W.A.: Cake-cutting algorithms: Be fair if you can. AK Peters Ltd. (1998)
21. Steinhaus, H.: The problem of fair division. Econometrica 16, 101–104 (1948)
22. Stromquist, W.: How to cut a cake fairly. American Mathematical Monthly 87(8), 640–644 (1980)
23. Stromquist, W.: Envy-free cake divisions cannot be found by finite protocols. The Electronic Journal of Combinatorics 15, R11 (2008)
24. Woeginger, G.J., Sgall, J.: On the complexity of cake-cutting. Discrete Optimization 4, 213–220 (2007)

On Multi-dimensional Envy-Free Mechanisms

Ahuva Mu'alem

Social and Information Sciences Laboratory,
California Institute of Technology, Pasadena, CA, 91106
ahumu@yahoo.com

Abstract. We study the problem of *fairness design*. Specifically, we focus on approximation algorithms for indivisible items with supporting envy-free bundle prices. We present the first polynomial-communication envy-free profit-maximizing combinatorial auctions for general bidders. In this context, envy-free prices can be interpreted as anonymous non-discriminatory prices. Additionally, we study the canonical makespan-minimizing scheduling problem of unrelated machines, in an envy-free manner. For the special case of related machines model we show that tight algorithmic bounds can be achieved.

Keywords: Mechanism Design, Anonymous Prices, Scheduling.

1 Introduction

Fair division of goods has been a central problem in economic theory. In such scenarios each bidder would like to get a fair share from her point of view. At the same time, it might be the case that the social designer has a certain global goal in mind, and thus the designer seeks a fair partition that is as close as possible to the global goal. Several concepts of fairness were studied over the years. Envy-free allocations introduced by Foley [8]. An allocation is called *envy-free* if every bidder likes his own bundle at least as well as that of anyone else.

A contemporary motivation to study fair allocations stems from the emerging technology of Computational Grids. Computational grids offer users simple access to tremendous computer resources for solving large scale computing problems. A typical grid is composed of *shared* resources owned by different organizational entities that may varied over time. The on-line nature and the different degrees of contributions and consumptions suggest new fundamental fairness issues (e.g., [1]).

In this paper we study envy-free allocations for indivisible goods with supporting *bundle* prices. We first study Combinatorial Auctions for general bidders with the goal of profit maximization. In this scenario, a collection of indivisible goods needs to be allocated concurrently, and bidders have preferences about various combinations of items, and not just on single items or single subsets of goods. In this context, envy-free prices can be interpreted as anonymous non-discriminatory prices.

F. Rossi and A. Tsoukis (Eds.): ADT 2009, LNAI 5783, pp. 120–131, 2009.

The second common scenario studied in this paper is a minimization problem. Suppose a new project with several tasks is just arrived to a company. The challenge is to find a fair allocation of the tasks among the employees such that the last task of the project finishes as soon as possible. A mechanism is envy-free in this setting, if no employee prefers the set of tasks and the payment assigned to some other employee.

Overview and results. In Section 2, we briefly state a known characterization of envy-free mechanisms for multi-dimensional domains in terms of local-efficient bundle-assignments [11]. Informally, an envy-free allocation must be locally optimum with respect to the social welfare, so that the overall social welfare cannot be improved by exchanging the allocated bundles among the agents.

In Section 3, we study envy-free profit-maximization Combinatorial Auctions with general bidders. For a very restricted special case of the unit-demand setting, Guruswami et al. [10] showed that finding optimal envy-free prices is APX-hard. We describe an envy-free mechanism that requires polynomial communication and achieves $O(\min\{n, \sqrt{k}\log k\})$-approximation with respect to profit, where k is the number of items and n is the number of bidders. We then show that any envy-free profit mechanism with approximation ratio better than 2, requires exponential communication.

Section 4 studies envy-free scheduling mechanisms. We focus on the scheduling problem extensively studied by Lenstra, Shmoys, and Tardos [16]. This NP-hard optimization problem was formulated as a mechanism design problem by Nisan and Ronen in their seminal paper on Algorithmic Mechanism Design [19]: There are k tasks that are to be scheduled on m non-identical machines ("unrelated machines"). The total cost of a subset of tasks on machine i is the additive sum of the costs of the individual tasks on that machine. The global goal is minimizing the makespan of the chosen schedule. I.e., assigning the tasks to the machines in a way that minimizes the finishing time of the last task. Nisan and Ronen considered this global goal in the context of truthfulness (assuming agents are selfish and thus should be incentivized to report their true costs).

We consider minimizing the makespan in the context of envy-free design. Specifically, using the characterization, we derive general bounds on the approximability of deterministic envy-free mechanisms that seek to minimize the makespan on unrelated machines. We exhibit a lower bound of $2 - \frac{1}{m}$ and an upper bound of $\frac{m+1}{2}$ for the best approximation ratio achievable by any envy-free mechanism. For $m = 2$ the result is tight. However, our upper bound is not known to be computationally efficient for any $m \geq 2$. This leaves several interesting open problems. Similar upper bound and lower bound were achieved independently in a recent work by Hartline et al. [12]. We also show that any envy-free mechanism for minimizing the makespan with supporting *item prices* cannot achieve approximation ratio better than m.

In Section 5, we consider the NP-hard problem of minimizing the makespan of related parallel machines [14]. We show that the envy-freeness constraint does not impose any further computational burden. Specifically, we show that there exists a poly-time computable deterministic envy-free mechanism that achieves the approximation ratio of $1 + \epsilon$ with respect to the optimal makespan. We also show that in this case, resale of tasks among agents can be prevented in quasi-poly-time (without using verifications).

Related Work. Envy-free profit-maximization approximations for combinatorial auctions were first studied by Guruswami et al. [10]. They showed $O(\log n)$-approximation for unit demand bidders, and $O(\log n + \log k)$-approximation for *single-minded bidders* with unlimited supply, where n is the number of bidders and k is the number of items. In the unlimited supply setting the number of copies of each item is as large as the number of bidders. The latter result was extended to an $O(\log n + \log k)$-approximation for general bidders with unlimited-supply, by Balcan, Blum and Mansour [3].

Recently, Cheung and Swamy [5] obtained $O(\sqrt{k} \log u_{\max})$-approximation for single-minded bidders with limited-supply, where $u_{\max}$ is the maximum number of item supply, by using a LP-based technique. Achieving an approximation ratio better then $\sqrt{k}$ is NP-hard, even if $u_{\max} = 1$ [9]. The paper [17] studies *envy-free allocations without money* from a computational point of view. In this setting envy-free allocations might not exist, and thus they consider approximations for the minimum envyness. None of these papers studies the deterministic envy-free profit maximization pricing of combinatorial auctions for general bidders when supply is limited (see [5] for a recent detailed overview and references therein).

The fundamental purely algorithmic scheduling problem of *minimizing the makespan of unrelated machines* is studied in [16]. This paper presents a non-trivial 2-approximation poly-time algorithm. They also showed that the problem cannot be approximated in poly-time within a factor less than $\frac{3}{2}$.

A seminal paper by Nisan and Ronen [19] defines the notion of *algorithmic mechanism design* [20]. In this paper each machine is treated as a strategic agent. The paper proves that not only is it impossible to minimize the makespan in a *truthful* manner, but that any approximation ratio better than 2 cannot be achieved by a truthful deterministic mechanism. They also showed that there is computationally efficient truthful mechanism that achieves an approximation ratio of m. Their result is tight for $m = 2$. The lower bound was recently improved (from 2 to 2.61 [15]).

Hochbaum and Shmoys describe a PTAS for *minimizing the makespan of related parallel machines* [14].[1] In this NP-hard problem, the type of each machine can be described by a single number (single-parameter type). This problem was first studied from algorithmic mechanism design perspective in [2]. Archer and Tardos designed a 3-approximation mechanism based on a randomized rounding of the optimal fractional solution [2]. Recently, Dhangwatnotai et al. [7] presented a truthful *randomized* $1 + \epsilon$-approximation mechanism.

---

[1] A PTAS is an $(1 + \epsilon)$-approximation algorithm that runs in poly-time, assuming ϵ is a fixed constant.

2   Characterizing Envy-Free Mechanisms

This section characterizes envy-free mechanisms. The characterization is stated in terms of the *local efficiency* of the social choice function and applies to every domain of valuations.

The Setting. We consider a finite set K of k indivisible items and a set N of n agents.[2] We assume that agents value combinations of items. Formally, each agent $i \in N$ has a valuation function $v_i()$ that describes his valuation for each subset S of items, i.e. $v_i(S)$ is the maximum finite amount of money i is willing to pay for S. An allocation $a = (a_1, ..., a_n)$ is a partition of items among the agents. Formally, a_i denotes the subset of items allocated to agent i, $a_1 \cup a_2 \cup \cdots \cup a_n \subseteq K$ (observe that not all items need to be allocated), and $a_i \cap a_j = \emptyset$, whenever $i \neq j$. The set of all possible allowed allocations is denoted by A. Every valuation $v_i \in V_i$ satisfies the following three conditions: *No externalities* meaning that the valuation of agent i depends only on his allocated bundle. *Free disposal* meaning that the valuation is nondecreasing with the set of allocated items (for every S and T, $S \subseteq T$ implies $v_i(S) \leq v_i(T)$). *Normalization* meaning that the value of the empty bundle is always zero. V_i denotes the domain of all possible valuations of agent i.

Agents have *quasi-linear utilities*, and so $v_i(a_i) - y$ is the overall utility agent i can obtain from the subset of goods a_i and paying the price y.

A *social choice function* $f : V \rightarrow A$ maps an n-tuple of valuations $v = (v_1, v_2, \ldots, v_n) \in V_1 \times V_2 \times \cdots \times V_n = V$ to an outcome $a \in A$. In this setting, a social choice function is simply an allocation rule.

A mechanism defines an allocation and a set of prices for every possible valuation of the agents. Formally, a *mechanism* is a tuple $M = (f, p)$, where f is a social choice function and the pricing function $p_i : V_i \rightarrow R$ assigns a payment to each agent $i \in N$. Intuitively, the social choice function f represents the global goal of the mechanism designer, and the payment p determines the fairness of f.

Definition 1 (Envy-Free Mechanism). *Let $M = (f, p)$ be a mechanism. Let $i, j \in N$ and let $v = (v_1, ..., v_n)$ be an n-tuple of valuations. Denote by $a \in A$ the allocation f outputs for v. The mechanism M is said to be envy-free if for every agents i, j and valuation v it holds that:*

$$v_i(a_i) - p_i(v) \geq v_i(a_j) - p_j(v),$$

where p is non-negative individually-rational payment.

We say that a social choice function $f : V \rightarrow A$ is envy-free achievable if there exists a non-negative individually-rational payment p such that the mechanism $M = (f, p)$ is envy-free.[3]

---

[2] For the characterization theorem we might relax this requirement and assume that K is either finite or infinite set of items.

[3] We consider direct revelation mechanisms. However, the agents in our setting are non-strategic, they always report their true valuations. Observe also that we use the notion of "achievability" rather than the notion of "implementability".

Example 1. Suppose we have two agents, and one item. Consider the following allocation rule: the agent with the highest value wins the item. Additionally, the winner pays the average of both values and the other agent pays zero.

This mechanism is envy-free: Suppose without loss of generality the first agent won the item. Then $v_1 \geq v_2$. From the fact that $v_1 - \frac{v_1+v_2}{2} \geq 0 \geq v_2 - \frac{v_1+v_2}{2}$, we immediately get that that agent 1 does not envy agent 2, and vice versa.

Definition 2. *For arbitrary allocation $a \in A$ and valuation $v \in V$, let*

$$\Psi(v, \ a) = \Sigma_{i=1}^{n} \ v_i(a)$$

denotes the social-welfare of the allocation a with respect to v.

2.1   Locally-Efficient Bundle Assignments and Allocations

In order to be able to state the characterization theorem we proceed with some definitions. Let $a \in A$ be an arbitrary feasible allocation. We shall consider the following associated allocations based on a:

Definition 3 (Bundle Allocation based on a and β). *Let $\beta : N \to N$ be an arbitrary function. Let $a = (a_1, ..., a_n) \in A$ be an arbitrary feasible allocation. We say that the allocation a^β is the bundle-allocation based on a and β, if agent i in a^β is allocated all bundles a_k with $\beta(k) = i$.*

We also consider a special interesting case, in which each agent gets exactly one bundle of $a \in A$, based on a given permutation:

Definition 4 (Bundle Assignment based on a and π). *Let $\pi : N \to N$ be an arbitrary permutation. Let $a = (a_1, ..., a_n) \in A$ be an arbitrary allocation. We say that the allocation a^π is the bundle-assignment based on a and π, if the bundle allocated to agent i in a^π is exactly the bundle a_k, where $\pi(k) = i$.*

Definition 5 (Locally-Efficient Bundle Assignment). *An allocation $a = (a_1, ..., a_n)$ is said to be locally-efficient bundle assignment with respect to $v = (v_1, ..., v_n)$, if for every permutation $\pi : N \to N$ it holds that:*

$$\Psi(v, \ a) \geq \Psi(v, \ a^\pi).$$

Example 2. The allocation rule $f^*(v) \in \text{argmax}_{a \in A} \ \Psi(v, \ a)$ which maximizes the social-welfare, and $f^\emptyset$ that always allocates an empty bundle for every agent, clearly produce locally-efficient bundle assignments.

Consider the following 2 agents and 2 identical items setting with: $v_1(1) = v_1(2) = 1.5$, $v_2(1) = v_2(2) = 2$. The allocation a in which agent 2 gets both items and agent 1 gets the empty bundle is a locally-efficient bundle assignment. Similarly, the allocation a' in which every agent gets exactly one item is also a locally-efficient bundle assignment. Additionally, since bundles cannot be reconfigured, $a' \neq a^\beta$ for every β. However, there is β' such that $a = a'^{\beta'}$.

2.2   Characterizing Envy-Freeness

We now state a characterization of envy-free bundle pricing mechanisms in terms of locally-efficient bundle assignments. A similar proof technique in different contexts were used in [23,24,18].

Theorem 1. *[11] A deterministic social choice function $f : V \to A$ is envy-free achievable if and only if the allocation $f(v)$ is a locally-efficient bundle assignment w.r.t. v, for every $v \in V$.*

In this paper we shall extensively use both the necessary and the sufficient condition of this theorem. Intuitively, instead of considering the interplay between the the allocation rule and the payment rule, the above characterization allows us to focus on the allocation rule alone in order to prove or disprove the envy-free achievability. Additionally, based on the characterization, the envy-free achievability can be decided in poly-time, more formally:

Cost Minimization Problems. The characterization theorem is stated for the case each agent wishes to maximize his value. The characterization apply also for cost minimization problems, for which the agents would like to minimize their costs. Technically, the inequality is reversed to the other direction in the definition 5 when considering cost minimization settings. It is easy to see that the canonical payment in this context has the property that the true cost of each agent is covered by the mechanism.

3   Profit-Maximizing Combinatorial Auctions

In this section we consider the problem of maximizing the seller's profit in a combinatorial auction with general agents. We shall show an envy-free mechanism that requires polynomial communication and achieves $O(\min\{n, \sqrt{k}\log k\})$-approximation with respect to the maximal envy-free profit. This result uses the characterization theorem and is built on the work of Blumrosen and Nisan [4] for general bidders and Guruswami et al. [10] for unit demand bidders. We then show that a envy-free mechanism with approximation ratio better than 2 w.r.t. the optimal profit requires exponential communication.

Theorem 2. *There exists an envy-free profit maximizing mechanism for Combinatorial Auctions with general agents that achieves an $O(\min\{n, \sqrt{k}\log k\})$-approximation and requires polynomial communication.*

If $n < \sqrt{k}$ then we can use the simple algorithm that always allocate the grand bundle to the agent with the highest value and setting the price to be this value. This gives an n-approximation to the profit. We now need to describe the allocation algorithm for the general case. Our first building block is the BN-algorithm for Combinatorial Auctions with general agents by Blumrosen and Nisan [4]. The BN-algorithm uses polynomial number of demand queries (and thus polynomial communication) to achieve $4\sqrt{k}$-approximation ratio with respect to the overall social welfare. The BN-algorithm constructs two allocations and outputs the one with the highest social welfare. The first allocation is simply the allocation that gives all the goods to the agent with the highest value. The second allocation is based on a "greedy" procedure that in step i allocates a subset of the remaining unallocated goods S to an unallocated agent j satisfying

$$S_i \in \text{argmax } S, j \text{ is unallocated subset of goods and agent, respectively} \frac{v_j(S)}{|S|}.$$

The next algorithmic step is to convert the output a of the BN-algorithm into a locally-efficient bundle assignment a^{π^*}, where π^* is the permutation such that $\Psi(v, a^{\pi^*}) \geq \Psi(v, a^{\pi})$ for every π. Clearly, this cannot decrease the achieved social welfare as $\Psi(v, a^{\pi^*}) \geq \Psi(v, a)$, and can be done in polynomial communication. By the characterization theorem 1, a^{π^*} is envy-free achievable.

We now need to describe the pricing method. If the the chosen allocation a in the previous step gives all the goods to one agent then this agent will pay his bid for the grand bundle, and all other agents pay zero. Clearly this is an envy-free pricing and the achieved profit P in this case is exactly the welfare achieved by the BN-algorithm. Otherwise, a is based on the greedy method. Our second building block is the envy-free pricing method designed for unit demand agents by Guruswami et al. [10], based on a careful poly-time selection of *reserved* prices. The achieved envy-free pricing is within factor of $2 \ln n$ for the maximum profit. In particular, the profit P in this case is within a factor of $2 \ln n$ of the social welfare $\Psi(v, a^{\pi^*})$. It is immediate to see that the factor in their proof is $2 \ln(\min\{n, k\})$, since the number of possible non-trivially allocated agents is at most $\min\{n, k\}$. Additionally, it is not hard to see that we can use this pricing method for our setting as well (since as in the unit demand setting, an agent in our setting can only envy an entire bundle that is allocated to another agent defined by a^{π^*}, and not any other arbitrary bundle).

Proof. Clearly, the maximum envy-free profit P^* is at most the maximum social welfare W^*, and thus:

$$\frac{P^*}{4\sqrt{k}} \leq \frac{W^*}{4\sqrt{k}} \leq \Psi(v, a) \leq \Psi(v, a^{\pi^*}) \leq 2P \ln(\min\{n, k\}).$$

All together we get the desired bound.   □

The following proposition shows that any envy-free mechanism needs an exponential communication in the worst-case to produce approximation ratio better than 2 for the profit. This result uses the construct of Nisan and Segal [21] for efficient allocations.

Proposition 1. *Any envy-free profit maximizing mechanism for Combinatorial Auctions that achieves an approximation ratio better than 2 requires exponential communication.*

3.1   Item Prices

Item prices is a special case of bundle prices. If item prices are defined, then the price of a bundle is the total prices of the items in the bundle. Thus an agent can envy a bundle or any sub-bundle allocated to some other agent. Envy-free mechanisms with item prices for combinatorial-auction can extract only a small profit regardless of any computational considerations (equivalently, any auction with anonymous item-price extracts a small profit in the worst-case).

Proposition 2. *Any deterministic envy-free mechanism for profit maximization with supporting item prices cannot achieve approximation ratio better than k.*

4   The Envy-Free Approximability of Unrelated Machines

The Unrelated Machines Model was studied in the seminal paper of Algorithmic Mechanism Design [16,19]. For this model we prove that not only it is impossible to minimize the makespan in an envy-free manner, but that any approximation ratio better than $2 - \frac{1}{m}$ cannot be achieved by any envy-free *deterministic* mechanism. We then present an envy-free mechanism that achieves $\frac{m+1}{2}$-approximation. Our mechanism make use on the optimal allocation w.r.t. makespan, and thus is not computationally efficient. For $m = 2$ our result is tight. Similar upper and lower bounds were achieved recently and independently by Hartline et al. [12].

The Setting. The unrelated machine scheduling setting $(R||C_{\max})$ is a special case of the combinatorial auction setting. There are k tasks that are to be scheduled on m machines.[4] Every machine i is an agent with a nonnegative valuation function $v_i()$. Formally, $v_i(\{j\})$ (or simply $v_i(j)$) specifies the *cost* of task j on machine i. One can think of the cost of task j on machine i as the time it takes i to complete j. The total cost of a set of tasks S on machine i is the *additive* sum of the costs of the individual tasks on that machine. Formally, $v_i(S) = \Sigma_{j \in S} v_i(j)$ for every S. In the unrelated machines setting these costs can be arbitrary (every $(k \cdot m)$-tuple of non-negative costs is feasible), and thus it is a multi-dimensional scheduling problem.

   Let $a \in A$ be an arbitrary allocation of tasks to the machines ("scheduling"). The load of machine i is its cost $v_i(a_i) = \Sigma_{j \in a_i} v_i(j)$. Let $r(a, v) = \max\{v_1(a_1), v_2(a_2), ..., v_m(a_m)\}$, be the load of the most loaded machine. To simplify notation we shall use the notation $r(a)$ instead of $r(a, v)$, when v is clear from the context.

   The global goal is minimizing the makespan. I.e., it is a minmax goal: assign the tasks to machines so that the last task finishes as soon as possible (each task is assigned to exactly one machine). Formally, fix an arbitrary $v \in V$. The allocation $\hat{a}$ is *optimal w.r.t. the makespan* if $r(\hat{a}) \leq r(a)$ for every $a \in A$.

---

[4] We chose m to be the number of machines to be consistent with the formulation of Nisan and Ronen. Recall that we used n previously to denote the number of agents, whereas here the agents are the machines.

4.1 Lower Bound

Theorem 3. *Any envy-free mechanism cannot achieve an approximation ratio better than $2 - \frac{1}{m}$ with respect to the makespan.*

4.2 Upper Bound

In general the optimal allocation w.r.t. the makespan is not locally efficient, and in particular is not envy-free achievable. We will thus need to modify this optimal allocation in a careful way. We need the following:

Definition 6 (The Function β^*). *Let $a = (a_1, ..., a_m) \in A$ be an arbitrary allocation. Define the function $\beta^* : [m] \to [m]$ as follows.*

Let $\beta^*(j) \in \operatorname{argmin}_{i=1,...,m} v_i(a_j), j \in [m]$. That is, $\beta^*(j)$ is the machine with the minimal cost for the bundle a_j (breaking ties arbitrarily).

Intuitively, a^{β^*} is defined by independent "bundle-auctions": in each step $j = 1..m$ we allocate the bundle a_j of some given allocation a to the lowest cost machine for this bundle (independently of the history of former steps).

Fact: If the valuation of each machine is additive, then $b = a^{\beta^*}$ is a locally-efficient bundle assignment. That is, for every permutation π it holds that:

$$\Psi(v,\ b) \leq \Psi(v,\ b^\pi).$$

Definition 7 (The Permutation π^*). *Let $a = (a_1, ..., a_m) \in A$ be an arbitrary allocation. Define π^* to be a permutation such that a^{π^*} is locally-efficient bundle assignment. If there is more than one permutation, then arbitrarily choose one.*

Clearly, $a^{\pi^*} \neq a^{\beta^*}$ in general. We now describe our algorithm.

Algorithm 1 (Bundle-Local-Search). *Input: $v = v_1, v_2, ..., v_m$.*

Let $\hat{a}$ be the optimal allocation with respect to the makespan of v.

- *If the makespan of $\hat{a}^{\pi^*}$ is at most $\frac{m+1}{2}$ times the makespan of $\hat{a}$, then output $\hat{a}^{\pi^*}$.*
- *Otherwise, output $\hat{a}^{\beta^*}$.*

Informally, we permute the bundles dictated by $\hat{a}$ to achieve the locally lowest cost social welfare. If the resulted makespan has not increased by much, then we output this allocation. Otherwise, we re-assign each bundle $\hat{a}_i$ to the machine with the minimal cost for this bundle.

Theorem 4. *Algorithm bundle-local-search guarantees an approximation ratio of $\frac{m+1}{2}$ with respect to the makespan. Moreover, algorithm bundle-local-search is envy-free achievable. All together we get that there exists an envy-free $\frac{m+1}{2}$-approximation mechanism for minimizing the makespan on unrelated machines.*

4.3 Item Prices

In this subsection we shall see that for minimizing the makespan, any envy-free mechanism with item prices cannot achieve approximation ratio better than

m w.r.t to the makespan (regardless of any computational considerations), and this tight. Recall that if item prices are defined, the price of a bundle is the total prices of the items in the bundle. The following proposition is applicable for both the unrelated machines model and the related parallel machines model (see the next section for exact definition).

Proposition 3. *Any deterministic envy-free mechanism for minimizing the makespan with supporting item prices cannot achieve approximation ratio better than m, and this is tight.*

5 Near-Optimality of Related Parallel Machines

In this section we study an interesting case in which the envy-freeness constraint do not impose a further computational burden. Specifically, we consider minimizing the makespan of related parallel machines in an envy-free manner. We show that there exists a *poly-time* black-box procedure that takes an allocation and converts it to an "envy-free" allocation without any loss in the approximation ratio, and specifies supporting envy-free bundle prices. We then show that there exists a poly-time envy-free $(1 + \epsilon)$-approximation mechanism for every fixed ϵ, using the deterministic PTAS by Hochbaum and Shmoys [14]. We shall start by the formal setting.

The Setting. The related parallel machine scheduling setting $(Q||C_{\max})$ is a special case of the unrelated model $(R||C_{\max})$. In this model each task j has a load $l_j > 0$. Additionally, every machine i represents an agent with type t_i. It takes $t_i \cdot l_j$ time units to perform task j on machine i. More formally, $v_i(\{j\}) = t_i \cdot l_j$. The total cost of a set of tasks on machine i is the additive sum of the costs of the individual tasks on that machine. In this setting machines are comparable: if machine i_1 is faster than machine i_2 on task j, then machine i_1 is always faster than machine i_2. For convenience we use the notation $l(S) = \Sigma_{j \in S} \, l_j$, to denote the total load of a subset of tasks S. We also assume without loss of generality that $t_1 \geq t_2 \geq \cdots \geq t_m$.

5.1 The Envy-Free Mechanism

Lemma 1. *f is envy-free achievable if and only if it allocates more load to the fastest machines.*

Lemma 2. *Let f be a deterministic c-approximation algorithm. We can assume without loss of generality that f always allocates more load to the fastest machines (this will only improve the makespan).*

Definition 8 (Frugal Payment). *Suppose $f(v) = a$, and a allocates higher load to faster machines. The frugal payment to each machine is defined recursively as follows: $p_1(v) = l(a_1) \cdot t_1$, and $p_i(v) = p_{i-1}(v) + (l(a_i) - l(a_{i-1})) \cdot t_i$, for $i = 2, ..., m$.*

Intuitively, the average payment for a unit of load decreases for faster machines in the above payment rule.

Theorem 5. *Any c-approximation deterministic algorithm w.r.t. makespan for related parallel machines model can be converted in poly-time to a c-approximation envy-free mechanism using the frugal payments.*

Based on the deterministic PTAS by Hochbaum and Shmoys [14] and theorem 5 we can show the following theorem. The theorem also suggests that the resale of items ("tasks") among the agents can be prevented without using verification (see next subsection for a proof).

Theorem 6. *There exists a poly-time computable envy-free $(1+\epsilon)$-approximation mechanism for minimizing the makespan on related parallel machines for every fixed ϵ. Moreover, resale among the agents can be prevented (in quasi-poly-time).*

5.2   Frugal Payments and Preventing Resales

In many cases, a presence of a secondary resale market is not very desirable. Firstly, the secondary market might have a strong impact on the designer's goal (in our case, the makespan objective). Second, the existence of a secondary market indicates that there is a potential profit that was not fully extracted by the mechanism.

In what follows we shall first briefly address the issue of frugality. This will be helpful for preventing resales. Additionally, we show that our deterministic envy-free mechanism is not truthful (and thus it is different from the randomized truthful mechanism in [7])[5]. Specifically we show that the frugal payment method might encourage an agent to pretend to be slower. Obviously, laziness cannot be easily detected (unlike the case that a machine pretend to be faster).

Proposition 4. *The frugal payment provides the cheapest individually-rational envy-free payment.*

Proposition 5. *Any envy-free $(1 + \epsilon)$-mechanism supported by the frugal payment is not truthful.*

Acknowledgments. I would like to thank anonymous referees, Liad Blumrosen, Federico Echenique, David Kempe, John Ledyard, Debasis Mishra, Mohamed Mostagir, Mahyar Salek and Michael Schapira for helpful discussions.

References

1. Amar, L., Mu'alem, A., Stoesser, J.: On the importance of migration for fairness in online grid markets. In: GRID (2008)
2. Archer, A., Tardos, E.: Truthful mechanisms for one-parameter agents. In: FOCS, pp. 482–491 (2001)

---

[5] Archer and Tardos [2] showed that the optimal allocation w.r.t. the makespan is truthfully implementable. We thus show here that the frugal payment is different from the truthful payment for the optimal allocation with respect to the makespan.

3. Balcan, N., Blum, A., Mansour, Y.: Item pricing for revenue maximization. In: EC (2008)
4. Blumrosen, L., Nisan, N.: On the computational power of iterative auctions I: Demand queries. In: EC (2005)
5. Cheung, M., Swamy, C.: Approximation algorithms for single-minded envy-free profit-maximization problems with limited supply. In: FOCS (2008)
6. Christodoulou, G., Koutsoupias, E., Vidali, A.: A characterization of 2-player mechanisms for scheduling. In: Halperin, D., Mehlhorn, K. (eds.) Esa 2008. LNCS, vol. 5193, pp. 297–307. Springer, Heidelberg (2008)
7. Dhangwatnotai, P., Dobzinski, S., Dughmi, S., Roughgarden, T.: Truthful approximation schemes for single-parameter agents. In: FOCS (2008)
8. Foley, D.: Resource allocation and the public sector. Yale Economics Essays 7, 45–98 (1967)
9. Grigoriev, A., van Loon, J., Sitters, R., Uetz, M.: How to sell a graph: Guidelines for graph retailers. In: Fomin, F.V. (ed.) WG 2006. LNCS, vol. 4271, pp. 125–136. Springer, Heidelberg (2006)
10. Guruswami, V., Hartline, J.D., Karlin, A.R., Kempe, D., Kenyon, C., McSherry, F.: On profit-maximizing envy-free pricing. In: SODA, pp. 1164–1173 (2005)
11. Haake, C.-J., Raith, M.G., Su, F.E.: Bidding for envy-freeness: A procedural approach to n-player fair-division problems. Social Choice and Welfare 19(4), 723–749 (2002)
12. Hartline, J., Ieong, S., Schapira, M., Zohar, A.: Private communication (2008)
13. Heydenreich, B., Mller, R., Uetz, M.J., Vohra, R.: Characterization of revenue equivalence. Econometrica 77(1), 307–316 (2009)
14. Hochbaum, D.S., Shmoys, D.B.: A polynomial approximation scheme for scheduling on uniform processors: Using the dual approximation approach. SIAM J. Comput. 17(3), 539–551 (1988)
15. Koutsoupias, E., Vidali, A.: A lower bound of 1+phi for truthful scheduling mechanisms. In: Kučera, L., Kučera, A. (eds.) MFCS 2007. LNCS, vol. 4708, pp. 454–464. Springer, Heidelberg (2007)
16. Lenstra, J.K., Shmoys, D.B., Tardos, E.: Approximation algorithms for scheduling unrelated parallel machines. In: FOCS (1987)
17. Lipton, R.J., Markakis, E., Mossel, E., Saberi, A.: On approximately fair allocations of indivisible goods. In: EC (2004)
18. Mishra, D., Talman, D.: Characterization of walrasian equilibria of the assignment model, working paper (2007)
19. Nisan, N., Ronen, A.: Algorithmic mechanism design. Games and Economic Behavior 35, 166–196 (2001)
20. Nisan, N., Roughgarden, T., Tardos, E., Vazirani, V.V. (eds.): Algorithmic Game Theory. Cambridge University Press, Cambridge (2007)
21. Nisan, N., Segal, I.: The communication requirements of efficient allocations and supporting prices. Journal of Economic Theory 129(1), 192–224 (2006)
22. Papadimitriou, C.H., Steiglitz, K.: Combinatorial Optimization: Algorithms and Complexity. Dover Publications (1998)
23. Quinzii, M.: Core and competitive equilibria with indivisibilities. International Journal of Game Theory 13(1), 41–60 (1984)
24. Rochet, J.C.: A necessary and sufficient condition for rationalizability in a quasi-linear context. Journal of Mathematical Economics 16, 191–200 (1987)

Stable Rankings in Collective Decision Making with Imprecise Information

Ignacio Contreras[1], Miguel A. Hinojosa[1], and Amparo M. Mármol[2]

[1] Pablo de Olavide University
[2] Sevilla University, Seville, Spain

Abstract. The objective of the paper is to propose procedures to construct global ranking of a set of alternatives in situations in which each member of a group is able to provide imprecise information on his/her preferences about the relative importance of the criteria that have to be taken into account.

We first propose an approach based on the assumption that the final evaluation depends on the complete group since no possibility exists that the group might split into coalitions that search for more favourable solutions for the the coalitions members. To this end, the partial information on criteria weights provided by each individual is transformed into ordinal information on alternatives, and then the aggregation of individual preferences is addressed within a distance-based framework.

In a second approach, the possibility of coalition formation is considered, and the goal is to obtain rankings in which the disagreements of all the coalitions are taken into account. These rankings will exhibit an additional property of collective stability in the sense that no coalition will has the incentive to abandon the group and begin a separate evaluation process.

This last approach may be of interest in political decisions where different sectors have to be incorporated into a joint evaluation process with the desire to obtain a consensus across all possible subgroups.

Keywords: Group decision making, decision analysis with multiple criteria, imprecise information.

1 Introduction

In many group decision problems a set of alternatives must be evaluated on the basis of different and conflicting criteria which have to be taken into account in the final decision. In general, each of the group members has a particular view about the relative importance of the different criteria. The aim of the collective decision-making process is either to identify the best or most preferred alternative(s) from a set or to generate a ranking of alternatives in accordance with these individual preferences about criteria.

Recent research in the field of group decision making incorporates the possibility of dealing with imprecise preference information and permits the procedures to be applied in contexts where the group members are unable or unwilling to provide a precise representation of their preferences over alternatives. See for

F. Rossi and A. Tsoukis (Eds.): ADT 2009, LNAI 5783, pp. 132–143, 2009.
© Springer-Verlag Berlin Heidelberg 2009

instance [1], where a class of flexible weight indices for ranking alternatives is proposed, and [2], where a preference aggregation method based on the estimation of utility intervals is presented.

The case of imprecise information has also been addressed for collective decisions in the multicriteria framework. A detailed revision of group decision models with imprecise information can be found, for instance, in [3]. Recent contributions are also [4], [5], [6], [7], [8], [9], [10] and [11].

In this paper we propose multicriteria collective decision procedures which consist of the construction of compromise solutions by using a distance function on the set of rankings when the group wants to rank a set of alternatives. These procedures are especially well suited for group settings where each member of the group individually provides partial information about his/her preferences with respect to the criteria under consideration, whilst not having information about the preferences of the other agents.

The final objective is the construction of a global ranking of the alternatives that combines as accurately as possible the different evaluations of the alternatives with respect to criteria by taking into account the partial information sets provided by the agents.

The first approach proposed is based on the assumption that the final evaluation depends on the complete group since no possibility exists that the group might split into coalitions in order to seek more favorable solutions for the coalition members.

In a second approach, this coalition possibility is considered, and the goal is to obtain rankings in which the disagreements of all the coalitions are taken into account.

This latter approach might be of interest in political decisions where different sectors have to be incorporated into a joint evaluation process with the desire to obtain a consensus across all possible subgroups.

The rest of the paper is organized as follows. In Section 2, we introduce the collective decision-making model to be addressed. In Section 3, the procedure to obtain a final ranking of alternatives is presented, and an illustrative example is provided. In Section 4 the set of group rankings obtained by the procedure described in Section 3 is refined by taking into account the disagreement of all the coalitions. Section 5 is devoted to conclusions.

2   The Model

Let us consider a multicriteria group decision problem in which M alternatives, $X = \{x_1, \ldots, x_M\}$, have been evaluated with respect to N criteria. The evaluations of each alternative with respect to each criteria are represented by a matrix $A \in R^{N \times M}$, whose elements are denoted a_{ij} with $i = 1, \ldots, N, j = 1, \ldots, M$. These evaluations are assumed to be objective, in the sense that they do not depend on the assessment of the agents, but are measured independently. Hence, a_{ij} represents the cardinal value or the score given to alternative x_j with respect to the i-th criterion.

There are K Decision Makers (DMs), each of whom offers some information about his/her preferences with respect to the relative importance of the criteria.

We assume that DMs' preferences can be represented by means of an additive function. Thus, the k-th decision makers aggregated value associated to alternative x_h is given by,

$$V^k(x_h) = \sum_{i=1}^{N} w_i^k a_{ih}, \qquad (1)$$

where w^k denotes a vector of weights that represents the relative importance of the criteria for agent k.

In contrast to classic approaches which consist of the elicitation of weights, these parameters need not be be completely determined beforehand. We allow imprecision by permitting the values of the criteria weights for each agent to vary in partial information sets, $\Phi^k \subseteq R^N$, $k = 1, \ldots, K$.

A partial information set for an agent consists of those vectors of weights that the agent will accept as reasonable for the importance of the criteria. By convention the criteria weights are normalized to add up to one, hence, $\Phi^k \subseteq \{w^k \in R^N, \sum_{i=1}^{N} w_i^k = 1, w_i^k \geq 0, i = 1, \ldots, N\}$, for $k = 1, \ldots, K$. In particular, we will explore the cases where preference information is given by means of linear relations between the weighting coefficients. In this case, Φ^K are polyhedral sets described by linear constraints on the criteria weights.

The process of constructing the information set for each DM can be carried out in a sequential way (see [10]). The DMs can provide information by stating linear relations on the weights. For instance, they can provide partial or complete ordinal information on the importance of the criteria. Another example of representation of information by means of linear relations, which is also easily interpretable by the DM, is when the DM declares a preference of alternative x_h to x_j. This implies that x_h should not be ranked below x_j in any ranking of alternatives induced by the individual preferences of this particular decision-maker. Hence, a relation $\sum_{i=1}^{N} w_i^k a_{ih} - \sum_{i=1}^{N} w_i^k a_{ij} > 0$ must be incorporated into Φ^k.

3   Rankings Minimizing Global Disagreement

The main idea of the procedure proposed here is the achievement of a final consensus or compromise solution between DMs from the individual rankings of alternatives induced by the partial information sets. Implicit in this problem is the existence of a measure of global agreement or disagreement between rankings. Therefore, the approach implies the introduction of a distance function on the set of rankings in order to determine the ranking that minimizes the total distance across DMs. A detailed study of models based on distance functions can be seen in [8].

Let R_h and R_k be two priority vectors that represent two individual rankings, with r_{hj} denoting the position assigned to alternative x_j by DM h. As standard, the first category is assigned to the most preferred alternative and when ties occur, the average of the values corresponding to the tied alternatives is assigned.

The following distance, which is based on the L_1-metric, is considered on the set of priority vectors:

$$d(R_k, R_s) = \sum_{j=1}^{M} |r_{kj} - r_{sj}| \tag{2}$$

In this setting, a consensus vector, R_G, is a priority vector that satisfies

$$\begin{aligned} min \ &\sum_{k=1}^{K} d(R_k, R_G) \\ s.t. \quad &R_G \in \mathcal{R} \end{aligned} \tag{3}$$

where $\mathcal{R}$ is the set of vectors that represent ordinal rankings. Notice that neither the rankings for each DM, not the consensus vector, R_G, need to be unique.

It is important to point out that the consideration of partial information sets to represent DMs' preferences implies that, from an individual point of view, agent k would accept an evaluation of alternative x_h consisting of $V^k(x_h) = \sum_{i=1}^{N} w_i^k a_{hi}$ if $w^k \in \Phi^k$. Hence, every ranking of alternatives that can be achieved from the dominance relations induced by any $w^k \in \Phi^k$ is considered acceptable by the k-th DM. As a consequence, different rankings of alternatives (at least one, otherwise, $\Phi^k = \emptyset$) can be induced by each DM's preferences. Therefore, the procedure has to include not only an objective for the group in order to determine the ranking that best agrees with the individual preferences, but also a selection criterion to choose a ranking for each DM which represents the individual preferences (those rankings that best agree with the group order).

A dominance relationship between alternatives can be derived from (1). For a fixed vector of weights, $w^k \in \Phi^k$, we can say that alternative x_j strictly dominates alternative x_h under the preference structure of the k-th DM, if $V^k(x_j) - V^k(x_h) > 0$. The consensus ranking is induced here by the ordinal positions of the alternatives, determined by comparing the aggregated values $V^k(x_i)$ for the different values of $w^k \in \Phi^k$.

Hence, an intermediate step consisting of the elicitation of an individual ranking for each DM is required. To this end, we consider the following set of constraints, where δ_{ij} and γ_{ij} are binary variables:

$$\begin{aligned} V^k(x_h) - V^k(x_j) + \delta_{hj}^k B &\geq 0, \forall h \neq j, \\ V^k(x_h) - V^k(x_j) + \gamma_{hj}^k B &\geq \varepsilon, \forall h \neq j. \end{aligned} \tag{4}$$

Here B is a large number and ε is a discriminating factor between alternatives, such that we say alternative x_h strictly dominates x_j if $V^k(x_h) - V^k(x_j) \geq \varepsilon$.

The values of variables δ_{hj}^k will be equal to one each time that x_h does not strictly dominate x_j, i.e., whenever $V^k(x_j) > V^k(x_h)$. In contrast, γ_{hj}^k will be one whenever $V^k(x_j) \geq V^k(x_h)$, that is, each time x_h is not preferred to x_j in the k-th DM preferences. These variables are included in the expression in order to reflect the ties between alternatives.

The ranking position induced by the aggregate values $V^k(x_h)$ can be obtained as the sum of the binary variables divided by 2 (this fraction represents the number of alternatives that dominate x_h), plus one. That is,

$$r_{kh} = \sum_{h \neq j} \frac{\delta_{hj}^k + \gamma_{hj}^k}{2} + 1, \forall h = 1, \ldots, M. \tag{5}$$

To guarantee that the vectors $R_k = (r_{k1}, \ldots, r_{kM})$ represent priority vectors, we incorporate the following set of constraints

$$\delta_{hj}^k + \delta_{jh}^k \leq 1, \forall h \neq j, \tag{6}$$

$$\delta_{hj}^k + \gamma_{jh}^k = 1, \forall h \neq j. \tag{7}$$

Constraints from (6) to (7) are necessary in order to induce the ranking of alternatives from the values $V^k(x_i)$. These constraints guarantee that the values of the binary variables δ_{hj}^k and γ_{kj}^k are correct, in the sense that only when x_j strictly dominates x_h, then $\delta_{hj}^k = 1$ holds, and consequently $\delta_{jh}^k = 0$. The values of γ_{hj}^k also depend on the value assigned to δ_{hj}^k and represent ties between alternatives. It is worth noting that in each constraint we have to consider not only the dominance relation of x_i over x_j but also the relation of x_j over x_i.

Finally, the following requirement is included

$$\sum_{j=1}^{M} r_{kj} = \frac{M(M+1)}{2}, \forall k = 1, \ldots, K. \tag{8}$$

In addition to the above constraints, the condition in (8) assures that vector R_k is contained in set S, that is to say, represents a priority vector.

In order to obtain the compromise solution, a set of variables corresponding to the group has to be considered: a weighting vector, aggregate values and a ranking of alternatives for the group. Hence, a new agent labelled as the subindex G in the set of DMs[1] is included in the model.

To determine the group solution we have to deal with the distance defined in (2). The minimization of the sum of these nonlinear functions, as stated in (3) can be reduced to a linear programming model by considering a Goal Programming formulation. By taking into account the following change of variables proposed in [9],

$$\begin{aligned} \alpha_{kj} &= \tfrac{1}{2} \left[|r_{kj} - r_{Gj}| - (r_{kj} - r_{Gj}) \right] \\ \beta_{kj} &= \tfrac{1}{2} \left[|r_{kj} - r_{Gj}| + (r_{kj} - r_{Gj}) \right]; \end{aligned} \tag{9}$$

we will include the following set of constraints in the model to measure the distance between individual priority vectors and the compromise ranking.

---

[1] Subindexes k, that represent DMs, will now vary from 1 to K plus G, i.e. $k = 1 \ldots, K, G$.

$$r_{kj} - r_{Gj} + \alpha_{kj} - \beta_{kj} = 0, \forall k = 1, \ldots, K; j = 1, \ldots, N. \tag{10}$$

Therefore, the distance between R_k and R_G can be represented equivalently by the following expression

$$d(R_k, R_G) = \sum_{j=1}^{M} (\alpha_{kj} + \beta_{kj}). \tag{11}$$

All the specifications defined above yield the following model

$$
\begin{aligned}
min \ & \sum_{k=1}^{K} \sum_{j=1}^{M} (\alpha_{kj} + \beta_{kj}) \\
s.t. \ & V^k(x_h) = \sum_{i=1}^{N} w_i^k a_{ih}, & \forall h, k \\
& V^k(x_h) - V^k(x_j) + \delta_{hj}^k B \geq 0, & \forall h \neq j, \forall k \\
& V^k(x_h) - V^k(x_j) + \gamma_{hj}^k B \geq \varepsilon, & \forall h \neq j, \forall k \\
& \delta_{hj}^k + \delta_{jh}^k \leq 1, & \forall h \neq j, \forall k \\
& \delta_{hj}^k + \gamma_{hj}^k + \gamma_{jh}^k \leq 2, & \forall h \neq j, \forall k \\
& \delta_{hj}^k + \gamma_{hj}^k + \gamma_{jh}^k \geq 1, & \forall h \neq j, \forall k \\
& \delta_{hj}^k + \gamma_{jh}^k = 1, & \forall h \neq j, \forall k \\
& r_{kh} = \sum_{h \neq j} \frac{\delta_{hj}^k + \gamma_{hj}^k}{2} + 1, & \forall h, k \\
& \sum_{h=1}^{M} r_{kh} = \frac{M(M+1)}{2} & \forall k \\
& r_{kh} - r_{Gh} + \alpha_{kh} - \beta_{kh} = 0, & \forall h, k \neq G \\
& w^k \in \Phi^k, & \forall k \\
& \delta_{hj}^k, \gamma_{hj}^k \in \{0, 1\}, & \forall h, j, k \\
& B, \varepsilon \geq 0.
\end{aligned}
\tag{12}
$$

By solving (12), the compromise ranking r_G is obtained. The solution to the problem also provides a ranking of alternatives for each DM. This is the ranking which permits total disagreement to be minimized from among those induced by the individual preferences. Note that it does not necessarily coincide with the ranking that minimizes the DM individual disagreement. This ranking is determined by means of the aggregate values $V^k(x_i)$, hence the vector of weights which best agrees with the group solution is selected for each DM.

The aggregated values for the group, $V^G(x_h)$, are only considered for computational purposes in order to induce a ranking of alternatives for the group in the same format in which individual rankings have been constructed. They have no interpretation from a cardinal point of view since the goal of the procedure is only the determination of the group's ranking.

Some desirable properties of our procedure in the context of social choice processes which are a direct consequence of its construction are: feasibility, anonymity, neutrality and no dictatorship. That is to say, a compromise solution is always obtainable, all the group members are treated equally, all the alternatives are treated equally, and no DM exists whose individual preferences determine the consensus ranking.

3.1   Illustrative Example

This section illustrates the proposed procedure for group decision-making problems. We have considered an example in which four decision-makers want to decide between five alternatives denoted by $\{x_1, \ldots, x_5\}$. The alternatives have been evaluated with respect to four different criteria. Table 1 shows the scores of the alternatives with respect to each criterion. These evaluations have been normalized so that the sum with respect to each criterion add up to one.

Table 1. Matrix of utilities

Alternative	Crit. 1	Crit. 2	Crit. 3	Crit. 4
x_1	0.5776	0.1692	0.0481	0.1612
x_2	0.2054	0.2900	0.0340	0.2070
x_3	0.1143	0.1692	0.3107	0.2200
x_4	0.0717	0.3052	0.4489	0.1765
x_5	0.0310	0.0664	0.1583	0.2353

Four agents are to evaluate this information. They have provided partial information about the relative importance they assign to each criteria. Agent 1 considers that the importance of the criteria is ranked in order of decreasing magnitude. In addition, this agent states that the weight of criterion 5 is not less than 1% of the total value. If $\Lambda^+ = \{w \in R^5, \sum_{k=1}^{5} w_k = 1, w_k \geq 0, k = 1, \ldots, 5\}$, then the partial information set for agent 1, can be formalized as

$$\Phi^1 = \{w^1 \in \Lambda^+, \, w_1^1 \geq w_2^1 \geq w_3^1 \geq w_4^1 \geq 0.01\}$$

The remaining agents preference information about the criteria are represented in the following information sets and can be interpreted in a similar way to that above.

$$\Phi^2 = \{w^2 \in \Lambda^+, \, w_2^2 \geq w_1^2 + w_3^2, \, w_4^2 \geq w_1^2 + w_3^2, \, w_1^2 \geq 0.01, \, w_3^2 \geq 0.01\},$$
$$\Phi^3 = \{w^3 \in \Lambda^+, \, w_4^3 \geq 2w_3^3 \geq 4w_2^3 \geq 8w_1^3 \geq 0.08\},$$
$$\Phi^4 = \{w^4 \in \Lambda^+, \, w_3^4 \geq w_1^4 + w_2^4 + w_4^4, \, w_1^4 \geq 0.01, \, w_2^4 \geq 0.01, \, w_3^4 \geq 0.01\}.$$

The solution to problem (12) yields a level of disagreement equal to 8 units, although the collective ranking is not unique. The following table (Table 2)

Table 2. Group rankings

Priority vector (R_G)	$d(R_1, R_G)$	$d(R_2, R_G)$	$d(R_3, R_G)$	$d(R_4, R_G)$
(5, 4, 2, 1, 3)	6	2	0	0
(2, 4, 3, 1, 5)	0	2	6	0
(5, 3, 2, 1, 4)	6	0	2	0
(4.5, 3, 2, 1, 4.5)	5	0	3	0
(5, 3.5, 2, 1, 3.5)	6	1	1	0
(3.5, 3.5, 2, 1, 5)	3	1	4	0

summarizes the group solution that minimizes the total disagreement, represented by their respective priority vectors, and the individual disagreements each solution provides.

It is interesting to note that the individual disagreement of each solution can vary from one collective ranking to another. Note that we only fix the total disagreement at its minimum level which, in this case, is 8.

4   Stable Rankings across Coalitions

Unfortunately, the procedure described in Section 3 does not always provide a unique group ranking which minimizes the level of disagreement of the DMs. In this section we present a procedure to refine the set of group rankings inspired by cooperative game theory approaches.

The individual disagreement associated to the agents for each collective ranking can be seen as an allocation of the total disagreement. The procedure described in Section 3 may provide several collective rankings which minimize the total disagreement and, in addition, there may be different sets of individual rankings associated to the same colletive ranking, therefore several allocations among the DMs of the minimum level of disagreement may exist.

Let $x = (x_1, x_2, \ldots, x_K)$ denote any of such allocations, that is to say, for each $k = 1, 2, \ldots, K$, $x_k = d(R_k, R_G)$, where R_G is one of the group rankings that minimizes the total disagreement and R_k is one of the associated individual rankings. Thus, when we choose a collective ranking, R_G, and a set of associated individual rankings, R_k, $k = 1, 2, \ldots, K$, we are assigning a disagreement $x_S = \sum_{k \in S} x_k$ to each subgroup (or coalition) $S \subseteq \mathcal{K} = \{1, \ldots, K\}$. On the other hand, the procedure described in Section 3 can be applied, not only to the whole set of DMs, but also to each coalition $S \subset \mathcal{K}$. In this way a minimum level of disagreement, D_S, is obtained for each coalition $S \subseteq \mathcal{K}$, by taking into account the preferences of the DMs in S. We will say that D_S is a reference of the disagreement for coalition S.

For each coalition $S \subseteq \mathcal{K}$, we consider the difference $x_S - D_S$, which is a measure of the dissatisfaction of coalition S with the collective ranking R_G (and the associated individual rankings) which provides allocation x. The idea is to choose among all the collective rankings for which the the minimum level of disagreement for the whole set of DMs is achieved, those that lexicographically minimize the maximum dissatisfaction of the coalitions. If any vector $x = (x_1, x_2, \ldots, x_K)$, $x_k \geq 0$, $k = 1, 2, \ldots, K$, $x_\mathcal{K} = D_\mathcal{K}$ could be chosen as a allocation of the collective disagreement, then the problem to solve would be Problem 13.

$$lex - min \ max_{S \subset \mathcal{K}} \ x_S - D_S$$
$$s.t. \qquad x_\mathcal{K} = D_\mathcal{K} \tag{13}$$

This problem has, as unique solution, the pre-nucleolus (see [15]) of the cooperative coalitional game defined by the references of the disagreement.

Example in Section 3.1 (continued). *The references of the disagreement in this case are shown in Table 3.*

Table 3. References of the disagreement for the coalitions

Coalition (S)	Disagreement (D_S)	Coalition (S)	Disagreement (D_S)
$\mathcal{K} = \{1,2,3,4\}$	8	$\{2,3\}$	2
$\{1,2,3\}$	8	$\{2,4\}$	0
$\{1,2,4\}$	2	$\{3,4\}$	0
$\{1,3,4\}$	6	$\{1\}$	0
$\{2,3,4\}$	2	$\{2\}$	0
$\{1,2\}$	0	$\{3\}$	0
$\{1,3\}$	6	$\{4\}$	0
$\{1,4\}$	0		

In the first step to solve Problem 13, we solve the following linear problem:

$$min \ \mu$$
$$s.t. \ x_S - D_S \leq \mu \quad \forall S \subset \mathcal{K} \tag{14}$$
$$x_{\mathcal{K}} = D_{\mathcal{K}}.$$

The solution of this problem is $\mu = 4$ and the constraints corresponding to coalitions $\{1,2\}$ and $\{3,4\}$ are active for each allocation x associated to the solution (it is worth noting that for at least one of the coalitions this property is fulfilled). The next step consists of solving Problem 15.

$$min \ \mu$$
$$s.t. \ x_1 + x_2 = 4$$
$$x_3 + x_4 - 2 = 4 \tag{15}$$
$$x_S - D_S \leq \mu \quad \forall S \subset \mathcal{K}, S \neq \{1,2\}, \{3,4\}$$
$$x_{\mathcal{K}} = D_{\mathcal{K}}.$$

This new problem has a unique allocation $x^* = (\frac{8}{3}, \frac{4}{3}, \frac{10}{3}, \frac{2}{3})$, associated to the optimal solution, $\mu = \frac{10}{3}$. x^* is called the pre-nucleolus of the cooperative coalitional game of references of the disagreements.

Unfortunately, in general, a collective ranking, R_G, and a set of associated individual rankings, R_k, $k = 1, 2, \ldots, K$, for which the corresponding allocation $x = (d(R_k, R_G))_{k \in \mathcal{K}}$ coincides with the pre-nucleolus of the game, do not always exist. Nevertheless, it is always possible to find rankings for which allocation $x = (d(R_k, R_G))_{k \in \mathcal{K}}$ approximates x^* by solving Problem 16.

$$lex - min \ max_{S \subset \mathcal{K}} \ x_S - D_S$$
$$s.t. \quad d(R_k, R_G) = x_k, \quad \forall k \in \mathcal{K}$$
$$x_{\mathcal{K}} = D_{\mathcal{K}} \tag{16}$$
$$R_k \in \mathcal{R}(\phi^k)$$
$$R_G \in \mathcal{R}(\phi^G)$$

Moreover, in most cases this collective ranking is unique.

Table 4. Individual weighting vectors

	Crit 1	Crit 2	Crit 3	Crit 4
w^1	0,330	0,330	0,329	0,010
w^2	0,050	0,438	0,231	0,281
w^3	0,010	0,020	0,323	0,647
w^4	0,108	0,392	0,500	0,000

Table 5. Individual rankings

	\multicolumn{5}{c}{Rank positions}
	DM 1	DM 2	DM 3	DM 4	Group
x_1	2	4	5	3.5	3.5
x_2	4	3	4	3.5	3.5
x_3	3	2	2	2	2
x_4	1	1	1	1	1
x_5	5	5	3	5	5
Disagreement	3	1	4	0	

Example in Section 3.1 (continued). *The solution of Problem 16 provides a vector of individual disagreements $x = (3, 1, 4, 0)$, which is as close as possible to the pre-nucleolus, by taking into account that the constraint in Problem 16 only considers distances between rankings.*

Table 4 and Table 5 summarize the individual results. In Table 4 the weighting vector selected for each individual partial information set is included. Table 5 summarizes the individual ranking induced from these weighting vectors (note that this ranking is constructed throughout the aggregated values $V^k(x_i)$), and the individual distances from the group ranking R_G.

Therefore, in this case, the group solutions establish that alternative x_4 is ranked at the first position, followed by x_3, alternatives x_1 and x_2 tie in the third position and the least-valued alternative is x_5.

$$\begin{pmatrix} x_3 \\ x_2 \\ x_1, x_2 \\ x_5 \end{pmatrix}$$

5   Conclusions

We have proposed a compromise method for collective decision problems which is especially suited for situations where the members of the group provide imprecise information about their preferences with respect to the criteria, and there is no flow of information between these members. In this context the group members are not encouraged to misrepresent their true preferences in order to manipulate the group decision. Hence, the result derived could be considered a fair representation of the group evaluation of all the alternatives.

Imprecise information is formalized by means of linear relations between criteria weights. This way of providing preferential information is one of the most easily interpreted by the DMs, and includes interesting particular cases such as those in which they only provide ordinal information on the weights.

The basic procedure relies on the transformation of the partial information about criteria into ordinal information about alternatives, expressed through rankings. A model is constructed that provides a set of compromise rankings for the group and an additional step enables the achievement of a unique compromise ranking which is stable across coalitions.

The approach presented here uses a measure of agreement based upon the L_1-metric and, therefore, emphasizes the sum of individual disagreements with respect to every alternative. However, the use of other metrics could also be considered in order to analyze this class of collective decision making problems.

Acknowledgements. This research has been partially financed by the Spanish Ministry of Education and Science project SEJ2007-62711/ECON and by Consejera de Innovacin, Ciencia y Tecnologia (Junta de Andaluca) project P06-SEJ-01801.

References

1. Contreras, I., Hinojosa, M.A., Mármol, A.M.: A class of flexible weight indices for ranking alternatives. IMA Journal of Management Mathematics 16, 71–85 (2005)
2. Wang, Y.M., Yang, J.B., Xu, D.L.: A preference aggregation method through the estimation of utility interval. Computers and Operations Research 32, 2027–2049 (2005)
3. Dias, L.C., Climaco, J.N.: Dealing with imprecise information in group multicriteria decisions: A methodology and a GDSS architecture. European Journal of Operational Research 160(2), 291–307 (2005)
4. Salo, A.A.: Interactive decision aiding for group decision support. European Journal of Operational Research 84, 134–149 (1995)
5. Kim, S.H., Ahn, B.S.: Group decision making procedure considering preference strength under incomplete information. Computers and Operations Research 12, 1101–1112 (1997)
6. Kim, S.H., Choi, S.H., Kim, J.K.: An interactive procedure for multi-attribute group decision making with incomplete information: Range-based approach. European Journal of Operational Research 118, 139–152 (1999)
7. Malakooti, B.: Ranking and Screening Multiple Criteria Alternatives with Partial Information and Use of Ordinal and Cardinal Strengths of Preferences. IEEE Transactions on Systems, Man, and Cybernetics Part A 30, 355–367 (2000)
8. Salo, A., Hämäläinen, R.P.: Preference ratios in multiattribute evaluation (PRIME) elicitation and decision procedures under incomplete information. IEEE Transactions on Systems, Man and Cybernetics 31, 533–545 (2001)
9. Valadares Tavares, L.: A model to support the search for consensus with conflicting rankings: Multitrident. International Transactions in Operational Research 11, 107–115 (2004)

10. Climaco, J.N., Dias, L.C.: Negotiation Processes with Imprecise Information on Multicriteria Additive Models. Group Decision and Negotiation 15, 171–184 (2006)
11. Contreras, I., Mármol, A.M.: A lexicographical compromise method for multiple criteria group decision problems with imprecise information. European Journal of Operational Research 181(3), 1530–1539 (2007)
12. Cook, W.D., Kress, M.: An extreme-point approach for obtaining weighted ratings in qualitative multicriteria decision making. Naval Research Logistic 43, 519–531 (1996)
13. González-Pachón, J., Romero, C.: Distanced based consensus methods: a goal programming approach. Omega 27, 341–347 (1999)
14. Mármol, A.M., Puerto, J., Fernández, F.R.: Sequential incorporation of imprecise information in multiple criteria decision processes. European Journal of Operational Research 137, 123–133 (2002)
15. Schmeidler, D.: The nucleolus of a characteristic function game. SIAM Journal of Applied Mathematics 17, 1163–1170 (1969)

Finding Best k Policies

Peng Dai[1] and Judy Goldsmith[2]

[1] Computer Science & Engineering University of Washington, Seattle WA 98195-2350
daipeng@cs.washington.edu
http://www.cs.washington.edu/homes/daipeng
[2] Univ. of Kentucky, Dept. of Comp. Sci. Lexington, KY, USA 40506-0046
goldsmit@cs.uky.edu
http://www.cs.uky.edu/~goldsmit

Abstract. An optimal probabilistic-planning algorithm solves a problem, usually modeled by a Markov decision process, by finding its optimal policy. In this paper, we study the k *best policies* problem. The problem is to find the k best policies. The k best policies, $k > 1$, cannot be found directly using dynamic programming. Naïvely, finding the k-th best policy can be Turing reduced to the optimal planning problem, but the number of problems queried in the naïve algorithm is exponential in k. We show empirically that solving k best policy problem by using this reduction requires unreasonable amounts of time even when $k = 3$. We then provide a new algorithm, based on our theoretical contribution to prove that the k-th best policy differs from the i-th policy, for some $i < k$, on exactly one state. We show that the time complexity of the algorithm is quadratic in k, but the number of optimal planning problems it solves is linear in k. We demonstrate empirically that the new algorithm has good scalability.

1 Introduction

Markov Decision Processes (MDPs) [1] are a powerful and widely-used formulation for modeling probabilistic planning problems [2,3]. For instance, NASA researchers use MDPs to model the Mars rover decision making problems [4,5]. MDPs are also used to formulate military operations planning [6] and coordinated multi-agent planning [7], *etc.*

An optimal planner typically takes an MDP model of a problem and outputs an optimal plan. This is not always sufficient. In many cases, a planner is expected to generate more than one solution.

Furthermore, in the modeling phase, not every aspect of nature can be easily factored in a problem representation. For the case of NASA rover, for example, there are many safety constraints that need to be satisfied [5]. An optimal plan might be very close to a risky value—but another may not have many risks and so it is better to prefer the slightly suboptimal one. Similarly there are many decision criteria—probability of reaching the goal, expected reward, expected risk, various preferences, *etc.* Combining them into a single criterion is hard, and multi-objective planning is too slow [8,9]. Thus, a good alternative is to

F. Rossi and A. Tsoukis (Eds.): ADT 2009, LNAI 5783, pp. 144–155, 2009.

look for many suboptimal plans given a single criterion and later pick one that looks the best according to all criteria.

In this paper, we look at the k *best policies* problem. Given an MDP model, the problem is to find the k best policies, ranked by the expected value of the initial state, tie-broken by the "closeness" to a better policy, followed by lexical order of the policies. The classical optimal planning problem is a special case of the k best policy problem where $k = 1$. The optimal planning problem can be solved by *dynamic programming*, as the property of the optimality of sub-problems holds. The k best policy problem be directly solved by dynamic programming. However, finding the k-th best policy can be brute-force reduced to exponentially many instances of the optimal planning problem. Our experiments show that solving the k best policy problem this way requires unreasonable time even when $k = 3$.

A very similar problem has been explored by Nielsen, et al. [10,11,12]. Nielsen and Kristensen observed that the problem of finding optimal *history-dependent* policies (maps from the state space crossed with the time step to the action space) can be modeled as finding "a minimum weight hyperpath" in directed hypergraphs. A vertex in the hypergraph represents a state of the MDP at a particular time; the hypergraphs are, therefore, acyclic. They present an elegant and efficient algorithm for finding the k best time-dependent policies for an MDP. However, their algorithm cannot handle MDPs with probabilistic cycles, therefore its usefulness is limited.

Our new solution to the k best policy problem follows from the property: *The k-th best policy differs from a better policy on exactly one state.* We propose an original algorithm for the k best policy problem that leverages this property. We demonstrate both theoretically and empirically that the new algorithm has low complexity and good scalability.

2   Background

2.1   Markov Decision Processes

AI researchers often use MDPs to formulate probabilistic planning problems. An MDP is defined as a four-tuple $\langle \mathcal{S}, \mathcal{A}, T, C \rangle$, where $\mathcal{S}$ is a finite set of discrete states, $\mathcal{A}$ is a finite set of all applicable actions, T is the transition matrix describing the domain dynamics, and C denotes the cost of action transitions.

The agent executes its actions in discrete time steps called *stages*. At each stage, the system is at one distinct state $s \in \mathcal{S}$. The agent can pick any action a from a set of *applicable actions* $Ap(s) \subseteq \mathcal{A}$, incurring a cost of $C(s, a)$. The action takes the system to a new state s' stochastically, with probability $T_a(s'|s)$.

The *horizon* of an MDP is the number of stages for which costs are accumulated. We focus our attention on a special set of MDPs called *stochastic shortest path* (SSP) problems. The horizon in such an MDP is indefinite and the costs are accumulated with no discounting. There are an initial state s_0, and a set of sink *goal states* $\mathcal{G} \subseteq \mathcal{S}$. Reaching any state $g \in \mathcal{G}$ terminates the execution. The cost of the execution is the sum of all costs along the path from s_0 to g. Any infinite horizon discounted reward MDP can easily be converted to an undiscounted SSP [13].

To solve the MDP we need to find an *optimal policy* ($\pi^* : \mathcal{S} \rightarrow \mathcal{A}$), a probabilistic execution plan that reaches a goal state with the minimum expected cost. We evaluate any policy π by a *value function*.

$$V_\pi(s) = C(s, \pi(s)) + \sum_{s' \in \mathcal{S}} T_{\pi(s)}(s'|s)V_\pi(s').$$

Any optimal policy must satisfy the following system of *Bellman equations*:

$$V^*(s) = 0 \quad \text{if } s \in \mathcal{G} \text{ else} \tag{1}$$
$$V^*(s) = \min_{a \in Ap(s)} [C(s, a) + \sum_{s' \in \mathcal{S}} T_a(s'|s)V^*(s')].$$

The corresponding optimal policy can be extracted from the value function:

$$\pi^*(s) = argmin_{a \in Ap(s)}[C(s, a) + \sum_{s' \in \mathcal{S}} T_a(s'|s)V^*(s')].$$

2.2 Dynamic Programming

We define a *sub-problem* of an MDP with state space $\mathcal{S}' \subseteq \mathcal{S}$ to be a self-contained MDP with state space $\mathcal{S}'$ and associated action transitions. We define the *sub-policy* of a policy π given a sub-problem with state space $\mathcal{S}' \subseteq \mathcal{S}$ to be the mapping from all $s \in \mathcal{S}'$ to $\pi(s)$. An optimal policy satisfies the following necessary and sufficient condition: for any sub-problem, the corresponding sub-policy is also optimal. Many optimal MDP algorithms are based on dynamic programming. Its usefulness was first proved by a simple yet powerful algorithm called *value iteration* (VI) [1]. Value iteration first initializes the value function arbitrarily. Then the values are updated iteratively using an operator called *Bellman backup* to create successively better approximations per state per iteration. Value iteration stops updating when the value function converges (one future backup can change a state value by at most ϵ, a pre-defined threshold).

Another algorithm, named *policy iteration* (PI) [14], starts from an arbitrary policy and iteratively improves the policy. Each iteration of PI consists of two sequential steps. The first step, *policy evaluation*, finds the value function of the current policy. Values are calculated by solving the system of linear equations (in the original PI algorithm), or by iteratively updating the value functions in the VI manner till convergence (modified policy iteration [15]). The second step, *policy improvement*, updates the current policy by choosing a greedy action per state by a one step lookahead, based on the value function calculated in the policy evaluation step. PI stops when the policy improvement step doesn't change the policy.

3 k Best Policy Problem

Classical dynamic programming successfully finds *one* optimal policy of an MDP in time polynomial in $|\mathcal{S}|$ and $|\mathcal{A}|$ [16,17]. In this paper, we find the k best policies of an MDP. We first give the formal definition of the k best policy problem. Then we introduce the main theoretical contribution of the paper by proving a very strong result about the k-th best policy.

Let M be an MDP, π a policy for M. We define the *policy graph* of M given π, denoted by G_π, to be a graph constructed by: (1) the set of states (vertices) that are reachable from s_0 given π, and (2) their corresponding transitions in π (edges).

Let s and s' be states of M. We say that s' is a *policy descendant* of s with respect to π if there is a path from s to s' in G_π or if $s = s'$. We define $Policydesc(s, \pi)$ to be the set of all policy descendant states of s under policy π. We assume that, for every state $s \in \mathcal{S}$, there are at least two possible actions for s.

Note that, for a given MDP and a given value function, there may be multiple policies with that value function. We define a notion of "best among equals", namely, the "closest" to better policies followed by a lexicographic ordering, so that the notion of "best policy" is well defined.

Lemma 1. *Using value iteration, we can find an optimal value function for M, and the optimal $V_{\pi^*}(s_0)$. We can then find the lexicographically least policy, π_1, that has that value for $V_{\pi_1}(s_0) = V_{\pi^*}(s_0)$.*

The proof of Lemma 1 is straightforward. Given the value function, for each state, we choose the lexicographically first action that achieves the desired value. (If $\mathcal{A} = \{a_0, a_1, \ldots, a_j\}$, the lexicographically first action satisfying a property is the lowest-numbered a_i with that property.) Once we have the *best* policy, we then need to define an ordering on policies so that we may define the k-th best.

Definition 1. *Given two policies π and π', we can consider them as vectors of length $|\mathcal{S}|$ over alphabet $|\mathcal{A}|$, and define the Hamming distance $Ham(\pi, \pi')$ to be the number of states on which π and π' differ. We also define $<_{lex}$ to be the lexicographic ordering on such vectors.*

Finally, we define an order on policies.

Definition 2. *Given an MDP M and a dynamic list of p best policies generated so far $\{\pi_1 \ldots, \pi_p\}$, the next best policy is computed based on the following ordering $\prec$ on the rest of the policies for M.*

$$\pi \prec \pi' \textbf{ if } \quad V_\pi(s_0) < V_{\pi'}(s_0)$$
$$\textbf{else if } \ min_{j \leq p} Ham(\pi_j, \pi) < min_{j \leq p} Ham(\pi_j, \pi')$$
$$\textbf{else if } \ \pi <_{lex} \pi'.$$

Intuitively, two policies with the same initial state value are first compared by how "close" each one is to some better policy, followed by lexicographic order if they are equally close.

Theorem 1. *Let M be an MDP, and let $\{\pi_1, \ldots, \pi_k\}$ be the k best policies for M, in order. Let $k \geq 1$. Then there is some $m < k$ such that π_k differs from π_m on exactly one state.*

The proof sketch to Theorem 1 is provided in the Appendix.

4   Algorithm

Consider the k-th ($k > 1$) best policy of an MDP M, called π_k. The necessary and sufficient condition of the optimality on sub-problems does not hold. With the loss of the optimality on sub-problems, dynamic programming is not immediately applicable. However, we can reduce it to many optimal planning problems, each solved by dynamic programming. Before illustrating the reduction, we present the high-level idea of our first algorithm in Algorithm 1. We call it k *best naïve* algorithm (KBN), as it is a brute force algorithm that doesn't use Theorem 1. KBN is based on the following observation: The $k + 1$-st best policy must differ from each of the k best policies on at least one state. We can enumerate the possible sets of state/action pairs the new policy must avoid, and find an optimal policy for each thus-constrained MDP, then take the best of those policies.

Algorithm 1. k best naïve (KBN)

1: **Input:** M (an MDP), k
2: find best policy π_1 by VI
3: $\Pi \leftarrow \{\pi_1\}$
4: **for** $i \leftarrow 2$ to k **do**
5:    $\pi_i \leftarrow$ best policy that differs from any policy $\pi \in \Pi$ by at least one state
6:    $\Pi \leftarrow \Pi \cup \{\pi_i\}$
7: return $\pi_1, \ldots, \pi_k$

For instance, given the best and second best policies, π_1 and π_2, to find π_3, we say that either it differs from π_1 on s_0 and from π_1 on s_0, or from π_1 on s_0 and from π_1 on s_1, or.... In this case, we solve $|\mathcal{S}|^2$ many optimal planning problems. To find the k-th best policy, we solve $|\mathcal{S}|^k$ many. Each newly-computed policy will be compared with the best policy computed so far, so that the number of comparisons is linear in the number of policies computed. Suppose we use VI to solve those optimal planning problems, KBN has a complexity $|\mathcal{S}|^k \times O(VI)$, an exponential function of k.

Some of these combinations of constraints may constrain away all actions for a particular state, so do not yield a next-best policy. However, the next best policy must be among those computed, and will be the best such.

Using Theorem 1, we have a new algorithm, called k *best improved* (KBI). The KBI pseudo-code is shown in Algorithm 2. KBI keeps a set of candidate policies $\mathcal{P}$, which is initially empty. We first find the optimal policy by value iteration. To find the i-th best policy, we generate $k - i + 1$ distinct policies as candidates. These candidates (1) must not be duplicates of any policy in $\mathcal{P}$, and (2) each differs from π_{i-1} on exactly one state. We have the following theorem.

Theorem 2. *The i-th best policy must be an element of $\mathcal{P}$.*

Proof. As we know from Theorem 1 that the i-th ($i \leq k$) best policy is exactly one state different from one of $\pi_1, \ldots, \pi_{i-1}$, say, π_j, where $j < i$. Therefore, it must have been generated when π_{j+1} was computed. Since it is the i-th best policy, it would

---

Algorithm 2. k best improved (KBI)

1: **Input:** M (an MDP), k
2: find best policy π_1 by VI
3: $\mathcal{P} \leftarrow$ empty set
4: **for** $i \leftarrow 2$ to k **do**
5:   generate distinct $k-i+1$ best policies that each differs from π_{i-1} on exactly one state and differs from $\{\pi_1, \ldots, \pi_{i-1}\}$ and insert them into $\mathcal{P}$ in order, discarding duplicates
6:   $\pi_i \leftarrow$ the best policy in $\mathcal{P}$
7:   delete π_i from $\mathcal{P}$
8: **return** $\pi_1, \ldots, \pi_k$

---

have been amongst the $i - j$-th best of those policies that are one state different from π_j, so it belongs to the $k - j$ best policies added to $\mathcal{P}$ at stage $j + 1$.

Thus, we find the i-th best policy by picking the best policy in $\mathcal{P}$. There are $(|\mathcal{A}| - 1) \times |\mathcal{S}|$ policies that are exactly one state different from π_i. Finding the best $k-i$ of them has a complexity $|\mathcal{A}| \times |\mathcal{S}| \times O(policy\ evaluation)$, plus the complexity of keeping the list $\mathcal{P}$ in sorted order ($O(k^2 \log k)$). KBI computes these policies $k - 1$ times, so its complexity is $(k - 1) \times |\mathcal{A}| \times |\mathcal{S}| \times O(policy\ evaluation)$, a linear function of k. (Note that the sorting term is dominated by $|\mathcal{A}| \times |\mathcal{S}| \times O(policy\ evaluation)$.)

5 Experiments

We address the following three questions in our experiments: (1) How does KBI compare with KBN on different problems and k values? (2) Does KBI scale well on large k values? (3) How different are the k best policies from the optimal policy?

We implemented KBN and KBI in C. We performed all experiments on a 2.2GHz Dual-Core Intel(R) Core(TM)2 Processor with 6GB memory. We picked problems from three domains, namely Racetrack [18], Single-arm pendulum (SAP) and Double-arm pendulum (DAP) [19]. We used a threshold value of $\epsilon = 10^{-6}$.

5.1 Comparing KBI and KBN

We compare KBN and KBI on a suite of six problems of various sizes. The running times of both algorithms when $k = 2$ are listed in Table 1. We see that KBI outperforms KBN on all problems. In four problems, the speedup is an order of magnitude. According to our analysis in the Algorithm section, when k increases by 1, the running time of KBN increases by a factor of $|\mathcal{S}|$, so for cases $k = 3$ and $k = 4$ we take the expectations of its running time based on its performance on the same problem when $k = 2$. Even for small k values, the running times of KBN are prohibitively high. For example, in SAP 2 problem, its expected running time is approximately one thousand hours for $k = 3$ and tens of millions of hours for $k = 4$.

Table 1. Running time (seconds) of KBN and KBI in various problems with different k values. The running time of KBN on $k > 2$ are expectations. KBI outperforms KBN on most problems by an order of magnitude even when $k = 2$.

Domain	States $|S|$	$k = 2$ KBN	$k = 2$ KBI	$k = 3$ KBN (expected)	$k = 3$ KBI	$k = 4$ KBN (expected)	$k = 4$ KBI
DAP 1	625	0.90	0.44	10^2	0.87	10^5	1.32
Racetrack 1	1,847	0.56	0.07	10^3	0.14	10^6	0.21
SAP 1	2,500	12.39	2.58	10^4	4.93	10^7	7.29
SAP 2	10,000	461.87	66.15	10^6	131.30	10^{10}	196.46
DAP 2	10,000	944.14	333.97	10^6	665.89	10^{10}	1001.23
Racetrack 2	21,371	11.10	2.02	10^5	4.03	10^9	6.02

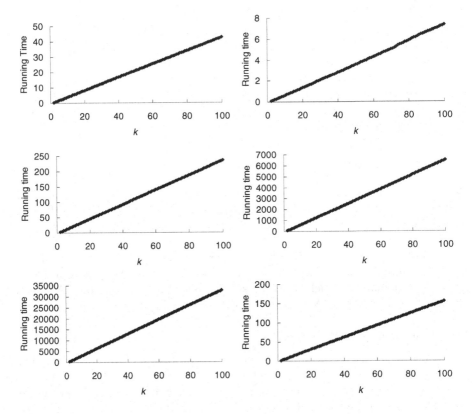

Fig. 1. Running time (seconds) of KBI when $k = 2, \ldots, 100$ on DAP 1, Racetrack 1, SAP 1, SAP 2, DAP 2, Racetrack 2 problems (left to right, top to bottom). The running times increase linearly in k for all problems.

5.2   The Scalability of KBI

In this experiment we investigate whether the KBI algorithm scales to large k values. We run KBI for $k = 100$ on the same set of problems, and record the elapsed times when it finishes generating the i-th best policy ($i = 2, \ldots, k$) of each problem. Figure 1 clearly shows that, for all problems KBI spends times linear in k when calculating k-th best policies. This experiment indicates that KBI has good scalability.

5.3   How k Best Policies Differ from the Optimal

We are also curious to know how the k best policies differ from the optimal policy. We analyze the list of k best policies calculated in the previous experiment, and compare the total number of different states, d, between each of these policies and the optimal policy π_1 for each problem. When d is small for a problem, it means that the k best policies are very similar to the optimal policy. This shows

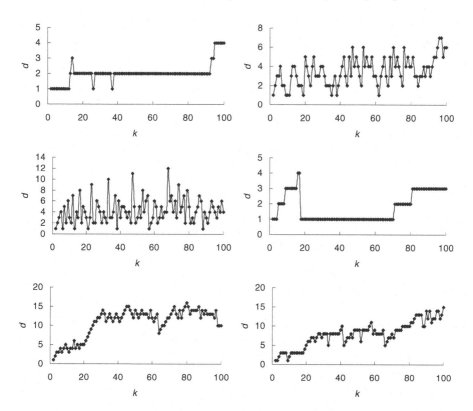

Fig. 2. The total number of different states between the k-th best policy and the optimal policy when $k = 2, \ldots, 100$ on DAP 1, Racetrack 1, SAP 1, SAP 2, DAP 2, Racetrack 2 problems (left to right, top to bottom). All k best policies are quite close to their π_1's.

that zmany good policies can be generated by a few small changes to the optimal policy. In other words, changes to few states can have very little impact on the optimality of the rest of the policy. When d is large, the optimal policy is more tightly coupled. When a sub-optimal action is chosen for a state, in order to get a good sub-optimal plan, changes to other states are usually also required.

We plot the d values for the k best policies on the same set of problems in Figure 2. These problems have relatively low d values (< 20 for all k). This shows that the k best policies are always quite close the the optimal policies. Some problems have relatively higher d values than others, namely SAP 1, DAP 2, and Racetrack 2, which means they have relatively tightly coupled optimal policies. As these problems are from diverse domains and of different sizes, it seems that the tightness of coupling of the optimal policies is probably problem-dependent.

6   Conclusions

This paper makes several contributions. First, we introduce the k best policy problem, and argue for its importance. Second, we prove a strong and useful theorem that the k-th best policy differs from some $m(< k)$-th best policy on exactly one state. Without that result, the brute-force algorithm for solving the k best policy problem (KBN) has time complexity exponential in k. Third, we propose a new algorithm, named k best policy improved (KBI), based on our theorem. We show that the time complexity of KBI is dominated by a computation linear in k. Fourth, we demonstrate that KBI outperforms KBN by an order of magnitude when $k = 2$ in most cases. The KBN algorithm does not scale to larger k values, as its running time increases exponentially in k. On the other hand, the running time of KBI increases only linearly in k. This makes KBI suitable for problems for which we want a long list of best policies. Fifth, we notice that the k best policies for different MDPs are quite similar to the optimal policies, though some problems' optimal policies are more tightly coupled than others'.

This is just the beginning of work on k best policies. There is much to be done in improving the algorithms, and in looking at applications-driven variants.

Acknowledgments

Dai was partially supported by Office of Naval Research grant N00014-06-1-0147. Goldsmith was partially supported by NSF grant ITR–0325063. We thank Mausam for helpful discussions on the problem.

References

1. Bellman, R.: Dynamic Programming. Princeton University Press, Princeton (1957)
2. Boutilier, C., Dean, T., Hanks, S.: Decision-theoretic planning: Structural assumptions and computational leverage. J. of Artificial Intelligence Research 11, 1–94 (1999)

3. Bonet, B., Geffner, H.: Planning with incomplete information as heuristic search in belief space. In: ICAPS, pp. 52–61 (2000)
4. Bresina, J.L., Dearden, R., Meuleau, N., Ramkrishnan, S., Smith, D.E., Washington, R.: Planning under continuous time and resource uncertainty: A challenge for AI. In: UAI, pp. 77–84 (2002)
5. Bresina, J.L., Jónsson, A.K., Morris, P.H., Rajan, K.: Activity planning for the mars exploration rovers. In: ICAPS, pp. 40–49 (2005)
6. Aberdeen, D., Thiébaux, S., Zhang, L.: Decision-theoretic military operations planning. In: ICAPS, pp. 402–412 (2004)
7. Musliner, D.J., Carciofini, J., Goldman, R.P., Durfee, E.H., Wu, J., Boddy, M.S.: Flexibly integrating deliberation and execution in decision-theoretic agents. In: ICAPS Workshop on Planning and Plan-Execution for Real-World Systems (2007)
8. Galand, L., Perny, P.: Search for compromise solutions in multiobjective state space graphs. In: ECAI, pp. 93–97 (2006)
9. Bryce, D., Cushing, W., Kambhampati, S.: Probabilistic planning is multiobjective! Technical Report ASU CSE TR-07-006 (June 2007)
10. Nielsen, L.R., Kristensen, A.R.: Finding the k best policies in finite-horizon mdps. European Journal of Operational Research 175(2), 1164–1179 (2006)
11. Nielsen, L.R., Pretolani, D., Andersen, K.A.: Finding the k shortest hyperpaths using reoptimization. Oper. Res. Lett. 34(2), 155–164 (2006)
12. Nielsen, L.R., Andersen, K.A., Pretolani, D.: Finding the k shortest hyperpaths. Computers & OR 32, 1477–1497 (2005)
13. Bertsekas, D.P., Tsitsiklis, J.N.: Neuro-Dynamic Programming. Athena Scientific (1996)
14. Howard, R.: Dynamic Programming and Markov Processes. MIT Press, Cambridge (1960)
15. Puterman, M.: Markov Decision Processes: Discrete Stochastic Dynamic Programming. John Wiley, New York (1994)
16. Littman, M.L., Dean, T., Kaelbling, L.P.: On the complexity of solving Markov decision problems. In: UAI, pp. 394–402 (1995)
17. Bonet, B.: On the speed of convergence of value iteration on stochastic shortest-path problems. Mathematics of Operations Research 32(2), 365–373 (2007)
18. Barto, A., Bradtke, S., Singh, S.: Learning to act using real-time dynamic programming. Artificial Intelligence J. 72, 81–138 (1995)
19. Wingate, D., Seppi, K.D.: Prioritization methods for accelerating MDP solvers. JMLR 6, 851–881 (2005)
20. Munos, R., Moore, A.: Influence and variance of a Markov chain: Application to adaptive discretization in optimal control. In: CDC (1999)
21. Bertsekas, D.P., Tsitsiklis, J.N.: An analysis of stochastic shortest path problems. Mathematics of Operations Research 16(3), 580–595 (1991)

Appendix

In order to prove Theorem 1, we consider the effects of changing a policy one state at a time.

Lemma 2. *Let M be an MDP, and π and π' be two policies for M that differ only on state s. Suppose that $V_\pi(s) \leq V_{\pi'}(s)$. Then $V_\pi(s_0) \leq V_{\pi'}(s_0)$. More strongly, if $s \in Policydesc(s_0, \pi)$ (which implies $s \in Policydesc(s_0, \pi')$) and $V_\pi(s) < V_{\pi'(s)}$, then $V_\pi(s_0) < V_{\pi'(s_0)}$.*

Proof. We know the values of $V_\pi(s)$ and $V_{\pi'}(s)$ are two unknown constants with $V_\pi(s) \leq V_{\pi'}(s)$. We write the two systems of linear equations with respect to π and π' by ignoring variables $V_\pi(s)$ and $V_{\pi'}(s)$ on the left hand side, and replacing them with their values whenever they are on the right hand side. We find the two systems of equations have the same set of coefficients, but the one given π has smaller or equal constant values on the right hand sides. If we solve the equations by factoring out all the variables on the right hand side iteratively, the same process as replacing a variable by its corresponding state's *influence* [20], we finally get the same value for all states where s is not a policy descendant given π', since all states' influences are the same in π and π', and a better value in π for all states where s is a policy descendant given π', since the influence of s on them is decreased (due to a smaller value), where the influence of other states remain unchanged. We call this property *monotonicity of influence*. This implies $V_\pi(s_0) \leq V_{\pi'}(s_0)$. Here, we actually proved a more general result, namely that $\forall s' \in \mathcal{S}[V_\pi(s') \leq V_{\pi'}(s')]$.

Lemma 3. *Let M be an MDP, and π and π' be two policies for M that differ only on state s. Suppose that $V_\pi(s_0) < V_{\pi'}(s_0)$. Then $V_\pi(s) < V_{\pi'}(s)$. More strongly, $\forall s' \in \mathcal{S}, [V_\pi(s') \leq V_{\pi'}(s')]$.*

Proof (Sketch). Suppose that $V_\pi(s) \geq V_{\pi'}(s)$.

We divide the states in $Policydesc(s_0, \pi')$ into two subsets: (1) *policy ancestors of s given π'*, the set of states where s is a policy descendant given π', and (2) *non-policy ancestors of s given π'*, the complement of (1).

We claim that the values of the non-policy ancestors of s given π' are the same as those given π. This is because the values of those states do not depend on s or any policy ancestors of s given π', so their values are not influenced by any potential value changes caused by s. For policy ancestors of s given π', their values cannot be improved, by the monotonicity of influence. Because their coefficients remain unchanged while the constants (values of non-policy ancestors of s given π' and value of s) are equal or larger. This contradicts the assumption that $V_\pi(s_0) < V_{\pi'}(s_0)$. Now, we know that $V_\pi(s) < V_{\pi'}(s)$. From Lemma 2 we have that $\forall s' \in Policydesc(s_0, \pi') [V_\pi(s) \leq V_{\pi'}(s)]$.

Lemma 4. *Let M be an MDP, and π and π' be two policies for M that differ only on two states s^1 and s^2. Suppose that $V_\pi(s_0) \leq V_{\pi'}(s_0)$. Consider the following two policies π^1, π^2 obtained from by starting with π by replacing exactly one distinct action each from $\pi(s)$, $s \in \{s^1, s^2\}$, with the corresponding $\pi'(s)$. Without loss of generality, suppose $\pi^i(s^i) = \pi'(s^i)$. Then π^1 and π^2 cannot both have larger initial state values than π' does.*

Proof (Sketch). For either s^i, if s^i is not a policy descendant of s_0 given π or π', then $V_{\pi'}(s_0) = V_{\pi^i}(s_0)$, and we're done.

Now suppose $V_{\pi'}(s_0) < V_{\pi^i}(s_0)$ for $i = 1, 2$. From Lemma 3, we have

$$\forall s' \in \mathcal{S}[V_{\pi'}(s') \leq V_{\pi^1}(s')], \text{ and } V_{\pi'}(s^2) < V_{\pi^1}(s^2), \tag{2}$$

$$\forall s' \in \mathcal{S}[V_{\pi'}(s') \leq V_{\pi^2}(s')], \text{ and } V_{\pi'}(s^1) < V_{\pi^2}(s^1). \tag{3}$$

There are three cases. Case 1: Neither s^1 nor s^2 is a policy descendant of the other given π. From Equation 2 we know $V_{\pi'}(s^2) < V_{\pi^1}(s^2) = V_\pi(s^2)$, as the values of all policy descendants of s_2 given π^1 and π are the same, and $\pi^1(s^2) = \pi(s^2)$. From Equation 3 we know $V_{\pi'}(s^1) < V_{\pi^2}(s^1) = V_\pi(s^1)$ for the same reason. Then from the monotonicity of influence together with all derived inequalities, we know $V_{\pi'}(s_0) < V_\pi(s_0)$. A contradiction.

Case 2: s^2 is a policy descendant of s^1 given π, but s^1 is not a policy descendant of s^2 given π (or vice versa). From Equation 2 we first know $V_{\pi'}(s^2) < V_{\pi^1}(s^2) = V_\pi(s^2)$. From Equation 3, and $V_{\pi'}(s^2) < V_\pi(s^2)$, by the monotonicity of influence we know $V_{\pi'}(s^1) < V_\pi(s^1)$. Then, from the monotonicity of influence together with all derived inequalities, we know $V_{\pi'}(s_0) < V_\pi(s_0)$. A contradiction.

Case 3: s^1 and s^2 are both policy descendants of each other given π'. From both Equations 2 and 3 and the monotonicity of influence we can prove $V_{\pi'}(s^1) < V_\pi(s^1)$ and $V_{\pi'}(s^2) < V_\pi(s^2)$. Then from the monotonicity of influence together with all derived inequalities, we know $V_{\pi'}(s_0) < V_\pi(s_0)$. A contradiction.

Lemma 5. *Let M be an MDP, and π and π' be two policies for M that differ only on m states $s^1, s^2, \ldots, s^m$, $m > 1$. Suppose that $V_\pi(s_0) = V_{\pi'}(s_0)$. Consider the 2^m distinct policies π^T, $T \subseteq \{s^1, s^2, \ldots, s^m\}$ that agree with π on all states not in T, and agree with π' on T. Then for at least one such T of size 1, $V_{\pi^T}(s_0) \leq V_{\pi'}(s_0)$.*

This Lemma can be proved inductively from Lemma 5.

Note that a fundamental assumption underlying dynamic programming algorithms for MDPs is: If M is a MDP and π a non-optimal policy (in the sense of having a non-optimal value function), then there is some $s \in \mathcal{S}$ and $a \in \mathcal{A}$ such that $v_\pi(s) > C(s, a) + \gamma \sum_{s' \in \mathcal{S}} T_a(s'|s) \cdot v_\pi(s')$. Bertsekas and Tsitsiklis showed that this holds for stochastic shortest path problems, when $\gamma = 1$ [21]. Their proof can be extended.

Lemma 6. *If If $V_\pi(s_0)$ is not optimal, there must be an $s^+ \in Policydesc(s_0, \pi)$ and $a \in \mathcal{A}$ such that $v_\pi(s^+) > C(s^+, a) + \sum_{s' \in \mathcal{S}} T_a(s'|s^+,) \cdot v_\pi(s')$. If we let $\pi'(s) = \pi(s)$ for $s \neq s^+$, and let $\pi'(s^+) = a$, then $V_{\pi'}(s_0) < V_{\pi(s_0)}$.*

Proof (Theorem 1). Let M be an MDP, and $\Pi_i = \{\pi_1, \ldots, \pi_i\}$ be the list of i best policies, for $i \leq k$. We claim that, for $k > 1$, there is some $j < k$ and state s such that π_j differs from π_k exactly on s.

If $V_{\pi_k}(s_0) = V_{\pi_1}(s_0)$, the theorem follows from Lemma 5.

If $V_{\pi_k}(s_0) > V_{\pi_1}(s_0)$, the theorem follows from Lemma 6.

New Hybrid Recommender Approaches: An Application to Equity Funds Selection

Nikolaos F. Matsatsinis and Eleftherios A. Manarolis

Department of Production Engineering and Management Decision Support Systems
Laboratory, Technical University of Crete, Crete, Greece
{nikos,manarolis}@ergasya.tuc.gr

Abstract. Recommender Systems and Multicriteria Decision Analysis remain two separate scientific fields in spite of their similarity in supporting the decision making process and reducing information overload. In this paper we present a novel algorithmic framework, which combines features from Recommender Systems literature and Multicriteria Decision Analysis to alleviate the sparsity problem and the absence of multidimensional correlation measures. We apply the introduced framework for recommending Greek equity funds to a set of simulation generated investors. The proposed framework treats MCDA's algorithm UTADIS as a content - based recommendation technique which, in conjunction with collaborative filtering results in two Hybrid Recommendation approaches. The resulting approaches manage to outperform the separate application of the UTADIS and collaborative filtering methods in terms of recommendation accuracy.

Keywords: Hybrid Recommender, Collaborative Filtering, UTADIS, equity funds.

1 Introduction

The development of recommender systems is a phenomenon of the last decade. A vast amount of commercial applications have been presented during this period of time among which Amazon.com, Netflix, Last.Fm and others, being arguably the most familiar commercial examples. The main purpose of these applications is to help the active user deal with the increasing information overload by supporting and facilitating the decision making process. The user is thus absolved from having to consider every alternative separately in order to decide which is best for him. Consequently the decision making process is accelerated.

Our model describes an interdisciplinary investment recommendation system in which we are exploring the possibility of combining and improving two methods (Recommender Systems - Collaborative Filtering and Multicriteria Decision Analysis – UTADIS) that even though they confront the same problem (recommending items and/ or classify them into predefined classes) they have been applied separately. In brief, collaborative filtering recommender systems are based on the philosophy of searching correlations among users or items and take advantage of these correlations in generating recommendations. On the other hand UTADIS and utility based recommenders base their recommendations on a user's utility function which describes his behavioral pattern.

F. Rossi and A. Tsoukis (Eds.): ADT 2009, LNAI 5783, pp. 156–167, 2009.
© Springer-Verlag Berlin Heidelberg 2009

The academic contribution of this paper extends from the decision science in general and its applications of recommenders and MCDA to the financial research sector, with portfolio selection and management serving as the major focusing area.

The augmentation of Collaborative Filtering algorithm with features from an MCDA algorithm that is designed to base its recommendations on multiple criteria is considered to introduce rationale in Collaborative Filtering recommendations.

We introduce the shift from the modeling of the imaginary average user – a utopian aspiration of doubtful usefulness - towards per user application of the UTADIS method. The financial sector and investment recommendation in particular is not excluded from the need for personalization. We try to address the literature gap present, regarding applications of Recommender Systems in mutual fund portfolio selection and the lack of a personalized approach concerning investing preferences.

In the following section we present a brief literature review on user to user memory based collaborative filtering systems. We also refer to the limitations presented during their application. The philosophy of the Multicriteria Decision Analysis' method UTADIS is presented in Section 3, along with some of its applications in the portfolio recommendation problem. In section 4 we thoroughly present the proposed algorithmic framework that resulted in the introduction of the hybrid approaches Hybrid 1 and Hybrid 2. In the fifth section we apply the new approaches to a set of Greek equity funds and measure their performance. In the final section the conclusions are presented along with the extent to which the initial goal is achieved. Future research proposals in the field are also included.

2   Recommender Systems

Recommender systems, having their roots in approximation theory, data mining, forecasting theory, emerged as a separate research filed in the mid 1990's. In most of the cases they aim at calculating the possibility of an item, not having been rated from the active user, being one of interest to him. In literature there are many alternative classifications for the recommender systems and the implemented algorithms [1-3].

We briefly present collaborative filtering algorithm which is considered the most popular and is also applied in our approach.

2.1   Collaborative Filtering

Collaborative filtering approaches are assigned the definition of a user "neighborhood" composed of users with similar tastes. Recommendations are produced based on the notion that items liked by the active user's neighborhood will also be liked by the active user. Collaborative filtering systems have the potential to provide filtering based on complex attributes, such as quality, taste, or aesthetics [4], thus being applicable to a wide field range. Collaborative filtering algorithm takes into consideration previous knowledge regarding user tastes, provided by the user explicitly or implicitly, in defining behavioral similarities. In order to measure similarity among users collaborative filtering algorithms use statistical metrics such as Pearson's linear correlation [5], cosine similarity [4], [6] and distance measures adjusted for measuring correlation e.g. Chebyshev distance [7].

Following user correlation measurement a "neighborhood" is defined by selecting users with behavior similar to the behavior of the active user. In order to generate recommendations the collaborative filtering algorithm calculates the weighted average or the adjusted weighted average of the neighbors' ratings.

In the years following the first collaborative filtering systems there have been numerous papers focusing on limitations resulting from their application [8-10]. New users need to rate a relatively large amount of items before the algorithm can provide them with accurate recommendations and this is due to the fact that correlation measures based on a small amount of data can often be misleading. This issue is commonly referred to as cold - start problem [11]. A similar issue emerges when a new item is introduced (first rater problem).

In addition to the cold start and first rater problems, data sets used in recommender systems are sparse. This refers to the general lack of sufficient ratings to generate accurate recommendations. In commercial applications it is common phenomenon for the available ratings to be less than 1% of the total items (e.g. Amazon) [6]. When applying collaborative filtering approaches to such a small amount of data it is precarious to generate recommendations when recommendation accuracy is considered. In extension to this problem we find the lack of available peers for users with "uncommon" tastes. Correlation measures are led to false neighborhood considerations, thus generating inaccurate recommendations. This is also referred to as gray sheep problem [9].

Furthermore collaborative filtering is prone to critique because of its function as black box, which prevents the user form understanding the process that generates a recommendation and its context. This consequently reduces recommendation explainability and possibly the user's confidence [12]. The augmentation of the recommendation generation process with a model that contributes ratings that result from each item's valuation on a set of criteria is considered introduce rationale. In order to serve this functionality and alleviate the previously presented issues in this paper we combine Multicriteria Decision Analysis presented in the next section with simple collaborative filtering.

2.2 Hybrid Recommenders

Hybrid recommenders combine two or more recommendation techniques in order to achieve better recommendations. In most of the cases collaborative filtering is the element that is combined with another method. The study of hybrid recommender systems requires an adequate classification framework. Burke provides us this framework that classifies hybrids according the way the participating techniques are combined [1].

3   MCDA

Multicriteria Decision Analysis is a field from Operational Research that has known rapid development in the last 40 years. This set of techniques resulted from the realization that single criterion analysis in decision aiding contradicts real life decision making. The aim of Multicriteria decision analysis is to represent the decision maker's way of thinking in order to support him in the decision making process by indentifying pareto optimal solutions.

There is an extensive literature available of MCDA applications that model and solve financial problems. In their literature review Steuer and Na study and classify 265 papers that deal with financial decisions [13]. Among those there are 77 that focus exclusively on portfolio management. The problem solved in these studies is similar to the one at hand, meaning the selection of investment alternatives that maximize an investor's utility without being limited by the dipole "risk - return".

None of the applications studied by Steuer and Na utilizes the additional information generated by the correlations present among users in the investment community. These studies focus merely on the behavioral replication of single user preferences.

3.1 UTADIS

UTADIS (UTilites Additives Discriminantes) [14, 15], which is a MCDA preference disaggregation approach, is often used in addressing multicriteria problems including country risk analysis, financial institutions' credit risk analysis, portfolio selection etc. The methodology analyses the set of evaluation criteria present in the decision making process, leading to a nonlinear mathematical framework combining these criteria in an accurate representation of decision maker's previous behavior. The method is similar to other commonly used statistical and econometric classification techniques (e.g., discriminant analysis, logit, probit, etc.) as well non parametric techniques (neural networks, machine learning, fuzzy sets etc).

The similarities between preference disaggregation and machine learning in particular are summarized in that they assign labels (classes) to the examples (decision alternatives) under consideration, based on a set of attributes that define the set of examples. The fundamental difference though, is that machine learning applications do not consider ordered classes nor ordered attributes (decision criteria) in contrast to UTADIS which is an algorithm dealing with monotone problems. Among the exceptions to this rule we find a decision tree algorithm presented by Potharst and Bioch that addresses monotone learning problems [16], the technique presented by Frank and Witten for replacing nominal attributes constituting the input space with binary attributes, thus treating them as ordinal [17] and the approach for ordinal classification presented by Frank and Hall, which in combination with the previous technique could also address monotone problems [18]. Apart from the above and due to the different approach adopted, machine learning seems to lack applications for ordinal classification problems (sorting).

In order to briefly present UTADIS formulation let the training sample, on which the utility function is calculated, consist of m alternatives described over a set of g evaluation criteria. The alternatives under consideration are classified into C ordered classes (C_{k-1} is preferred to C_k) based on the decision maker's stated opinion. The additive utility model, which is developed through the UTADIS method has the following form:

$$U(\boldsymbol{g}) = \sum_{i=1}^{n} p_i u_i(g_i) \tag{1}$$

where, $U(\boldsymbol{g})$ is the total utility for an alternative on the set of the evaluation criteria $\boldsymbol{g}=\{g_1,...,g_n\}$ and $u_i(g_i)$ is the marginal utility of the alternative on the evaluation criterion g_i. The global utility provides an overall evaluation of the performance of the alternative, whereas the marginal utilities provide the partial evaluations of the alternative's performance on each individual evaluation criterion. Details on the solution of the linear programming (LP) problem are given in [15].

The majority of applications utilizing MCDA's preference disaggregation techniques attempt to model someone's expertise in the sector the problem originated from in order to generate commonly accepted solutions for the entire users community. In this paper UTADIS is applied on a per user basis in an effort to separately model a user's taste and consequently come up with a set of utility functions able to provide personalized recommendations.

Utility based algorithms including UTADIS present limitations that Burke discusses thoroughly in his work [1]. In particular Burke argues that utility based systems require immediate interaction with the user in order to obtain his utility function, while in the meantime they present a static character with limited ability to adjust to users evolving preferences. UTADIS on the contrary does not require from the user to explicitly input his utility function since it is able to obtain it by analyzing his previous decisions, such as every technique from MCDA's preference disaggregation. As a result of the indirect user involvement in calculating a utility function the second limitation presented by Burke does not hold for UTADIS either, since it is possible to run the UTADIS algorithm and obtain an updated utility function once a critical amount of new ratings and/ or items have been imported in the system.

4   Proposed Approach

In this paper we follow the guidelines provided by Adomavicius and Kwon on the calculation of multicriteria user to user correlation measurement to introduce a two stage approach [7]. In the first stage UTADIS generates additive utility functions for every user based on the alternative classifications available from the simulation process. These utility functions are then combined with the common collaborative filtering scheme in two ways in order to enhance correlation's computation. Figure 1 graphically represents our approach.

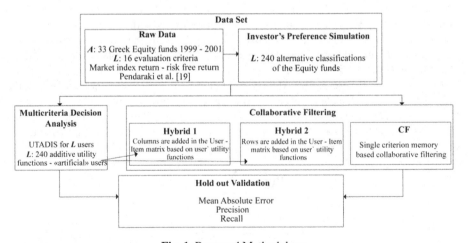

Fig. 1. Proposed Methodology

4.1 Data Set

The data set is divided into two subsets. The first subset has been also used in previous work [19] and it constitutes of 33 equity funds, which are analyzed for a three year period 1999-2001. Each equity fund is divided into three alternatives, one for every year of the analysis, thus resulting in 99 investment alternatives. The increase in the set of the alternatives evidently increases data sparsity but on the other hand allows a more realistic approach of investment habits (active investing).

The buying prices of the equity funds were used to calculate a set of indexes that are considered as participating in the investment decision. Furthermore in order to obtain a benchmark we used data to measure market performance, namely the Athens Stock Exchange index. The calculated indexes are: equity fund's annual return, average return, standard deviation of return, geometric mean of excess return over benchmark, value at risk, Sharpe ratio, Modigliani index, information ratio, beta, Treynor index, Jensen's alpha coefficient, Treynor and Mazuy's alpha coefficient, Treynor and Mazuy's gamma coefficient, Henriksson and Merton's alpha coefficient, Henriksson and Merton's gamma coefficient, Treynor and Black ratio. The second subset is composed of investment behaviors. The transformation of these behaviors into classifications for the alternatives is necessary in order to be able to apply both collaborative filtering and UTADIS algorithms. Since it was not possible to obtain real investment behavior, due to the lack of such publicly available data set and constraints regarding its purchase (increased cost - customer confidentiality), we were forced to simulate this behavior following the procedure described below.

Simulation of Investment Behavior. The simulation process is divided in two phases. In the first phase each of the 16 indexes that form the decision criteria is selected sequentially. Every equity fund is classified in one of the three predefined classes based on its performance on the selected index. The item's per criterion classification is based on 5 predefined scenarios. In order to form each scenario we need to identify 2 threshold points that separate the 3 classification categories. If an alternative's performance is above the r_1 threshold which is the separation point between the C_1 and C_2 categories ($C_1 \succ C_2$) then the alternative is classified in the C_1 category or it is considered to have obtained a rating of 1 and so on. By the end of the first phase we have managed to construct a set S numbering 80 (16 criteria x 5 scenarios) alternative classifications.

In the next phase of the simulation we came up with 3 additional classifications for every classification $s_i \in S$ that was generated in the first phase, simply by adding noise. This increased the simulation fairness and resulted in a set L of 240 alternative classifications. At this point we should clarify that classifying an alternative in one of the three possible categories is considered identical to giving an alternative a rating in the three point scale and from now on there will be no semantic distinction. Furthermore the ratings assigned to every alternative are considered the expression of an individual's investing preferences. For the following steps of the analysis and the application of the collaborative filtering and UTADIS algorithms we consider that these alternative classifications represent 240 different user investment behaviors for the 99 available investment alternatives, thus resulting in a total of 23760 available ratings.

In order to assign meaningful interpretation to the three classification classes we should underline that they consist of buying and selling activities of particular individuals in the market. Rating an alternative in C_1 denotes buying suggestion for the specific equity fund. Items classified in C_2 require further investigation, while the C_3 class denotes suggestion for selling or avoid buying the relevant equity fund.

4.2 Hybrid 1

For the Hybrid 1 approach each user's additive utility function takes part in the generation of a rating for the complete set of alternatives. The calculated rating is then integrated in the user - item matrix. This process leads to the introduction of "artificial" users that behave exactly as the corresponding utility function suggests. In a similar manner Sarwar et al. used feature augmentation to improve email filtering [20]. They introduced filtering bots that acted as artificial users by evaluating emails based on a set of criteria and then contributed these ratings into the system for the collaborative filtering algorithm to utilize. Our approach further improves the approach proposed by Sarwar et al. since the behavioral pattern of every introduced "artificial" user in our approach is determined by a utility-based model that is developed through the UTADIS algorithm and is based on (simulated) observed investment behavior. Due to the utilization of UTADIS it is possible to develop non linear utility functions that, in conjunction with the previous feature, can equip the recommender system with more credible recommendations. The number of columns in the user - item matrix is increased by L, where L is the number of additive utility functions generated from UTADIS thus doubling in width. The graphical representation of the modified user – item matrix is given in Figure 2. Ratings available before the integration are shown in the light pattern, while the dark pattern indicates data generated by the additive utility functions. Following Burke's classification of hybrid recommender techniques our approach can be classified as feature augmentation hybrid, since the ratings resulting from UTADIS perform as input to simple memory based collaborative filtering.

Every "artificial" user is considered to have rated the entire item set based on the intrinsic utility function of the real user corresponding to him. This means that, for example, user UTADIS$_1$ rates items based on user U_1 utility function as it has been calculated during the MCDA part of the introduced framework. The incorporated columns significantly alleviate the sparsity problem (including cold-start and first rater problems), since there exist no missing data, allowing a more thorough computation of user to user similarity. Correlation among users is also more cohere, especially in sparse data set applications, due to the introduction of multicriteria ratings that reduce inconsistency by being subject to a mathematical model that takes into consideration an item's performance on predefined evaluation criteria.

Future studies could change the leverage for the ratings contributed by "artificial" users following a certain criterion such as the fitting index of the user's calculated utility function and the utility function implied by the real data.

4.3 Hybrid 2

The second modification of the common collaborative filtering approach is applied by incorporating every item's marginal utilities - which were generated based on each item's performance on the evaluation criteria - in the user - item matrix. The proposed technique is classified according to the classification scheme for hybrid recommender by Burke, as a "feature combination". This is because apart from importing the actual ratings provided by every user to the collaborative filtering algorithm, we also incorporate the deeper knowledge generated through the application of the UTADIS algorithm regarding user investment preferences. The Hybrid 2 technique can be also classified as "meta – level" hybrid, since during the conjunction of the participating techniques the whole model generated by UTADIS is imported in the collaborative filtering algorithm through every alternative's marginal utilities. The graphical representation of the modified user - item matrix is also given in Figure 2.

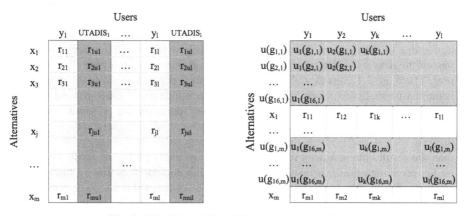

Fig. 2. Hybrid 1 and Hybrid 2 user – item matrices

In the same way as in the case of graphical representation of the Hybrid 1 technique the previously available information (ratings $r_{i,l}$) are shown in light pattern, while the rest of the matrix originates from the application of UTADIS ($u_l(g_{i,j,l})$). The display of the ratings generated during simulation and the marginal utilities in a single matrix is only to improve comprehension. Taking under consideration that marginal utilities are on different scale than real ratings we are forced to regard them as two separate user – item matrixes in order to calculate correlation among users. The correlations that result from these two matrices are equally leveraged in order to obtain the final correlation measure.

The recommendation process though utilizes only the real ratings excluding the imported data. This is contrast to the Hybrid 1 approach where the imported ratings participated in the recommendation process.

5 Application Results

Complying with the previously described framework we initially solved the UTADIS algorithm, including post - optimality analysis, for every available user and every formed partition (40,800 runs), in order to import the generated information in the user - item matrix. Based on the expanded matrices we estimated the Pearson correlation coefficients for the Hybrid 1 and Hybrid 2 approaches. We selected the 10 most similar users to the active user in order to generate ratings by utilizing the adjusted weighted average formula. The performance on both the training and test samples was recorded for every partition of the available data. In order to measure performance we used 3 metrics; Mean Absolute Error (MAE), precision (P) and recall (Re). Thorough presentation of these metrics can be found in previous papers dealing with recommendation accuracy measurement [21]. Following the precision measurement we conducted the Wilcoxon signed rank test to determine if the differences observed are statistically significant. We chose the Wilcoxon signed rank test because the examined sets as a whole were not validated as following normal distribution when the Lilliefors normality test was ran.

In the presented results we decided to exclude the precision and recall metrics since precision was highly correlated to MAE, in fact there was no change in the best performer for any partition examined, and there was no evidence of statistical significance for the obtained results. Recall on the other hand should be given a special notice due to the fact that "good" items generated by the simulation are only a small part of the data set which in turn results in a small number of relevant alternatives per user. There are 2.12 relevant alternatives on average per user in the test sample, which is the reason that prevents us from obtaining an accurate recall measurement.

The MAE results for every method and every partition are presented in Table 1 along with the additional information from the statistical significance testing. We marked with ● every result that was significantly different (worse) at the 5% level than the result obtained from the best performer of every partition. Table 2 elaborates on the error measurement results and presents the comparisons among the examined methods. Every cell number refers to the number of times that its corresponding column method outperforms (significantly) the method in the associated row.

Table 1. Mean Absolute Error/ partition (test sample), ● indicates statistically significant difference over the performance of the best performing method (marked in bold)

	20%	40%	60%	80%	90%
Hybrid 1	0.2229	**0.2221**	**0.2320**	**0.2777**	0.2777
Hybrid 2	**0.2216**	0.2327 ●	0.2564 ●	0.2978 ●	0.2893 ●
CF	0.2239	0.2337 ●	0.2567 ●	0.2928 ●	**0.2699**
UTADIS	0.2522 ●	0.2539 ●	0.2661 ●	0.2977 ●	0.3058 ●

Table 2. Pair-wise comparison of mean absolute error performance, number indicates how often method in column (significantly) outperforms method in row

	Hybrid 1	Hybrid 2	CF	UTADIS
Hybrid 1	–	1 (0)	1 (0)	0
Hybrid 2	4 (4)	–	1 (1)	1 (1)
CF	3 (3)	3 (1)	–	0
UTADIS	5 (5)	4 (2)	5 (3)	–
Total	12 (12)	8 (3)	7 (4)	1(1)

It is evident that the Hybrid 1 approach is the winner among the examined methods in our paper, since it managed to (significantly) improve recommendation accuracy in comparison to the majority of the partitions. UTADIS seems to be the worst performer but the augmentation of collaborative filtering algorithm with some of its features and the resulting multicriteria correlation measurement still managed to give encouraging results.

Based on the overall results from Table 2 we can safely state that the incorporation of MCDA features in collaborative filtering improves performance in terms of (statistically significant) MAE. For relatively sparse data single criterion memory based collaborative filtering surpassed the introduced Hybrid 2 approach even when considering the statistical significance of the result. For more cohere datasets though, Hybrid 2 demonstrates higher recommendation accuracy.

6 Conclusions

The major obstacle that we faced in realizing this work was the absence of a database containing real life data on investment behavior (user rankings, evaluation criteria). We were forced to simulate investor's behavior and reuse a dataset from a previous work in the field of mutual fund recommendation [19]. We used 16 evaluation criteria to apply UTADIS, a preference disaggregation model that generates a utility function that is both able to reproduce classifications that an active user has provided and recommend alternatives of interest based on the user's generated utility function. We integrated UTADIS in collaborative filtering algorithm thus introducing new hybrid approaches that benefit from the utilization of features from both academic fields and manage to outperform their separate application in the majority of metrics examined.

We showed that optimal performance depends on the level of sparsity and through this the need for extensive analysis and thorough presentation of an algorithm's results over different partitions of the initial data is mandatory. The analysis showed that the Hybrid 1 approach performed better than the rest of the examined methods in the majority of data sparsity levels, while Hybrid 2 managed to outperform collaborative filtering only when less sparse data sets were considered. Still the integration of marginal utilities in the similarity measurement which was introduced through Hybrid 2 approach did not perform as expected and the same applies to UTADIS, which was used in our approach for augmenting collaborative filtering.

Future work could focus on applying goal programming and non linear programming techniques allowing the thorough analysis of the parameters of each approach. This would benefit potential e-commerce applications because it provides the ability to select the approach and underlying parameters that could guarantee prediction accuracy for every sparsity level.

The search for a framework to combine the proposed techniques with MCDA, collaborative filtering or even other recommendation techniques could lead into new hybrids further improving recommendation accuracy. Such a hybrid would emerge from the solution of a linear problem that leverages the participating techniques based on their previously achieved recommendation accuracy. The introduced framework is suitable for application in fields that traditionally attract recommendation algorithms attention and present less computational and conceptual complexity than portfolio selection. Furthermore the demonstrated improvement in the overall performance of the UTADIS algorithm suggests that the new approaches are applicable to numerous research fields previously dominated by operational research among which country risk analysis, financial institutions' credit risk analysis, assessment of corporate performance and viability, stock evaluation, business failure prediction and others.

References

1. Burke, R.: Hybrid recommender systems: Survey and experiments. User Model. User-Adapt. Interact. 12, 331–370 (2002)
2. Resnick, P., Varian, H.R.: Recommender Systems. Communications of the ACM 40, 56–58 (1997)
3. Terveen, L., Hill, W.: Human-Computer Collaboration in Recommender Systems. In: Carroll, J. (ed.) Human Computer Interaction in the New Millenium, pp. 487–509. Addison-Wesley, New York (2001)
4. Breese, J.S., Heckerman, D., Kadie, C.: Empirical Analysis of Predictive Algorithms for Collaborative Filtering. In: 14th Conference Uncertainty in Artificial Intelligence (1998)
5. Resnick, P., Iakovou, N., Sushak, M., Bergstrom, P., Riedl, J.: GroupLens: An open architecture for Collaborative Filtering of Netnews. In: Computer Supported Cooperative Work Conf. (1994)
6. Sarwar, B., Karypis, G., Konstan, J., Riedl, J.: Item-based Collaborative Filtering Recommendation Algorithms. In: 10th International World Wide Web Conference (2001)
7. Adomavicius, G., Kwon, Y.: New Recommendation Techniques for Multicriteria Recommendation Systems. IEEE Intelligent Systems, 48–55 (2007)
8. Balabanovic, B., Shoham, Y.: Fab: Content-based, Collaborative Recommendation. Comm. ACM, 66–72 (1997)
9. Claypool, M., Gokhale, A., Miranda, T., Murnikov, P., Netes, D., Sartin, M.: Combining Content-based and Collaborative Filters in online Newspaper. In: ACM SIGIR 1999 Workshop Recommender Systems: Algorithms and Evaluation (1999)
10. Lee, W.S.: Collaborative Learning for Recommender Systems. In: International Conference Machine Learning (2001)
11. Schein, J.B., Popescul, A., Ungar, L.H., Pennock, D.M.: Methods and Metrics for Cold-start Recommendations. In: 25th Annual International ACM SIGIR Conference (2002)
12. Jonathan, L.H., Joseph, A.K., John, R.: Explaining collaborative filtering recommendations. In: Proceedings of the 2000 ACM conference on Computer supported cooperative work. ACM, Philadelphia (2000)
13. Steuer, R.E., Na, P.: Multiple criteria decision making combined with finance: A categorized bibliographic study. European Journal of Operational Research 150, 496–515 (2003)
14. Devaud, J.M., Groussaud, G., Jacquet-Lagreze, E.: UTADIS: Une méthode de construction de fonctions d'utilité additives rendant compte de jugements globaux. In: European Working Group on Multicriteria Decision Aid, Bochum (1980)
15. Doumpos, M., Zopounidis, C.: Multicriteria Decision Aid Classification Methods. Kluwer Academic Publishers, Dordrecht (2002)
16. Potharst, R., Bioch, J.C.: Decision trees for ordinal classification. Intell. Data Anal. 4, 97–111 (2000)
17. Frank, E., Witten, I.H.: Making Better Use of Global Discretization. In: Proceedings of the Sixteenth International Conference on Machine Learning. Morgan Kaufmann Publishers Inc., San Francisco (1999)

18. Frank, E., Hall, M.: A Simple Approach to Ordinal Classification. In: Flach, P.A., De Raedt, L. (eds.) ECML 2001. LNCS (LNAI), vol. 2167, p. 145. Springer, Heidelberg (2001)
19. Pendaraki, K., Zopounidis, K., Doumpos, M.: On the construction of mutual fund portfolios: A multicriteria methodology and an application to the Greek market of equity mutual funds. European Journal of Operational Research 2, 462–481 (2005)
20. Sarwar, B.M., Konstan, J.A., Borchers, A., Herlocker, J., Miller, B., Riedl, J.: Using filtering agents to improve prediction quality in the GroupLens research collaborative filtering system. In: Proceedings of the 1998 ACM conference on Computer supported cooperative work. ACM, Seattle (1998)
21. Herlocker, J.L., Konstan, J.A., Terveen, L.G., Riedl, J.T.: Evaluating Collaborative Filtering Recommender Systems. ACM Transactions on Information Systems, 5–53 (2004)

A Prescriptive Approach for
Eliciting Imprecise Weight Statements
in an MCDA Process

Mona Riabacke[1], Mats Danielson[1], Love Ekenberg[1], and Aron Larsson[1,2]

[1] Dept. of Computer and Systems Sciences, Stockholm University and
Royal Institute of Technology, Forum 100, SE-164 40 Kista, Sweden
[2] Dept. of Information Technology and Media, Mid Sweden University,
SE-851 70 Sundsvall, Sweden,
{mona,mad,lovek,aron}@dsv.su.se

Abstract. In this article, we discuss decision making involving multiple objectives (MCDA) and especially the lack of more prescriptively useful elicitation methods for weights within MCDA. We highlight the discrepancy between how elicitation is handled in current decision analysis applications and the abilities of real decision-makers to provide what is required from them. Based on theory and highlighted problems with current methods, we propose a novel approach for weight elicitation which relaxes the need for numeric preciseness from decision-makers and reduces some of the practical issues related to such processes. The method is tested in a comparative study, as well as employed in a real-life case study.

Keywords: Multi-criteria decision making, Elicitation process, Imprecise criteria weights.

1 Introduction

Research on quantitative decision making has proceeded from the study of decision theory founded on single criterion decision making towards decision support for more realistic decision making situations with multiple, often conflicting, criteria, and more than one decision-maker. In particular, Multi-Criteria Decision Making (MCDA) stands out as a promising category within decision support methods. MCDA can provide the decision-makers with a better understanding of the trade-offs involved in a decision, e.g., between economic, social and environmental objectives (criteria). The number of MCDA applications has increased during the last decade, but behavioural issues have not received much attention within this field of research, yet the identification of such problems and the call for research on behavioural issues have been recognized for a long time [1].

Traditional decision analysis methods are heavily influenced by normative theories and expected utility models for the guidance of rational choice. A widely discussed practical difficulty involved in the use of such models for decision making is the difficulty of assessing precise probabilities and utilities (cf., e.g., [2]). In decisions involving multiple objectives, there is also the need to make value tradeoffs to indicate

F. Rossi and A. Tsoukis (Eds.): ADT 2009, LNAI 5783, pp. 168–179, 2009.

the relative desirability of achievement levels on one objective in comparison to others. The relative importance of the different criteria is a central concept in MCDA and many methods for deriving these weights from preference statements exist. Like probability and utility elicitation, the elicitation of weights is a cognitively demanding task (cf., e.g., [3]) and the elicited values can be heavily dependent on the method of assessment (cf., e.g., [4]). Inconsistencies in assessed weights can occur both within a method for eliciting weights as well as between different weighting methods. However, it seems difficult to reach definite conclusions from comparative studies between different methods for weight elicitation (cf., e.g., [5, 6]), which may be due to the inherent complexity of evaluating the quality of results.

1.1 Problem Background

Although elicitation has been an area of concern for quite some time (cf., e.g., [7, 8, 9, 10, 11]), there are still no generally accepted methods for elicitation and the process of eliciting adequate quantitative information from people is still one of the major challenges facing research within the field. The current demand for numeric precision within elicitation is unrealistic for several reasons. People have problems judging exact values (cf., e.g., [10]), and their preferences and beliefs are not naturally represented in this fashion. Several studies have pointed at the difficulties in expressing such values with numeric preciseness (cf., e.g., [12]), which poses problems when the required values are point estimates. Barron and Barrett [13] state that the elicitation of exact weights demands an exactness which may not exist in the mind of the decision-maker, and von Winterfeldt and Edwards [14] argue that "the precision of numbers is illusory". As pointed out in [15], there is a need to adapt the elicitation process to the behaviours of real users in a prescriptive manner. Also, the heuristics and biases programme initiated by Tversky and Kahneman [7] illustrates many of the systematic deviations from traditional theoretical expectations inherent in our thinking, judgment and memory, which cause problems for elicitation processes. Moreover, the framing of the problem (its formulation) often has great impact on people's preferences [8]. Preferences can vary depending on subtle differences in presentation (of mathematically equivalent choices), like the order of the probabilities presented or on the sizes of the probabilities involved (cf., e.g., [16]). Consequently, an important area to focus on within applied MCDA is the development and/or improvement of prescriptive techniques and methods for elicitation, better suited for real-life usage by decision-makers/experts.

The aim of this paper is, thus, to find a more realistic weight elicitation method, adapted for real-life problems and deployable for usage by a wider spectrum of people. Such a method must be relatively simple to use, transparent and possible to incorporate without too much facilitation into a natural decision analytical process.

1.2 Methods for Eliciting Weights in MCDA

In the literature, there have been a number of methods suggested for assessing criteria weights using exact values. These range from relatively simple ones, like the commonly used direct rating (DR) and point allocation (PA) methods (for a comparison of the two methods, cf., e.g., [17]), to somewhat more advanced procedures, such as the often used SMART [18] or SWING [14] methods. There are several weighting methods that appear to be minor variants of one another, but the small differences have shown to have important effects for inference and decision making [19]. Trade-off methods (where the subject is asked to state what trade-offs he or she is willing to do for certain changes in values) have also been proposed for weight elicitation, but, e.g., [20] concludes that trade-off methods have a tendency to give greater weight to the most important attribute in comparison to methods like DR and SWING.

As many reports of the difficulties with eliciting precise weights from decision-makers exist, some other approaches, less reliant on great precision from the decision-makers have been suggested. In such methods, ordinal and imprecise preference information is used to determine criteria weights and/or values of alternatives. For instance, the decision-maker could be asked to state importance weights on a semantic scale (e.g. very much more important, much more important, moderately more important etc.) like the AHP method [21]. However, the correctness of the conversion from the semantic scale to the numeric scale used by Saaty [21] as a measure for preference strength has been questioned by, e.g., [3], and the AHP method requires pair wise comparisons of all criteria (which can be very time consuming). Also, the use of verbal terms, in general, has been criticised, since words can have very different (numerical) meanings to different people (cf., e.g., [11]).

Another, more indirect approach is to let decision-makers simply rank the different criteria (i.e. supply the ordinal values), and thereafter use surrogate weights that are consistent with the supplied rankings (i.e. convert ordinal weights to cardinal weights). Advantages with this approach are that it is effort-saving [22], allows decision-makers to be more vague, it is less cognitively demanding and that groups are more likely to agree on ranks than on more precise weights [23]. Several proposals on how to convert the rankings into numerical weights exist, e.g., rank sum (RS) weights and rank reciprocal (RR) weights [24], and centroid (ROC) weights [25]. Barron and Barrett [13] found the latter superior to the other two on the basis of simulation experiments.

In some decision analysis applications, preferential uncertainties and incomplete information is handled by using intervals (cf., e.g., [26, 27]), where a range of possible values is represented by an interval. This approach is also claimed to put less demands on the decision-maker, and is suitable for group decision making as individual differences in preferences and judgments can be represented by value intervals (cf., e.g., [28]). In Mustajoki et al. [29], the authors propose an Interval SMART/SWING method, in which they generalize the SMART and SWING methods (for point estimates) into a method that allows interval judgements to represent imprecision. Another system, which also allows the use of intervals (to represent incomplete information about consequences as well as preferences), is a generic multi-attribute analysis (GMAA) system [28].

1.3   Problems with Existing Weight Elicitation Methods in Applied MCDA

The promising solutions offered today to aid decision making processes are seldom used in organizations (cf., e.g., [30, 10]) and there is an obvious need to reduce the current discrepancy regarding how elicitation is handled in current decision analysis applications and the ability of real decision-makers to provide what is required from them in order to increase the utilization of decision analytical methods in applied decision making. An often demanding task for users of decision analysis tools is to supply the input, and simpler methods for elicitation, which are preferred by users, are very important for the practical applicability of decision analysis (cf., e.g., [19, 22, 31, 32]). There is a great deal of uncertainty involved in elicitation, and when reviewing the literature in search of existing methods for weight elicitation that are satisfactory for real-life decision making problems, most of them fall short as they demand unrealistic numeric precision, are cognitively demanding and/or are difficult to employ without too much facilitation.

In an additive model, the weights reflect the importance of one dimension relative to the others. Most commonly, the degree of importance of an attribute depends on its spread (the range of the scale of the attribute), and this is why elicitation methods, like SMART [18], which do not consider the spread specifically have been criticized. Yet, with methods where ranges are explicitly considered during the elicitation of weights, several empirical studies imply that people still do not adjust weight judgments properly when there are changes in the ranges of the attributes (cf., e.g., [33]). However, if all criteria ranges are equal, e.g., [0, 10], and each step on each scale is considered equivalent, simpler and more practically useful methods can be used for weight elicitation. The weights are instead intuitive (importance) weights, reflecting the subjects' general attitude towards the attributes and an implicit range of outcomes.

A more realistic approach, which alleviates some of the problems related to the elicitation of exact values, is ranking. It allows decision-makers to be less precise and is less cognitively demanding on the decision-makers, but the current approaches of converting ordinal weights to cardinal weights may not produce surrogate weights that are adequate representations of the decision-makers' preferences. Although promising, it would be desirable to complement the supplied ranking somehow with preference relation information.

In conclusion, there seems to be a need for elicitation methods that do not require formal decision analysis education, are not too cognitively demanding by forcing people to express unrealistic preciseness or to state more than they can, do not require too much time, and are relatively easy to understand and use without losing trustworthiness.

2 A Prescriptive Method for Weight Elicitation

In order to address some of the major hindrances above for real-life usage of existing weight elicitation methods as described in the previous section, and to solve problems associated with the required exactness in subjective values, we propose an elicitation method which can be seen as a further development of ranking using surrogate weights. The method is subsequently tested in a comparative study (see Section 2.2) as well as employed in a real-life case (see Section 2.3).

2.1 Weight Elicitation Using Distance Functions

The proposed method consists of two stages. In the user interaction stage, information on the decision-maker's ranking of the criteria in the criteria set is elicited. The idea is that instead of using a conversion method (e.g., ROC weights) to receive surrogate (cardinal) weights, the decision-maker expresses the magnitude of the differences between the ranked criteria. After the user interaction, the information is interpreted in the second stage. In the following, we denote the set of criteria $\mathbf{G} = \{G_1, ..., G_N\}$ where each criterion $G_i \in \mathbf{G}$ is, during the elicitation, associated with a scale position (value) x_i and a criterion weight variable w_i. The first stage consists of three steps that can be iterated until convergence.

In the first step, only ordinal information is collected. The decision-maker (or other stakeholder) states the priority order of the criteria by ranking them from most important to least important. Without loss of generality, we assume that G_1 is more important than G_2 which is more important than G_3 and so forth. If one or more criteria are considered equally important, they will be ranked at the same level.

In the second step, the decision-maker is asked to assess the difference in importance between the most and the least important criteria. The idea behind the step is that it is easier to assess extremal than intermediate criteria, i.e. that the decision-maker is more likely to have opinions on the most and least important criteria than those in between. This was confirmed in the comparative study. Denote the highest visible user score x_H and the lowest x_L. x_H is set to 100 at the outset. The decision-maker is thus asked to express strength in his/her differences between the most and least important criteria by setting x_L.

In the third step, after the ranking of the criteria, they are distributed equally along a slider. The slider initially indicates that there is no information on the magnitude of the differences between their relative importances. This is displayed as the criteria being equal, i.e. that the magnitude of the difference between criterion G_1 and criterion G_2 is equal to the magnitude of the difference between G_2 and G_3, and so on. We call these initial differences default distances. Thereafter, the decision-maker will be able to adjust the distances between the criteria in order to express his or her cardinal importance information between them, e.g., he or she may feel that two criteria are closer together or more apart importance-wise, which is indicated by decreasing or increasing the distance between the criteria (see Fig. 1).

In this fashion, we receive preference strength information regarding the differences between the criteria that more adequately represents the weights. This approach, when using a graphical user interface such as the one in Fig. 1, allows the user to interactively adjust the distances between the criteria until the user is satisfied and thinks that the distances represent an adequate representation of his/her preferences with respect to priority. The user may also be less precise in their statements since no exact numbers are required. We further circumvent the problem of having the user understand the meaning of such values at an early stage. Instead of distributing weights, distances between weights are distributed. During prescriptive decision analysis, perceptions change and evolve, and the representation of these perceptions should be dynamic (cf., e.g., [34]). Beliefs and preferences are not static, and the decision-maker's view on what is important (and the relations) for the decision may change during the progress of the decision process. At a later stage, when the understanding of the problem has increased, the decision-maker may be asked to redo the procedure. This completes the user interaction stage of the elicitation process.

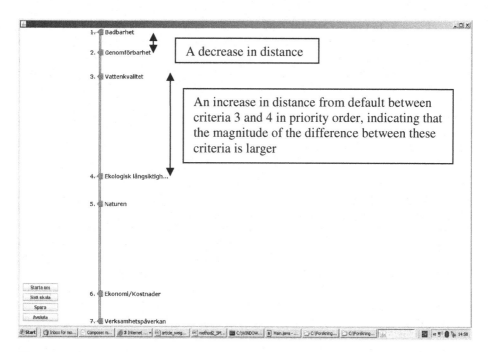

Fig. 1. A priority order with seven criteria, with adjustments made regarding the magnitude of the differences between the criteria

The interpretation stage is performed in different ways depending on the expressive power of the decision analytic method used. Underlying all interpretations is the scaling of the user statements from the slider. Assuming normalised criteria weights that sum to one, i.e. $\Sigma w_i = 1$ and $0 \le w_i \le 1$ for all $i \le N$, the user information ($x_i \le x_H$ for all $i \le N$) from the first stage is mapped onto a normalised scale such that $\Sigma x_i \rightarrow 1$; $x_H \rightarrow x_H / \Sigma x_i$; $x_L \rightarrow x_L / \Sigma x_i$; and $0 \rightarrow 0$. The middle two are the endpoints of the visible part of the user slider, see Fig. 1. If the expressive power of the analysis method only permits fixed numbers, the normalised slider weights are the output of the elicitation process. This is not the ideal situation and it is advised to use a decision analysis method capable of handling imprecision in the weight information. Such imprecision can be of two kinds, either in the form of intervals or additionally also allowing comparisons between weights. If imprecision is handled by allowing intervals, each user statement x_i on the slider is interpreted as an interval such that $w_i \in [x_i / \Sigma x_i - a_i, x_i / \Sigma x_i + b_i]$, where $0 < a_i \le 1$ and $0 < b_i, \le 1$ are proportional imprecision constants. These constants could, e.g., be interpreted as reflecting the degree of confidence in the weights ($w_1, ..., w_i, ..., w_N$) which are results of the user's statements, i.e. all x_i. While this is an improvement over having fixed numbers, it does still not fully capture the idea of not requiring more information than is actually available during the elicitation.

The preferred usage of the decision-maker information is through a representation using both intervals and comparative statements in order to represent the information actually given by the decision-maker. In step one of the user interaction stage, an ordinal ranking is obtained. Using comparative statements, this is represented as

$w_i \geq w_j$ for $i < j$. In the last step of the user interaction stage, the decision-maker then slides the different criteria until he/she is satisfied with their relative positions. The end result is then viewed relative to the initial equidistant positions. For each pair G_i and G_j of criteria, the user could have modified the distance between them in three ways relative to the initial distances: By **A**) increasing the distance between them, by **B**) decreasing it, or by **C**) having kept the same distance.

In case **A**), the interpretation is that the two criteria weights differ *at least* with a certain value, yielding the additional inequality constraint

$$w_i \geq w_j + \text{proj}_{d_{ij}} [D_L(d_{12}, ..., d_{(N-1)N})] \qquad (1)$$

where d_{ij} is the distance $|x_i - x_j|$ between criteria G_i and G_j on the slider as indicated by the decision-maker (indicating the magnitude of the difference between them), and $D_L: [0, x_H - x_L]^N \rightarrow [0,1]^N$ is a multi-variate difference function which maps a vector of distances to a vector of *lower* bounds of differences between each w_i and w_j. Let **d** be the vector of default differences prior user manipulation, and let **d*** be the vector of differences as indicated (manipulated) by the decision maker, then let $\mathbf{d_{diff}} = \mathbf{d^*} - \mathbf{d}$. Then positive components of $\mathbf{d_{diff}}$ should be positive components of $D_L(d_{12}, ..., d_{(N-1)N})$ with the ratios between these components preferably preserved (maintaining the preference strength information as indicated by the decision maker). It should further be required that D_L should yield consistent constraints, i.e. the conditions $\Sigma w_i = 1$ and $0 \leq w_i \leq 1$ for all $i \leq N$ must be possible to satisfy simultaneously together with the additional constraints. These are the basic requirements on D_L yielding a reasonable quantitative interpretation of the user's statements, however this quantitative interpretation is by means of sets of feasible weight distributions and not on a single one. Verifying consistency can either be done by consistency checks performed in an employed decision tool (see, e.g., [35] which is the case at present), or embedded as a property of the function D_L. It should be noted however, that D_L is context dependent and the process of defining it may vary between cases.

In case **B**), the interpretation is that the two criteria weights differ *at most* with a certain value, yielding the additional inequality constraint

$$w_i \leq w_j + \text{proj}_{d_{ij}} [D_U(d_{12}, ..., d_{(N-1)N})] \qquad (2)$$

Here, $D_U: [0, x_H - x_L]^N \rightarrow [0,1]^N$ is a difference function similar to D_L which maps a vector of distances to a vector of *upper* bounds of differences between w_i and w_j. In case **C**), the interpretation is that there is no more information available between the two criteria weights. Thus, $w_i \geq w_j$ is kept as the only constraint.

2.2 Comparative Study

In order to test the appropriateness of the proposed weight elicitation method (which in theory could reduce many of the problems discussed in section 1.3, it was compared with two simplified multi-attribute rating approaches often used in practical analysis, *SMART* and *Direct Rating*, in a comparative study.

In this study, the five participants all used computers in their daily work, were in the age range of 28-62, had driver's licences, and had bought at least one car in their past. The participants were interviewed individually for an hour each on two occasions (one week apart). They were asked to picture themselves about to make a decision on buying a new car and that they after extensive research had identified a number of possible car types within the price range they could manage. Furthermore, they were given the seven criteria that they could want to consider for the evaluation of the possible alternatives. Their weights regarding these criteria were elicited using all three methods (in a varied order) on both occasions. After using each method, as well as in the end of each occasion, they were interviewed about their preferences regarding method, perceived pros and cons with each method, their beliefs in results, as well as their perceived effort. The number of subjects in the comparative study was kept down in order to facilitate an iterative improvement of the method prior to the case study (Section 2.3). Still, the study gave some valuable results, especially about the subjects' confidence in their own preference statements, and about how to improve the graphical interface to the proposed elicitation method.

From the study, we confirmed that the participants were most sure of their preferences concerning the most important criteria and the least important criterion, and were less sure of their preferences regarding the criteria in the middle (or slightly downwards) of the ranking order. Their preferences regarding the middle criteria often differed between the two occasions. Several of the participants preferred SMART (out of the 3 methods they tried). However, the "free" scale upwards gave very different results (from the same participants) between the two occasions, e.g., the most important criterion got 50 points in the first test and 100 points in the re-test, or 20 in the first test and 90 in the re-test. Furthermore, the participants all valued the possibility to interactively adjust their statements during the "thought process" in the application. Finally, the proposed elicitation method was also the method that generated the most consistent results between the two occasions.

The comparative study confirms that reflections about the problem and explicit considerations about one's beliefs on more than one occasion is perceived to contribute to better quality of results and increased insight, which is an important goal. The participants asked for some modifications of the graphical interface in order to improve understanding of the proposed elicitation method. In spite of some criticism of the first version of the proposed method, it generated the most consistent results between the two occasions, i.e. the internal consistency was the highest of the three methods. Regarding convergent validity, the SMART and DR methods differed (many empirical studies report the same findings), and as importance preferences are subjective values it is difficult to determine which method actually generates the most accurate results. The free scale upwards of the SMART method also seems like an element that can influence the convergent validity of the method a great deal as people in the comparative study seemed highly affected by the "available information" (see, the availability heuristic, [7]). In order to test the proposed elicitation method further, it was employed in a real-life decision making situation where an MCDA application was used to guide the decision process.

2.3 Employment of the Weight Elicitation Method in a Real-Life Case

A lot of the advances within decision analysis have been theoretical developments, which are implemented in practical applications without much empirical testing. If empirical studies on new or existing decision support techniques have been done, they are seldom applied to real-life cases. Brown and Vari [36] suggest that we need to use real-life cases as test beds in order to retrieve more prescriptive (practically useful) methods. There is, thus, a need to examine new decision support techniques in real-life decision making situations, and hence, the proposed elicitation method (which theoretically could reduce many of the major hindrances to real-life usage) was also employed in a real-life case and tested with real decision-makers.

A debated decision that the governing politicians in Örebro (a municipality in Sweden) faced was how to improve the water quality of Svartån, a river running through the city of Örebro. The problems with Svartån had been debated for long, and the decision was multi-facetted in nature. For several years, there had been unacceptably high amounts of intestinal bacteria in the water, and the different spots for bathing along Svartån had been deemed unsuitable according to EU regulations. The primary goal of the decision-makers was to make it possible to swim in the Svartå river by the year 2010, but also to obtain a more sustainable (long-term) solution with an increased quality of the water in general. The decision-makers were seven politicians with different political standpoints, so the decision was a multi-criteria, multi-stakeholder problem. Thus, it was a real-life decision making problem suitable for the employment of the weight elicitation method described in section 2.1.

In the Svartå case, we tested this paper's approach to eliciting weights as part of the MCDA model used to aid the decision making process (after initial testing in the small comparative study). Decomposed scaling, where the weights and the partial value functions are assessed separately, was used. The seven main criteria were identified collectively by the politicians, and thereafter the weights were elicited individually from the politicians on two occasions. The first occasion was early on in the process, right after the identification of the top-level criteria, in order to initiate and motivate the decision-makers' reflections about their own beliefs. The second occasion was later on in the process, when the decision-makers had understood the problem, the different options, and their own beliefs better. The identification of the sub-criteria and the value assessments were initially performed by civil servants. The decision-makers thereafter continuously confirmed and/or adjusted these assessments as the work proceeded in workshop settings (lead by facilitators), where the participants were politicians and civil servant representatives.

2.4 Results

In the real-life case, all politicians found the elicitation method clear and easy to use. To give a priority order for the seven main criteria was considered fairly easy by all politicians, since they had thought about the problem quite extensively and had a rather fair idea of what they considered important. However, several of the participants commented on the fact that the order of some of the middle criteria was less definite (in their minds) than the most important criteria. They also completed their task within a time frame acceptable to them, and quite easily adjusted the distances between criteria to indicate the magnitudes of the differences between the criteria. Some of the decision-makers found the step to express the scale by giving the least important criterion a value in comparison to the most important criterion (x_H, which was set to 100 points) to be the most difficult. Yet, they all completed this step within minutes of thought.

In the analysis of the Svartå problem, the individual decision maker information could be studied using a representation of both intervals and comparative statements, but in the final group evaluation the simpler interpretation was used. The use of the simpler technique for interpreting the decision-makers' preferences was motivated by the aim of promoting understanding of the process. Intervals were instead used to represent the group's preferences, and the robustness of the decision-makers' preferences was studied by allowing for different widths of weight intervals (encompassing the weights derived from the elicitation data). In this case, a simple method was chosen for aggregating their different individual weight statements representing conflicting priorities (several methods have been proposed in the literature for such aggregations, cf., e.g., [37]). For each criterion, the individual weight distributions were collected by forming an interval (which embraced all decision-makers' weights). This aggregation technique was used in the evaluation - in addition to evaluating the decision using the individual weights for each politician - resulting in two out of the seven alternatives being clearly superior. The information obtained from the proposed elicitation method could thus be interpreted in both the simpler and more advanced way depending on the needs and wants of the decision-makers.

2.5 Discussion

Numbers are incorrectly associated with precision to people [14], which is somewhat paradoxical when they are used to describe uncertain measures, like preferences in this case. In both the comparative and the real-life case studies, people's preferences between some of the criteria were shown to be somewhat dynamic and we have found that the proposed weight elicitation method is a more realistic (and thus more prescriptively useful) way of eliciting decision-makers' preferences. The possibility to simply rank and thereafter interactively adjust the preferences regarding their differences seems to be less cognitively demanding and supports the creative, dynamic and cyclic modelling approach suggested by prescriptive analysts (cf., e.g., [34]). This interactive part stimulated thoughts on importance relations among the criteria and the users often spoke out loud during this step, explaining their thoughts of mind (often re-positioning criteria in the priority order as well). Moreover, the use of intervals to represent these preferences could be a way to cover the dynamicity of preference entities, although the elicitation data could also be represented by fixed numbers if the expressive power of the decision analytic method cannot handle intervals.

Participants in both studies felt that providing a priority order, as well as expressing the relation between the criteria, made them formulate their importance preferences in more detail than before the final elicitation step.

In conclusion, the goal was to find a more prescriptively useful weight elicitation method, which would reduce some of the problems with existing methods. The promising results of the comparative study promoted the use of the proposed method for weight elicitation in a real-life case, where it proved to be easily understood, did not require too much time, did not require the users to express more than they could, and was not perceived as too demanding, consequently limiting the cognitive load on the users. Thus, the proposed method seems to be a more prescriptively useful method than most methods offered today. There is reason to believe that the prescriptive approach is a direction we should move in, in order to increase the utilization of decision analysis applications in real-life decision making.

References

1. Wallenius, J., Dyer, J.S., Fishburn, P.C., Steuer, R.E., Zionts, S.: Multiple Criteria Decision Making, Multiattribute Utility Theory: Recent Accomplishments and What Lies Ahead. Management Science 54(7), 1336–1349 (2008)
2. Fox, J.: Probability, Logic and the Cognitive Foundations of Rational Belief. J. Appl. Logic. 1, 197–224 (2003)
3. Belton, V., Stewart, T.: Multiple Criteria Decision Analysis: An Integrated Approach. Kluwer Academic Publishers, UK (2002)
4. Pöyhönen, M., Hämäläinen, R.: On the Convergence of Multiattribute Weighting Methods. European Journal of Operational Research 129, 106–122 (2001)
5. Doyle, J.R., Green, R.H., Bottomley, P.A.: Judging Relative Importance: Direct Rating and Point Allocation are Not Equivalent. Org. Behav. & Human Dec. Proc. 70, 65–72 (1997)
6. Borcherding, K., Eppel, T., von Winterfeldt, D.: Comparison of Weighting Judgments in Multiattribute Utility Measurement. Management Science 37(12), 1603–1619 (1991)
7. Tversky, A., Kahneman, D.: Judgment Under Uncertainty: Heuristics and Biases. Science 185(4157), 1124–1131 (1974)
8. Tversky, A., Kahneman, D.: The Framing of Decisions and the Psychology of Choice. Science 211, 453–458 (1981)
9. Keeney, R.L., Raiffa, H.: Decisions with Multiple Objectives: Preferences and Value Trade-offs. Wiley, NY (1976)
10. Shapira, Z.: Risk taking: A Managerial Perspective. Russel Sage Foundation, NY (1995)
11. Kirkwood, C.W.: Strategic Decision Making: Multiobjective Decision Making with Spreadsheets. Wadsworth Publishing Company, US (1997)
12. Riabacke, A., Påhlman, M., Larsson, A.: How Different Choice Strategies Can Affect the Risk Elicitation Process. IAENG Intern. J. of Comp. Science 32(4), 460–465 (2006)
13. Barron, F.H., Barrett, B.E.: Decision Quality Using Ranked Attribute Weights. Management Science 42(11), 1515–1523 (1996)
14. von Winterfeldt, D., Edwards, W.: Decision Analysis and Behavioural Research. Cambridge University Press, Cambridge (1986)
15. Riabacke, A., Påhlman, M., Baidya, T.: Risk Elicitation in Precise and Imprecise Domains – A Comparative Study, Sweden and Brazil. In: Proc. of the Int. Conf. on Computational Intelligence for Modelling, Control and Automation (2006)
16. Påhlman, M., Riabacke, A.: A Study on Framing Effects in Risk Elicitation. In: Proceedings of the Int. Conf. on Computational Intelligence for Modelling, Control & Autom. (2005)
17. Bottomley, P.A., Doyle, J.R., Green, R.H.: Testing the Reliability of Weight Elicitation Methods: Direct Rating versus Point Allocation. J. Marketing Res. 37, 508–513 (2000)
18. Edwards, W.: How to Use Multiattribute Utility Measurement for Social Decisionmaking. IEEE Transactions on Systems, Man & Cybernetics 7(5), 326–340 (1977)
19. Bottomley, P.A., Doyle, J.R.: A Comparison of Three Weight Elicitation Methods: Good, Better, and Best. Omega 29, 553–560 (2001)
20. Fischer, G.W.: Range Sensitivity of Attribute Weights in Multiattribute Value Models. Org. Behav. & Human Dec. Proc. 62(3), 252–266 (1995)
21. Saaty, T.L.: The Analytic Hieararchy Process. McGraw-Hill, NY (1980)
22. Katsikopoulos, K.V., Fasolo, B.: New Tools for Decision Analysis. IEEE Transactions on Systems, Man, and Cybernetics – Part A: Systems and Humans 36(5), 960–967 (2006)
23. Barron, F.H., Barrett, B.E.: The Efficacy of SMARTER: Simple Multi-Attribute Rating Technique Extended to Ranking. Acta Psychologica 93, 23–36 (1996)

24. Stillwell, W.G., Seaver, D.A., Edwards, W.: A Comparison of Weight Approximation Techniques in Multiattribute Utility Decision Making. Org. Behav. & Human Performance 28(1), 62–77 (1981)
25. Barron, F.H.: Selecting a Best Multiattribute Alternative with Partial Information about Attribute Weights. Acta Psychologica 80(1-3), 91–103 (1992)
26. Walley, P.: Reasoning with Imprecise Probabilities. Chapman and Hall, London (1991)
27. Danielson, M., Ekenberg, L., Ekengren, A., Hökby, T., Lidén, J.: A Process for Participatory Democracy in Electronic Government. J. Multi-Criteria Dec. Anal. 15, 15–30 (2008)
28. Jiménez, A., Rios-Insua, S., Mateos, A.: A Generic Multi-Attribute Analysis System. Computers & Operations Research 33(4), 1081–1101 (2006)
29. Mustajoki, J., Hämäläinen, R., Salo, A.: Decision Support by Interval SMART/SWING - Incorporating Imprecision in the SMART and SWING Methods. Dec. Sciences 36 (2005)
30. March, J.G.: A Primer on Decision Making, How Decisions Happen. The Free Press, NY (1994)
31. Edwards, W., Barron, F.H.: SMARTS and SMARTER: Improved Simple Methods for Multiattribute Utility Measurement. Org. Behav. & Human Dec. Proc. 60, 306–325 (1994)
32. Stewart, T.J.: A Critical Survey on the Status of Multiple Criteria Decision Making Theory and Practice. Omega 20(5-6), 569–586 (1992)
33. von Nitzsch, R., Weber, M.: The Effect of Attribute Ranges on Weights in Multiattribute Utility Measurements. Management Science 39(8), 937–943 (1993)
34. French, S., Rios Insua, D.: Statistical Decision Theory. Oxford University Press Inc., NY (2000)
35. Danielson, M., Ekenberg, L., Idefeldt, J., Larsson, A.: Using a Software Tool for Public Decision Analysis: The Case of Nacka Municipality. Dec. Analysis 4(2), 76–90 (2007)
36. Brown, R., Vari, A.: Towards a Research Agenda for Prescriptive Decision Science: The Normative Tempered by the Descriptive. Acta Psych. 80, 33–47 (1992)
37. Matsatsinis, N.F., Samaras, A.P.: MCDA and Preference Disaggregation in Group Decision Support Systems. Euro. J. Oper. Research 130(2), 414–429 (2001)

Inverse Analysis from a Condorcet Robustness Denotation of Valued Outranking Relations

Raymond Bisdorff[1], Patrick Meyer[2,3], and Thomas Veneziano[1]

[1] University of Luxembourg
Faculty of Sciences, Technology, and Communication
Computer Science and Communications Research Unit
Interdisciplinary Lab for Intelligent and Adaptive Systems
6, rue Richard Coudenhove-Kalergi, L-1359 Luxembourg
{raymond.bisdorff,thomas.veneziano}@uni.lu
[2] Institut Télécom, Télécom Bretagne
UMR CNRS 3192 Lab-STICC
Technopôle Brest Iroise CS 83818 F-29238 Brest Cedex 3, France
patrick.meyer@telecom-bretagne.eu
[3] Université européenne de Bretagne

Abstract. In this article we develop an indirect approach for assessing criteria significance weights from the robustness of the significance that a decision maker acknowledges for his pairwise outranking statements in a Multiple Criteria Decision Aiding process. The main result consists in showing that with the help of a mixed integer linear programming model this kind of a priori knowledge is sufficient for estimating adequate numerical significance weights.

Keywords: inverse Multiple Criteria Decision Analysis, significance weights ellicitation, uncertainty, robustness.

1 Introduction

We consider a decision situation in which a finite set of decision alternatives is evaluated on a finite family of performance criteria. A decision maker is willing to pairwisely compare these alternatives according to the outranking paradigm. One considers indeed that an alternative *a outranks* an alternative *b* when a *significant majority* of criteria validates the fact that *a* is performing at least as good as *b* and there is no criterion where *b* seriously outperforms *a* [1]. To assess when such a significant majority of criteria validates an outranking situation requires a more or less precise numerical knowledge of the significance of each criterion in the multiple criteria preference aggregation. Two different approaches exist to specify theses values:

- either via *direct* preference information, where the criteria significance is first assessed and then the aggregated outranking situations are computed,
- or, via *indirect* preference information, where some a priori partial knowledge of the resulting aggregated outranking is used in order to infer plausible estimators of the criteria significance.

F. Rossi and A. Tsoukiàs (Eds.): ADT 2009, LNAI 5783, pp. 180–191, 2009.

In this article we exclusively concentrate on the indirect preference information approach. Similar approaches, mostly in the domain of Multiple Attribute Value Theory, already appeared in the literature where they are generally called *disaggregation/aggregation* or *ordinal regression* methods [2,3,4,5,6,7,8]. In analogy with corresponding techniques in inferential statistics, we prefer to group all indirect preference information modeling techniques under the generic term *inverse Multiple Criteria Decision Analysis*. The innovative a priori knowledge on which we focus our inverse analysis here is the robustness of the significant majorities that the decision maker acknowledges for his pairwise comparisons with respect to all potential significance weights, a fact we call the CONDORCET robustness of the outranking situation in the sequel of this article. The main result of our article is to show that this kind of a priori knowledge alone is sufficient for estimating numerical significance weights.

The article is organised as follows: in the next section, we define the CONDORCET robustness denotation of valued outranking relations and then, in Section 3 we briefly detail the way of computing it. Afterwards, in Section 4 we present a mathematical model for estimating the significance weights followed by some brief remarks on practical application issues.

2   Defining the CONDORCET Robustness Denotation of Valued Outranking Relations

Let $A = \{x, y, z, \ldots\}$ be a finite set of $n > 1$ potential decision alternatives and $F = \{g_1, \ldots, g_m\}$ a coherent finite family of $m > 1$ criteria.

The alternatives are evaluated on each criterion on real performance scales to which an indifference q_i and a preference p_i discrimination threshold (for all g_i in F) is associated [1]. The performance of alternative x on criterion g_i is denoted x_i.

In order to characterise a local *at least as good as* situation [9,10] between any two alternatives x and y of A, with each criterion g_i is associated a double threshold order S_i whose numerical representation is given by:

$$S_i(x,y) = \begin{cases} 1 & \text{if} \quad x_i + q_i \geqslant y_i \,, \\ 0 & \text{if} \quad x_i + p_i \leqslant y_i \,, \\ 0.5 & \text{otherwise.} \end{cases}$$

Furthermore, we associate with each criterion $g_i \in F$ a rational *significance weight* w_i which represents the contribution of g_i to the *overall warrant or not* of the *at least as good as* preference situation between all pairs of alternatives. Let $W = \{w_i : g_i \in F\}$ be the set of relative significance weights associated with F such that $0 < w_i < 1$ $(\forall g_i \in F)$ and $\sum_{g_i \in F} w_i = 1$ and let $\mathcal{W}$ be the set of such significance weights sets.

The *overall valued outranking relation*, denoted $\widetilde{S}^w$, aggregating the partial *at least as good as* situations, is given by :

$$\widetilde{S}^w(x,y) = \sum_{w_i \in W} w_i \cdot S_i(x,y), \ \forall(x,y) \in A \times A.$$

$\widetilde{S}^W(x, y)$ is thus evaluated in the rational interval $[0, 1]$ with the following semantics [9]:

- $\widetilde{S}^W(x, y) = 1$ indicates that all criteria warrant unanimously the "*at least as good as*" preference situation between x and y;
- $\widetilde{S}^W(x, y) > 0.5$ indicates that a majority of criteria warrant the "*at least as good as*" preference situation between x and y;
- $\widetilde{S}^W(x, y) = 0.5$ indicates a balanced situation where the criteria warranting the "*at least as good as*" preference situation between x and y are exactly as significant as those who do not warrant this situation;
- $\widetilde{S}^W(x, y) < 0.5$ indicates that a majority of criteria do not warrant the "*at least as good as*" preference situation between x and y;
- $\widetilde{S}^W(x, y) = 0$ indicates that all criteria warrant unanimously the negation of the "*at least as good as*" preference situation between x and y.

Let $\succsim_w$ be the preorder[1] on F associated with the natural $\geqslant$ relation on the set of significance weights W. $\sim_w$ induces r ordered equivalence classes $\Pi_1^W \succ_w \ldots \succ_w \Pi_r^W$ $(1 \leq r \leq m)$. The criteria of an equivalence class have the same significance weight in W and for $i < j$, those of Π_i^W have a higher significance weight than those of Π_j^W. Let $\mathcal{W}_{\succsim_w} \subset \mathcal{W}$ denote the set of all significance weights sets that are preorder-compatible with $\succsim_w$.

Let $W \in \mathcal{W}$. The CONDORCET *robustness denotation*[2] [12] of $\widetilde{S}^W$, denoted $[\![\widetilde{S}^W]\!]$, is defined, for all $(x, y) \in A \times A$, as follows:

$$
[\![\widetilde{S}^W]\!](x, y) = \begin{cases}
3 & \text{if } \widetilde{S}^V(x, y) = 1 \ \forall V \in \mathcal{W}; \\
2 & \text{if } [\widetilde{S}^V(x, y) > 0.5 \ \forall V \in \mathcal{W}_{\succsim_w}] \wedge [\exists V' \in \mathcal{W} : \widetilde{S}^{V'}(x, y) < 1]; \\
1 & \text{if } [\widetilde{S}^W(x, y) > 0.5] \wedge [\exists V' \in \mathcal{W}_{\succsim_w} : \widetilde{S}^{V'}(x, y) \leqslant 0.5]; \\
0 & \text{if } \widetilde{S}^W(x, y) = 0.5; \\
-1 & \text{if } [\widetilde{S}^W(x, y) < 0.5] \wedge [\exists V' \in \mathcal{W}_{\succsim_w} : \widetilde{S}^{V'}(x, y) \geqslant 0.5]; \\
-2 & \text{if } [\widetilde{S}^V(x, y) < 0.5 \ \forall V \in \mathcal{W}_{\succsim_w}] \wedge [\exists V' \in \mathcal{W} : \widetilde{S}^{V'}(x, y) > 0]; \\
-3 & \text{if } \widetilde{S}^V(x, y) = 0 \ \forall V \in \mathcal{W};
\end{cases}
$$

with the following semantics:

- $[\![\widetilde{S}^W]\!](x, y) = \pm 3$ if all criteria *unanimously warrant* (resp. *do not warrant*) the outranking situation between x and y;
- $[\![\widetilde{S}^W]\!](x, y) = \pm 2$ if a *significant majority* of criteria *warrants* (resp. *does not warrant*) the outranking situation between x and y for all $\succsim_w$-compatible weights sets;
- $[\![\widetilde{S}^W]\!](x, y) = \pm 1$ if a significant majority of criteria *warrants* (respectively *does not warrant*) this outranking situation for W but not for all $\succsim_w$-compatible weights sets;

---

[1] As classically done, $\succ_w$ denotes the asymmetric part of $\succsim_w$, whereas $\sim$ denotes its symmetric part.

[2] The simple majority validated outranking relation $S^W(x, y)$ such that $\widetilde{S}^W(x, y) > 0.5$ is generally called the CONDORCET *relation* (see Barbut [11]), in honours of the Marquis DE CONDORCET (1743–1794) who first promoted social choice procedures based on pairwise simple majority votings.

- $[\![\widetilde{S}^W]\!](x, y) = 0$ if the total significance of the warranting criteria is *exactly balanced* by the total significance of the not warranting criteria for W.

The careful reader may have noticed that, in the presence of veto thresholds as defined in [10], if a veto situation occurs in the comparison of a couple of alternatives, the associated CONDORCET robustness denotation is -3, as the overall outranking relation $\widetilde{S}^W$ equals 0, disregarding the criteria significance weights.

3   Computing the CONDORCET Robustness Denotation

In this section, we briefly explain how to obtain the CONDORCET robustness denotation. Further details can be found in [12].

Let us consider the following numerical example to illustrate our purpose throughout this paper.

Example. Consider a set $A = \{a, b, c, d, e\}$ of five decision alternatives and a consistent family F of three *cardinal* criteria $\{g_1, g_2, g_4\}$ measuring performances on rational scales from 0.0 to 100.0 and two *ordinal* criteria $\{g_3, g_5\}$ measuring performances on a discrete ordinal scale from 0 to 10. Criterion g_2 is a *cost*–type criterion on which performances have to be minimised, whereas the four other are benefit–type criteria, i.e. the higher the performance is the better a decision alternative is considered.

Table 1 presents the randomly generated performances of the alternatives on each criterion. Notice the significance weights set W shown in the third column which induces the significance ordering $\{g_1\} \succ_w \{g_4\} \succ_w \{g_3\} \succ_w \{g_5\} \succ_w \{g_2\}$.

Let us start by presenting the notation which allows us to detail the construction of the CONDORCET robustness denotation associated with a valued outranking relation $\widetilde{S}^W$ and a significance weights set W.

Let $c_k^W(x, y)$ be the sum of "*at least as good as*" characteristics $S_i(x, y)$ for all criteria $g_i \in \Pi_k^W$, and $\overline{c_k^W}(x, y)$ the sum of the negation $(1.0 - S_i(x, y))$ of these characteristics. Furthermore, let $C_k^W(x, y) = \sum_{i=1}^k c_i^W(x, y)$ be the cumulative sum of "*at least as good as*" characteristics for all criteria having significance at

Table 1. Performance table

crit. (F)	pref. dir.	weights (W)	decision alternatives (A)					thresholds indiff. pref.	
			a	b	c	d	e	indiff.	pref.
g_1	max	5/15	70.9	61.8	90.2	31.2	33.1	5.0	8.0
g_2	min	1/15	20.9	17.1	76.3	69.2	35.5	3.0	6.0
g_3	max	3/15	1	4	6	8	6	0	1
g_4	max	4/15	17.3	46.3	24.5	40.6	68.2	6.0	7.0
g_5	max	2/15	2	1	8	2	6	0	1

least equal to the one associated to Π_k^W, and let $\overline{C_k^W}(x,y) = \sum_{i=1}^k \overline{c_i^W}(x,y)$ be the cumulative sum of the negation of these characteristics, for all k in $\{1, \ldots, r\}$.

In the absence of ± 3 denotations, the following proposition gives us a test for the presence of a ± 2 denotation:

Proposition 1 (Bisdorff [12])

$$[\![\widetilde{S}^W]\!](x,y) = 2 \iff \begin{cases} \forall k \in 1, \ldots, r : C_k^W(x,y) \geqslant \overline{C_k^W}(x,y); \\ \exists k \in 1, \ldots, r : C_k^W(x,y) > \overline{C_k^W}(x,y). \end{cases}$$

The negative -2 denotation corresponds to similar conditions with reversed inequalities.

The ± 2 denotation test of Proposition 1 corresponds in fact to the verification of *stochastic dominance*-like conditions (see [12]).

A ± 1 CONDORCET robustness denotation, corresponding to the observation of a weighted majority (resp. minority) in the absence of the ± 2 case, is simply verified as follows:

$$[\![\widetilde{S}^W]\!](x,y) = \pm 1 \iff \left((\widetilde{S}^W(x,y) \gtrless 0.5) \wedge [\![\widetilde{S}^W]\!](x,y) \neq \pm 2 \right).$$

Example. Back to the example, we can now compute the CONDORCET robustness denotation associated with the outranking relation. Let us detail these calculations for the following two couples, (b,c) and (a,d). Recall that the significance ordering is given by a five-class preorder $\{g_1\} \succ_w \{g_4\} \succ_w \{g_3\} \succ_w \{g_5\} \succ_w \{g_2\}$.

We can easily verify via Table 2 that $[\![\widetilde{S}^W]\!](b,c) = -2$. Besides we can see that $[\![\widetilde{S}^W]\!](a,d) \neq \pm 2$. Since $\widetilde{S}^W(a,d) = 0.53 > 0.5$, we finally have $[\![\widetilde{S}^W]\!](a,d) = 1$. Table 3 presents the outranking relation $\widetilde{S}^W$ and its corresponding CONDORCET robustness denotation $[\![\widetilde{S}^W]\!]$ for all pairs of alternatives of $A \times A$.

The issue we address in this paper is now the following. Consider that we have given a performance table as shown in Table 1, but without any explicit significance weights information, as well as a CONDORCET robustness denotation $[\![\widetilde{S}^W]\!]$ similar to the one shown in the right part of Table 3, with W unknown. *Is it possible to infer from these information alone the apparent significance weights of the criteria?* In other words, *may we compute on the basis of the given information a preorder $\succsim$ on the criteria and a numerical instance W^* of a $\succsim$-compatible*

Table 2. Cumulative sums for couples (b,c) and (a,d)

	$c_i^W(b,c)$	$\overline{c_i^W}(b,c)$	$C_i^W(b,c)$	$\overline{C_i^W}(b,c)$	$c_i^W(a,d)$	$\overline{c_i^W}(a,d)$	$C_i^W(a,d)$	$\overline{C_i^W}(a,d)$
Π_1^W	0	1	0	1	1	0	1	0
Π_2^W	1	0	1	1	0	1	1	1
Π_3^W	0	1	1	2	0	1	1	2
Π_4^W	0	1	1	3	1	0	2	2
Π_5^W	1	0	2	3	1	0	3	2

Table 3. Global outranking with CONDORCET robustness denotation

A	$\tilde{S}^w$					$[\![\tilde{S}^w]\!]$				
	a	b	c	d	e	a	b	c	d	e
a	-	.50	.07	.53	.40	-	0	-2	1	-1
b	.53	-	.33	.67	.40	1	-	-2	2	-1
c	.93	.67	-	.47	.67	2	2	-	-1	2
d	.60	.60	.53	-	.53	1	1	1	-	1
e	.60	.60	.53	.80	-	1	1	1	2	-

weights set which satisfies the given CONDORCET robustness denotation $[\![\tilde{S}^w]\!]$, i.e. W^* and $\succsim$ are such that $[\![\tilde{S}^{w^*}]\!] = [\![\tilde{S}^w]\!]$?

4 Inferring the Criteria Significance Weights

To solve this estimation problem we are going to formulate a mixed integer linear programming model.

We start with characterising a constraint model for every possible CONDORCET robustness denotation except the ±3 ones. Indeed, we may ignore unanimous positive and negative (±3) robustness denotations as they concern in fact the trivial pairwise comparison of alternatives that are either Pareto dominating or Pareto dominated. Their aggregated outranking situation is thus always unanimously warranted (resp. not warranted), independently of any particular criteria significance weights. These denotations therefore do not contain any specific information for inferring the particular significance of an individual criterion.

We denote $A^2_{\pm 2}$ (resp. $A^2_{\pm 1}$ or A^2_0) the set of pairs (x, y) of alternatives such that $[\![\tilde{S}^w]\!](x, y) = \pm 2$ (resp. ±1 or 0).

As the criterion significance weights are supposed to be rational, we may without loss of generality restrict our estimation problem to integer weight sets. Thus every criterion may get an integer significance weight $w_i \in [1, M]$, where M denotes the maximal admissible value. Limiting our purpose to genuine decision aid situations, we may choose this bound in practical applications to be equal to the number m of criteria.

We denote $P_{m \times M}$ a Boolean $(0, 1)$-matrix, with general term $[p_{i,u}]$, that characterises row-wise the number of weight units allocated to criterion g_i. Formally, the row i represents the decomposition of the weight associated to g_i into M bits in a unary base (little-endian) and thus $\sum_{u=1}^{M} p_{i,u} = w_i$.

The fact that every criterion g_i of F must have a strictly positive significance may thus be expressed with the help of the following constraint:

$$\sum_{g_i \in F} p_{i,1} = m.$$

At least one weight unit is allocated to every criterion, i.e. $p_{i,1} = 1$ for all $g_i \in F$. As an example, if g_i has an integer weight of 3 and if we decide that $M = 5$, then the ith row of $P_{m \times 5}$ is given by $(1, 1, 1, 0, 0)$.

The required cumulative semantics of $P_{m \times M}$ is therefore achieved with the following set of constraints:

$$p_{i,u} \geqslant p_{i,u+1} \quad (\forall i = 1, ..., m, \ \forall u = 1, ..., M - 1).$$

4.1   Constraints for $[\![\widetilde{S}^w]\!](x, y) = \pm 2$ Conditions

Let us now translate Proposition 1 to a computable set of constraints.

Corollary 1
When considering integer weights, Proposition 1 may be reformulated as:

$$[\![\widetilde{S}^w]\!](x, y) = 2 \iff \begin{cases} \forall u \in 1, ..., \max w_i : C_u'^w(x, y) \geqslant \overline{C_u'^w}(x, y); \\ \exists u \in 1, ..., \max w_i : C_u'^w(x, y) > \overline{C_u'^w}(x, y); \end{cases}$$

where $C_u'^w(x, y)$ (resp. $\overline{C_u'^w}(x, y)$) is the sum of all $S_i(x, y)$ (resp. $1 - S_i(x, y)$) such that the significance weight $w_i \leq u$. The negative -2 denotation corresponds again to similar conditions with reversed inequalities.

Proof. We easily verify that all constraints from Proposition 1 are present in the corollary (for the set $U = \{u / \exists w_i \in W, w_i = u\}$ of indexes). For all other values of u the constraints are redundant.   □

This leads to the property that $p_{i,u} = 1 \iff w_i \geq u$ and we directly obtain:

$$C_u'^w(x, y) = \sum_{g_i \in F} \left(p_{i,u} \cdot S_i(x, y) \right).$$

In order to model now the $[\![\widetilde{S}^w]\!](x, y) = \pm 2$ conditions, we introduce for all pairs $(x, y) \in A_{\pm 2}^2$ the following set of constraints:

$$\sum_{g_i \in F} \left(p_{i,u} \cdot [S_i(x, y) - \overline{S_i}(x, y)] \right) \geqslant b_u(x, y) \quad (\forall u = 1, ..., M),$$

where $\overline{S_i}$ is the negation $(1 - S_i)$ of the criterion's double threshold order characteristic function, and where the $b_u(x, y)$ are Boolean $(0, 1)$ variables for each pair of alternatives and each equi-significance level u in $\{1, ..., M\}$. Note that the negative -2 denotation again corresponds to a similar inequation with a reversed inequality and negative $b_u(x, y)$. These binary variables allow us to impose at least one case of strict inequality for each $(x, y) \in A_{\pm 2}^2$ as required in Corollary 1 via the following constraints:

$$\sum_{u=1}^{m} b_u(x, y) \geqslant 1, \quad \left(\forall (x, y) \in A_{\pm 2}^2 \right).$$

4.2   Constraints for $[\![\widetilde{S}^w]\!](x, y) = \pm 1$ Conditions

In order to introduce the $[\![\widetilde{S}^w]\!](x, y) = \pm 1$ conditions, we may formulate for all pairs $(x, y) \in A_{\pm 1}^2$ the following set of constraints:

$$\sum_{g_i \in F} \Big(\sum_{u=1}^{M} p_{i,u} \Big) \cdot \pm \big(S_i(x, y) - \overline{S}_i(x, y) \big) \geqslant 1 \ \forall (x, y) \in A_{\pm 1}^2, \tag{1}$$

where the factor $(\sum_{u=1}^{M} p_{i,u})$ represents the integer value of the estimated weight w_i of criterion g_i.

Recall that a CONDORCET robustness denotation of ± 2 represents an outranking situation which is validated (or non-validated) for all possible weights sets compatible with the given significance preorder. Such a situation therefore represents a robust validation by the decision maker, and should as such be considered highly trustful. Consequently, if the decision maker imposes a ± 1 or 0 CONDORCET robustness, this can be considered as more anecdotical. In practical situations, it might happen that the CONDORCET robustness given by the decision maker might not be compatible with the underlying problem. To avoid not finding any solution, we relax Constraints (1) by adding positive slack variables which have to be minimised in order to satisfy best possibly the constraints:

$$\sum_{g_i \in F} \Big(\sum_{u=1}^{M} p_{i,u} \Big) \cdot \pm \big(S_i(x, y) - \overline{S}_i(x, y) \big) \pm s^{\pm 1}(x, y) \geqslant 1 \ \forall (x, y) \in A_{\pm 1}^2.$$

4.3   Constraints for $[\![\widetilde{S}^w]\!](x, y) = 0$ Conditions

Similarly as in the previous section, for all pairs $(x, y) \in A_0^2$, we formulate the corresponding set of soft equality constraints:

$$\sum_{g_i \in F} \Big(\sum_{u=1}^{M} p_{i,u} \Big) \cdot \big(S_i(x, y) - \overline{S}_i(x, y) \big) + s_+^0(x, y) - s_-^0(x, y) = 0.$$

4.4   Objective Function

Finally, our overall objective is to determine a significance weights set W^* which:

- satisfies all the $[\![\widetilde{S}^{w^*}]\!](x, y) = \pm 2$ constraints,
- respects the $[\![\widetilde{S}^{w^*}]\!](x, y) = \pm 1$ and $[\![\widetilde{S}^{w^*}]\!](x, y) = 0$ constraints as well as possible, and
- gives the smallest possible weights w_i $(g_i \in F)$ (which, in practice, tends to use the least possible number of equi-significance classes).

Therefore, we introduce the following objective function which is to be minimised:

$$K_1\Big(\sum_{g_i \in F}\sum_{u=1}^{M} p_{i,u}\Big) \tag{2}$$

$$-K_2\Big(\sum_{u=1}^{M}\Big(\sum_{(x,y)\in A_{\pm 2}^2} b_u(x,y)\Big)\Big) \tag{3}$$

$$+K_3\Big(\sum_{(x,y)\in A_{\pm 1}^2} s^{\pm 1}(x,y)\Big) + K_4\Big(\sum_{(x,y)\in A_0^2}(s_+^0(x,y)+s_-^0(x,y))\Big) \tag{4}$$

where $K_1 ... K_4$ are parametric constants used for the correct hierarchical ordering of the four sub-goals. Note that (3) is not necessary for solving our problem, but it guarantees the strictest possible enforcing of the $[\![\widetilde{S}^w]\!](x,y) = \pm 2$ constraints with strict inequalities.

In summary, we obtain the following linear mixed integer program which covers all positive, negative and zero CONDORCET robustness denotations:

MILP

Variables:

$$p_{i,u} \in \{0,1\} \qquad\qquad \forall g_i \in F, \ \forall u = 1,..,M$$
$$b_u(x,y) \in \{0,1\} \qquad\qquad \forall(x,y)\in A_{\pm 2}^2, \forall u = 1,..,M$$
$$s^{\pm 1}(x,y) \geqslant 0 \qquad\qquad \forall(x,y)\in A_{\pm 1}^2$$
$$s_+^0(x,y)\geqslant 0 , \ s_-^0(x,y)\geqslant 0 \qquad\qquad \forall(x,y)\in A_0^2$$

Parameters:

$$K_i > 0 \qquad\qquad \forall i = 1...4$$

Objective function:

$$\min \quad K_1\Big(\sum_{g_i\in F}\sum_{u=1}^{M} p_{i,j}\Big) - K_2\Big(\sum_{u=1}^{M}\sum_{(x,y)\in A_{\pm 2}^2} b_u(x,y)\Big)$$
$$+K_3\Big(\sum_{(x,y)\in A_{\pm 1}^2} s^{\pm 1}(x,y)\Big) + K_4\Big(\sum_{(x,y)\in A_0^2}(s_+^0(x,y)+s_-^0(x,y))\Big)$$

Constraints:

s.t.
$$\sum_{g_i\in F} p_{i,1} = m$$

$$p_{i,u} \geqslant p_{i,u+1} \qquad\qquad \forall g_i \in F, \ \forall u = 1,..,M-1$$

$$\sum_{g_i\in F}\Big(p_{i,u}\cdot \big[S_i(x,y)-\overline{S_i}(x,y)\big]\Big) \gtrless b_u(x,y) \ \forall(x,y)\in A_{\pm 2}^2, \ \forall u = 1,..,M$$

$$\sum_{u=1}^{M} b_u(x,y) \geqslant 1 \qquad\qquad \forall(x,y)\in A_{\pm 2}^2$$

$$\sum_{g_i\in F}\Big(\big(\textstyle\sum_{u=1}^{M} p_{i,u}\big)\cdot \pm (S_i(x,y)-\overline{S_i}(x,y))\Big)$$
$$\pm s_{\pm}^1(x,y) \geqslant 1 \qquad\qquad \forall(x,y)\in A_{\pm 1}^2, \ \forall u = 1,..,M$$

$$\sum_{g_i\in F}\big(\textstyle\sum_{u=1}^{M} p_{i,u}\big)\cdot (S_i(x,y)-\overline{S_i}(x,y))$$
$$+s_+^0(x,y)-s_-^0(x,y) = 0 \qquad\qquad \forall(x,y)\in A_0^2, \ \forall u = 1,..,M$$

Table 5. Optimal MILP solution for our example with estimated significance weights

F	$p^*_{i,u}$ 1 2 3 4 5	W^*	W
g_1	1 1 1 1 1	5/13	5/15
g_2	1 0 0 0 0	1/13	1/15
g_3	1 1 1 0 0	3/13	3/15
g_4	1 1 1 0 0	3/13	4/15
g_5	1 0 0 0 0	1/13	2/15

Let $P^* = [p^*_{i,u}]$ be an optimal solution of the MILP model. We may calculate the estimated significance weights as the row sum of $[p^*_{i,u}]$, i.e. $w^*_i = \sum_{u=1}^{M} p^*_{i,u}$ for all criteria $g_i \in F$ and thus recover the corresponding significance preorder $\succsim_{w^*}$.

Example. Let us reconsider our example. Solving MILP with Cplex 11.0 gives the optimal P^* matrix shown in Table 5[3]. The resulting estimated normalised weights are: $w^*_1 = 0.385$, $w^*_2 = 0.077$, $w^*_3 = 0.231$, $w^*_4 = 0.231$ and $w^*_5 = 0.077$, whereas the *real* weights that we initially generated are : $w_1 = 0.333$, $w_2 = 0.067$, $w_3 = 0.200$, $w_4 = 0.267$ and $w_5 = 0.133$.

All constraints related to the 6 pairs $(x,y) \in A^2_{\pm 2}$ are positively verified, as well as those concerning the 13 pairs $(x,y) \in A^2_{\pm 1}$ and the pair $(a,b) \in A^2_0$. Therefore we get with $K_i = 1\ \forall i = 1...4$ the optimal value of -22 for the objective function $(13 - 7 \cdot 5 + 0 + 0)$.

The original linear significance order: $\{g_1\} \succ_w \{g_4\} \succ_w \{g_3\} \succ_w \{g_5\} \succ_w \{g_2\}$ is reconstructed as a three-level significance preorder: $\{g_1\} \succ_{w^*} \{g_3, g_4\} \succ_{w^*} \{g_2, g_5\}$. Recomputing the corresponding overall outranking relation we obtain the estimated $\widetilde{S}^{W^*}$ relation (see Table 6), which admits an identical CONDORCET robustness denotation as the original $\widetilde{S}^W$ relation.

The example that we detailed through this article illustrates the fact that the reconstruction from the CONDORCET robustness denotation alone of the significance weights set following the original valued outranking relation is in general

Table 6. Global outranking relation with inferred significance weights

A	estimated $\widetilde{S}^{W^*}$ $x_1\ \ x_2\ \ x_3\ \ x_4\ \ x_5$	original $\widetilde{S}^W$ $x_1\ \ x_2\ \ x_3\ \ x_4\ \ x_5$
x_1	- .50 .08 .54 .46	- .50 .07 .53 .40
x_2	.54 - .31 .69 .46	.53 - .33 .67 .40
x_3	.92 .69 - .46 .69	.93 .67 - .40 .67
x_4	.54 .54 .54 - .62	.53 .53 .53 - .53
x_5	.54 .54 .54 .77 -	.53 .53 .53 .80 -

---

[3] Cplex 11.0 solves this tiny mixed integer linear program with 26 MIP simplex iterations, 0 branch-and-bound nodes and 4 Gomory cuts.

not unique and not completely faithful. Several admissible significance preorders and numerical weights sets might indeed support the same robustness denotation and some tuning of the MILP objective function may be necessary depending on the decision aid goal we intend to follow in order to get a useful result.

4.5 Practical Application Issues

If we apply the MILP model with Cplex 11.0, associated with an AMPL front end modeler on more or less real-sized random multiple criteria decision problems (20 alternatives evaluated on 13 criteria) we observe quite reasonable solving times on an 6 threaded standard application server. Depending on the maximal value M allowed for an individual criterion significance weight we indeed obtain average computation times of 2.5 seconds for $M = 7$ up to 2 minutes for $M = 13$.

As already mentioned, for a given value of M, the MILP might have some non zero slacks. In such a case, as our purpose is here to find a solution without any slacks, we need to increase the value of M to reduce the slacks. In practice, we simply reiterate the resolution, with M slightly incremented. Notice that 1 more unit for M produces m new binary variables (from $p_{1,M+1}$ to $p_{m,M+1}$), increasing significantly the computation time as we have noticed before.

Furthermore, a great number of problems may be solved using values M much lower than the number of criteria such that the number of columns of matrix P is generally overestimated. Consequently, to limit the expected computation time, we recommend to set the initial maximal admissible value M to the requested depth of the estimated significance preorder and to increase it only if necessary.

5   Conclusion

In this paper we have presented an innovative method to determine significance weights of criteria in Multiple Criteria Decision Aid, while guaranteeing a high degree of robustness and therefore a high reliability of the outranking relation.

In the future we plan to examine the questioning of a decision maker in order to obtain valuable information for the determination of the significance weights from robustness affirmations. This involves the analysis of the decision maker's responses as well as the study of the interactive use of the algorithm presented in this paper. In particular we intend to restrict the decision maker's intervention on a few pairs of alternatives and infer the outranking relation for those remaining.

References

1. Roy, B., Bouyssou, D.: Aide Multicritère à la Décision: Méthodes et Cas. Economica, Paris (1993)
2. Jacquet-Lagrèze, E., Siskos, Y.: Assessing a set of additive utility functions for multicriteria decision making: the UTA method. European Journal of Operational Research 10, 151–164 (1982)
3. Mousseau, V., Słowinski, R.: Inferring an Electre TRI model from assignment examples. Journal of Global Optimization 12(2), 157–174 (1998)

4. Mousseau, V., Dias, L.C., Figueira, J., Gomes, C., Clímaco, J.N.: Resolving inconsistencies among constraints on the parameters of an MCDA model. European Journal of Operational Research 147(1), 72–93 (2003)

5. Siskos, Y., Grigoroudis, E., Matsatsinis, N.F.: Uta methods. In: Figueira, J., Greco, S., Ehrgott, M. (eds.) Multiple Criteria Decision Analysis: State of the Art Surveys, pp. 297–344. Springer, Boston (2005)

6. Grabisch, M., Kojadinovic, I., Meyer, P.: A review of capacity identification methods for Choquet integral based multi-attribute utility theory: Applications of the Kappalab R package. European Journal of Operational Research 186, 766–785 (2008)

7. Greco, S., Mousseau, V., Słowinski, R.: Ordinal regression revisited: multiple criteria ranking using a set of additive value functions. European Journal of Operational Research 191(2), 415–435 (2008)

8. Meyer, P., Marichal, J.-L., Bisdorff, R.: Disaggregation of bipolar-valued outranking relations. In: An, L.T.H., Bouvry, P., Tao, P.D. (eds.) MCO. Communications in Computer and Information Science, vol. 14, pp. 204–213. Springer, Heidelberg (2008)

9. Bisdorff, R.: Logical foundation of multicriteria preference aggregation. In: Bouyssou, D., et al. (eds.) Aiding Decisions with Multiple Criteria, pp. 379–403. Kluwer Academic Publishers, Dordrecht (2002)

10. Bisdorff, R., Meyer, P., Roubens, M.: Rubis: a bipolar-valued outranking method for the best choice decision problem. 4OR: A Quarterly Journal of Operations Research 6(2), 143–165 (2008)

11. Barbut, M.: Médianes, Condorcet et Kendall. Mathématiques et Sciences Humaines 69, 9–13 (1980)

12. Bisdorff, R.: Concordant outranking with multiple criteria of ordinal significance. 4OR 2(4), 293–308 (2004)

Directional Decomposition of Multiattribute Utility Functions

Ronen I. Brafman[1] and Yagil Engel[2]

[1] Department of Computer Science, Ben Gurion University, Israel
brafman@cs.bgu.ac.il
[2] Industrial Engineering & Management, Technion, Israel
yagile@ie.technion.ac.il

Abstract. Several schemes have been proposed for compactly representing multiattribute utility functions, yet none seems to achieve the level of success achieved by Bayesian and Markov models for probability distributions. In an attempt to bridge the gap, we propose a new representation for utility functions which follows its probabilistic analog to a greater extent. Starting from a simple definition of marginal utility by utilizing reference values, we define a notion of conditional utility which satisfies additive analogues of the chain rule and Bayes rule. We farther develop the analogy to probabilities by describing a directed graphical representation that relies on our concept of conditional independence. One advantage of this model is that it leads to a natural structured elicitation process, very similar to that of Bayesian networks.

1 Introduction

Specifying a multi-variate utility function is known to be a difficult task, and often considered a bottleneck in implementation of intelligent systems. It requires quantifying one's preferences – a non-trivial cognitive task which involves contemplating a large number of questions about the relative desirability of uncertain outcomes, or gambles. Furthermore, the very personal and subjective nature of utility information makes it harder to reuse and learn, unlike probabilistic knowledge, which can often be learned from data and reused for various instances of a system. Yet, the preference and utility elicitation tasks must be carried out when analyzing decision problems. A number of attempts have been made to aid this elicitation process by structuring it so that either the type of questions that must be answered is simpler and/or the number of questions is smaller. Often, this process is aided by some graphical structure that captures some properties of the utility function.

The level of success of current formalisms is not clear, partly because assessing the benefit of various utility elicitation processes is difficult. Even given that current models provide significant theoretical simplifications, the cognitive burden imposed by the elicitation process may still be prohibitive for practical applications. Given the well recognized practical benefits yielded by probabilistic graphical models, it is likely that much more can be done for utilities, too. In this work we attempted to follow the footsteps of probability theory more closely than before, by defining a notion of conditional utility that is closer in form to its probabilistic analog. We then show how this concept

F. Rossi and A. Tsoukis (Eds.): ADT 2009, LNAI 5783, pp. 192–202, 2009.

leads naturally to milestones such as the chain rule and Bayes rule analogies, and finally to a graphical representation based on a directed acyclic graph. While our new method of representing and eliciting utilities bears certain similarities to existing methods, as detailed below, it offers an elicitation process – both for qualitative structure and numeric values – that is clear, simple, and intuitive. Furthermore, it provides immediate computational benefits, and several promising direction for future research that are based on the close resemblance to probabilistic models. We believe that this method can become an essential part of the toolkit of decision analysts and an important component in real-world decision support systems.

In the remainder of this paper we define a new notion of conditional utility and utilize it to define *utility difference networks*. We explain their elicitation process and compare them to existing formalism for representing structured utility functions. Finally, we discuss a few open questions.

2 Background and Related Work

Let Θ denote the space of possible outcomes, with $\preceq$ a preference relation (weak total order) over Θ. Let $\Gamma = \{a_1, \ldots, a_n\}$ denote a set of attributes describing Θ. Each attribute $a \in \Gamma$ has a domain $\mathcal{D}(a)$, so that $\Theta \subseteq \prod_{i=1}^{n} \mathcal{D}(a_i)$. We use prime signs and superscripts to denote specific assignment for an attribute, and a concatenation of assignment symbols (as in $a_i' a_j''$) means that each of the attributes gets a respective value. We use γ and γ_i to denote subsets of Γ, and the same notation as before to denote assignments to all the attributes in the set. For example, if $\gamma_1 = \{a_i, a_j\}$, then $\gamma_1^0 = a_i^0 a_j^0$. Finally, we use $\mathcal{D}(\gamma)$ to denote the set of all possible assignments to γ, that is the projection of Θ over $\prod_{a_i \in \gamma} \mathcal{D}(a_i)$.

Definition 1. *Let* $\gamma_1, \gamma_2 \subset \Gamma$. *$\gamma_1$ and γ_2 are* conditionally additive independent (CAI) *given their complement* $\Gamma \setminus (\gamma_1 \cup \gamma_2)$, *if preferences over lotteries on* Γ *depend only on their marginal conditional probability distributions over* γ_1 *and* γ_2.

Graphical models have been employed for the representation of decomposed utility, as early as by [9, 13]. However, the first representation that relies on *conditional independence*, and thus follow the footsteps of probabilistic models, can be attributed to Bacchus and Grove [1]. These authors show that conditional additive independence has a perfect map, meaning that given a set of attributes and a preference order, there exists a graph whose node separation expresses the exact set of independence conditions. Further, they show that the utility function decomposes to a sum over lower dimensional functions, each defined over a maximal clique of the graph. This decomposition is a special type of *generalized additive independence (GAI)*, a global independence condition introduced originally by Fishburn [7].

Definition 2. *Let* $\gamma_1, \ldots, \gamma_g \subseteq \Gamma$ *such that* $\bigcup_{i=1}^{g} \gamma_i = \Gamma$. *$\gamma_1, \ldots, \gamma_g$ are called* generalized additive independent *(GAI) if preferences over lotteries on* Γ *depend only on their marginal distributions over* $\gamma_1, \ldots, \gamma_g$.

An (expected) utility function $u(\cdot)$ can be decomposed additively according to its (possibly overlapping) GAI sub-configurations.

Theorem 1 ([7]). *Let* $\gamma_1, \ldots, \gamma_g$ *be GAI. Then there exist functions* $f_1, \ldots, f_g$ *such that*

$$u(a_1, \ldots, a_m) = \sum_{r=1}^{g} f_r(\gamma_r).$$ (1)

Bacchus and Grove revived this notion and named it GAI. This opened the way to an increasing body of research on representation and reasoning with GAI. Boutilier et al. [2] introduce UCP networks, which is a directed form of CAI-maps. The directionality though is obtained from identifying preferential independence conditions over sets of attributes, that is exogenously to the GAI decomposition. Gonzales and Perny [8] introduce GAI nets, which is a graphical representation for GAI, where nodes represent subsets of attributes, and nodes are connected if their respective subsets intersect. Braziunas and Boutilier [3] provide a method of elicitation that takes advantage of the locality property of GAI.

CAI and GAI require comparisons of probability distributions and preferences over lotteries. In applications in which uncertainty is not a crucial element (e.g., electronic commerce applications), it is not required and usually not desired to involve probabilities in user interaction. Engel and Wellman [5] extend the work of [4] and introduce *conditional difference independence (CDI)*. Intuitively, attributes x and y are CDI of each other if any difference in value over assignments to x does not depend on the current assignment of y, for any possible assignment to the rest of the variables. CDI is very similar to CAI, and therefore has a perfect map as well.

Definition 3. [1] *Let* $\gamma_1, \gamma_2 \subset \Gamma$. γ_1 *and* γ_2 *are* conditionally difference independent *given* $\gamma_3 = \Gamma \setminus (\gamma_1 \cup \gamma_2)$, *denoted as* $CDI(X, Y)$, *if*

$$\forall \text{ assignments } \hat{\gamma}_3, \gamma_1', \gamma_1'', \gamma_1', \gamma_2''$$
$$u(\gamma_1' \gamma_2' \hat{\gamma}_3) - u(\gamma_1'' \gamma_2' \hat{\gamma}_3) = u(\gamma_1' \gamma_2'' \hat{\gamma}_3) - u(\gamma_1'' \gamma_2'' \hat{\gamma}_3)$$

Our new concept of independence and graphical model most closely resemble CDI. However, in comparison to CDI, it introduces several benefits: (i) it is directional, allowing for a more intuitive elicitation process and (ii) the independence condition is weaker, meaning it can be applied in some cases wherein which CDI does not hold.

Another direction of research relied on other types of utility independence. CUI networks [6] is a graphical model that relies on the concept of *conditional utility independence* [10], which intuitively requires the (cardinal) preference order over a subset of the attributes to be independent of another subset of attributes. Earlier works by Shoham [12] and La Mura and Shoham [11] are also seeking utility representation that is similar to a probability distribution. Shoham [12] proposes a redefinition of utility function as a set function, over additive factors in the domain that together contribute to the decision maker's well being. La Mura and Shoham [11] propose only a redefinition of the utility independence concept, which is a multiplicative version of difference independence (that is, refers to utility ratios rather than differences). In non-probabilistic settings, and

---

[1] Difference independence and CDI are defined given a preference order over preference differences, and its numeric representation is a measurable value function. For brevity of presentation we describe it in terms of utilities.

especially in situations in which decision outcomes can be measured against monetary differences (as in purchasing), we believe that utility differences are more natural to elicit than ratios.

A common drawback of most previous models is that most focus is given to the process of data elicitation, whereas the process of *structure elicitation*, in which the independence structure is identified, is usually left to domain experts. This is particularly true for GAI based representations, as UCP and GAI networks, because there is no explicit and intuitive process for identifying and/or verifying GAI conditions. Our novel model, in contrast, has the benefit of an intuitive and incremental structure elicitation process.

3 Reference and Conditional Utility

There are inherent differences between probability distributions and utility functions, which make any analogy between the two problematic. Arguably, the most primal difference is the fact that probability distribution is a set function, defined over events that encapsulate a set of atomic outcomes. In contrast, there is no meaning for the utility of a set of atomic outcomes. For probability distributions, there is a natural definition for a function over a subspace of the world on which the problem is defined. Technically, if the world is represented by a set of attributes Γ, one can define a probability distribution over some $\gamma \subset \Gamma$ by summing over the atomic outcomes that hold for any assignment to γ, thus marginalize out the irrelevant parameters (namely, $\Gamma \setminus \gamma$).

Whereas there is no meaning for marginalizing parameters of a utility function, a similar effect can be achieved by fixing those parameters on some *reference value*. For probabilities, we ask the question *what is the probability of outcomes in γ when we don't know the value of $\Gamma \setminus \gamma$*. While we do not have an exact analogy for utilities, with reference values we get ask: *what is the utility of outcomes in γ when the value of $\Gamma \setminus \gamma$ is fixed on the reference*. The idea of using a reference value has been exploited in previous works [7, 3, 6], however it was never taken quite as far in driving the analogy to probabilities.

Let $a_1^0 \ldots a_n^0 \in \Theta$ denote a predetermined complete assignment, which we call *the reference assignment*. The reference assignment allows us to define a utility function over a subspace of the joint domain. Let $\overline{\gamma} = \Gamma \setminus \gamma$.

Definition 4. *The reference utility function is defined as follows*

$$u_r(\gamma) = u(\gamma \overline{\gamma}^0)$$

The next step is to define the notion of conditioning, within a subspace of the domain.

Definition 5. *The conditional utility function is defined as follows*

$$u_r(\gamma_1 | \gamma_2) = u_r(\gamma_1 \gamma_2) - u_r(\gamma_1^0 \gamma_2)$$

where $\overline{\gamma} = \Gamma \setminus \{\gamma_1 \cup \gamma_2\}$.

This definition has a direct rooting in the definition of conditional probabilities. The definition of the latter is

$$p(\gamma_1 | \gamma_2) = \frac{p(\gamma_1 \gamma_2)}{\gamma_2}$$

As common in probabilistic reasoning, we take a log of the definition in order to replace multiplication with additivity. This results exactly in Definition 5.

Given that, it is not surprising that the utility function exhibits an additive decomposition which is similar to the multiplicative decomposition of a probability function. We first have to normalize the utility function (henceforth) such that $u(\Gamma^0) = 0$.

Theorem 2 (The chain rule).

$$u(\Gamma) = \sum_{i=1}^{n} u_r(a_i|\{a_j\}_{j=1}^{i-1})$$

Proof. By definitions of conditional utility and reference utility,

$$u_r(a_i|\{a_j\}_{j=1}^{i-1}) = u_r(a_1 \ldots a_{i-1}a_i) - u_r(a_1 \ldots a_{i-1}a_i^0) =$$
$$u(a_1 \ldots a_{i-1}a_i a_{i+1}^0 \ldots a_n^0) - u(a_1 \ldots a_{i-1}a_i^0 a_{i+1}^0 \ldots a_n^0)$$

Summing over $i = 1, \ldots, n$ on both sides yields the desired result, because: (1) the negative term for $i = 1$ is $u(a_1^0 \ldots a_n^0) = 0$, (2) the negative term for i cancels out with the positive term for $i - 1$ (both are $u(a_1 \ldots a_{i-1}a_i^0 a_{i+1}^0 \ldots a_n^0)$), and (3) the positive term for $i = n$ is $u(\Gamma)$. □

Finally, it is easy to see that this definition obeys an additive adaptation of Bayes rule. Again, taking log over the probabilistic equation we obtain the following

Theorem 3 (Bayes Rule Analog).

$$u_r(\gamma_1|\gamma_2) = u_r(\gamma_2|\gamma_1) + u_r(\gamma_1) - u_r(\gamma_2)$$

3.1 Conditional Independence

The chain rule by itself does not provide significant computational value, because the last term ($i = n$) includes the left-hand side of the equation $u(\Gamma)$. The idea, similar to the one employed to achieve compact probability functions, is that the conditional utility function $u_r(a_i|a_1, \ldots, a_{i-1})$ may not depend on all of the attributes $a_1, \ldots, a_{i-1}$, but only on some subset of them, in which case the terms considered by the chain rule have lower dimensionality. This is formalized as follows.

Definition 6. γ_1 *is said to be* conditionally independent *of* γ_2 *given* γ_3 ($CDI_r(\gamma_1, \gamma_2|\gamma_3)$) *if for any* $\gamma_3' \in \mathcal{D}(\gamma_3)$,

$$u_r(\gamma_1|\gamma_2\gamma_3') = u_r(\gamma_1|\gamma_3')$$

When $\gamma_3 = \Gamma \setminus \gamma_1 \cup \gamma_2$, then $CDI_r(\gamma_1, \gamma_2|\gamma_3)$ is equivalent to γ_1 and γ_2 being CDI. Therefore, CDI_r is a generalization of CDI. The novelty of this definition is that it refers to a subset of the attributes. Whereas in previous independence concepts the conditional set must always be "the rest of the attributes", here we specifically select a conditional set, and can ignore the attributes which are not relevant to γ_1.

Table 1. Utility for each assignment to attributes x, y, and z

x^0yz	x^0y^0z	x^0yz^0	$x^0y^0z^0$
9	6	6	3
xyz	xy^0z	xyz^0	xy^0z^0
12	7	8	5

As an example, consider the values in Table 1, which provides the value for the eight different instantiations of three boolean attributes, x, y, and z. The difference between the two values in each column corresponds to the difference in x given difference instantiations of yz.

We see that CDI(x, y) does not hold because $u(xyz) - u(x^0yz) \neq u(xy^0z) - u(x^0y^0z)$ (according to the two left columns). In our terms it means that CDI$_r(x, y|z)$ does not hold. However, CDI$_r(x, y|)$ does hold, because the difference is equal for the reference value z^0 (see the two right columns): $u(xyz^0) - u(x^0yz^0) = u(xy^0z^0) - u(x^0y^0z^0)$ (or, equivalently, $u_r(x|y) = u_r(x|)$).

4   Utility Difference Networks

Loyal to the Bayes-net analogy, we seek a directed graphical structure, with a node for each attribute, and the following property: each attribute is conditionally directional independent, given its parents, of all its other non-descendants. Let $Pa(a)$ denote the parents of a node a in a graph, and let $Dn(a)$ denote its descendants. Furthermore, let $Co(a) = \Gamma \setminus \{a\} \cup Pa(a) \cup Dn(a)$.

Definition 7. *A utility difference network is a DAG $G = (V, E)$, with V corresponding to a set of attributes Γ, and for any $a \in \Gamma$, CDI$_r(a, Pa(a)|Co(a))$.*

The utility computation from the directed graph is again very similar to how probabilities are computed from a Bayes-net. The following theorem is a direct result of the chain rule and Definition 7.

Theorem 4. *The utility function can be computed from the utility difference network as follows*

$$u(\Gamma) = \sum_{i=1}^{n} u_r(a_i|Pa(a_i))$$

Previous graphical models usually assume that the model is given, obtained by some domain expert. In particular, how to identify a GAI decomposition remains an unsolved question (except for the case that the GAI structure is a result of a collection of CAI or CDI conditions). Note also that while each pairwise CDI condition requires in theory the verification of order of $\exp(n)$ equalities for utility differences (because a verification is required for each instantiation of the rest of the attributes), with our new notion of conditional independence we only need to consider the independent attributes and the conditioning set. However, we note that when creating a full network this is not a significant advantage, because the number of queries for the *last* variable in the ordering will reach the same order of magnitude as in CDI.

4.1 Elicitation

The process of obtaining a utility difference network structure is similar in spirit to that used for Bayesian networks. It is summarized by the following procedure. As is the case in Bayesian network, the result depends on the *variable ordering* that is used by the procedure, and the choice of variable ordering is usually based on heuristic assessments. Intuitively, we would like to place the most important variables first, because these are the variables that are likely to have many connections to other variables. By keeping them on top we avoid having to represent all of these dependencies as parents of the same variable. Furthermore, it makes intuitive sense to have the important variables first, so the dependence between other variables are conditioned on them.

For each variable in its turn, we find a set of parents: those attributes that are required in order to render the current variable independent of the rest. We use the notation $\Gamma^i = \{a_1, \ldots, a_i\}$, and (x, y) refers to a directed arc from $x \in V$ to $y \in V$.

algorithm $\underline{ProcGetStructure}(\Gamma)$
    input: Γ, ordered as $\{a_1, \ldots, a_n\}$
    output: a utility difference network over Γ
for i=1 to n:
    find minimal $\hat{\Gamma}^i \subseteq \Gamma^{i-1}$ such that $\text{CDI}_r(a_i, \Gamma^{i-1} \setminus \hat{\Gamma}^i | \hat{\Gamma}^i)$
    For each $x \in \hat{\Gamma}^i$, Add (x, a_i) to E
    return $G = (\Gamma, E)$

The data in the nodes of a utility difference network is in form of conditional utility function, that is obtained by querying a user for preference differences. For example, the node a with parents γ requires the function $u_r(a|\gamma)$, which is obtained by queries for the differences $u_r(a\gamma) - u_r(a^0\gamma)$.

4.2 Example

In order to demonstrate the difference between CDI and CDI_r, we consider the hard-drive example used by Engel and Wellman [5], and we show their CDI-map of the problem in Figure 1a. The example describes various decision criteria that a procurement department of a company evaluates when purchasing some quantity of new hard-drives. Below, each attribute is listed with a designated attribute name (the first letter), and its (sometimes arbitrary) domain.

RPM (R) 3600, 4200, 5400 RPM
Transfer rate (T) 3.4, 4.3, 5.7 MBS
Volume (V) 60, 80, 120, 160 GB
Supplier ranking (S) 1, 2, 3, 4, 5
Quality rating (Q) (of the HD brand) 1, 2, 3, 4, 5
Delivery time (D) 10, 15, 20, 25, 30, 35 days
Warranty (W) 1, 2, 3 years
Insurance (I) (for the case the deal is signed but not implemented) α, β, γ
Payment timeline (P) 10, 30, 90 days

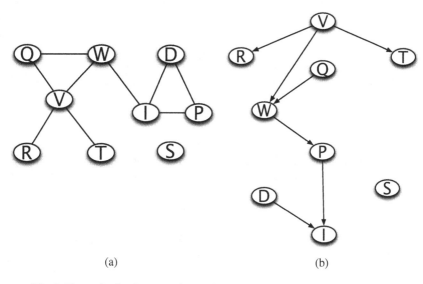

Fig. 1. Networks for the example: (a) CDI-map (b) utility difference network

Intuitively, the most salient criterion is volume, and the other important ones are RPM, warranty, and quality; hence the ordering prefix V, R, W, Q seems sensible. The rest of the ordering is less crucial, and we use I, D, P, T. First, V is placed as a root. When R is considered next, we find that it depends on V because for high volume hard-drives the marginal utility of improving RPM is higher. We find that W depends on V, because larger hard-drives tend to fail more. Now when considering Q, we might find that there is similar dependence between Q and V. However, if the reference value of W is the maximum value (3 years), we might find that given the reference value of W (and the rest of the variables), Q and V are independent because the longer warranty alleviates (substitutes) quality concerns. Further down in the network we might see a similar effect: a convenient value for insurance alleviates the dependency of delivery terms on the payment terms. In addition, we find the CDI conditions as described by Engel and Wellman [5]. For example, given any fixed value of V the marginal value of improving the quality rating does not depend on the RPM. Also, S is CDI and therefore CDI_r of all the rest. We obtain the DAG depicted in Figure 1b. The utility difference network does not achieve lower dimension than the corresponding CDI-map, however it provides directionality that can be exploited for a more natural elicitation process.

5 Discussion

CDI is a stronger condition than CDI_r, and as such the locality property it achieves is stronger. To see this, consider the elicitation of data for an attribute x. In a CDI-map, the marginal utility over x is independent of the value of any node outside its local neighborhood. Therefore, the marginal utility of x can be elicited using local queries that involve differences over x and given a fixed value of the neighbors, without even knowing the value of the rest of the variables. In utility difference network the marginal utility of x may in some cases depend on the value of a non-neighbor y. For example, if y is an ancestor of x, the marginal value of x is independent of y given any value of x's parents, but only *given that the rest of the variables are fixed on the reference value*. It is possible that there is some instantiation of the rest of the variables, under which the independence is lost. Therefore, elicitation must specify explicitly that the the rest of the variables are fixed on the reference value.

In fact, this can be seen as an advantage of utility difference network. In CDI-map, such x and y will necessarily have an edge connecting them, whereas in a utility difference network such edge can be omitted. In that sense, utility difference network refines CDI. Furthermore, if x is fully CDI of y, this can be exploited in utility difference networks as well; as long as a local query makes sense to the user there is no need to indicate explicitly that the rest of the variables are fixed on the reference value.

A promising direction coming out of this representation is in introducing a form of Bayesian Learning for utilities. Consider a digital camera manufacturer that wishes to obtain information about customers' preferences. The company may be able to observe some limited set of choices made by customers. Perhaps we can also assume that single dimensional utilities (e.g., how much worths an improvement in a single attribute, all else being equal) is easy to estimate, or elicit. The company can now use the evidence (customer's choices) and Theorem 3, in order to obtain data regarding future choices of the customer. For example, the customer may have chosen to pay an extra \$60 for a camera with $10_\times$ zoom and 6_{mgp}, over one with $7_\times$ zoom and 6_{mgp}. Now (given the single dimensional data, in the form of reference utilities over each attribute) the company can compute the amount that the customer is willing to pay to get 6_{mgp} over 4_{mgp}, given that the zoom is $10_\times$.

Practical problem with this direction are yet to be resolved: this assumes that the outcomes above differ only in these two attributes, and in addition the rest of the attributes are fixed on the reference value (or, alternatively, difference independence holds between the two attributes we considered and any other attribute). Furthermore, we should theoretically be able to infer information about a customer only according to choices made by that customer. It is possible though that in some cases heuristic information can be inferred across different customers.

6   Conclusions

We propose a new representation scheme for utility functions. Starting from a definition of utility for a subspace of the domain, with respect to reference value of the rest of the attributes, we proceed with a definition of conditional utility as the marginal utility of an attribute, conditioned on some other attributes, and relative to the reference value. We show that conditional utility accommodates the logarithmic adaptations of the chain rule and Bayes rule, and develop the analogy to probabilities further by describing a directed graphical representation that relies on a concept of conditional independence.

In comparison with previous directed models [2, 6], we believe that our representation is simpler and easier to construct. Utility Difference Networks can be considered an adaptation of CDI-maps into a DAG, and though does not provide reduction of dimensionality, we believe that it has the potential to benefit the field in a similar way to how Bayesian Networks facilitated probabilistic reasoning in comparison to Markov Networks.

There are several direction to explore following this work. One is the form of Bayesian Learning proposed in Section 5. Furthermore, the fact that conditional utilities satisfy the chain rule and Bayes rule, implies that it may be possible to perform utility inference using algorithms similar to those that are used for belief propagation. Partial information obtained from observing the agent behavior, possibly coupled with observations about single dimensional preferences of similar users, can be used to infer other preferences. Value conditioning stemming from partial user choices or product constraints can be reasoned with, much like evidence in belief propagation, yielding estimates of the utility of various choices for other attributes.

Acknowledgements

Ronen Brafman is partially supported by ISF grant 8254320, by the Paul Ivanier Center for Robotics Research and Production Management, and by the Lynn and William Frankel Center for Computer Science. Yagil Engel is supported in part by an Aly Kaufman fellowship at the Technion.

References

[1] Bacchus, F., Grove, A.: Graphical models for preference and utility. In: Eleventh Conference on Uncertainty in Artificial Intelligence, Montreal, pp. 3–10 (1995)

[2] Boutilier, C., Bacchus, F., Brafman, R.I.: UCP-networks: A directed graphical representation of conditional utilities. In: Seventeenth Conference on Uncertainty in Artificial Intelligence, Seattle, pp. 56–64 (2001)

[3] Braziunas, D., Boutilier, C.: Local utility elicitation in GAI models. In: Twenty-first Conference on Uncertainty in Artificial Intelligence, Edinburgh, pp. 42–49 (2005)

[4] Dyer, J.S., Sarin, R.K.: Measurable multiattribute value functions. Operations Research 27, 810–822 (1979)

[5] Engel, Y., Wellman, M.P.: Generalized value decomposition and structured multiattribute auctions. In: Eighth ACM Conference on Electronic Commerce, pp. 227–236 (2007)

[6] Engel, Y., Wellman, M.P.: CUI networks: A graphical representation for conditional utility independence 31, 83–112 (2008)

[7] Fishburn, P.C.: Interdependence and additivity in multivariate, unidimensional expected utility theory. International Economic Review 8, 335–342 (1967)

[8] Gonzales, C., Perny, P.: GAI networks for utility elicitation. In: Ninth International Conference on Principles of Knowledge Representation and Reasoning, Whistler, BC, Canada, pp. 224–234 (2004)

[9] Gorman, W.M.: The structure of utility functions. Review of Economic Studies 35, 367–390 (1968)

[10] Keeney, R.L.: Utility independence and preferences for multiattributed consequences. Operations Research 19(4), 875–893 (1971)

[11] Mura, P.L., Shoham, Y.: Expected utility networks. In: Fifteenth Conference on Uncertainty in Artificial Intelligence, Stockholm, pp. 366–373 (1999)

[12] Shoham, Y.: A symmetric view of probabilities and utilities. In: Fifteenth International Joint Conference on Artificial Intelligence, Nagoya, Japan, pp. 1324–1329 (1997)

[13] Tatman, J.A., Shachter, R.D.: Dynamic programming and influence diagrams 20, 365–379 (1990)

The Possible and the Necessary for Multiple Criteria Group Decision

Salvatore Greco[1], Vincent Mousseau[2], and Roman Słowiński[3]

[1] Faculty of Economics, University of Catania, Italy
salgreco@unict.it
[2] Ecole Centrale Paris, Industrial Engineering Laboratory, France
vincent.mousseau@ecp.fr
[3] Institute of Computing Science, Poznań University of Technology, Poznań,
and Systems Research Institute, Polish Academy of Sciences, 00-441 Warsaw, Poland
roman.slowinski@cs.put.poznan.pl

Abstract. We introduce the principle of robust ordinal regression to group decision. We consider the main multiple criteria decision methods to which robust ordinal regression has been applied, i.e., UTA^{GMS} and GRIP methods, dealing with choice and ranking problems, $UTADIS^{GMS}$, dealing with sorting (ordinal classification) problems, and $ELECTRE^{GMS}$, being an outranking method applying robust ordinal regression to well known ELECTRE methods. In this way, we obtain corresponding methods for group decision: UTA^{GMS}-GROUP, $UTADIS^{GMS}$-GROUP and $ELECTRE^{GMS}$-GROUP.

Keywords: Robust ordinal regression, Multiple criteria choice, sorting and ranking, Additive value functions, Outranking methods, Multiple criteria group decision.

1 Introduction

In Multiple Criteria Decision Analysis (MCDA - for a recent state-of-the-art see [6]), an alternative a, belonging to a finite set of alternatives $A = \{a, b, \ldots, j, \ldots, m\}$ ($|A| = m$), is evaluated on n criteria $g_i : A \to \mathbf{R}$ belonging to a consistent family $G = \{g_1, g_2, \ldots, g_i, \ldots, g_n\}$ ($|G| = n$). From here on, to designate an i-th criterion, we will use interchangeably g_i or i ($i = 1, \ldots, n$). For the sake of simplicity, but without loss of generality, we suppose that evaluations on each criterion are increasing with respect to preference, *i.e.*, the more the better, defining a marginal weak preference relation as follows:

a is at least as good as b with respect to criterion i $\Leftrightarrow$ $g_i(a) \geq g_i(b)$.

There are two main approaches to construction of decision models in MCDA: Multi-Attribute Utility Theory (MAUT) [16], [5], and the outranking approach [22], [23], [8].

The purpose of MAUT is to represent preferences of a Decision Maker (DM) on a set of alternatives A by an overall value (utility) function

$$U(g_1(a), \ldots, g_n(a)) : \mathbf{R}^n \to \mathbf{R}$$

F. Rossi and A. Tsoukis (Eds.): ADT 2009, LNAI 5783, pp. 203–214, 2009.

such that:

$$a \text{ is at least as good as } b \iff U(g_1(a), \ldots, g_n(a)) \geq U(g_1(b), \ldots, g_n(b)).$$

The goal of the outranking approach is to represent preferences of a DM on a set of alternatives A by a pairwise comparison function

$$S(g_1(a), g_1(b), \ldots, g_n(a), g_n(b)) \colon \mathbf{R}^{2n} \to \mathbf{R}$$

such that:

$$a \text{ is at least as good as } b \iff S(g_1(a), g_1(b), \ldots, g_n(a), g_n(b)) \geq 0$$

Each decision model requires specification of some parameters. For example, using multiple attribute utility theory, the parameters are related to the formulation of marginal value functions $u_i(g_i(a))$, $i = 1, \ldots, n$, while using the outranking approach, the parameters can be weights, indifference, preference, and veto thresholds for each criterion g_i, $i = 1, \ldots, n$.

Recently, MCDA methods based on indirect preference information and on the *disaggregation paradigm* [15] are considered more interesting, because they require less cognitive effort from the DM in order to express preference information. The DM provides some holistic preferences on a set of reference alternatives A^R, and from this information the parameters of a decision model are induced using a methodology called *ordinal regression*. The resulting decision model consistent with the provided preference information is used to evaluate the alternatives from set A (*aggregation stage.*) Typically, ordinal regression has been applied to MAUT models, so in these cases we speak of *additive ordinal regression*. For example, additive ordinal regression has been applied in the well-known method called UTA (see [14]). The ordinal regression methodology has been applied, moreover, to some nonadditive decision models. In this case, we speak of *nonadditive ordinal regression* and its typical representatives are the UTA like-methods substituting the additive value function by the Choquet integral (see [4], [17], [1]), and the DRSA methodology using a set of decision rules as the decision model (see [9]).

Usually, among the many sets of parameters of a decision model representing the preference information given by the DM, only one specific set is considered. We say that the set of parameters or the decision model is *compatible* with the preference information given by the DM if it is consistent with the preference information given by the DM. For example, from among the many compatible value functions only one function is selected to rank the alternatives from set A. Since the choice of one among many compatible sets of parameters is arbitrary to some extent, recently *robust ordinal regression* has been proposed with the aim of taking into account all compatible sets of parameters. The first robust ordinal regression method is the recently proposed generalization of the UTA method, called UTA^{GMS} [11]. In UTA^{GMS}, instead of only one compatible additive value function composed of piecewise-linear marginal functions, all compatible additive value functions composed of general monotonic marginal functions are taken into account.

As to the preference information, the UTA^{GMS} method requires from a DM to make some pairwise comparisons on a set of reference alternatives $A^R \subseteq A$. The set of all compatible decision models defines two relations in set A: the *necessary* weak preference relation, which holds for any two alternatives $a, b \in A$ if and only if all compatible value functions give to a a value greater than the value given to b, and the *possible* weak preference relation, which holds for this pair if and only if at least one compatible value function gives to a a value greater than the value given to b.

Recently, an extension of UTA^{GMS} has been proposed and called the *GRIP* method [7]. The *GRIP* method builds a set of all compatible additive value functions, taking into account not only a preorder on a set of alternatives, but also the intensities of preference among some reference alternatives. This kind of preference information is required in other well-known MCDA methods, such as $MACBETH$ [3] and AHP [24], [25]. Both UTA^{GMS} and $GRIP$ apply the robust ordinal regression to the multiple attribute additive model and, therefore, we can say that these methods apply the *additive robust ordinal regression*. In the literature, the *nonadditive robust ordinal regression* has been proposed, applying the approach of robust ordinal regression to a value function having the form of Choquet integral in order to represent positive and negative interactions between criteria [2]. The robust ordinal regression approach can be applied also to the outranking approach [10].

In this paper, we wish to consider the robust ordinal regression in a group decision context. Therefore, we consider a set of decision makers $\mathcal{D} = \{d_1, \ldots, d_p\}$ with each own preferences, and we use robust ordinal regression to investigate spaces of consensus between them. The article is organized as follows. Section 2 is devoted to presentation of the general scheme of robust ordinal regression for choice and ranking problems within MAUT, as well as basic principles of UTA^{GMS} and $GRIP$ methods. In section 3, robust ordinal regression for group choice and ranking problems is introduced within MAUT, and the UTA^{GMS}-GROUP method is presented. Section 4 presents a general scheme of robust ordinal regression for sorting problems within MAUT, as well as basic principles of $UTADIS^{GMS}$. In section 5, robust ordinal regression for group sorting problems is introduced within MAUT, and the $UTADIS^{GMS}$-GROUP method is presented. Section 6 presents a general scheme of robust ordinal regression within the outranking approach, as well as basic principles of $ELECTRE^{GMS}$. In section 7, robust ordinal regression for group decision problems is introduced within the outranking approach, and the $ELECTRE^{GMS}$-GROUP method is presented. The last section contains conclusions.

2   The Robust Ordinal Regression Approach for Choice and Ranking Problems within MAUT

MAUT provides a theoretical foundation for preference modeling using a real-valued utility function, called value function, aggregating evaluations of alternatives on multiple criteria. The value function is intended to be a preference model of a particular DM. It is also a decision model, since it gives scores to alternatives which permit to order them from the best to the worst, or to choose the best alternative with the highest score. Its most popular form is additive:

$$U(a) = \sum_{i=1}^{n} u_i(g_i(a)), \tag{1}$$

where $u_i(g_i(a))$, $i = 1, \ldots, n$, are real-valued marginal value functions.

Ordinal regression has been known for at least fifty years in the field of multidimensional analysis. It has been applied within MAUT, first to assess weights of an additive linear value function [27], [21], and then to assess parameters of an additive piece-wise linear value function [14]. The latter method, called UTA, initiated a stream of further developments, in both theory and applications [26].

Recently, two new methods, UTA^{GMS} [11] and $GRIP$ (Generalized Regression with Intensities of Preference) [7], have generalized the ordinal regression approach of the UTA method in several aspects, the most important of which is that they are taking into account all additive value functions (1) compatible with the preference information, while UTA is using only one such function.

2.1   The Preference Information Provided by the Decision Maker

The DM is expected to provide the following preference information:

- a partial preorder $\succeq$ on $A^R \subseteq A$ whose meaning is: for $x, y \in A^R$

$$x \succeq y \Leftrightarrow x \text{ is at least as good as } y,$$

- a partial preorder $\succeq^*$ on $A^R \times A^R$, whose meaning is: for $x, y, w, z \in A^R$,

$$(x, y) \succeq^* (w, z) \Leftrightarrow x \text{ is preferred to } y \text{ at least as much as } w \text{ is preferred to } z,$$

- a partial preorder $\succeq_i^*$ on $A^R \times A^R$, whose meaning is: for $x, y, w, z \in A^R$, $(x, y) \succeq_i^* (w, z) \Leftrightarrow x$ is preferred to y at least as much as w is preferred to z on criterion g_i, $i = 1, \ldots, n$.

2.2   Possible and Necessary Rankings

A compatible value function is able to restore the preference information expressed by the DM on A^R and $A^R \times A^R$. Each compatible value function induces, moreover, a complete preorder on the whole set A. In particular, for any two solutions $x, y \in A$, a compatible value function orders x and y in one of the following ways: $x \succ y$, $y \succ x$, $x \sim y$. With respect to $x, y \in A$, it is thus reasonable to ask the following two questions:

- Are x and y ordered in the same way by *all* compatible value functions?
- Is there *at least one* compatible value function ordering x at least as good as y (or y at least as good as x)?

In the answer to these questions, UTA^{GMS} and $GRIP$ produce two rankings on the set of alternatives A, such that for any pair of alternatives $a, b \in A$:

- in the *necessary* ranking (partial preorder), a is ranked at least as good as b if and only if, $U(a) \geq U(b)$ for *all* value functions compatible with the preference information,
- in the *possible* ranking (strongly complete and negatively transitive relation), a is ranked at least as good as b if and only if, $U(a) \geq U(b)$ for *at least one* value function compatible with the preference information.

The necessary ranking can be considered as *robust* with respect to the preference information. Such robustness of the necessary ranking refers to the fact that any pair of alternatives compares in the same way whatever the additive value function compatible with the preference information. Indeed, when no preference information is given, the necessary ranking boils down to the dominance relation, and the possible ranking is a complete relation.

As $GRIP$ is taking into account additional preference information in form of comparisons of intensities of preference between some pairs of reference alternatives, the set of all compatible value functions restoring the whole preference information is also used to produce four types of relations on $A \times A$, such that for any four alternatives $a, b, c, d \in A$:

- the *necessary* relation $(a, b) \succeq^{*^N} (c, d)$ (partial preorder) holds (a is preferred to b *necessarily* at least as much as c is preferred to d), if and only if $U(a) - U(b) \geq U(c) - U(d)$ for *all* compatible value functions,
- the *possible* relation $(a, b) \succeq^{*^P} (c, d)$ (strongly complete and negatively transitive relation) holds (a is preferred to b *possibly* at least as much as c is preferred to d), if and only if $U(a) - U(b) \geq U(c) - U(d)$ for *at least one* compatible value functions,
- the *necessary* relation $(a, b) \succeq_i^{*^N} (c, d)$ (partial preorder) holds (on criterion i, a is preferred to b *necessarily* at least as much as c is preferred to d), if and only if $u_i(a) - u_i(b) \geq u_i(c) - u_i(d)$ for *all* compatible value functions $(i = 1, \ldots, n)$,
- the *possible* relation $(a, b) \succeq_i^{*^P} (c, d)$ (strongly complete and negatively transitive relation) holds (on criterion i, a is preferred to b *possibly* at least as much as c is preferred to d), if and only if $u_i(a) - u_i(b) \geq u_i(c) - u_i(d)$ for *at least one* compatible value functions $(i = 1, \ldots, n)$.

3   Robust Ordinal Regression for Group Decision about Choice and Ranking: The UTA^{GMS}-GROUP Method

The UTA^{GMS}-GROUP method applies the robust ordinal regression approach to the case of group decision, in which several DMs cooperate to make a collective decision. DMs share the same "description" of the decision problem (the same set of alternatives, evaluation criteria and performance matrix). Each DM provides his/her own preference information, composed of pairwise comparisons of some reference alternatives. The collective preference model accounts for the preference expressed by each DM. Although in the considered framework it is also possible to handle preference information about intensity of preference, we will skip this preference information for the lack of space.

Let us denote the set of DMs by $\mathcal{D} = \{d_1, \ldots, d_p\}$. For each DM $d_h \in \mathcal{D}' \subseteq \mathcal{D}$, we consider all compatible value functions. Four situations are interesting for a pair $(a, b) \in A$:

- $a \succeq_{\mathcal{D}'}^{N,N} b$: $a \succeq^N b$ for *all* $d_h \in \mathcal{D}'$,

- $a \succeq_{\mathcal{D}'}^{N,P} b$: $a \succeq^N b$ for *at least one* $d_h \in \mathcal{D}'$,

- $a \succeq_{\mathcal{D}'}^{P,N} b$: $a \succeq^P b$ for *all* $d_h \in \mathcal{D}'$,

- $a \succeq_{\mathcal{D}'}^{P,P} b$: $a \succeq^P b$ for *at least one* $d_h \in \mathcal{D}'$.

4   Robust Ordinal Regression for Sorting Problems: The $UTADIS^{GMS}$ Method

Robust ordinal regression has also been proposed for sorting problems in the new $UTADIS^{GMS}$ method [12], considering an additive value function (1) as a preference model. Let us remember that sorting procedures consider a set of k predefined preference ordered classes $C_1, C_2, \ldots, C_k$, where $C_{h+1} \gg C_h$ ($\gg$ a complete order on the set of classes), $h = 1, \ldots, k - 1$. The aim of a sorting procedure is to assign each alternative to one class or to a set of contiguous classes. The robust ordinal regression takes into account a *value driven sorting procedure*, that is, it uses a value function U to decide the assignments in such a way that if $U(a) > U(b)$ then a is assigned to a class not worse than b.

We suppose the DM provides preference information in form of possibly imprecise assignment examples on a reference set of alternatives $A^R \subseteq A$, i.e. for $a^R \in A^R$ the DM defines a desired assignment $a^R \to [C_{L^{DM}(a^R)}, C_{R^{DM}(a^R)}]$, where $[C_{L^{DM}(a^R)}, C_{R^{DM}(a^R)}]$ is an interval of contiguous classes $C_{L^{DM}(a^R)}$, $C_{L^{DM}(a^R)+1}, \ldots, C_{R^{DM}(a^R)}$. An assignment example is said to be precise if $L^{DM}(a^R) = R^{DM}(a^R)$, and imprecise, otherwise.

Given a value function U, a set of assignment examples is said to be *consistent with U* iff

$$\forall a^R, b^R \in A^R, U(a^R) \geq U(b^R) \Rightarrow R^{DM}(a^R) \geq L^{DM}(b^R). \qquad (2)$$

Given a set A^R of assignment examples and a corresponding set $\mathcal{U}_{A^R}$ of compatible value functions, for each $a \in A$ we define the *possible assignment* $C_P(a)$ as the set of indices of classes C_h for which there exist *at least one* value function $U \in \mathcal{U}_{A^R}$ assigning a to C_h, and the *necessary assignment* $C_N(a)$ as set of indices of classes C_h for which *all* value functions $U \in \mathcal{U}_{A^R}$ assign a to C_h.

5 Robust Ordinal Regression for Group Decision about Sorting: The $UTADIS^{GMS}$-GROUP Method

Given a set of DMs $\mathcal{D} = \{d_1, \ldots, d_p\}$, for each DM $d_r \in \mathcal{D}' \subseteq \mathcal{D}$ we consider the set of all compatible value functions $\mathcal{U}_{A^R}^{d_r}$. Given a set A^R of assignment examples, for each $a \in A$ and for each DMs $d_r \in \mathcal{D}'$ we define his/her possible and necessary assignments as

$$C_P^{d_r}(a) = \{h \in H \text{ such that } \exists U \in \mathcal{U}_{A^R}^{d_r} \text{ assigning } a \text{ to } C_h\} \qquad (3)$$

$$C_N^{d_r}(a) = \{h \in H \text{ such that } \forall U \in \mathcal{U}_{A^R}^{d_r}, \mathcal{U} \text{ is assigning } a \text{ to } C_h\} \qquad (4)$$

Moreover, for each subset of DMs $D' \subseteq \mathcal{D}$, we define the following assignments:

$$C_{P,P}^{D'}(a) = \bigcup_{d_r \in \mathcal{D}'} C_P^{d_r}(a) \qquad (5)$$

$$C_{N,P}^{D'}(a) = \bigcup_{d_r \in \mathcal{D}'} C_N^{d_r}(a) \qquad (6)$$

$$C_{P,N}^{D'}(a) = \bigcap_{d_r \in \mathcal{D}'} C_P^{d_r}(a) \qquad (7)$$

$$C_{N,N}^{D'}(a) = \bigcap_{d_r \in \mathcal{D}'} C_N^{d_r}(a). \qquad (8)$$

Possible and necessary assignments $C_P^{d_r}(a)$ and $C_N^{d_r}(a)$ are calculated for each decision maker $d_r \in \mathcal{D}$ using UTADIS^{GMS}, and then the four assignments $C_{P,P}^{D'}(a)$, $C_{N,P}^{D'}(a)$, $C_{P,N}^{D'}(a)$ and $C_{P,P}^{D'}(a)$ can be calculated for all subsets of decision makers $\mathcal{D}' \subseteq \mathcal{D}$.

6 Robust Ordinal Regression for Outranking Methods

Outranking relation is a non-compensatory preference model used in the $ELECTRE$ family of multiple criteria decision aiding methods [22]. Its construction involves two concepts known as concordance and discordance. Outranking relation, usually denoted by S, is a binary relation on a set A of actions. For an ordered pair of actions $(a, b) \in A$, aSb means "a is at least as good as b". The assertion aSb is considered to be true if the coalition of criteria being in favor of this statement is strong enough comparing to the rest of criteria, and if among the criteria opposing to this statement, there is no one for which a is significantly worse than b. The first condition is called concordance test, and the second, non-discordance test.

Let us denote by k_j the weight assigned to criterion g_j, $j = 1, \ldots, n$; it represents a relative importance of criterion g_j within family G of n criteria. The indifference, preference and veto thresholds on criterion g_j are denoted by q_j, p_j and v_j, respectively. For consistency, $v_j > p_j > q_j \geq 0$, $j = 1, \ldots, n$. In all formulae that follow, we suppose, without loss of generality, that all these thresholds are constant, that preferences are increasing with evaluations on particular criteria, and that criteria are identified by their indices.

The concordance test involves calculation of concordance index $C(a, b)$. It represents the strength of the coalition of criteria being in favor of aSb. This coalition is composed of two subsets of criteria:

- subset of criteria being clearly in favor of aSb, i.e., such that $g_j(a) \geq g_j(b) - q_j$,
- subset of criteria that do not oppose to aSb, while being in an ambiguous position with respect to this assertion; these are those criteria for which a weak preference relation bQa holds; i.e., such that $g_j(b) - p_j \leq g_j(a) < g_j(b) - q_j$.

Consequently, the concordance index is defined as

$$C(a, b) = \frac{\sum_{j=1}^{n} \phi_j(a, b) \times k_j}{\sum_{j=1}^{n} k_j}, \tag{9}$$

where, for $j = 1, \ldots, n$,

$$\phi_j(a, b) = \begin{cases} 1, & \text{if } g_j(a) \geq g_j(b) - q_j, \\ \frac{g_j(a) - [g_j(b) - p_j]}{p_j - q_j}, & \text{if } g_j(b) - p_j \leq g_j(a) < g_j(b) - q_j, \\ 0, & \text{if } g_j(a) < g_j(b) - p_j. \end{cases} \tag{10}$$

$\phi_j(a, b)$ is a marginal concordance index, indicating to what extend criterion g_j contributes to the concordance index $C(a, b)$. As defined by (10), $\phi_j(a, b)$ is a piecewise-linear function, non-decreasing with respect to $g_j(a) - g_j(b)$.

Remark that $C(a, b) \in [0, 1]$, where $C(a, b) = 0$ if $g_j(a) \leq g_j(b) - p_j$, $j = 1, \ldots, n$ (b is strictly preferred to a on all criteria), and $C(a, b) = 1$ if $g_j(a) \geq g_j(b) - q_j$, $j = 1, \ldots, n$ (a outranks b on all criteria).

The result of the concordance test for a pair $(a, b) \in A$ is positive if $C(a, b) \geq \lambda$, where $\lambda \in [0.5, 1]$ is a cutting level, which has to be fixed by the DM.

Once the result of the concordance test has been positive, one can pass to the non-discordance test. Its result is positive for the pair $(a, b) \in A$ unless "a is significantly worse than b" on at least one criterion, i.e., if $g_j(b) g_j(a) < v_j$ for $j = 1, \ldots, n$.

It follows from above that the outranking relation for a pair $(a, b) \in A$ is true, and denoted by aSb if both the concordance test and the non-discordance test are positive. On the other hand, the outranking relation for a pair $(a, b) \in A$ is false, and denoted by $aS^c b$, either if the concordance test or the non-discordance test is negative. Knowing S or S^c for all ordered pairs $(a, b) \in A$, one can proceed to exploitation of the outranking relation in set A, which is specific for the choice, or sorting or ranking problem, as described in [8].

Experience indicates that elicitation of preference information necessary for construction of the outranking relation is not an easy task for a DM. In particular, the inter-criteria preference information concerning the weights of criteria and the veto thresholds are difficult to be expressed directly.

For this reason, some aggregation-disaggregation procedures have been proposed in the past to assist the elicitation of the weights of criteria and all the thresholds required to construct the outranking relation [19], [20], [18]. A robust ordinal regression approach to outranking methods has been presented in [10]. Below, we briefly sketch this proposal.

We assume that the preference information provided by the DM is a set of pairwise comparisons of some reference actions. The set of reference actions is denoted by A^R, and it is usually, although not necessarily, a subset of set A. The comparison of a pair of actions $(a, b) \in A^R$ states the truth or falsity of the outranking relation, denoted by aSb or aS^cb, respectively. It is worth stressing that the DM does not need to provide all pairwise comparisons of reference actions, so this comparison can be confined to a small subset of pairs.

We also assume that the intra-criterion preference information concerning indifference and preference thresholds $p_j > q_j \geq 0$, $j = 1, \ldots, n$, is given. The last assumption is not unrealistic because these thresholds are relatively easy to provide by an analyst who is usually aware what is the precision of criteria, and how much difference is non-significant or relevant.

In order to simplify calculations of the ordinal regression, we assume that the weights of criteria sum up to one, i.e. $\sum_{j=1}^{n} k_j = 1$. Thus, (9) becomes

$$C(a, b) = \sum_{j=1}^{n} \phi_j(a, b) \times k_j = \sum_{j=1}^{n} \psi_j(a, b), \tag{11}$$

where the marginal concordance index $\psi_j(a, b) = \phi_j(a, b) \times k_j$ is a monotone non-decreasing function with respect to $g_j(a) - g_j(b)$, such that $\psi_j(a, b) \geq 0$ for $g_j(a) - g_j(b) \geq -q_j$, $j = 1, \ldots, n$, and $\sum_{j=1}^{n} \psi_j(a, b) = 1$ in case $g_j(a) - g_j(b) = \beta_j - \alpha_j$ for all $j = 1, \ldots, n$, α_j and β_j being the worst and the best evaluation on criterion g_j, respectively.

The ordinal regression constraints defining the set of concordance indices $C(a, b)$, cutting levels λ and veto thresholds v_j, $j = 1, \ldots, n$, compatible with the pairwise comparisons provided by the DM have the following form:

$$\left. \begin{array}{l} C(a, b) = \sum_{j=1}^{n} \psi_j(a, b) \geq \lambda \text{ and } g_j(b) - g_j(a) \leq v_j - \epsilon, j = 1, \ldots, n, \\ \quad \text{if } aSb, \text{ for } (a, b) \in A^R, \\ C(a, b) = \sum_{j=1}^{n} \psi_j(a, b) \leq \lambda + \epsilon + M_0(a, b) \text{ and } g_j(b) - g_j(a) \leq v_j - \delta M_j(a, b), \\ M_j(a, b) \in \{0, 1\}, \sum_{j=0}^{n} M_j(a, b) \leq n, \ j = 1, \ldots, n, \\ \quad \text{if } aS^cb, \text{ for } (a, b) \in A^R, \\ 1 \geq \lambda \geq 0.5, \quad v_j \geq p_j, \ j = 1, \ldots, n, \\ \psi_j(a, b) \geq 0 \text{ if } g_j(a) - g_j(b) \geq -q_j, \text{ for all } (a, b) \in A^R, \ j = 1, \ldots, n, \\ \sum_{j=1}^{n} \psi_j(a, b) = 1 \text{ if } g_j(a) - g_j(b) = \beta_j - \alpha_j \text{ for all } (a, b) \in A^R, \ j = 1, \ldots, n, \\ \psi_j(a, b) \geq \psi_j(c, d) \text{ if } g_j(a) - g_j(b) \geq g_j(c) - g_j(d), \\ \quad \text{for all } a, b, c, d \in A^R, \ j = 1, \ldots, n, \end{array} \right\} E(A^R)$$

where ϵ is a small positive value and δ is a big positive value. Remark that $E(A^R)$ are constraints of a 0-1 mixed linear program.

Given a pair of actions $(x, y) \in A$, the following values are useful to build necessary and possible outranking relations:

$$d(x, y) = Min\left\{ \sum_{j=1}^{n} \psi_j(x, y) - \lambda \right\}, D(x, y) = Max\left\{ \sum_{j=1}^{n} \psi_j(x, y) - \lambda \right\}.$$

subject to constraints $E(A^R)$, where $\psi_j(a, b) \geq \psi_j(c, d)$ if $g_j(a) - g_j(b) \geq g_j(c) - g_j(d)$, for all $a, b, c, d \in A^R \cup \{x, y\}$, $j = 1, \ldots, n$, and $g_j(y) - g_j(x) \geq v_j$, $j = 1, \ldots, n$.

Given a pair of actions $(x, y) \in A$, x *necessarily* outranks y, which is denoted by $xS^N y$, if and only if $d(x, y) \geq 0$. $d(x, y) \geq 0$ means that for all compatible outranking models x outranks y. Analogously, given a pair of actions $(x, y) \in A$, x *possibly* outranks y, which is denoted by $xS^P y$, if and only if $D(x, y) \geq 0$. $D(x, y) \geq 0$ means that for at least one compatible outranking model x outranks y. The necessary and the possible outranking relations are to be exploited as usual outranking relations in the context of choice, sorting and ranking problems.

7   Robust Ordinal Regression for Outranking Methods in Group Decision Problems

The above approach can be adapted to the case of group decision. In this case, several DMs cooperate in a decision problem to make a collective decision. DMs share the same "description" of the decision problem (the same set of actions, evaluation criteria and performance matrix). Each DM provides his/her own preference information, composed of pairwise comparisons of some reference actions. The collective preference model accounts for the preference expressed by each DM.

Let us denote the set of DMs by $\mathcal{D} = \{d_1, \ldots, d_p\}$. For each DM $d_r \in D' \subseteq D$, we consider all compatible outranking models. Four situations are interesting for a pair $(x, y) \in A$:

- $x\ S^{N,N}(D')\ y$: $xS^N y$ for all $d_r \in D'$,
- $x\ S^{N,P}(D')\ y$: $xS^N y$ for at least one $d_r \in D'$,
- $x\ S^{P,N}(D')\ y$: $xS^P y$ for all $d_r \in D'$,
- $x\ S^{P,P}(D')\ y$: $xS^P y$ for at least one $d_r \in D'$.

8   Conclusions

In this article we presented basic principles of robust ordinal regression for group decision. After recalling the robust ordinal regression methods within MAUT for choice and ranking problems (UTA^{GMS} and $GRIP$), for sorting problems (($UTADIS^{GMS}$), as well as ordinal regression methods within the outranking approach ($ELECTRE^{GMS}$), we extended all these methods to group decision introducing UTA^{GMS}-Group, $UTADIS^{GMS}$-GROUP and $ELECTRE^{GMS}$-GROUP.

References

1. Angilella, S., Greco, S., Lamantia, F., Matarazzo, B.: Assessing non-additive utility for multicriteria decision aid. European Journal of Operational Research 158, 734–744 (2004)
2. Angilella, S., Greco, S., Matarazzo, B.: Nonadditive robust ordinal regression: aggregation-disaggregation approach considering all compatible fuzzy measures in a multiple criteria decision model based on the Choquet integral. Forthcoming in European Journal of Operational Research (2008)

3. Bana e Costa, C.A., Vansnick, J.C.: MACBETH: An interactive path towards the construction of cardinal value functions. International Transactions in Operational Research 1(4), 387–500 (1994)
4. Choquet, G.: Theory of capacities. Annales de l'Institut Fourier 5, 131-295 (1953)
5. Dyer, J.S.: Multiattribute Utility Theory. In: Figueira, J., Greco, S., Ehrgott, M. (eds.) Multiple Criteria Decision Analysis: State of the Art Surveys, pp. 265–295. Springer, New York (2005)
6. Figueira, J., Greco, S., Ehrgott, M. (eds.): Multiple Criteria Decision Analysis: State of the Art Surveys. Springer, New York (2005)
7. Figueira, J., Greco, S., Słowiński, R.: Building a set of additive value functions representing a reference preorder and intensities of preference: GRIP method. European Journal of Operational Research 195, 460–486 (2009)
8. Figueira, J., Mousseau, V., Roy, B.: ELECTRE methods. In: Figueira, J., Greco, S., Ehrgott, M. (eds.) Multiple Criteria Decision Analysis: State of the Art Surveys, ch. 4, pp. 133–162. Springer, Heidelberg (2005)
9. Greco, S., Matarazzo, B., Słowiński, R.: Rough sets theory for multicriteria decision analysis. European Journal of Operational Research 129, 1–47 (2001)
10. Greco, S., Mousseau, V., Słowiński, R.: The necessary and the possible in a perspective of robust decision aiding. Presented at 66th Meeting of the European Working Group on Multiple Criteria Decision Aiding, Marrakech, Morocco, October 18-20 (2007)
11. Greco, S., Mousseau, V., Słowiński, R.: Ordinal regression revisited: Multiple criteria ranking with a set of additive value functions. European Journal of Operational Research 191, 415–435 (2008)
12. Greco, S., Mousseau, V., Słowiński, R.: Multiple criteria sorting with a set of additive value functions. European Journal of Operational Research (submitted, 2009)
13. Kiss, L., Martel, J.M., Nadeau, R.: ELECCALC - an interactive software for modelling the decision maker's preferences. Decision Support Systems 12, 757–777 (1994)
14. Jacquet-Lagrèze, E., Siskos, Y.: Assessing a set of additive utility functions for multicriteria decision-making, the UTA method. European Journal of Operational Research 10, 151–164 (1982)
15. Jacquet-Lagrèze, E., Siskos, Y.: Preference disaggregation: 20 years of MCDA experience. European Journal of Operational Research 130, 233–245 (2001)
16. Keeney, R.L., Raiffa, H.: Decision with Multiple Objectives - Preferences and value Tradeoffs. Wiley, New York (1976)
17. Marichal, J.L., Roubens, M.: Determination of weights of interacting criteria from a reference set. European Journal of Operational Research 124, 641–650 (2000)
18. Mousseau, V., Dias, L.: Valued outranking relations in ELECTRE providing manageable disaggregation procedures. European Journal of Operational Research 156, 467–482 (2004)
19. Mousseau, V., Słowiński, R.: Inferring an ELECTRE TRI model from assignment examples. Journal of Global Optimization 12, 157–174 (1998)
20. Mousseau, V., Słowiński, R., Zielniewicz, P.: A user-oriented implementation of the ELECTRE-TRI method integrating preference elicitation support. Computers and Operations Research 27, 757–777 (2000)
21. Pekelman, D., Sen, S.K.: Mathematical programming models for the determination of attribute weights. Management Science 20, 1217–1229 (1974)
22. Roy, B.: The outranking approach and the foundations of ELECTRE methods. Theory and Decision 31, 49–73 (1991)

23. Roy, B., Bouyssou, D.: Aide Multicritère à la Décision: Méthodes et Cas. Economica, Paris (1993)
24. Saaty, T.L.: The Analytic Hierarchy Process. McGraw-Hill, New York (1980)
25. Saaty, T.L.: The Analytic Hierarchy and Analytic Network Processes for the Measurement of Intangible Criteria and for Decision-Making. In: Figueira, J., Greco, S., Ehrgott, M. (eds.) Multiple Criteria Decision Analysis: State of the Art Surveys, pp. 345–407. Springer, New York (2005)
26. Siskos, Y., Grigoroudis, V., Matsatsinis, N.: UTA methods. In: Figueira, J., Greco, S., Ehrgott, M. (eds.) Multiple Criteria Decision Analysis: State of the Art Surveys, pp. 297–343. Springer, New York (2005)
27. Srinivasan, V., Shocker, A.D.: Estimating the weights for multiple attributes in a composite criterion using pairwise judgments. Psychometrika 38, 473–493 (1973)

Preferences in an Open World

Ulrich Junker

uli.junker@free.fr

Abstract. We consider constructive approaches to decision making which allow incomplete preference orders over multiple criteria. Whereas additional preferences may be acquired during the decision making process, the set of criteria is usually kept fixed. In this paper, we study the addition of new criteria and examine how this may refine or even reverse the existing preferences. We identify essential changes in the preference order and show that these changes provide a compact representation of preference relations in an open world.

1 Introduction

We consider constructive approaches to decision making which allow incomplete preference orders over multiple criteria. Whereas additional preferences may be acquired during the decision making process, the set of criteria is usually kept fixed. We believe that this assumption is not realistic and that the discovery of new dimensions or the merging of different viewpoints may lead to the discovery of new preferences that could not be formulated within the more restricted viewpoints.

In this paper, we propose a more realistic preference acquisition model. The key notion is that of enlarging given viewpoints. The notion of a viewpoint is used by [2] to describe an independent way of analyzing, evaluating, and comparing alternative actions. In collaborative decision making, different agents want to make a common decision by comparing the outcomes of the actions. As different agents may prefer different outcomes, each agent has a particular viewpoint and these viewpoints have different preference relations. In multi-criteria decision making, multiple outcomes of the actions may be compared independently of each other. Each of these outcomes constitutes a criterion for evaluating and comparing the actions. For example, we may compare the available hotels for a night stop by criteria such as their price and their distance to the airport. Each of these criteria can constitute an independent viewpoint. In this case, any trade-off between the two criteria price and distance is as good as the other those trade-offs. However, multiple criteria can also be combined into a single viewpoint. If the decision maker has preferences between different price- and distance-combinations, then we encounter a viewpoint which maps the different actions to a combinatorial outcome space defined by price and distance.

In the same way as preferences may be acquired or discovered incrementally, it may also happen that new criteria are discovered during such an incremental decision making process. We thus enlarge an existing viewpoint to a new

F. Rossi and A. Tsoukis (Eds.): ADT 2009, LNAI 5783, pp. 215–224, 2009.

viewpoint by adding a new dimension to the original outcome space. For example, our decision maker may suddenly discover that the hotels around the airport differ much in quality. As a consequence, the two-dimensional outcome space over price and distance is replaced by a three-dimensional outcome space defined by price, distance, and quality. We can then lift the preferences on the original outcome space to the new outcome space in a ceteris-paribus manner. However, the enlarged viewpoint may refine and even reverse these preferences lifted from the original viewpoint. These changes between expected and actual preference orders provide crucial information for preference representation and reasoning.

The paper is organized as follows: section 2 and 3 recall background in decision making in terms of the concept of a viewpoint. Section 4 introduces the new concept of enlarging a viewpoint. Sections 5 and 6 study compact representations in terms of transitive reductions.

2   Viewpoints with Incomplete Preferences

We consider classic decision-making problems where a single decision has to be chosen from a set of actions $\mathcal{A}$. A *viewpoint* constitutes an independent way to analyze the actions and to evaluate and compare the outcomes of the actions [2]. We restrict our discussion to viewpoints where actions have a deterministic outcome. A viewpoint captures weak preferences between outcomes, i.e. certain outcomes are at least as preferred as other outcomes. We follow a constructive approach to decision making where the preference relation between outcomes is incomplete initially and can successively be refined, for example by eliciting information from the decision maker. However, we suppose that the preference relation between outcomes is reflexive and transitive, i.e. an incomplete pre-order $\succsim$. This permits to model indifference, strict preferences, and incomparability between outcomes:

Definition 1. *A viewpoint v for a set of actions $\mathcal{A}$ is characterized by an outcome space Ω_v, a criterion $z_v : \mathcal{A} \to \Omega_v$ which maps actions to their outcomes, and a pre-order $\succsim_v$ over Ω_v which defines weak preferences between outcomes.*

Two different viewpoints for the same set of actions may differ in their preferences, their criteria, and their outcome spaces. The first difference is encountered in collaborative decision making where multiple agents want to make a common decision, but compare the same outcomes of the actions differently. A good example is the choice of a restaurant by a group of people who have different preferences about the kind of food that is served. Hence, each agent has a separate viewpoint and these viewpoints map the actions to the same outcomes, but differ in their preference relations. The second difference is obtained if two viewpoints map the actions to different outcomes from the same outcome space. This occurs in robust decision making involving multiple scenarios. For example, optimistic scenarios may assume small delays for given flights, whereas pessimistic scenarios may assume large delays for the same flights. Each scenario constitutes a viewpoint and the viewpoints map the actions to the outcomes differently, but use the same preferences for comparing them. The third difference is encountered in multi-criteria decision making and multi-objective optimization [4] where each viewpoint defines a distinct outcome space or dimension. For example, there may be a viewpoint considering the costs of a trip, another viewpoint for the comfort, and a third viewpoint for the interest of the trip.

We now discuss how to make rational decisions from a viewpoint with incomplete preferences. We recall that a pre-order $\succsim_v$ over Ω_v can be split into a strict preference relation $\succ_v$ and an indifference relation $\sim_v$. An outcome ω_1 is strictly preferred to an outcome ω_2 iff ω_1 is weakly preferred to ω_2, but not vice versa. The outcomes ω_1 and ω_2 are indifferent iff ω_1 is weakly preferred to ω_2 and ω_2 is weakly preferred to ω_1. The strict relation $\succ_v$ is a strict partial order and the indifference relation is an equivalence relation. Furthermore, two outcomes ω_1 and ω_2 are called comparable iff either ω_1 is weakly preferred to ω_2 or ω_2 is weakly preferred to ω_1. A pre-order $\succsim_v$ is called complete iff all pairs of outcomes are comparable.

In a constructive approach to decision making, preference relations are initially incomplete and are successively refined. This can be achieved by eliciting preferences actively from a user, by learning user preferences passively, or by constructing these preferences according to higher-level reflections. We thus encounter a sequence of viewpoints having the same criteria, but more and more refined preference relations with the purpose of reaching complete relations in the end of this construction process. We impose some constraints on the way a viewpoint w may extend the preference relation of a viewpoint v. Firstly, we suppose that preferences are added, but not removed, meaning that the pre-order $\succsim_w$ is a superset of the pre-order $\succsim_v$. If we allow the addition of arbitrary preferences, we may end in viewpoints that are indifferent among all outcomes. Whereas the addition of new preferences preserves previous indifference relations between outcomes, it need not preserve a strict preference between an outcome ω_1 and ω_2 since a weak preference between ω_2 and ω_1 may be added. In order to preserve strict preferences, we forbid the addition of preferences among comparable outcomes:

Definition 2. *A viewpoint w is an extension of a viewpoint v iff these viewpoints are defined for the same actions $\mathcal{A}$, agree in their outcome spaces and criteria, i.e. $\Omega_v = \Omega_w$ and $z_v = z_w$, and the preference relations $\succsim_w$ and $\succsim_v$ agree on all outcomes that are comparable with respect to $\succsim_v$.*

If the preference relation of a viewpoint v is incomplete, we can define an extension with fewer incomparable outcomes by choosing a pair of incomparable outcomes ω_1 and ω_2 and by making them comparable. We can either establish a strict preference or an indifference. We thus obtain three ways of extending $\succsim_v$. We either add the pair (ω_1, ω_2) or the pair (ω_2, ω_1), or both pairs to $\succsim_v$, and determine the transitive closure of the result, which will be the pre-order $\succsim_w$ of a new viewpoint w which has the same outcome space and criterion as v. It can be shown that this new viewpoint w is an extension of v.

An incomplete viewpoint can thus be made complete by a sequence of steps where each step transforms at least one pair of incomparable outcomes into a pair of comparable outcomes. It can therefore be shown that complete extensions of a viewpoint always exist. Moreover, incomplete viewpoints have more than one complete extension.

Viewpoints with complete pre-orders satisfy the postulates of rational decision making. Complete viewpoints have an outcome that is at least as preferred as all outcomes. A rational decision of a complete viewpoint is an action that produces this optimal outcome. An incomplete viewpoint has multiple complete extensions. We say that a rational decision of an incomplete viewpoint is an action that leads to an optimal outcome in some complete extension of the viewpoint. It can be shown that an outcome is optimal in some complete extension of a viewpoint iff it is non-dominated in the viewpoint, i.e. there is no strictly preferred outcome in the viewpoint:

Proposition 1. *Let v be a viewpoint. An outcome $\omega^* \in \Omega_v$ is optimal in some complete extension w of v, i.e. $\omega^* \succsim_w \omega$ for all $\omega \in \Omega_v$, iff there is no outcome $\omega \in \Omega_v$ s.t. $\omega \succ_v \omega^*$.*

Thanks to this, it is not necessary to construct the complete extensions of a viewpoint to find its rational decisions. It is sufficient to find the non-dominated outcomes of the incomplete viewpoints and to determine the decisions that lead to these outcomes.

3   Combinatorial Viewpoints

So far we have considered examples for multiple viewpoints that have same criteria or that have distinct criteria. In this section, we explore a further possibility, namely that the criteria of two viewpoints partially overlap. This possibility may arise for combinatorial outcome spaces which are the Cartesian product of several 'dimensions' $\Omega_{v,1}, \ldots, \Omega_{v,n}$. Two viewpoints partially overlap if their outcome spaces share some dimension. We introduce the notion of a combinatorial viewpoint that exhibits this structure:

Definition 3. *A combinatorial viewpoint v for a set of actions $\mathcal{A}$ is characterized by n_v outcome spaces $\Omega_{v,1}, \ldots, \Omega_{v,n_v}$, n_v criteria $z_{v,i} : \mathcal{A} \to \Omega_{v,i}$ that map the actions to these outcome spaces, and a weak preorder $\succsim_v$ over the Cartesian product $\Omega_{v,1} \times \ldots \times \Omega_{v,n_v}$ of the outcome spaces.*

A combinatorial viewpoint v corresponds to a standard viewpoint that has the combinatorial outcome space $\Omega_v := \Omega_{v,1} \times \ldots \times \Omega_{v,n_v}$ and the criterion $z_v : \mathcal{A} \to \Omega_v$ that maps an action a to the vector $z(a) := (z_{v,1}(a), \ldots, z_{v,n_v}(a))$. Moreover, a standard viewpoint corresponds to a combinatorial viewpoint of single dimension.

There are many examples for combinatorial viewpoints in the real-world. In cooperative decision making, multiple agents may be regrouped in different interest groups. Each interest group constitutes a viewpoint. Each agent who is a member of the group defines a dimension of this combinatorial viewpoint. As the same agent can belong to multiple interest groups, these groups may have dimensions in common. Combinatorial viewpoints are also encountered in multi-criteria decision making as multiple criteria can be combined together.

This discussion shows that combinatorial viewpoints can be combined into larger viewpoints. In certain cases, they can also be decomposed into smaller viewpoints. Sometimes, it may be necessary to permute the dimensions and criteria of a viewpoint. We therefore define a three operations on viewpoints, namely decomposition, aggregation, and permutation.

If a combinatorial viewpoint v has a group of criteria that are preferentially independent from the other criteria, we can extract this group of criteria and define a smaller viewpoint for it. Let I be a set of indices from $1, \ldots, n_v$. If this set consists of the indices $i_1, \ldots, i_k$ and these indices are listed in increasing order, we denote the subspace for these indices by $\Omega_I := \Omega_{v,i_1} \times \ldots \times \Omega_{v,i_k}$. A vector α from Ω_I may replace the elements of a vector β from Ω at indices I. We denote this by $r_I(\alpha, \beta)$. The i-th element of the resulting vector is equal to α_j if i is equal to i_j and it is equal to β_i if i is not in I. The criteria $z_{v,i_1}, \ldots, z_{v,i_k}$ are preferentially independent from the remaining criteria iff all α, β in Ω_I and all γ, δ in Ω satisfy $r_I(\alpha, \beta) \succsim_v r_I(\beta, \gamma)$ if and only if $r_I(\alpha, \delta) \succsim_v r_i(\beta, \delta)$. We can therefore safely define a viewpoint w that has the outcome spaces $\Omega_{v,i_1}, \ldots, \Omega_{v,i_k}$, the criteria $z_{v,i_1}, \ldots, z_{v,i_k}$ and the preference order $\succsim_w$ such that $\alpha \succsim_w \beta$ iff $r_I(\alpha, \gamma) \succsim_v r_I(\beta, \gamma)$ for all $\gamma \in \Omega$.

Two combinatorial viewpoints v and w can be aggregated into a viewpoint u with $n_v + n_w$ dimensions by using the Pareto-dominance. The resulting Pareto-aggregation is new viewpoint that has all the outcome spaces $\Omega_{v,1}, \ldots, \Omega_{v,n_v}$, $\Omega_{w,1}, \ldots, \Omega_{w,n_w}$ and the criteria $z_{v,1}, \ldots, z_{v,n_v}, z_{w,1}, \ldots, z_{w,n_w}$. Its pre-order $\succsim_u$ is the weak Pareto-dominance order for the pre-orders $\succsim_v$ and $\succsim_w$. Given an outcome $\omega := (\omega_1, \ldots, \omega_{n_u})$ in Ω_u, we denote the sub-vector $(\omega_i, \ldots, \omega_j)$ in Ω_u from index i to j by $\omega_{[i,j]}$. An outcome ω^* from Ω_u is weakly preferred to an outcome ω from Ω_u, i.e. $\omega^* \succsim_u \omega$ iff $\omega^*_{[1,n_v]} \succsim_v \omega_{[1,n_v]}$ and $\omega^*_{[n_v+1,n_u]} \succsim_v \omega_{[n_v+1,n_u]}$. Note that the criteria of viewpoint v are preferentially independent from the criteria of viewpoint w in the Pareto-aggregagtion and vice versa. Moreover, the weak Pareto-dominance order is the smallest relation that has this property.

Sometimes it is necessary to permute the criteria of the viewpoint. Let π be a permutation of the indices $1, \ldots, n_v$. Applying this permutation to the viewpoint v results into a new viewpoint $\pi(v)$. It has the outcome spaces $\Omega_{v,\pi_1}, \ldots, \Omega_{v,\pi_{n_v}}$, the criteria $z_{v,\pi_1}, \ldots, z_{v,\pi_{n_v}}$ and a weak preorder $\succsim_{\pi(v)}$ such that $(\omega^*_{\pi_1}, \ldots, \omega^*_{\pi_{n_v}}) \succsim_{\pi(v)} (\omega_{\pi_1}, \ldots, \omega_{\pi_{n_v}})$ iff $(\omega^*_1, \ldots, \omega^*_{n_v}) \succsim_v (\omega_1, \ldots, \omega_{n_v})$.

4   Enlargement of Viewpoints

The field of multi-criteria decision aiding makes the assumption that all criteria are identified before the preference elicitation starts. The identification of these criteria or objectives is a substantial part of the decision analysis process as explained in the second chapter of Raiffa's and Keeney's seminal work [6].

The criteria define the world in which preferences will be elicited, learned, or constructed. The usual assumption is that this world is closed, meaning that neither new values, nor new dimensions can be added to the outcome space. Nevertheless, the discovery of new dimensions happens quite often if the decision maker deals with a new domain. In this case, the decision making process is unstructured and ill-framed and the decision maker learns step by step which criteria are important. If new criteria are discovered, we say that the current viewpoint is enlarged. We enlarge a viewpoint by adding new dimensions and criteria to it:

Definition 4. *A combinatorial viewpoint w for the set of actions $\mathcal{A}$ is an enlargement of a combinatorial viewpoint v for the set of actions $\mathcal{A}$ iff $n_w \geq n_v$ and $\Omega_{v,i} = \Omega_{w,i}$ and $z_{v,i} = z_{w,i}$ for all $i = 1, \ldots, n_v$.*

The addition of new dimensions raises new questions concerning the preference elicitation. What should happen to the existing preferences that are defined over the existing dimensions when a new dimension is discovered? We may assume that the original criteria are preferentially independent from the new criteria. This assumption defines a default preference relation $\succsim_{v \to w}$ for an enlargement, namely $\alpha \succsim_{v \to w} \beta$ iff $\alpha_{[1,n_v]} \succsim_v \beta_{[1,n_v]}$ and $\alpha_{[n_v+1,n_w]} = \beta_{[n_v+1,n_w]}$. This relation is a pre-order.

However, the user may give additional preference information that either refines or overrides this default preference relation. Indeed, we can suppose that the discovery of a new dimension triggers a profound preference revision process which revisits the preferences of the original viewpoint. The decision maker can then state whether this preference holds for all, for some, or no value of the new dimension. In the first case, the decision maker sticks to the default order. In the second case, she should list the values of the third dimension for which the preference still holds. In the third case, she should completely reverse the preference. Moreover, the decision maker may not only modify the existing preferences, but also add new ones. Hence, an enlargement adds preferences to the default preference relation and removes others from it.

The added preferences are obtained as set difference of the enlargement preferences and the default preference relation, i.e. $A := \{(\alpha, \beta) \in \Omega_w^2 \mid \alpha \succsim_w \beta$ and $\beta \not\succsim_{v \to w} \alpha\}$. The removed preferences are obtained as set difference of the the default preference relation and the enlargement preferences, i.e. $R := \{(\alpha, \beta) \in \Omega_w^2 \mid \alpha \succsim_{v \to w} \beta$ and $\beta \not\succsim_w \alpha\}$. By definition, these two sets are disjoint and the pre-order $\succsim_w$ is obtained from the pre-order $\succsim_{v \to w}$ by removing R and by adding A. Hence, the sets of added and removed preferences capture the important changes in the preference relation due to the revision process that is initiated by the discovery of a new dimension.

We can also define enlargements by combining two viewpoints v and w. The combination has the outcome spaces and criteria of the viewpoints v and w, but may have an arbitrary preference relation. The weak Pareto dominance relation based on $\succsim_v$ and $\succsim_w$ provides a default preference relation for the combination if no other preference information for the combination is given. The sets of added and removed preferences can be defined for this default preference relation as before.

5   Compact Representations by Transitive Reductions

In the remainder of the paper, we investigate compact representations for preference relations of viewpoints, starting with that of standard viewpoints. Compact representations can be obtained by exploiting the structure of a concept. Preorders can be split into a strict partial order and an equivalence relation. Equivalence relations can be represented in form of a set of equivalence classes. For each equivalence class, a representative is chosen and the classes are represented by mapping the elements of the considered outcome space to this representative. Strict partial orders can be represented by a directed acyclic graph such that the transitive closure of this graph is equal to the pre-order. The essential notion here is that of a transitive reduction of a graph.

A transitive reduction of a binary relation R over Ω is a minimal subset of R that has the same transitive closure as R [1]. As the transitive reduction is minimal it does not contain a pair (α, γ) if it already contains the pairs (α, β) and (β, γ). With other words, the transitive reduction is anti-transitive. Formally, a relation R over Ω is *anti-transitive* iff for all $\alpha, \beta, \gamma \in \Omega$ the following property holds:

$$(\alpha, \beta) \in R \text{ and } (\beta, \gamma) \in R \text{ implies } (\alpha, \gamma) \notin R \tag{1}$$

A transitive reduction establishes anti-transitivity, but unfortunately a transitive reduction need not be unique. For example, the relation $\{(a, b), (a, c), (b, a), (b, c), (c, a), c, b)\}$ has two transitive reductions, namely $\{(a, b), (b, c), (c, a)\}$ and $\{(a, c), (c, b), (b, a)\}$. Moreover the definition of a transitive reduction is not symmetrical to that of a transitive closure. The transitive closure R^+ of a binary relation R is defined as the smallest superset of this relation that is transitive. In analogy, we define the *strong transitive reduction* R^- of a binary relation R as a greatest subset of this relation that is anti-transitive. This set is obtained by removing all transitive links from R:

$$R^- = R - \{(\alpha, \beta) \in R \mid \text{there is } \beta \in \Omega \text{ s.t. } (\alpha, \beta) \in R \text{ and } (\beta, \gamma) \in R\} \tag{2}$$

As a consequence of this correspondence, the strong transitive reduction always exists and is unique. The strong transitive reduction of an anti-transitive relation is equal to this relation. Furthermore the strong transitive reduction is monotonic in the following sense: Let R_1 and R_2 be two binary relations over Ω. If $R_1 \subseteq R_2$ then $R_1^- \subseteq R_2^-$. The strong transitive reduction of a strict partial order is sufficient to represent the strict partial order:

Proposition 2. *Let $\succ$ be a strict partial order over Ω and $\succ^-$ be its strong transitive reduction. The transitive closure of $\succ^-$ is equal to $\succ$.*

Moreover, the strong transitive reduction of a strict partial order is equal to the transitive reduction of the strict partial order. If a binary relation is anti-transitive and acyclic, then the strong transitive reduction of its transitive closure is equal to this binary relation.

The strong transitive reduction of a pre-order only represents the strict part of this pre-order and removes indifference:

Proposition 3. *Let $\succsim$ be a pre-order over Ω and $\succsim^-$ be its strong transitive reduction. The transitive closure of $\succsim^-$ is equal to the strict part $\succ$ of $\succsim$.*

Usually, the available preference information will not be given in form of a pre-order, but in form of some binary relation R over Ω. The pre-order is then obtained as the reflexive transitive closure R^* of this binary relation. According to the properties above, the strong transitive reduction R^- extracts the strict preferences from R. Indeed, the transitive closure of R^- is the strict part of R^*.

The strong transitive reduction can be computed by standard graph algorithms. In a first step, the strongly connected components of R can be computed by Tarjan's algorithms. Each strongly connected component is an equivalence class in the indifference relation of R^*. After this, the graph without strongly connected components is constructed. This graph is a directed acyclic graph and represents the strict preferences. An algorithm for computing the transitive reduction can then be used. As the graph is acyclic its transitive closure is a strict partial order, meaning that the resulting transitive reduction is equal to the strong transitive reduction.

We can thus determine a compact representation for viewpoint preferences by using standard graph algorithms. The notion of a strong transitive reduction gives a clear characterization of this representation.

6   Compact Representations for Enlargements

We now seek compact representations for enlargements. An enlargement w of a combinatorial viewpoint v adds a new dimension. The default preference order for the enlargement is the pre-order $\succsim_{v\rightarrow w}$ which simply assumes that the previous criteria are preferentially independent of the new criteria and that this is all what is known. However, the decision maker adds preferences to this default order and removes other preferences from it in order to establish the new pre-order of w. Can we find compact representations of these change sets?

We are now interested to transform the strong transitive reduction $\succsim_{v\rightarrow w}^-$ of the default pre-order $\succsim_{v\rightarrow w}$ into the strong transitive reduction $\succsim_w^-$. We neglect here the question of the change of indifference classes as this is a relatively simple task. Existing classes may be split into disjoint subsets and some of these subsets may be merged into new classes. In the remainder of the section we focus on the question of how to represent changes in the strict preferences.

Instead of using the original change sets, we now consider the changes in the strong transitive reductions of the pre-order of the enlargement and the default preference order. The removed preferences are $\Delta_{v-w} := \{(\alpha, \beta) \in \Omega_w^2 \mid \alpha \succsim_{v\rightarrow w}^- \beta$ and $\beta \not\succsim_w^- \alpha\}$ and the added preferences are $\Delta_{w-v} := \{(\alpha, \beta) \in \Omega_w^2 \mid \alpha \succsim_w^- \beta$ and $\beta \not\succsim_{v\rightarrow w}^- \alpha\}$.

Both sets are anti-transitive and acyclic since they are subsets of a strong transitive reduction. Moreover, the added preferences Δ_{w-v} do not belong to the strong reduction of the default preference order $\succsim_{v\rightarrow w}^-$ and the removed preferences Δ_{v-w} belong to it. It is possible that the set of added preferences Δ_{w-v} contains a pair (α, β) although $\alpha \succ_{v\rightarrow w} \beta$ holds. In this case, the set of removed preferences Δ_{v-w} must contain a preference from each chain $\gamma_1, \ldots, \gamma_k$ s.t. $\alpha = \gamma_1$, $\gamma_i \succ_{v \to w}^- \gamma_{i+1}$ for $i = 1, \ldots, k-1$, and $\gamma_k = \beta$ in order to respect the anti-transitivity of $\succsim_w^-$. Furthermore it is possible that the set of added preferences Δ_{w-v} contains a pair (α, β) although $\beta \succ_{v \to w} \alpha$ holds. In this case, the set of removed preferences Δ_{v-w} must contain a preference from each chain $\gamma_1, \ldots, \gamma_k$ s.t. $\beta = \gamma_1$, $\gamma_i \succ_{v \to w}^- \gamma_{i+1}$ for $i = 1, \ldots, k-1$, and $\gamma_k = \alpha$ in order to respect the acyclicity of $\succ_w^-$.

Vice versa, if two binary relations Δ_1 and Δ_2 respect those properties, then adding Δ_1 to the default preference order and removing Δ_2 from it results in an anti-transitive and acyclic relation R. The transitive closure of this relation is a strict partial order and the strong transitive reduction of this strict partial order is equal to R. This leads to the following representation theorem which describes how to modify the preference relation of an existing viewpoint. We here suppose that the viewpoint v is an enlargement that has the default preference relation:

Proposition 4. *Let v be a combinatorial viewpoint, $\succ_v^-$ the strong transitive reduction of its strict preferences, and Δ_1 and Δ_2 be two binary relations over Ω_v such that the following properties hold:*

1. *Δ_1 and Δ_2 are anti-transitive and acyclic.*
2. *Δ_1 is a subset of $\succ_v^-$ and Δ_2 is disjoint to $\succ_v^-$.*
3. *If $\alpha \succ \beta$ for an $(\alpha, \beta) \in \Delta_2$ then for each chain $\gamma_1, \ldots, \gamma_k$ satisfying $\alpha = \gamma_1$, $\gamma_i \succ_v^- \gamma_{i+1}$ for $i = 1, \ldots, k-1$ and $\gamma_k = \beta$ there exists a j s.t. $(\gamma_j, \gamma_{i+1}) \in \Delta_1$.*
4. *If $\alpha \succ \beta$ for an $(\beta, \alpha) \in \Delta_2$ then for each chain $\gamma_1, \ldots, \gamma_k$ satisfying $\alpha = \gamma_1$, $\gamma_i \succ_v^- \gamma_{i+1}$ for $i = 1, \ldots, k-1$ and $\gamma_k = \beta$ there exists a j s.t. $(\gamma_j, \gamma_{i+1}) \in \Delta_1$.*

Then $\{(\alpha, \beta) \in \Omega_v^2 \mid \alpha \succ_v^- \beta\} - \Delta_1 \cup \Delta_2$ is an anti-transitive and acyclic relation.

This proposition imposes conditions on the preferences that a decision maker may add to and remove from a default preference relation when discovering a new dimension. It ensures that the current viewpoint can be enlarged to a new viewpoint while keeping the representation of preferences compact and while minimizing the changes.

7 Conclusion

This paper considered constructive decision making processes which not only support incremental preference elicitation, but also the discovery of new criteria during this process. If this happens, a new dimension is added to the outcome space. By default, the existing preferences are interpreted as ceteris-paribus preferences on this new space meaning that they hold independent of the value of the new dimension. This interpretation results into a default preference relation on the new outcome space. As a new dimension has been discovered, the existing preferences may be revisited within a preference revision process. As a consequence, the decision maker may remove or modify existing preferences and add new ones. We have elaborated conditions on these change sets that ensure the validity of the resulting preference relation and a compact representation.

In the final version of the paper, we will extend this treatment of enlargements of viewpoints which are due to the combination of multiple viewpoints which have been independent so far. This may be useful in group decision making when two groups merge together for the purpose of lobbying. The merger may revise some of the existing group preferences. Inversely, this more general notion of enlargement also permits a compact representation of combinatorial viewpoints. If the criteria of such a viewpoint can be partitioned into mutually independent factors, then is it possible to decompose the viewpoint completely into smaller parts. If preferential independence is not guaranteed in general, then it may still be reasonable to perform the decomposition and to compare its Pareto-aggregation with the original viewpoint. If the change sets are small, then the decomposition and the change sets may provide a more compact representation of the original viewpoint.

This work has been started as an attempt to provide a clear meaning to the preference reversal approach in [5], which seeks to elaborate optimization method under change sets. Different forms of preference revision are also studied in [3]. Furthermore, operations for the enlargement and compaction of outcome spaces are introduced in [7] for the purpose of defining similarity measures for preference orders. Further work is needed to establish the precise relationship between those approaches.

References

1. Aho, A.V., Garey, M.R., Ullman, J.D.: The transitive reduction of a directed graph. SIAM J. Comput. 1(2), 131–137 (1972)
2. Bouyssou, D., Marchant, T., Pirlot, M., Tsoukiàs, A., Vincke, P.: Evaluation and Decision Models with Multiple Criteria, vol. 86. Springer, Heidelberg (2006)
3. Chomicki, J., Song, J.: Monotonic and nonmonotonic preference revision. CoRR, abs/cs/0503092 (2005)
4. Ehrgott, M.: Multicriteria Optimization. Springer, Berlin (2000)
5. Junker, U.: Upside-down preference reversal: How to override ceteris paribus preferences? In: AAAI Workshop on Preference Handling in Artificifial Intelligence, pp. 118–122 (2007)
6. Keeney, R.L., Raiffa, H.: Decisions with Multiple Objectives. Wiley, Chichester (1976)
7. Wicker, A.W., Doyle, J.: Comparing preferences expressed by cp-networks. In: AAAI Workshop on Advances in Preference Handling, pp. 128–133 (2008)

Extending Argumentation to Make Good Decisions

Yannis Dimopoulos[1], Pavlos Moraitis[2], and Leila Amgoud[3]

[1] Dept. of Computer Science, University of Cyprus, Nicosia, Cyprus
yannis@cs.ucy.ac.cy
[2] Laboratory of Informatics Paris Descartes (LIPADE), Paris Descartes University,
Paris, France
pavlos@mi.parisdescartes.fr
[3] Paul Sabatier University, Toulouse, France
amgoud@irit.fr

Abstract. Argumentation has been acknowledged as a powerful mechanism for automated decision making. In this context several recent works have studied the problem of accommodating preference information in argumentation. The majority of these studies rely on Dung's abstract argumentation framework and its underlying acceptability semantics.

In this paper we show that Dung's acceptability semantics, when applied to a preference-based argumentation framework for decision making purposes, may lead to counter intuitive results, as it does not take appropriately into account the preference information. To remedy this we propose a new acceptability semantics, called *super-stable extension* semantics, and present some of its properties. Moreover, we show that argumentation can be understood as a multiple criteria decision problem, making in this way results from decision theory applicable to argumentation.

1 Introduction

In many decision making situations we are confronted with a set of alternatives or options each of which has its own advantages that can be expressed as different arguments supporting that alternative. For instance, in a car purchase scenario, one argument that supports small cars is that they have low running cost, while another argument that favors big cars is that they have better safety features. The final decision is usually based upon the preferences one has over the arguments, or more generally how arguments relate to each other.

It is therefore not surprising that during the last years, argumentation has been acknowledged as a powerful mechanism for automating the decision making process of autonomous agents. Several recent works (see e.g. [1,2,3,4,5]) have emphasized the role of agents' preferences in the evaluation of their arguments within a particular class of argumentation frameworks called *preference-based argumentation* frameworks. The majority of these frameworks are using the acceptability semantics of the Dung's abstract argumentation framework [6].

F. Rossi and A. Tsoukis (Eds.): ADT 2009, LNAI 5783, pp. 225–236, 2009.
© Springer-Verlag Berlin Heidelberg 2009

In [7], it was shown that preference-based argumentation under the stable extension semantics is essentially a method for making decisions that are supported by "good" or "strong" arguments. Roughly speaking, a set of arguments E is a stable extension if every argument of E is strictly preferred to any other argument that is not included in E.

In this work we show that the stable extensions semantics of Dung's framework when applied to decision making may lead to counterintuitive results and therefore fail to deliver the correct conclusions.

More precisely, we show that stable extensions consider as equally good two sets of arguments (and therefore the options they support), although for every argument of the second set, the first set contains a more preferred argument. One may understand that in this case the agent could randomly select an option that is supported either by an argument from the first or the second set and this could be a wrong decision if these arguments support incorrect conclusions. This problem relates to a similar problem identified independently by Horty in [8] in the context of the use of argumentation for defeasible reasoning. For this reason we propose a new semantics called *super-stable* extension which allows to fix this problem.

Finally, in this paper we show the correspondence between argumentation and multi-criteria decision making. Then we emphasize that an aggregation method like *regime* [9] can be an alternative approach for defining a ranking on the set of arguments supporting the options and consequently on the options themselves.

2   Basics of Argumentation

Argumentation is a reasoning model based on the following main steps: i) constructing *arguments* and counter-arguments, ii) defining the *strengths* of those arguments, and iii) defining the *justified conclusions*. Argumentation systems are built around an underlying logical language and an associated notion of logical consequence, defining the notion of argument. The argument construction is a monotonic process: new knowledge cannot rule out an argument but only gives rise to new arguments which may interact with the first argument. Arguments may be conflicting for different reasons.

Definition 1 (Argumentation system [6]). *An argumentation system is a pair $T = (\mathcal{A}, \mathcal{R})$. $\mathcal{A}$ is a set of arguments and $\mathcal{R} \subseteq \mathcal{A} \times \mathcal{A}$ is an attack relation. We say that an argument a attacks an argument b iff $(a, b) \in \mathcal{R}$.*

Among all the arguments, it is important to know which arguments to keep for inferring conclusions. In [6], different acceptability semantics were proposed. The basic idea behind these semantics is the following: for a rational agent, an argument a_i is acceptable if he can defend a_i against all attacks. All the arguments acceptable for a rational agent will be gathered in a so-called *extension*. An extension must satisfy a consistency requirement and must defend all its elements.

Definition 2 (Conflict-free, Defence [6]). *Let* $\mathcal{B} \subseteq \mathcal{A}$, *and* $a_i \in \mathcal{A}$.

- $\mathcal{B}$ *is* conflict-free *iff* $\nexists\, a_i, a_j \in \mathcal{B}$ *s.t.* $(a_i, a_j) \in \mathcal{R}$.
- $\mathcal{B}$ *defends* a_i *iff* $\forall\, a_j \in \mathcal{A}$, *if* $(a_j, a_i) \in \mathcal{R}$, *then* $\exists\, a_k \in \mathcal{B}$ *s.t.* $(a_k, a_j) \in \mathcal{R}$.

The main semantics introduced by Dung are summarized in the following definition.

Definition 3 (Acceptability semantics [6]). *Let* $\mathcal{B}$ *be a conflict-free set of arguments.*

- $\mathcal{B}$ *is* admissible *iff it defends any argument in* $\mathcal{B}$.
- $\mathcal{B}$ *is a* preferred extension *iff it is a maximal (w.r.t* $\subseteq$*) admissible extension.*
- $\mathcal{B}$ *is a* stable extension *iff it is a preferred extension that attacks any argument in* $\mathcal{A} \backslash \mathcal{B}$.

Example 1. *Let* $T = (\mathcal{A}, \mathcal{R})$ *be an argumentation theory where* $\mathcal{A} = \{\alpha_1, \alpha_2, \alpha_3, \alpha_4\}$ *is the set of the arguments and* $\mathcal{R} = \{(a_1, a_2), (a_2, a_1), (a_1, a_4), (a_2, a_3)\}$ *is the set of attacks. This argumentation theory has two stable extensions* $\mathcal{E}_1 = \{\alpha_1, \alpha_3\}$ *and* $\mathcal{E}_2 = \{\alpha_2, \alpha_4\}$.

3   Preference-Based Argumentation Framework: Properties and Limitations

In [10] the basic argumentation framework of Dung was extended into *preference-based argumentation theory (PBAT)*. The framework was further developed and studied in [7]. The basic idea of a PBAT is to consider two binary relations between arguments:

1. A *conflict* relation, denoted by $\mathcal{C}$, that is based on the logical links between arguments.
2. A *preference* relation, denoted by $\succeq$, that captures the idea that some arguments are stronger than others. Indeed, for two arguments $a, b \in \mathcal{A}$, $a \succeq b$ means that a is at least as good as b. The relation $\succeq$ is assumed to be a partial pre-order (that is *reflexive* and *transitive*). The relation $\succ$ denotes the corresponding strict relation. That is, $a \succ b$ iff $a \succeq b$ and $b \not\succeq a$.

The two relations are combined into a unique attack relation, denoted by $\mathcal{R}$, and the Dung's semantics are applied on the resulting framework. In what follows, we focus on a particular class of PBATs, presented in [7], where the conflict relation $\mathcal{C}$ is *irreflexive* and *symmetric*.

Definition 4 (Preference-based Argumentation Theory (PBAT)). *([3])*
Given an irreflexive and symmetric conflict relation $\mathcal{C}$ *and a preference relation* $\succeq$ *on a set of arguments* $\mathcal{A}$, *a* preference-based argumentation theory *(PBAT)* *on* $\mathcal{A}$ *is an argumentation system* $T = (\mathcal{A}, \mathcal{R})$, *where* $(a, b) \in \mathcal{R}$ *iff* $(a, b) \in \mathcal{C}$ *and* $b \not\succ a$.

It follows directly from the definition that if $(a, b) \in C$ and $a \succeq b$ and $b \not\succeq a$, then $(a, b) \in \mathcal{R}$. Moreover, if $(a, b) \in C$ and a, b are either indifferent or incompatible in $\succeq$, then $(a, b) \in \mathcal{R}$ and $(b, a) \in \mathcal{R}$. Also note that if $(a, b) \in C$, then either $(a, b) \in \mathcal{R}$ or $(b, a) \in \mathcal{R}$. Finally, if $(a, b) \in \mathcal{R}$ and $(b, a) \notin \mathcal{R}$, then $a \succ b$.

The following example illustrates some features of PBATs.

Example 2. *Let $\mathcal{A} = \{a, b, c, d\}$ be a set of arguments, and C the conflict relation on $\mathcal{A}$ defined as $C = \{(a, b), (b, a), (b, c), (c, b), (c, d), (d, c)\}$. Moreover, let the preference relation $\succeq$ contain transitive closure of the set of pairs $a \succeq b$, $b \succeq c$, $c \succeq d$, and $d \succeq c$. The corresponding PBAT is $T = (\mathcal{A}, \mathcal{R})$, where $\mathcal{R} = \{(a, b), (b, c), (c, d), (d, c)\}$. Theory T has two stable extensions, $E_1 = \{a, c\}$ and $E_2 = \{a, d\}$.*

In [3] the impact of the preference relation on an argumentation system was studied. After defining a relation $\triangleright$ on the powerset $2^{\mathcal{A}}$ of the arguments of a PBAT $T = (\mathcal{A}, \mathcal{R})$, it was shown that the stable extensions of T correspond to the most preferred elements of $2^{\mathcal{A}}$ wrt this relation.

Definition 5. *([3]) Let $T = (\mathcal{A}, \mathcal{R})$ be a PBAT built on an underlying pre-order $\succeq$. If $A_1, A_2 \in 2^{\mathcal{A}}$, with $A_1 \neq A_2$, then $A_1 \triangleright A_2$ iff one of following holds:*

- *$A_1 \supset A_2$*
- *for all a, b such that $a \in A_1 \setminus A_2$ and $b \in A_2 \setminus A_1$, it holds that $a \succ b$*

The exact correspondence between the relation $\triangleright$ and stable extensions is as follows.

Theorem 1. *([3]) Let $T = (\mathcal{A}, \mathcal{R})$ be a PBAT built on an underlying pre-order $\succeq$ and a conflict relation C. E is a stable extension of T iff there are no arguments $a, b \in E$ s.t. $(a, b) \in C$, and for all $A \in 2^{\mathcal{A}}$ such that $A \triangleright E$, there are $a_1, a_2 \in A$ such that $(a_1, a_2) \in C$.*

The example below illustrates the link between $\triangleright$ and stable extensions.

Example 3. *Let $T = (\mathcal{A}, \mathcal{R})$ be a PBAT with $\mathcal{A} = \{a, b, c\}$ and $\mathcal{R}$ composed from the conflict relation $C = \{(a, b), (b, a)(a, c), (c, a)\}$ and preference relation that contains the pairs $a \succ b$ and $a \succ c$, and marks all other pairs of arguments as indifferent. The relation $\triangleright$ on $2^{\mathcal{A}}$ induced by $\succeq$ contains the pairs $\{a\} \triangleright \{b, c\}$, $\{a\} \triangleright \{b\}$, $\{a\} \triangleright \{c\}$. Since the sets $\{a, b, c\}$, $\{a, b\}$, $\{a, c\}$ are ruled out by C, the set $E = \{a\}$ is the stable extension of T.*

One feature of the $\triangleright$ relation is that it may *not be transitive*. Consider for instance the theory of the previous example, and observe that $\{a, b\} \triangleright \{b, c\} \triangleright \{c\}$. However, it is not the case that $\{a, b\} \triangleright \{c\}$.

The second important observation, which is the main focus of this work, relates to the conclusions sanctioned by preference-based argumentation under the stable model semantics. The following example, borrowed from [8], shows clearly that these results can be counterintuitive even in simple cases.

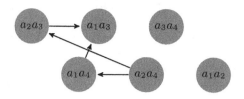

Fig. 1. The preference relation induced on sets of arguments

Example 4. *The story of the example is about conclusions that can been drawn regarding the financial situation of a person based on arguments built on information about her occupation and residence. Let's suppose that lawyers are, in general, considered to be wealthy, but a certain subclass, the pubic defenders, are considered not to be. Consider now an area in Paris -say, Passy - containing a large number of expensive private homes along with a much smaller number of middle-income rental properties. Thus the residents of Passy can be generally considered to be wealthy although the renters, are considered to not to be. Assume that Ann is a public defender (PDa), and therefore a lawyers (La), who rents in Passy (Ra), and is therefore a resident of Passy (Pa).*

If we assume that Wa represents the proposition that Ann is wealthy, the arguments that can be generated in an underlying propositional language from the above story are $a_1 = \{PDa, PDa \longrightarrow La, La \longrightarrow Wa\}$, $a_2 = \{PDa, PDa \longrightarrow \neg Wa\}$ $a_3 = \{Ra, Ra \longrightarrow Pa, Pa \longrightarrow Wa\}$, $a_4 = \{Ra, Ra \longrightarrow \neg Wa\}$.

From the above arguments we generate the PBAT $T=(\mathcal{A}, \mathcal{R})$, where $\mathcal{A} = \{a_1, a_2, a_3, a_4\}$. The attack relation $\mathcal{R}$ is composed from the conflict relation $C = \{(a_1, a_2), (a_2, a_1), (a_1, a_4), (a_4, a_1), (a_3, a_2), (a_2, a_3), (a_3, a_4), (a_4, a_3)\}$, and the preference relation $\succeq$ that is defined as $a_2 \succ a_1$ and $a_4 \succ a_3$, whereas all other pairs of arguments are incomparable.

Theory T has two extensions, namely $E_1 = \{a_1, a_3\}$ and $E_2 = \{a_2, a_4\}$. The first extension supports the conclusion that Ann is wealthy, whereas the the second that she is not. Intuitively however one would conclude that Ann is not wealthy. In other words we could argue that the second extension is more preferred than the first, as for every argument of the first it contains a more preferred argument.

The relation $\rhd$ on the subset of $\mathcal{A}$ with two elements is depicted in figure 1. Note again that $\rhd$ is not transitive. Indeed, it holds that $\{a_2, a_4\} \rhd \{a_2, a_3\}$ and $\{a_2, a_3\} \rhd \{a_1, a_3\}$, but $\{a_2, a_4\} \not\rhd \{a_1, a_3\}$. Although, $a_2 \succ a_1$ and $a_4 \succ a_3$, the stable model semantics does not render $\{a_2, a_4\}$ better than $\{a_1, a_3\}$, because $a_2 \not\succ a_3$ and $a_4 \not\succ a_1$.

The main purpose of this work is to provide a preliminary study of the problem of the conclusions sanctioned by the state-of-the-art argumentation, and identify possible solutions by borrowing ideas from decision theory.

4   Preference-Based Argumentation Revisited

As we noted in section 3, the stable model semantics can lead to counter-intuitive results. To remedy the situation we present a new semantics for preference-based argumentation called *super-stable* extensions semantics. The main idea is to only accept conclusions drawn under the stable model semantics from a PBAT T that correspond to the conclusions that are sanctioned by some other PBAT T' which is obtained from T by removing incomparability. Therefore, the new semantics may differ from the standard stable model semantics only on theories with incomparability. As we will discuss in the next section theories without incomparability always sanction the correct conclusion under the stable extension semantics.

Before we proceed to the definition of the new semantics, we recall some useful concepts. A relation $\succeq$ on a set S is *total* if for all $a, b \in S$ with $a \neq b$, $a \succeq b$ or $b \succeq a$. A *strict total order* on a set S is an asymmetric (hence irreflexive), transitive and total relation on S. The notion of an *extension* of a relation is used in decision theory and economics eg. ([11], [12]).

Definition 6. *A binary relation $\succeq_E$ on S is an extension of a pre-order $\succeq$ on S if $\succeq_E$ is a pre-order on S such that $\succeq_E \supseteq \succeq$ and for all $a, b \in S$ if $a \succ b$ then $a \succ_E b$. An extension of a pre-order $\succeq$ that is complete (ie., for all $a, b \in S$, $a \succeq b$ or $b \succeq a$) is called ordering extension of $\succeq$.*

Hansson [11] has shown that every pre-order has an ordering extension. Moreover, Donaldson and Weymark [12] proved that a pre-order is the intersection of its ordering extensions.

Definition 7. *A strict total order $\succ_s$ on a set S is a strict ordering of a total pre-order $\succeq$ if for all $a, b \in S$ if $a \succ b$ then $a \succ_s b$.*
A strict total order is a strict ordering of a pre-order if it is a strict ordering of one of its ordering extensions.

The following definition extends the notions of ordering extension and strict ordering to the case of PBATs.

Definition 8. *Let $T = (\mathcal{A}, \mathcal{R})$ be a PBAT on an underlying pre-order $\succeq$ and a conflict relation $\mathcal{C}$. The PBAT $T_o = (\mathcal{A}, \mathcal{R}_o)$, on an underlying relation $\succeq_o$ and the conflict relation $\mathcal{C}$, is a ordering completion of T if $\succeq_o$ is an ordering extension of $\succeq$. The PBAT $T_s = (\mathcal{A}, \mathcal{R}_s)$, on an underlying relation $\succ_s$ and the conflict relation $\mathcal{C}$, is a strict projection of T if $\succ_s$ is a strict ordering of $\succeq$.*

The following result that relates the stable extensions of a PBAT and the stable extensions of its ordering completions and strict projections is easily provable.

Proposition 1. *Let T be a PBAT, T_o one of its ordering completions, and T_s one of its strict projections. If E_o is a stable extension of T_o, then it is also a stable extension of T. Moreover, if E_s is a stable extension of T_s, then it is also a stable extension of T.*

However, it is not the case that a stable extension of T is a stable extension of some T_o built on a pre-order $\succ_o$ that is an ordering extension of $\succeq$. Consider for instance again the theory of example 4, and its stable extension $E_1 = \{a_1, a_3\}$. To see that there is no PBAT T_o which is an ordering completion of T and has E_1 as a stable extension, observe that for E_1 to be a stable extension it must be the case that $a_1 \succeq a_4$ and $a_3 \succeq a_2$. However, together with $a_2 \succ a_1$ these would mean that $a_3 \succeq a_4$, which is impossible given that $a_4 \succ a_3$.

We now proceed with the definition of the new semantics for preference-based argumentation. As noted earlier, the basic idea is to only accept a set of arguments as an extension of a PBAT T if this set is a stable extension of an ordering completion of T. More formally the concept is defined as follows.

Definition 9. *Let $T = (\mathcal{A}, \mathcal{R})$ be a PBAT built on an underlying pre-order $\succeq$ and conflict relation C. A stable extension E of T is a* super-stable *extension of T if it is the stable extension of an ordering extension of T.*

By the results of [11] we know that every pre-order has an ordering extension. Therefore, every PBAT has an ordering completion which is itself a PBAT. From [7], we know that every PBAT has a stable extension. By combining these two results we obtain the following property for super-stable extensions.

Proposition 2. *Every PBAT has a super-stable extension.*

5   Theories without Incomparability

If a PBAT T contains no incomparability, T is an ordering completion of itself. Therefore, any stable extension of T is by definition a super-stable extension of T. In this section we prove that for this class of theories a correspondence holds between their stable extensions and the stable extensions of their strict projections.

For a PBAT without incomparability, $T = (\mathcal{A}, \mathcal{R})$, we define the *level* of argument $a \in \mathcal{A}$, denoted by $l(a)$, recursively as follows

- $l(a) = 1$ for all a such that there is no $b \in \mathcal{A}$ s.t. $b \succ a$
- $l(a) = k$ for all a such that for all $a' \in \mathcal{A}$ s.t. $a' \succ a$ it holds that $l(a') < k$, and $\exists a'' \in \mathcal{A}$ s.t. $a' \succ a$ and $l(a') = k - 1$

The following lemma relates the level of arguments with relation $\succ$ and will be used in the proof of the main result of this section (proposition 3 below).

Lemma 1. *Let $T = (\mathcal{A}, \mathcal{R})$ be a PBAT without incomparability on an underlying pre-order $\succeq$. For every $a, b \in \mathcal{A}$, $l(a) < l(b)$ iff $a \succ b$.*

Proof. The property that for every $a, b \in \mathcal{A}$ if $a \succ b$ then $l(a) < l(b)$, follows directly from the definition of the level of an argument. We prove now that if $l(a) < l(b)$ then $a \succ b$. Assume $a, b \in \mathcal{A}$ with $l(a) < l(b)$. Clearly it can not be the case $b \succ a$, because then $l(b) < l(a)$. Assume that $a \succeq b$ and $b \succeq a$, and

$l(b) = m$ (hence, $l(a) < m$). Then, there must be $c \in \mathcal{A}$ s.t $l(c) = m - 1$ and $c \succ b$. Therefore, it must be the case that $c \succ a$, and hence $l(a) \geq m$. But this contradicts $l(a) < m$. Therefore it holds that $a \succ b$. □

A direct consequence of the previous lemma is that $l(a) \leq l(b)$ iff $a \succeq b$. Also note that since a super-stable extension of a PBAT $T = (\mathcal{A}, \mathcal{R})$ is also a stable extension of T, it holds that for all $a \notin E$ there exist $b \in E$ such that $(b, a) \in \mathcal{R}$. From the above we conclude that for all $a \notin E$ there exist $b \in E$ such that $(a, b) \in \mathcal{C}$ and $l(b) \leq l(a)$. Note the above property does not hold in general for theories that contain incomparability.

Proposition 3. *Let T be a PBAT without incomparability. Every stable extension of T is a stable extension of some strict projection of T.*

Proof. We prove the claim by defining a strict projection of T, $T_s = (\mathcal{A}, \mathcal{R}_s)$, for which E is a stable extension. Let $\succ_s^{E^+}$ be a strict projection on the arguments of E. Similarly, let $\succ_s^{E^-}$ be a strict projection on the arguments of $\mathcal{A} \setminus E$. Finally, let $\succ_s^{E^+,-}$ be the binary relation on $(E \times (\mathcal{A} \setminus E)) \cup ((\mathcal{A} \setminus E) \times E)$ such that for any pair of arguments $a \in E$ and $b \notin E$

- if $l(a) > l(b)$ then $b \succ_s^{E^+,-} a$
- if $l(a) \leq l(b)$ then $a \succ_s^{E^+,-} b$

where $l(a)$ is the level of argument a in theory T. Define $\succ_s = \succ_s^{E^+} \cup \succ_s^{E^-} \cup \succ_s^{E^+,-}$.

We first show that $\succ_s$ is strict total order. It is easy to verify that $\succ_s$ is asymmetric and total by construction. We show that $\succ_s$ is transitive. Let $a, b, c \in \mathcal{A}$, such that $a \succ_s b$ and $b \succ_s c$. We need to show that $a \succ_s c$. We proceed by case analysis. For the case where $a, b, c \in E$ or $a, b, c \in \mathcal{A} \setminus E$, transitivity follows by construction.

Assume now that $a \in E$ and $b, c \notin E$. Then, by construction, $l(a) \leq l(b)$. Moreover, since $b \succ_s c$, by lemma 1, it must be the case $l(b) < l(c)$. Hence, $l(a) < l(c)$, which, again by lemma 1, means that $a \succ_s c$. The case where $a \in E$, $b \notin E$, $c \in E$ is similar.

Now suppose that $a, b \in E$, and $c \notin E$. Since $a \succ_s b$, by lemma 1, we obtain that $l(a) < l(b)$. Moreover, by construction, it must hold that $l(b) \leq l(c)$. Hence, $l(a) < l(c)$, and therefore $a \succ_s c$.

The remaining cases where $a \notin E$ can be proved analogously.

Finally, we prove that E is a stable extension of $T_s = (\mathcal{A}, \mathcal{R}_s)$. First note that E is conflict free. Now assume that $a \notin E$ and define $D(a) = \{b | b \in E$ and $(b, a) \in \mathcal{R}\}$. Since E is a stable extension, it must be $D(a) \neq \emptyset$. Let $b \in D(a)$, and assume that $(b, a) \notin \mathcal{R}_s$. By construction, it must hold that $l(a) < l(b)$, which by lemma 1 implies $a \succ b$. This however contradicts $b \in D(a)$. Therefore, $(b, a) \in \mathcal{R}_s$, which means that E is a stable extension of T_s. □

By combining the definition of a super-stable extension with proposition 3 we obtain the following strong property regarding the super-stable extensions of theories without incomparability.

Proposition 4. *Every super-stable extension of a PBAT T that contains no incomparability is a stable extension of some strict projection of T.*

6 A Multi-criteria View of PBAT

In the previous sections we investigated how standard argumentation semantics can be extended to accommodate preference information on the arguments. In this section we change our perspective and explore a more direct link between argumentation and decision theory. More specifically, we interpret arguments as criteria and regard preferences as information on the relative importance of these criteria. Under this perspective argumentation can be understood as Multiple Criteria Decision Problem (MCDP). We start our analysis with a definition of the problem that leaves out some of its aspects that are not directly relevant to our purposes.

Definition 10. *A* Multiple Criteria Decision Problem (MCDP) *is a triple* $P = (A, K, \succeq)$ *where*

- $A = \{a_1, \dots, a_n\}$ *is the set of* attributes.
  A set of values is associated with each attribute, denoted by $v(a_1), \dots, v(a_n)$
- $K = \{\gg_{a_1}, \dots, \gg_{a_n}\}$ *is the set of* criteria. *A criterion* $\gg_{a_i}$ *is a pre-order associated with the values of an attribute* a_i
- $\succeq$ *is pre-order on the criteria*

An alternative *l wrt to a MCDP* $P = (A, K, \succeq)$ *is any* $l \in v(a_1) \times \dots \times v(a_n)$. *We denote the set of alternatives by* L_P.

In certain situations, a *solution* to a MCDP is a ranking relation $\trianglerighteq$ on the set of alternatives L, ie. $\trianglerighteq \subset L \times L$. Usually a solution to a MCDP has to satisfy certain properties [13].

The following definition shows that argumentation can be transformed in a meaningful way into a MCDP.

Definition 11. *Given a PBAT* $T = (\mathcal{A}, \mathcal{R})$, *where* $\mathcal{A} = \{a_1, \dots, a_n\}$, *we define its corresponding MCDP* $M_T = (A_T, K_T, \succeq_T)$ *as follows:*

- $A_T = \mathcal{A}$, *with* $v(a_i) = \{a_i^+, a_i^-\}$, *for each* $a_i \in A_T$.
- $K_T = \{\gg_1, \dots, \gg_n\}$, *where* $\gg_i$, *for* $1 \leq i \leq n$, *is defined as the preference* $a_i^+ \gg_i a_i^-$.
- $\succeq_T = \succeq$

The following is an example of a translation of a specific PBAT into a MCDP.

Example 5. *Consider the PBAT* $T = (\mathcal{A}, \mathcal{R})$, *where* $\mathcal{A} = \{a_1, a_2, a_3\}$, *and the underlying preference relation* $\succeq$ *defined as:* $a_1 \succ a_2$, $a_1 \succ a_3$, $a_2 \succeq a_3$, $a_3 \succeq a_2$. *The corresponding MCDP is defined as* $M_T = (A_T, K_T, \succeq_T)$, *where:*

- $A_T = \{a_1, a_2, a_3\}$, *with* $v(a_1) = \{a_1^+, a_1^-\}$, $v(a_2) = \{a_2^+, a_2^-\}$, $v(a_3) = \{a_3^+, a_3^-\}$.

- $K_T = \{\gg_1, \gg_2, \gg_3\}$, with $a_1^+ \gg_1 a_1^-$, $a_2^+ \gg_2 a_2^-$, $a_3^+ \gg_3 a_3^-$.
- $\succeq_T = \succeq$

Several methods have been proposed in the literature, the applicability of which in many cases depends on the features of the MCDP at hand. Among the methods for tackling MCDPs that appear in the literature and are applicable to the case we consider, it seems that the Regime method [9] is the closest to the spirit of the stable extensions semantics. Here we discuss a simplified version of the method as it appears in [14].

The Regime method works as follows. For any two alternatives A_i, A_j, let K^+ be the set of criteria according to which A_i is better than A_j, and K^- be the set of criteria according to which A_j is better than A_i. Regime ranks A_i better than A_j, denoted by $A_i \rhd_R^t A_j$, if $K^+ \neq \emptyset$ and there is an injective map from K^- to K^+ by which each criterion in K^- is mapped to a more important criterion in K^+. The set of optimal alternatives is then $\{A_o : \forall i \neg (A_i \rhd_R^t A_o)\}$.

We can easily define a preference order on the sets of arguments of a PBAT that captures the Regime method. To do this we associate to any set of arguments A, an alternative $A^t = \{a^+ | a \in A\} \cup \{a^- | a \notin A\}$.

Definition 12. Let $T = (\mathcal{A}, \mathcal{R})$ be a PBAT and $M_T = (A_T, K_T, \succeq_T)$ its corresponding MCDP. For any $A_1, A_2 \subseteq 2^{\mathcal{A}}$, it holds that $A_1 \rhd_R A_2$ if $A_1^t \rhd_R^t A_2^t$.

We can now define the notion of a *regime extension* of a PBAT by characterizing it in a way similar to the stable extensions.

Definition 13. A set of arguments E is a regime extension *of a PBAT* $T = (\mathcal{A}, \mathcal{R})$ *if there are no arguments* $a, b \in E$ *s.t.* $(a, b) \in \mathcal{C}$, *and for all* $E' \in 2^{\mathcal{A}}$ *such that* $E' \rhd_R E$, *there are* $a_1, a_2 \in E'$ *such that* $(a_1, a_2) \in \mathcal{C}$.

We now apply the previous definition to the story of 4, and observe that it yields the correct result.

Example 6. Consider again the theory T of example 4. The corresponding MCDP $M_T = (A_T, K_T, \succeq_T)$ can be defined as outlined above. Consider the two sets of arguments $E_1 = \{a_1, a_3\}$ and $E_2 = \{a_2, a_4\}$ of T which correspond to the alternatives $E_1^t = \{a_1^+, a_2^-, a_3^+, a_4^-\}$ and $E_2^t = \{a_1^-, a_2^+, a_3^-, a_4^+\}$. For the comparison $E_2^t \rhd_R^t E_1^t$ we have that $K^+ = \{a_2, a_4\}$, $K^- = \{a_1, a_3\}$, and the mapping $a_1 \rightarrow a_2$, $a_3 \rightarrow a_4$. For the comparison $E_1^t \rhd_R^t E_2^t$ we have that $K^+ = \{a_1, a_3\}$ $K^- = \{a_2, a_4\}$, but there is no suitable mapping. Therefore we conclude $E_2 \rhd_R E_1$. Moreover, E_2 is a regime extension of T.

7   Conclusion and Future Work

In this paper we pointed out that Dung's stable extensions semantics when applied in preference-based argumentation frameworks for decision making purposes lead to counter-intuitive conclusions. A similar problem was also identified by Horty in [8]. This mainly holds for argumentation theories where the preference relation used for defining the relative strength of individual arguments contains incomparability. To resolve this problem we proposed a new acceptability semantics called *super-stable extensions* which allows to capture the conclusions corresponding to the good decisions and to avoid the counter intuitive ones which could correspond to bad decisions. Moreover, we showed that preference-based argumentation can be understood as a multiple-criteria decision problem allowing to that way the exploration of the application of theoretical results of the decision theory in argumentation. Therefore, this work can been seen as an attempt to bring new ideas from decision theory to argumentation.

Our future work concerns the definition of a binary relation $\triangleright_{SS}$ on the sets of arguments of a PBAT that will be proved exactly the preference relation that is induced by the super-stable extensions semantics and to prove the correspondence between both. This will be the equivalent result of the one we proved in [7] between the preference $\triangleright$ and the stable extensions.

Acknowledgments. We thank Alexis Tsoukias for many fruitful discussions on several aspects of this work. We also thank Michael Maher for his helpful comments.

References

1. Kakas, A., Moraitis, P.: Argumentation based decision making for autonomous agents. In: Proc. 2nd International Joint Conference on Autonomous Agents and Multi-Agents systems, pp. 883–890 (2003)
2. Amgoud, L.: A general argumentation framework for inference and decision making. In: 21st Conference on Uncertainty in Artificial Intelligence, UAI 2005, pp. 26–33 (2005)
3. Amgoud, L., Dimopoulos, Y., Moraitis, P.: Making decisions through preference-based argumentation. In: Principles of Knowledge Representation and Reasoning, KR, pp. 113–123 (2008)
4. Bonet, B., Geffner, H.: Arguing for decisions: A qualitative model of decision making. In: Proceedings of the 12th Conference on Uncertainty in Artificial Intelligence, pp. 98–105 (1996)
5. Amgoud, L., Prade, H.: Explaining qualitative decision under uncertainty by argumentation. In: 21st National Conference on Artificial Intelligence, AAAI 2006, pp. 16–20 (2006)
6. Dung, P.M.: On the acceptability of arguments and its fundamental role in non-monotonic reasoning, logic programming and n-person games. Artificial Intelligence 77, 321–357 (1995)
7. Dimopoulos, Y., Moraitis, P., Amgoud, L.: Theoretical and Computational Properties of Preference-based Argumentation. In: Proc. of European Conference on AI, ECAI 2008 (2008)
8. Horty, J.: Argument construction and reinstatement in logics for defeasible reasoning. Artificial Intelligence and Law 9, 1–28 (2001)
9. Hinloopen, E., Nijkamp, P., Rietveld, P.: The regime method: A new multicriteria technique. In: Hansen, P. (ed.) Essays and Surveys on Multiple Criteria Decision Making, pp. 146–155. Springer, Heidelberg (1983)

10. Amgoud, L., Cayrol, C.: On the acceptability of arguments in preference-based argumentation framework. In: Proceedings of the 14th Conference on Uncertainty in Artificial Intelligence, pp. 1–7 (1998)
11. Hansson, B.: Choice structures and preference relations. Synthése 18, 443–458 (1968)
12. Donaldson, D., Weymark, A.: A quasiordering is the intersection of orderings. Journal of Economic Theory 178, 382–387 (1998)
13. Bouyssou, D., Marchant, T., Pirlot, M., Perny, P., Tsoukias, A., Vincke, P.: Multicriteria Evaluation and Decision Models: stepping stones for the analyst. Springer, Heidelberg (2006)
14. Moffett, A., Sarkar, S.: Incorporating multiple criteria into the design of conservation area networks: A minireview with recommendations. Diversity and Distributions 12, 125–137 (2006)

Building Consistent Pairwise Comparison Matrices over Abelian Linearly Ordered Groups

Bice Cavallo, Livia D'Apuzzo, and Massimo Squillante

University Federico II,
Naples, Italy
University of Sannio,
Benevento, Italy

Abstract. In the paper, algorithms are provided to check the consistency of pairwise comparison matrices and to build consistent matrices over abelian linearly ordered groups. A measure of consistency is also given; this measure improves a consistent index provided in a previous paper.

Keywords: Pairwise comparison matrices, consistency index, abelian linearly ordered group.

1 Introduction

Let $X = \{x_1, x_2, ..., x_n\}$ be a set of alternatives or criteria. An useful tool to determine a weighted ranking on X is a *pairwise comparison matrix* (PCM)

$$
A = \begin{pmatrix}
a_{11} & a_{12} & ... & a_{1n} \\
a_{21} & a_{22} & ... & a_{2n} \\
... & ... & ... & ... \\
a_{n1} & a_{n2} & ... & a_{nn}
\end{pmatrix}
\tag{1}
$$

which entry a_{ij} expresses how much the alternative x_i is preferred to alternative x_j. A condition of *reciprocity* is assumed for the matrix $A = (a_{ij})$ in such way that the preference of x_i over x_j expressed by a_{ij} can be exactly read by means of the element a_{ji}.

Under a suitable condition of *consistency* for $A = (a_{ij})$, X is totally ordered and the values a_{ij} can be expressed by means of the components w_i and w_j of a suitable vector $\underline{w}$, that is called *consistent vector* for the matrix $A = (a_{ij})$; then $\underline{w}$ provides the weights for the elements of X.

The shape of the reciprocity and consistency conditions depends on the different meaning given to the number a_{ij}, as the following well known cases show.

Multiplicative case: $a_{ij} \in]0, +\infty[$ is a preference ratio and the conditions of *multiplicative reciprocity* and *consistency* are given respectively by

$$
a_{ji} = \frac{1}{a_{ij}} \quad \forall\, i,j = 1, \ldots, n,
$$

$$
a_{ik} = a_{ij} a_{jk} \quad \forall\, i,j,k = 1, \ldots, n.
$$

F. Rossi and A. Tsoukis (Eds.): ADT 2009, LNAI 5783, pp. 237–248, 2009.

A consistent vector is a positive vector $\underline{w} = (w_1, w_2, ..., w_n)$ verifying the condition $\frac{w_i}{w_j} = a_{ij}$ and so perfectly representing the preferences over X.

Additive case: $a_{ij} \in] - \infty, +\infty[$ is a preference difference and *additive reciprocity* and *consistency* are expressed as follows

$$a_{ji} = -a_{ij} \quad \forall \, i, j = 1, \ldots, n,$$
$$a_{ik} = a_{ij} + a_{jk} \quad \forall \, i, j, k = 1, \ldots, n.$$

A consistent vector is a vector $\underline{w} = (w_1, w_2, ..., w_n)$ verifying the condition $w_i - w_j = a_{ij}$.

Fuzzy case: $a_{ij} \in [0, 1]$ measures the distance from the indifference that is expressed by 0.5; the conditions of *fuzzy reciprocity* and *fuzzy additive consistency* are

$$a_{ji} = 1 - a_{ij} \quad \forall \, i, j = 1, \ldots, n,$$
$$a_{ik} = a_{ij} + a_{jk} - 0.5 \quad \forall \, i, j, k = 1, \ldots, n.$$

A consistent vector is a vector $\underline{w} = (w_1, w_2, ..., w_n)$ verifying the condition $w_i - w_j = a_{ij} - 0.5$.

The multiplicative *PCMs* play a basic role in the Analytic Hierarchy Process, a procedure developed by T.L. Saaty at the end of the 70s [11], [12], [13].

In [2], [3], [5], [4], and [8], properties of multiplicative PCMs are analyzed in order to determine a qualitative ranking on the set of the alternatives and to find vectors representing this ranking. Additive and fuzzy matrices are investigated for instance by [1] and [10].

In the case of a multiplicative matrix, Saaty suggests that the comparisons expressed in verbal terms have to be translated into preference ratios a_{ij} taking value in $S^* = \{1, 2, 3, 4, 5, 6, 7, 8, 9, \frac{1}{2}, \frac{1}{3}, \frac{1}{4}, \frac{1}{5}, \frac{1}{6}, \frac{1}{7}, \frac{1}{8}, \frac{1}{9}\}$. The assumption of the Saaty scale restricts the Decision Maker's possibility to be consistent: indeed, if the Decision Maker (DM) expresses the following preference ratios $a_{ij} = 5$ and $a_{jk} = 3$, then he will not be consistent because $a_{ij}a_{jk} = 15 > 9$. Analogously for the fuzzy case, the assumption that $a_{ij} \in [0, 1]$, restricts the possibility to respect the fuzzy consistency: indeed, if the DM claims $a_{ij} = 0.9$ and $a_{jk} = 0.8$, then he will not be consistent because $a_{ij} + a_{jk} - 0.5 = 1.7 - 0.5 > 1$.

In order to unify the several approaches to PCMs and remove the above drawbacks, in [6] the authors introduce PCMs whose entries belong to an abelian linearly ordered group (*alo-group*) $\mathcal{G} = (G, \odot, \leq)$. In this way the reciprocity and consistency conditions are expressed in terms of the group operation $\odot$ and the drawbacks related to the consistency condition are removed; in fact the consistency condition is expressed by $a_{ik} = a_{ij} \odot a_{jk}$, thus, for each choice of $a_{ij} \in G$ and $a_{jk} \in G$, the result $a_{ij} \odot a_{jk}$ is an element of G. As a non trivial alo-group $\mathcal{G} = (G, \odot, \leq)$ has neither the greatest element nor the least element (see [6]), the Saaty set S* and the interval $[0, 1]$, embodied with the usual order $\leq$ on R, can be not structured as alo-groups.

In reference [6], the assumption of *divisibility* for $\mathcal{G}$ (see Section 2) allows us to introduce the notion of mean $m_\odot(a_1, ..., a_n)$ of n elements and associate a *mean vector* $\underline{w}_{m_\odot}$ to a PCM. In order to give a measure of consistency for a PCM over an alo-group $\mathcal{G}$, in [6] a notion of distance $d_\mathcal{G}$, linked to $\mathcal{G}$, is introduced. Then, the consistency index $I_\mathcal{G}(A)$ is defined as mean of the distances $d_\mathcal{G}(a_{ik}, a_{ij} \odot a_{jk})$, with $i < j < k$.

In this paper, we analyze the property of consistency for a PCM in order to:

1. provide an algorithm to check whenever or not a matrix is consistent;
2. provide an algorithm to build a consistent matrix by means of $n - 1$ comparisons;
3. provide a new consistency index linked to the index $I_\mathcal{G}(A)$ introduced in [6] but easier to compute.

2   Preliminaries on Alo-groups

Let us recall some notions and results related to an alo-group (see [6] for details). These results will be useful in the sequel to build a consistent PCM over an alo-group and to define in this context a new consistency index.

Let G be a non empty set provided with a total weak order $\leq$ and a binary operation $\odot : G \times G \to G$. $\mathcal{G} = (G, \odot, \leq)$ is called *alo-group*, if and only if $(G, \odot)$ is an abelian group and the the following implication holds:

$$a \leq b \Rightarrow a \odot c \leq b \odot c.$$

The above implication is equivalent to

$$a < b \Rightarrow a \odot c < b \odot c,$$

where $<$ is the strict simple order associated to $\leq$.

If $\mathcal{G} = (G, \odot, \leq)$ is an alo-group, then G is naturally equipped with the order topology induced by $\leq$ and $G \times G$ is equipped with the related product topology, and $\mathcal{G}$ is a *continuous* alo-group if and only if $\odot$ is continuous.

Let $\mathcal{G} = (G, \odot, \leq)$ be an alo-group, then we assume that: e denotes the *identity* of $\mathcal{G}$, $x^{(-1)}$ the *symmetric* of $x \in G$ with respect to $\odot$, $\div$ the *inverse operation* of $\odot$ defined by "$a \div b = a \odot b^{(-1)}$". It results $(a \div b)^{(-1)} = b \div a$. Moreover we define the *norm* of $a \in G$ by setting:

$$||a|| = a \vee a^{(-1)}. \tag{2}$$

Proposition 1. *[6] Let $\mathcal{G} = (G, \odot, \leq)$ be an alo-group. Then, the operation*

$$d_\mathcal{G} : (a, b) \in G^2 \to ||a \div b|| \in G \tag{3}$$

verifies the conditions:

1. $d_{\mathcal{G}}(a,b) \geq e$;
2. $d_{\mathcal{G}}(a,b) = e \Leftrightarrow a = b$;
3. $d_{\mathcal{G}}(a,b) = d_{\mathcal{G}}(b,a)$;
4. $d_{\mathcal{G}}(a,b) \leq d_{\mathcal{G}}(a,c) \odot d_{\mathcal{G}}(b,c)$.

Definition 1. *The operation $d_{\mathcal{G}}$ in* (3) *is a $\mathcal{G}$-metric or $\mathcal{G}$-distance.*

For a positive integer n, the (n)-*power* $x^{(n)}$ of $x \in G$ is defined as follows

$$x^{(1)} = x$$

$$x^{(n)} = \bigodot_{i=1}^{n-1} x_i \odot x_n = \bigodot_{i=1}^{n} x_i, \qquad x_i = x \quad i = 1, ..., n, \ n \geq 2$$

If $b^{(n)} = a$, then we say that b is the (n)-*root* of a and write $b = a^{(1/n)}$.

$\mathcal{G}$ is *divisible* if and only if for each positive integer n and each $a \in G$ there exists the (n)-root of a.

Definition 2. *Let $\mathcal{G} = (G, \odot, \leq)$ be a divisible alo-group. Then, the $\odot$- mean $m_{\odot}(a_1, a_2, ..., a_n)$ of the elements $a_1, a_2, ..., a_n$ of G is defined by*

$$m_{\odot}(a_1, a_2, ..., a_n) = \begin{cases} a_1 & n = 1, \\ (\bigodot_{i=1}^{n} a_i)^{(1/n)} & n \geq 2. \end{cases}$$

Definition 3. *An* isomorphism *between two alo-groups $\mathcal{G} = (G, \odot, \leq)$ and $\mathcal{G}' = (G', \circ, \leq)$ is a bijection $h : G \to G'$ that is both a lattice isomorphism and a group isomorphism, that is:*

$$x < y \Leftrightarrow h(x) < h(y)$$
$$h(x \odot y) = h(x) \circ h(y).$$

Proposition 2. *[6] Let $h : G \to G'$ be an isomorphism between the alo-groups $\mathcal{G} = (G, \odot, \leq)$ and $\mathcal{G}' = (G', \circ, \leq)$. Then,*

$$d_{\mathcal{G}'}(a', b') = h(d_{\mathcal{G}}(h^{-1}(a'), h^{-1}(b'))).$$

Moreover, $\mathcal{G}$ is divisible if and only if $\mathcal{G}'$ is divisible and, under the assumption of divisibility:

$$m_{\circ}(y_1, y_2, ..., y_n) = h\big(m_{\odot}(h^{-1}(y_1), h^{-1}(y_2), ..., h^{-1}(y_n))\big).$$

2.1   Real Alo-groups

An alo-group $\mathcal{G} = (G, \odot, \leq)$ is a *real* alo-group if and only if G is a subset of the real line R and $\leq$ is the total order on G inherited from the usual order on R.

Let $+$ and $\cdot$ be the usual addition and multiplication on R and $\otimes :]0, 1[^2 \to]0, 1[$ the operation defined by

$$x \otimes y = \frac{xy}{xy + (1-x)(1-y)},$$

then, examples of real divisible and continuous alo-groups are the following:

Multiplicative alo-group

$]0, +\infty[= (]0, +\infty[, \cdot, \leq)$; then $e = 1$, $x^{(-1)} = 1/x$, $x^{(n)} = x^n$ and $x \div y = \frac{x}{y}$.

So $d_{]0,+\infty[}(a, b) = \frac{a}{b} \vee \frac{b}{a}$ and $m.(a_1, ..., a_n)$ is the geometric mean: $\left(\prod_{i=1}^n a_i \right)^{\frac{1}{n}}$.

Additive alo-group

$\mathcal{R} = (R, +, \leq)$; then $e = 0$, $x^{(-1)} = -x$, $x^{(n)} = nx$, $x \div y = x - y$.

So $d_{\mathcal{R}}(a, b) = |a - b| = (a - b) \vee (b - a)$ and $m_+(a_1, ..., a_n)$ is the arithmetic mean: $\frac{\sum_i a_i}{n}$.

Fuzzy alo-group

$]0, 1[= (]0, 1[, \otimes, \leq)$; then $e = 0.5$, $x^{(-1)} = 1 - x$, $x \div y = \frac{x(1-y)}{x(1-y)+(1-x)y}$.

So $d_{]0,1[}(a, b) = \frac{a(1-b)}{a(1-b)+(1-a)b} \vee \frac{b(1-a)}{b(1-a)+(1-b)a}$.

Remark 1. *Our choice of the operation structuring the ordered interval $]0, 1[$ as an alo-group wants to obey the requests: 0,5 is the identity element and $1 - x$ is the symmetric of x. In this way the condition of reciprocity for a PCM over a fuzzy alo-group is given again by $a_{ji} = 1 - a_{ij}$, as defined in Section 1. In order to obtain an operation verifying the above requests, we apply the following result of [6]:*

Theorem 1. *Let G be a proper open interval of R and $\leq$ the total order on G inherited from the usual order on R, then the following assertions are equivalent:*

1. *$\mathcal{G} = (G, \odot, \leq)$ is a continuous alo-group;*
2. *there exists a continuous and strictly increasing function $\psi :]0, +\infty[\to G$ verifying the equality*

$$x \odot y = \psi(\psi^{-1}(x) \cdot \psi^{-1}(y)).$$

Setting $G =]0, 1[$ and

$$\psi : t \in]0, +\infty[\to \frac{t}{t + 1} \in]0, 1[, \tag{4}$$

that is a continuous and strictly increasing function between $]0, +\infty[$ and $]0, 1[$, we get

$$x \otimes y = \psi(\psi^{-1}(x) \cdot \psi^{-1}(y)).$$

Remark 2. *The operation $\otimes$ is the restriction to $]0, 1[^2$ of the uninorm (see [9]):*

$$U(x, y) = \begin{cases} 0, & (x, y) \in \{(0, 1), (1, 0)\}; \\ \frac{xy}{xy+(1-x)(1-y)}, & otherwise. \end{cases}$$

The multiplicative, the additive and the fuzzy alo-groups are isomorphic; in fact the bijection

$$h : x \in]0, +\infty[\to \log x \in R$$

is an isomorphism between $]0,+\infty[$ and $\mathcal{R}$ and ψ in (4) is an isomorphism between $]0,+\infty[$ and $]0,1[$. So, by Proposition 2, the mean $m_\otimes(a_1, ..., a_n)$ related to the fuzzy alo-group can be computed, by means of the function in (4), as follows:

$$m_\otimes(a_1, ..., a_n) = \psi\left(\left(\prod_{i=1}^{n} \psi^{-1}(a_i)\right)^{\frac{1}{n}}\right).$$

3 Consistent PCMs over a Divisible Alo-group

In this section, $\mathcal{G} = (G, \odot, \leq)$ is a divisible alo-group and $A = (a_{ij})$ in (1) is a PCM over $\mathcal{G}$, that is $a_{ij} \in G$, $\forall i, j \in \{1, \ldots, n\}$.
We assume that A is *reciprocal* with respect to $\odot$, that is :

$$a_{ji} = a_{ij}^{(-1)} \quad \forall\, i, j = 1, \ldots, n \tag{5}$$

so $a_{ii} = e$ for each $i = 1, 2, ..., n$ and $a_{ij} \odot a_{ji} = e$ for $i, j \in \{1, 2, ..., n\}$.

Definition 4. *[6], $A = (a_{ij})$ is a consistent matrix with respect to $\odot$, if and only if:*

$$a_{ik} = a_{ij} \odot a_{jk} \quad \forall i, j, k. \tag{6}$$

Moreover, $\underline{w} = (w_1, \ldots, w_n)$, with $w_i \in G$, is a consistent vector for $A = (a_{ij})$ if and only if $w_i \div w_j = a_{ij} \,\forall\, i,\, j=1,2,...,n$.

Let $\underline{a}_1, \underline{a}_2, \ldots, \underline{a}_n$ be the rows of A; then the *mean vector* associated to A is the vector

$$\underline{w}_{m_\odot}(A) = (m_\odot(\underline{a}_1), m_\odot(\underline{a}_1), \cdots, m_\odot(\underline{a}_n)). \tag{7}$$

If the matrix A is consistent then $\underline{w}_{m_\odot}(A)$ is a consistent vector. Indeed in [6], we prove the following

Proposition 3. *The following assertions related to $A = (a_{ij})$ are equivalent:*

i) *A is a consistent PCM;*
ii) *each column $\underline{a}^k$ is a consistent vector;*
iii) *the mean vector $\underline{w}_{m_\odot}$ is a consistent vector.*

Proposition 4. *[6] $A = (a_{ij})$ is a consistent matrix with respect to $\odot$, if and only if:*

$$a_{ik} = a_{ij} \odot a_{jk} \quad \forall\, i, j, k : i < j < k. \tag{8}$$

Proposition 5. *[6] $A = (a_{ij})$ is a consistent matrix with respect to $\odot$, if and only if:*

$$d_{\mathcal{G}}(a_{ik}, a_{ij} \odot a_{jk}) = e \quad \forall\, i, j, k : i < j < k$$

3.1   Checking the Consistency

We provide new characterizations of a consistent PCM that allows us to give an efficient algorithm to check the consistency of a PCM.

Proposition 6. *The following assertions are equivalent:*

1. *A is a consistent PCM with respect to $\odot$;*
2. *$a_{ik} = a_{i\,i+1} \odot a_{i+1\,k}$ $\forall i, k : i < k$;*
3. *$a_{ik} = a_{i\,i+1} \odot a_{i+1\,i+2} \odot \ldots \odot a_{k-1\,k}$ $\forall i, k : i < k$.*

Proof. 1. $\Rightarrow$ 2. It is straightforward because of Proposition 4.
    2. $\Rightarrow$ 3. By 2.:

$$a_{ik} = a_{i\,i+1} \odot a_{i+1\,k}$$

$$a_{i+1\,k} = a_{i+1\,i+2} \odot a_{i+2\,k}$$

$$\vdots$$

$$a_{k-2\,k} = a_{k-2\,k-1} \odot a_{k-1\,k}$$

Thus, by associativity of $\odot$, 3. is achieved.
    3. $\Rightarrow$ 1. By Proposition 4, it is enough to prove that 3. $\Rightarrow$ (8).
Let $i < j < k$. By 3., we have that:

$$a_{ik} = a_{i\,i+1} \odot \ldots a_{j-1\,j} \odot a_{j\,j+1} \ldots \odot a_{k-1\,k};$$

so, by associativity of $\odot$ and applying again 3., we have that:

$$a_{ik} = (a_{i\,i+1} \odot a_{i+1\,i+2} \odot \ldots \odot a_{j-1\,j}) \odot (a_{j\,j+1} \odot a_{j+1\,j+2} \odot \ldots \odot a_{k-1\,k}) = a_{ij} \odot a_{jk}.$$

Thus, the following corollary follows:

Corollary 1. $A = (a_{ij})$ *is a consistent matrix with respect to $\odot$, if and only if:*

$$d_{\mathcal{G}}(a_{ik}, a_{i\,i+1} \odot a_{i+1\,k}) = e \quad \forall\, i, k : i < k$$

Finally, in order to check whenever or not a matrix is consistent, we provide Algorithm 1, for which computational complexity order is equal to $O(n^2)$. In Algorithm 1, we assume that:

- i is the index of the rows of A, it is initialized to $i = 1$;
- k is the index of the columns of A, it is initialized to $k = i + 2$;
- n is the order of A;
- *ConsistentMatrix* is a boolean variable and the algorithm returns *ConsistentMatrix = true* if and only if the matrix is consistent. It is initialized to *true*, but *ConsistentMatrix = false* is immediately returned when an inconsistent triple $(a_{ik}, a_{i\,i+1}, a_{i+1\,k})$ occurs.

Algorithm 1. Checking consistency

$i = 1$;
ConsistentMatrix=true;
while $i \leq n - 2$ and ConsistentMatrix **do**
  $k = i + 2$;
  while $k \leq n$ and ConsistentMatrix **do**
    if $a_{ik} \neq a_{i\ i+1} \odot a_{i+1\ k}$ **then**
      ConsistentMatrix=false;
    end if
    $k = k + 1$;
  end while
  $i = i + 1$;
end while
return ConsistentMatrix;

3.2   Building a Consistent Matrix

Given $X = \{x_1, x_2, \ldots, x_n\}$ the set of alternatives, item 2. of Proposition 6 allow us to build a consistent PCM starting from a fixed alternative x_i and the $n - 1$ comparisons between x_i and x_j, for $j \neq i$. These comparisons are expressed by one of the following sequences:

1. $a_{i1}, \ldots a_{i\ i-1}, a_{i\ i+1}, \ldots a_{in}$,
2. $a_{1i}, \ldots a_{i-1\ i}, a_{i+1\ i}, \ldots a_{ni}$.

By item 2. of Proposition 6, we can also to build a consistent PCM starting from one of the following sequences:

3. $a_{12}, a_{23}, \ldots a_{n-1\ n}$,
4. $a_{21}, a_{32}, \ldots a_{n\ n-1}$.

For the fuzzy case, reference [7] builds a consistent matrix by means of sequence **3**. Here, we provide Algorithm 2 to build a consistent PCM starting from the sequence $a_{12}, a_{13}, \ldots, a_{1n}$; we use the equalities $a_{i+1\ j} = a_{i+1\ i} \odot a_{i\ j}$, $\forall i, j$ such that $i < j - 1$, obtained from item 2 of Proposition 6.

Example 1. *Let $\{x_1, x_2, x_3, x_4, x_5\}$ be a set of alternatives. We suppose that the DM prefers x_1 to each other alternative and expresses the following preference ratios (multiplicative case): $a_{12} = 2$, $a_{13} = 4$, $a_{14} = 5$ and $a_{15} = 6$. By means of Algorithm 2, we obtain:*

$$a_{21} = \frac{1}{2}, a_{31} = \frac{1}{4}, a_{41} = \frac{1}{5}, a_{51} = \frac{1}{6},$$

$$a_{11} = a_{22} = a_{33} = a_{44} = a_{55} = 1,$$

$$a_{23} = a_{21}a_{13} = 2, \ a_{32} = \frac{1}{2},$$

$$a_{24} = a_{21}a_{14} = \frac{5}{2}, \ a_{42} = \frac{2}{5},$$

Algorithm 2. Building a consistent matrix

for $j = 2, \ldots, n$ do
   $a_{j1} = a_{1j}^{(-1)}$
end for
for $i = 1, \ldots, n$ do
   $a_{ii} = e$
end for
for $i = 1, \ldots n - 2$ do
  for $j = i + 2 \ldots n$ do
    $a_{i+1\, j} = a_{i+1\, i} \odot a_{i\, j}$
    $a_{j\, i+1} = a_{i+1\, j}^{(-1)}$
  end for
end for

$$a_{25} = a_{21}a_{15} = 3, \; a_{52} = \frac{1}{3},$$

$$a_{34} = a_{32}a_{24} = \frac{5}{4}, \; a_{43} = \frac{4}{5},$$

$$a_{35} = a_{32}a_{25} = \frac{3}{2}, \; a_{53} = \frac{2}{3},$$

$$a_{45} = a_{43}a_{35} = \frac{6}{5}, \; a_{54} = \frac{5}{6},$$

and therefore:

$$A = \begin{pmatrix} 1 & 2 & 4 & 5 & 6 \\ \frac{1}{2} & 1 & 2 & \frac{5}{2} & 3 \\ \frac{1}{4} & \frac{1}{2} & 1 & \frac{5}{4} & \frac{3}{2} \\ \frac{1}{5} & \frac{2}{5} & \frac{4}{5} & 1 & \frac{6}{5} \\ \frac{1}{6} & \frac{1}{3} & \frac{2}{3} & \frac{5}{6} & 1 \end{pmatrix}$$

Example 2. *Let $\{x_1, x_2, x_3, x_4, x_5\}$ be a set of alternatives. We suppose that the DM prefers x_1 to each other alternative and expresses the following preferences (fuzzy case): $a_{12} = 0.6$, $a_{13} = 0.7$, $a_{14} = 0.8$ and $a_{15} = 0.9$. By means of Algorithm 2, we obtain:*

$$a_{21} = 0.4, a_{31} = 0.3, a_{41} = 0.2, a_{51} = 0.1,$$

$$a_{11} = a_{22} = a_{33} = a_{44} = a_{55} = 0.5,$$

$$a_{23} = \frac{a_{21}a_{13}}{a_{21}a_{13} + (1 - a_{21})(1 - a_{13})} = 0.609,$$

$$a_{32} = 0.391,$$

$$a_{24} = \frac{a_{21}a_{14}}{a_{21}a_{14} + (1 - a_{21})(1 - a_{14})} = 0.727,$$

$$a_{42} = 0.273,$$

$$a_{25} = \frac{a_{21}a_{15}}{a_{21}a_{15} + (1 - a_{21})(1 - a_{15})} = 0.857,$$

$$a_{52} = 0.143,$$

$$a_{34} = \frac{a_{32}a_{24}}{a_{32}a_{24} + (1 - a_{32})(1 - a_{24})} = 0.632,$$

$$a_{43} = 0.368,$$

$$a_{35} = \frac{a_{32}a_{25}}{a_{32}a_{25} + (1 - a_{32})(1 - a_{25})} = 0.794,$$

$$a_{53} = 0.206,$$

$$a_{45} = \frac{a_{43}a_{35}}{a_{43}a_{35} + (1 - a_{43})(1 - a_{35})} = 0.692,$$

$$a_{54} = 0.308,$$

and therefore:

$$A = \begin{pmatrix} 0.5 & 0.6 & 0.7 & 0.8 & 0.9 \\ 0.4 & 0.5 & 0.609 & 0.727 & 0.857 \\ 0.3 & 0.391 & 0.5 & 0.632 & 0.794 \\ 0.2 & 0.273 & 0.368 & 0.5 & 0.692 \\ 0.1 & 0.143 & 0.206 & 0.308 & 0.5 \end{pmatrix}$$

4   A New Consistency Index

Let T be the set $\{(a_{ij}, a_{jk}, a_{ik}), i < j < k\}$ and $n_T = |T|$. By Proposition 5 $A = (a_{ij})$ is inconsistent if and only if $d_{\mathcal{G}}(a_{ik}, a_{ij} \odot a_{jk}) > e$ for some triple $(a_{ij}, a_{jk}, a_{ik}) \in T$. Thus, in [6] the authors have provided the following definition of consistency index and the related results:

Definition 5. *The consistency index of A is given by:*

$$I_{\mathcal{G}}(A) = \begin{cases} d_{\mathcal{G}}(a_{13}, a_{12} \odot a_{23}) & n = 3, \\ \left(\bigodot_T d_{\mathcal{G}}(a_{ik}, a_{ij} \odot a_{jk}) \right)^{\left(\frac{1}{n_T} \right)} & n > 3. \end{cases}$$

with

$$n_T = \frac{n(n-2)(n-1)}{6}.$$

Proposition 7. *$I_{\mathcal{G}}(A) \geq e$ and A is consistent if and only if $I_{\mathcal{G}}(A) = e$.*

Proposition 8. *Let $\mathcal{G}' = (G', \circ, \leq)$ be a divisible alo-group isomorphic to $\mathcal{G}$ and $A' = (h(a_{ij}))$ the transformed of $A = (a_{ij})$ by means of the isomorphism $h : G \to G'$. Then $I_{\mathcal{G}}(A) = h^{-1}(I_{\mathcal{G}'}(A'))$.*

At the light of Corollary 1, it is reasonable to define a new consistency index, considering only the distances $d_{\mathcal{G}}(a_{ik}, a_{i\,i+1} \odot a_{i+1\,k})$, with $i < k - 1$; of course if $i = k - 1$ then $d_{\mathcal{G}}(a_{ik}, a_{i\,i+1} \odot a_{i+1\,k}) = e$. Let T^* be the set $\{(a_{i\,i+1}, a_{i+1\,k}, a_{ik}), i < k - 1\}$ and $n_{T^*} = |T^*|$, then we consider the following index:

$$I_{\mathcal{G}}^*(A) = \begin{cases} d_{\mathcal{G}}(a_{13}, a_{12} \odot a_{23}) & n = 3, \\ \left(\bigodot_{T^*} d_{\mathcal{G}}(a_{ik}, a_{i\,i+1} \odot a_{i+1\,k}) \right)^{\left(\frac{1}{n_{T^*}}\right)} & n > 3. \end{cases}$$

with

$$n_{T^*} = \frac{(n-2)(n-1)}{2}.$$

By Corollary 1, we have:

Proposition 9. *$I_{\mathcal{G}}^*(A) \geq e$ and A is consistent if and only if $I_{\mathcal{G}}^*(A) = e$.*

As for $n > 3$ it results $n_{T^*} < n_T$, the index $I_{\mathcal{G}}^*(A)$ is more easy to compute than the consistency index $I_{\mathcal{G}}(A)$, thus, we provide the following definition:

Definition 6. *A consistency index of A is given by $I_{\mathcal{G}}^*(A)$.*

Moreover, proposition analogous to Proposition 8 follows:

Proposition 10. *Let $\mathcal{G}' = (G', \circ, \leq)$ be a divisible alo-group isomorphic to $\mathcal{G}$ and $A' = (h(a_{ij}))$ the transformed of $A = (a_{ij})$ by means of the isomorphism $h : G \to G'$. Then $I_{\mathcal{G}}^*(A) = h^{-1}(I_{\mathcal{G}'}^*(A'))$.*

Example 3. *Let us consider*

$$A = \begin{pmatrix} 0.5 & 0.3 & 0.4 & 0.4 \\ 0.7 & 0.5 & 0.1 & 0.2 \\ 0.6 & 0.9 & 0.5 & 0.8 \\ 0.6 & 0.8 & 0.2 & 0.5 \end{pmatrix}$$

that is a PCM over the fuzzy alo-group $]0,1[$. By applying the function ψ^{-1}, with ψ in (4), to the entries of A, we get the matrix

$$A' = \begin{pmatrix} 1 & \frac{3}{7} & \frac{2}{3} & \frac{2}{3} \\ \frac{7}{3} & 1 & \frac{1}{9} & \frac{1}{4} \\ \frac{3}{2} & 9 & 1 & 4 \\ \frac{3}{2} & 4 & \frac{1}{4} & 1 \end{pmatrix}.$$

A' is a PCM over the multiplicative alo-group $]0, +\infty[$ and its consistency index is

$$I_{]0,+\infty[}^*(A') = \sqrt[3]{I_{]0,+\infty[}^*(A'_{123}) \cdot I_{]0,+\infty[}^*(A'_{124}) \cdot I_{]0,+\infty[}^*(A'_{234})}$$

$$= \sqrt[3]{14 \cdot \frac{56}{9} \cdot \frac{16}{9}} = 5.37$$

Applying Proposition 10, we can compute the consistency index of A by means of the isomorphism ψ in (4):

$$I_{]0,1[}^*(A) = \psi(I_{]0,+\infty[}^*(A')) = \frac{5.37}{6.37} = 0.84.$$

References

1. Barzilai, J.: Consistency measures for pairwise comparison matrices. J. MultiCrit. Decis. Anal. 7, 123–132 (1998)
2. Basile, L., D'Apuzzo, L.: Ranking and weak consistency in the a.h.p. context. Rivista di matematica per le scienze economiche e sociali 20(1), 99–110 (1997)
3. Basile, L., D'Apuzzo, L.: Weak consistency and quasi-linear means imply the actual ranking. International Journal of Uncertainty, Fuzziness and Knowledge-Based Systems 10(3), 227–239 (2002)
4. Basile, L., D'Apuzzo, L.: Transitive matrices, strict preference and intensity operators. Mathematical Methods in Economics and Finance 1, 21–36 (2006)
5. Basile, L., D'Apuzzo, L.: Transitive matrices, strict preference and ordinal evaluation operators. Soft Computing 10(10), 933–940 (2006)
6. Cavallo, B., D'Apuzzo, L.: A general unified framework for pairwise comparison matrices in multicriterial methods. International Journal of Intelligent Systems 24(4), 377–398 (2009)
7. Chiclana, F., Herera-Viedma, E., Alonso, S., Herera, F.: Cardinal Consistency of Reciprocal Preference Relations: A Characterization of Multiplicative Transitivity. IEEE Transaction on Fuzzy Systems 17(1), 14–23 (2009)
8. D'Apuzzo, L., Marcarelli, G., Squillante, M.: Generalized consistency and intensity vectors for comparison matrices. International Journal of Intelligent Systems 22(12), 1287–1300 (2007)
9. Fodor, J., Yager, R., Rybalov, A.: Structure of uninorms. International Journal of Uncertainty, Fuzziness and Knowledge-Based Systems 5(4), 411–427 (1997)
10. Herrera-Viedma, E., Herrera, F., Chiclana, F., Luque, M.: Some issue on consistency of fuzzy preferences relations. European Journal of Operational Research 154, 98–109 (2004)
11. Saaty, T.L.: A scaling method for priorities in hierarchical structures. J. Math. Psychology 15, 234–281 (1977)
12. Saaty, T.L.: The Analytic Hierarchy Process. McGraw-Hill, New York (1980)
13. Saaty, T.L.: Axiomatic foundation of the analytic hierarchy process. Management Science 32(7), 841–855 (1986)

Aggregating Interval Orders by Propositional Optimization

Daniel Le Berre, Pierre Marquis, and Meltem Öztürk*

Université Lille Nord de France, F-59000 Lille, France
Université d'Artois, CRIL, F-62307 Lens, France
CNRS, UMR 8188, F-62307 Lens, France
{leberre,marquis,ozturk}@cril.univ-artois.fr

Abstract. Aggregating preferences for finding a consensus between several agents is an important issue in many fields, like economics, decision theory and artificial intelligence. In this paper we focus on the problem of aggregating interval orders which are special preference structures allowing the introduction of tresholds for the indifference relation. We propose to solve this problem by first translating it into a propositional optimization problem, namely the Binate Covering Problem, then to solve the latter using a MAX-SAT solver. We discuss some properties of the proposed encoding and provide some hints about its practicability using preliminary experimental results.

Keywords: Interval orders, preference modelling and aggregation, propositional reasoning, Boolean optimization.

1 Introduction

Aggregating preferences for finding a consensus between several agents is an important issue in many fields, like economics, decision theory, and artificial intelligence. Given the preferences of a set of agents (or voters) over a set of alternatives (or candidates), where preferences are generally formulated as binary relations such as strict preference, indifference, etc., preference aggregation aims at determining a collective preference relation representing as much as possible the individual preferences.

However many works have shown through paradoxes and impossibility theorems that preference aggregation is not an easy task, the famous ones are Condorcet's paradox [3], Arrow's theorem [2] .

A common approach is to consider a preference relation as a complete preorder (i.e., a reflexive and transitive relation). In the above results each voter is supposed to present a complete preoder over the set of alternatives. However, such a model for preferences does not prove adequate to all situations, and other models (generalizing the complete preorder one) have been pointed out. In particular, different structures have been introduced for defining thresholds as in the famous example given by Luce [10] about a cup

---

* This work has been supported in part by two Projects: ANR-05-BLAN-0384 and PEPS-09 37. This support is gratefully acknowledged by the autors. The authors also thank anonymous referees for their helpful comments and suggestions.

F. Rossi and A. Tsoukis (Eds.): ADT 2009, LNAI 5783, pp. 249–260, 2009.

of coffee. Indeed, in contrast to the strict preference relation, the indifference relation induced by such structures is not necessarily transitive. Semiorders may form the simplest class of such structures and they appear as a special case of interval orders. The axiomatic analysis of what we call now interval orders has been given by Wiener [16], then the term "semiorders" has been introduced by Luce [10] and many results about their representations are available in the literature (for more details see [5,13]). Roughly speaking, within an interval order, alternative x_1 is strictly preferred to alternative x_2 if and only if the evaluation of x_1 is greater than the evaluation of x_2 plus a threshold. It is easy to see that preorders are special cases of interval orders where the value of threshold is fixed to zero.

In this paper we consider the interval order aggregation problem; to solve it, we propose a method based on the Kemeny distance which makes use of a translation into the Binate Covering Problem [4]. More precisely, we consider the case where the preferences of voters are interval orders and we try to find a final interval order which will be "as close as possible" to the set of voter's preferences. Let us note that having an interval order as a result of an aggregation is not a drawback for pointing out an undominated alternative since it is known that when the asymmetric part of a binary relation is transitive, which is the case of interval orders, there is always at least one such undominated alternative [14]. Moreover it is natural to ask an interval order as a result when preferences of voters are interval orders. Finally, as we will show it, even when the input preferences are preorders, focusing on interval orders as outputs is a way to get an aggregation which is closer to the given preferences than when preorders are targeted (just because the set of all interval orders over the! alternatives is a superset of the set of all preorders over the alternatives).

2   Aggregation as Optimisation

In this paper, we consider a finite set of alternatives A on which preference relations are applied ($|A| = n$), we represent with $a, b, c, ...$ specific elements of A and $x_1, x_2, ...$ or $x, y, z, ...$ variables ranging over the set A. We have a finite set of voters $V = \{v_1, ..., v_m\}$ ($|V| = m$). Voters express their preferences by the help of two binary relations represented in an explicit way as n^2-matrices: the notation aP_ib (resp. aI_ib) means that the voter v_i prefers strictly alternative a to b (resp. is indifferent between a and b). $\#p(a,b)$ (resp. $\#i(a,b)$) is the number of voters v_i for whom aP_ib (resp. aI_ib) holds. We call a *profile*, the set of voter's preference relations and denote it by $X = \{\langle P_1, I_1 \rangle, \langle P_2, I_2 \rangle, ..., \langle P_m, I_m \rangle\}$; its size is in $\mathcal{O}(m.n^2)$.

The result of the agregation is also expressed by two relations that we denote by P and I (P^{-1} represents the inverse of P: $\forall x, y \in A$, $xP^{-1}y$ iff yPx). aPb (resp. aIb) means that alternative a is preferred to alternative b (resp. a and b are indifferent) in the resulting order. We denote it as $f(\langle P_1, I_1 \rangle, \langle P_2, I_2 \rangle, ..., \langle P_m, I_m \rangle) = \langle P, I \rangle$.

The pair $\langle P, I \rangle$ is called a *preference structure* if and only if P is asymmetric, I is reflexive and symmetric, $P \cup I$ is complete and $P \cap I$ is empty. Such a pair is an interval order if and only if it is a preference structure and satisfies a property called Ferrers relation.[1]

---
[1] $\forall x, y, z, t \in A$, $xPy \wedge yIz \wedge zPt \Rightarrow xPt$.

Definition 1. *Let P and I be binary relations on $A \times A$, $\langle P, I \rangle$ is an interval order if and only if*

   i) $P \cup I \cup P^{-1} = A \times A$ *(completeness),*
   ii) $P \cap I = $ *(exclusivity),*
   iii) P *is asymmetric, I is symmetric and reflexive,*
   iv) $P.I.P \subset P$ *(Ferrers relation).*

The numerical representation of interval orders is as in the following:

Proposition 1. *[5] Let P and I be binary relations on $A \times A$, $\langle P, I \rangle$ is an interval order if and only if there exist a mapping g from A to $\mathbb{R}$ and a mapping q from $\mathbb{R}$ to $\mathbb{R}^+$ such that for any $x, y \in A$, we have:*
$$xPy \Leftrightarrow g(x) > g(y) + q(g(y)).$$
$$xIy \Leftrightarrow g(x) \leq g(y) + q(g(y)).$$

Interval orders are quasi-orders (i.e., orders with a transitive asymmetric part). Gibbard ([6]) has showed that Arrow's theorem can be generalized to the case of quasi-orders, hence we have this impossibility result for interval orders. Pirlot and Vincke ([13]) have focused also on this theorem with a special attention to interval orders. Before presenting this theorem we first need the following definitions in order to state it formally:

weak unanimity. an aggregation procedure satisfies the weak unanimity condition if
   and only if, for all voters $v_i \in V$ and for all $a, b \in A$, $aP_ib \Longrightarrow aPb$;
non-dictatorship. an aggregation procedure satisfies the non-dictatorship condition if
   and only if, for no voter $v_i \in V$ such that for all possible preferences of other voters
   and for all alternatives a and b $aP_ib \Longrightarrow aPb$;
independence of irrelevant alternatives. an aggregation procedure satisfies the independence of irrelevant alternatives condition if and only if $\forall (\langle P_1, I_1 \rangle, \ldots, \langle P_m,$
   $I_m \rangle), (\langle P_1', I_1' \rangle, \ldots, \langle P_m', I_m' \rangle), \forall a, b \in A,$
   $(\langle P_1, I_1 \rangle, \ldots, \langle P_m, I_m \rangle)/\{a, b\} = (\langle P_1', I_1' \rangle, \ldots, \langle P_m', I_m' \rangle)/\{a, b\} \Longrightarrow$
   $(\langle P, I \rangle), /\{a, b\} = (\langle P', I' \rangle)/\{a, b\}$
   where $\langle P, I \rangle$ is the result on $(\langle P_1, I_1 \rangle, \ldots, \langle P_m, I_m \rangle)$, and $\langle P, I \rangle/\{a, b\}$ is the
   restriction of $\langle P, I \rangle$ to $\{a, b\}$, etc.

Theorem 1 (Generalized Arrow's Theorem). *[13] If $|A| \geq 4$, if X is the set of all n-tuples of interval orders on A and if Y is the set of all interval orders on A, then there is no $(X$-$Y)$-aggregation procedure[2] satisfying simultaneously weak unanimity, non-dictatorship and independence conditions.*

Note that if all the considered relations are complete preorders, Theorem 1 is exactly Arrow's theorem with $|A| \geq 3$. We need four alternatives for interval orders because of the definition of Ferrers relation.

    There exist a number of papers addressing the aggregation issue for binary relations as an optimization problem. Typically, a 0/1 linear program is targeted. Contrastingly, in our approach, we associate to each profile of binary relations an instance of BCP, the

---

[2] X represents here the set of voter's preferences and Y the resulting order.

so-called *Binate Covering Problem* [4], where the set of constraints is not any set of 0/1 linear inequations but a SAT instance. This problem has been studied for decades by the circuit community where it is important for logic synthesis (minimizing the number of components needed to perform a given operation).

From a theoretical standpoint, like 0/1 linear programming, BCP is an NP-hard optimization problem (and the associated decision problem is in NP) (see e.g. [12]). In practice, each clause can be translated into an equivalent 0/1 linear inequation, but the converse does not hold. The specific format of the constraints considered in BCP (compared to 0/1 linear programs) enables us to take advantage of the power of existing MAX-SAT solvers in order to solve its instances in a more efficient way from the practical side.

To our knowledge there is a limited number of studies related to the aggregation of interval orders. Pirlot and Vincke [13] have shown that the schemes that work well for complete preorders such as lexicographic procedure or Borda's sum of ranks do not lean themselves easily to the generalization with interval orders. They proposed two types of aggregation procedures: one consisting in aggregating numerical representations into a "global evaluation" function, and the other inspired from pairwise comparison methods.

In this paper we propose a hybrid approach consisting in finding an interval order being optimal in the sense of minimal Kemeny distance [8] to the input profile. Intuitively, ranking the alternatives according to Kemeny's rule can be seen as the best compromise since on average it gives the "closest" social preference to the individual preferences. Our idea can be summarized as in the following:

1. Determine all pairwise comparisons for which all the voters have the same opinion and build a partial order that preserves those comparisons.
2. Search within the set of feasible interval orders in order to find a closest one to the input profile.

The first step can be easily achieved the following way:

$$\forall v_i \in V, \forall x, y \in A, xP_i y, \Longrightarrow xPy,$$
$$\forall v_i \in V, \forall x, y \in A, xI_i y, \Longrightarrow xIy.$$

The resulting $\langle P, I \rangle$ is a partial order.

From partial order to interval orders. Naturally this step provides in the majority of cases many interval orders. The worst case that we may expect is when the partial order provided in the first step is empty. In this case we have to find all the interval orders containing n objects (n being the cardinality of A). This case gives an idea on the number of interval orders that we may have. Stanley [15] has precised the number of interval orders with n elements; for this he has made use of relations between interval orders and hyperplanes arrangement. The coefficient of the following polynomial provides the number of interval orders:

$$z = \sum_{k \geq 0} c_k \frac{x^k}{k!}$$
$$z = 1 + x + 3\frac{x^2}{2!} + 19\frac{x^3}{3!} + 195\frac{x^4}{4!} + 2831\frac{x^5}{5!} + 53703\frac{x^6}{6!} + 1264467\frac{x^7}{7!} + \dots$$

z is the unique power series satisfying $\frac{z'}{z} = y^2$, $z(0) = 1$ where $1 = y(2 - e^{xy})$. The value c_k of the serie z is the number of interval orders on k alternatives. This number grows exponentially on the number of alternatives: for instance, with just 7 alternatives we have more than one million interval orders. However, we will see in the following that we do not need to represent those interval orders explicitly. We denote by $\langle P^{(1)}, I^{(1)} \rangle, \langle P^{(2)}, I^{(2)} \rangle, \ldots$ these interval orders.

Discriminating interval orders. In our approach, the distance of an interval order $\langle P^{(i)}, I^{(i)} \rangle$ to the input profile X, $D(\langle P^{(i)}, I^{(i)} \rangle, X)$, will be calculated as the sum of its distance to each voter's order $\langle P_j, I_j \rangle$.
Let us denote this distance by $d(\langle P^{(i)}, I^{(i)} \rangle, \langle P_j, I_j \rangle)$:

$$D(\langle P^{(i)}, I^{(i)} \rangle, X) = \sum_{\langle P_j, I_j \rangle \in X} d(\langle P^{(i)}, I^{(i)} \rangle, \langle P_j, I_j \rangle)$$

The distance d is computed using the difference between pairwise comparisons in the following way:

$$d(\langle P^{(i)}, I^{(i)} \rangle, \langle P_j, I_j \rangle) = \sum_{(x,y) \in A^2} \delta_{\langle P^{(i)}, I^{(i)} \rangle, \langle P_j, I_j \rangle}(x, y)$$

$$\delta_{(\langle P^{(i)}, I^{(i)} \rangle, \langle P_j, I_j \rangle)}(x, y) = \begin{cases} p2p \; if & (xP^{(i)}y \text{ and } yP_jx) \text{ or } (yP^{(i)}x \text{ and } xP_jy) \\ 0 \quad if & (xP^{(i)}y \text{ and } xP_jy) \text{ or } (xI^{(i)}y \text{ and } xI_jy) \\ p2i \; otherwise \end{cases}$$

Here $p2p$ and $p2i$ are nonnegative constant numbers. The rationale for this definition of d is to put a penalty when there is a discrepancy of preference relation between the comparison given by a voter and the one of the interval order. Naturally, a discrepancy of a strict preference (for instance xPy) to the inverse of this preference (yPx) is at least as problematic as a discrepancy of a strict preference (for instance xPy) to an indifference (xIy) for this reason we suggest that $p2p \geq p2i$. Even more one can impose the strict inequality ($p2p > p2i$) which will guarantee to have as a result aIb when the profile with three voters is aP_1b, aI_2b and bP_3a. Note that the distance used by Hudry ([7]) imposes $p2p = p2i = 1$ and provides as a result three interval orders (aPb, bPa and aIb) for this example.

We propose to represent the set of interval orders to be *implicitly* considered in the second step using propositional constraints (clauses). Then, computing the interval orders closest to the profile is encoded as minimizing an objective function. Accordingly, we reduce our interval order optimization problem to the BCP one.

3   Translation into the Binate Covering Problem

We first need propositional variables v_{xPy} and v_{xIy} to represent all pairs of the form xPy ($\forall x \neq y \in A$) and xIy ($\forall x, y \in A, x \leq y$). As a consequence, $n^2 - n + \frac{n \times (n-1)}{2} + n = \frac{3 \times n^2 - n}{2}$ variables must be considered. For instance, for 4 alternatives, we need 22 propositional variables. For 16 alternatives, we need 376 propositional variables.

3.1   Implicit Representation of Interval Orders

Structural constraints. The following constraints express that the result of the aggregation must be an interval order. They do not depend on the voters.

- $P \cup I$ is complete: $\forall x < y \in A\ v_{x P y} \vee v_{x I y} \vee v_{y P x}$,
- P is asymmetric: $\forall x < y \in A,\ \neg(v_{x P y} \wedge v_{y P x}) \equiv \neg v_{x P y} \vee \neg v_{y P x}$,
- P and I are exclusive: $\forall x \neq y \in A,\ \neg(v_{x P y} \wedge v_{x I y}) \equiv \neg v_{x P y} \vee \neg v_{x I y}$,
- I is symmetric by construction because a single propositional variable $v_{x I y}$ represents both xIy and yIx.
- I is reflexive: $\forall x \in A,\ v_{x I x}$ is forced to be true,
- $P \cup I$ is Ferrers: $\forall x, y, z, t \in A, x \neq y, z \neq t, x \neq t, y \neq t, x \neq z\ (v_{x P y} \wedge v_{y I z} \wedge v_{z P t}) \Rightarrow v_{x P t}$,

Note that we need to generate $2n(n-1) + n + n(n-1)(n-2)^2 = n(n^3 - 5n^2 + 10n - 5)$ structural constraints plus the unit clauses needed to preserve unanimity (see below). For 4 alternatives, it means at least 76 constraints. For 16 alternatives, it means at least 47536 constraints. The $O(n^4)$ space required by the above encoding is clearly dominated by the cost of ensuring Ferrers condition.

Unanimity constraints. Those additional constraints encode unanimity for both P and I. They are generated according to the votes. Since they force the truth value of some variables, they simplify in practice the computation of the best interval order.

- Unanimity for P: $\forall x \neq y \in A$, if $\#p(x, y) = |V|$ then xPy is forced to be true,
- Unanimity for I: $\forall x \neq y \in A$, if $\#i(x, y) = |V|$ then xIy is forced to be true.

3.2   Distance between Interval Orders and the Profile

The coefficient associated to each variable is computed according the individual penalty δ defined earlier and the number of voters that disagree with the interval order.

- $\forall x, y \in A$, satisfying $I(x, y)$ entails that voters that strictly prefer x to y or y to x disagree with that fact, with a simple individual penalty of $p2i$. As a consequence, the coefficient of the variables is exactly $p2i(\#p(x, y) + \#p(y, x))$,
- $\forall x, y \in A$, satisfying $P(x, y)$ entails that voters that are indifferent between x and y disagree with that fact with a simple penalty of $p2i$, while the voters that strictly prefer y to x disagree with that fact with an individual penalty of $p2p$. So the coefficient of those variables is exactly $p2i\#i(x, y) + p2p * \#p(y, x)$.

The objective function of the binate covering problem associated with X is denoted by $score_X(\langle P, I \rangle)$ and is

$$\sum_{x \leq y \in A} p2i(\#p(x, y) + \#p(y, x))v_{x I y} + \sum_{x \neq y \in A} (p2i\#i(x, y) + p2p * \#p(y, x))v_{x P y}.$$

Thus the space needed to represent the objective function is in $\mathcal{O}(n^2.log_2(m))$. Interestingly, the space needed by the encoding (constraints and objective function) is only logarithmic in the number of voters. This renders the approach feasible for a large number of voters. On the other hand, the space needed by the encoding is in $\mathcal{O}(n^4)$; considering that MAX-SAT solvers are currently able to solve *some* instances with millions of variables, it might be possible to solve aggregation problems up to roughly 40 alternatives (which leads to 2 millions of clauses using the above encoding).

The result of the aggregation step is any interval order which minimizes the value of the objective function. An important issue is to determine whether it makes sense to use of sophisticated SAT engine (or 0/1 linear program solver) to solve those specific BCP instances stemming from a translation from instances of the aggregation problem. [7] gave a positive answer to this query, by identifying the complexity of the following decision problem: SCORE:

Input: A finite profile X of binary relations $\langle P, I \rangle$ on A and a nonnegative integer k.
Question: Does there exist an interval order $\langle P, I \rangle$ on A such that $score_X(\langle P, I \rangle) \leq k$?

In a nutshell Hudry showed that SCORE is NP-complete as soon as the number of voters m is "sufficiently" large compared to the number n of alternatives, even in the restricted case when X consists of linear orders only, provided that $p2p = p2i = 1$. This justifies to take advantage of algorithms running in exponential time (as MAX-SAT solvers) in the worst case, since polynomial time ones are hardly expected.

Hudry's NP-hardness result extends easily to our framework when the parameters $p2p$ and $p2i$ are such that $p2p = p2i$ since linear orders are interval orders; on the other hand, the membership to NP of the SCORE probllem is obvious in our setting: in order to determine that an instance of this decision problem is positive, it is enough to guess a binary relation $\langle P, I \rangle$ on A (its size is $\mathcal{O}(n^2)$), then to check that it is an interval order (this can be easily achieved in polynomial time in the size of the relation), and finally to compute in polynomial time $score_X(\langle P, I \rangle)$ in order to compare it with k.

Our MAX-SAT algorithm for the BCP problem is a branch-and-bound algorithm. During the search, each time a (partial) assignment is found that satisfies all the constraints, the corresponding score is computed (each unassigned variable is set to 0) and a constraint which eliminates all the assignments leading to a greater bound is added, so that whenever a partial assignment leads to a score which is worse than this bound, a backtrack occurs. Its worst-case time complexity is simply exponential in the number of variables under consideration (hence linear in the size of X) and its space complexity is linear in the size of the constraints (hence quadratic in the size of X).

3.3   Examples

As a matter of illustration, let us consider the following examples. For these examples we suppose that $p2i = 1$ and $p2p = 2$.

Example 1. Consider first a case with 5 voters and 4 alternatives with the preferences of voters shown in Table 1.

These preferences of voters can be compactly represented in a matrix where $\forall x_i, x_j$, $P(x_i, x_j) = \alpha$ means that there are α voters who prefer alternative x_i to alternative x_j. Table 2 represents the matrix related to the previous example. Accordingly, this

Table 1. Pairwise comparisons on 4 alternatives given by 5 voters

V_1	a	b	c	d
a	I	P	P	P
b	P^{-1}	I	I	P
c	P^{-1}	I	I	P
d	P^{-1}	P^{-1}	P^{-1}	I

V_2	a	b	c	d
a	I	P	I	P
b	P^{-1}	I	P^{-1}	P
c	I	P	I	P
d	P^{-1}	P^{-1}	P^{-1}	I

V_3, V_4, V_5	a	b	c	d
a	I	P	P	P
b	P^{-1}	I	I	P
c	P^{-1}	I	I	P
d	P^{-1}	P^{-1}	P^{-1}	I

matrix contains all the information needed for running our aggregation procedure. Our method find as a result the following interval order: aPb, cPa, dPa, cPb, dPb, cPd (its distance to the profile is 12).

Table 2. The number of voters agreeing for a strict preference

P	a	b	c	d
a	0	5	1	2
b	0	0	0	2
c	3	4	0	5
d	3	3	0	0

Example 2. Here is a second example; Table 3 shows the pairwise comparisons given by three voters on three alternatives (a, b, c).

Table 3. The profile of Example 2

V_1	a	b	c
a	I	P	I
b	P^{-1}	I	P^{-1}
c	I	P	I

V_2	a	b	c
a	I	P	P
b	P^{-1}	I	I
c	P^{-1}	I	I

V_3	a	b	c
a	I	I	I
b	I	I	I
c	I	I	I

The result of our aggregation procedure provides a unique interval order as close as possible to the input profile. It is not a preorder (a is preferred to b and all the other comparisons are indifference), despite the fact that each preference relation in the input profile is a preorder.

3.4   More Than One Solution Is Often the Case

Clearly enough, there is no guarantee in general that a unique interval order $\langle P, I \rangle$ exists, leading to a minimal value s^* for the objective function $score_X(\langle P, I \rangle)$. This problem is inherent to the fact that voters may have different preferences, and it may happen in very simple scenarios, for instance when A consists of two alternatives a and b and X consists of two interval orders $\langle P_1, I_1 \rangle$ and $\langle P_2, I_2 \rangle$ on A so that aP_1b and aI_2b: in such a case, both $\langle P_1, I_1 \rangle$ and $\langle P_2, I_2 \rangle$ lead to the minimal value $s^* = p2i$, but not to the same sets of undominated alternatives. Nevertheless, this plurality is problematic since decisions made using only one of such optimal interval orders are not necessarily robust, in the sense that the choice of another optimal interval order could question them. Typically decisions are made by comparing alternatives or determining undominated ones. While robustness is a complex notion, a sufficient condition for a comparison to be robust is when it holds for every optimal interval order, and similarly an alternative is robustly undominated when it is undominated for all optimal interval orders. Formally, the following decision problems have to be considered: NEC-COMP(R):

Input: A finite profile X of binary relations $\langle P, I \rangle$ on A and two alternatives a, b from A.

Question: Is it the case that every interval order $\langle P, I \rangle$ on A satisfying $score_X(\langle P, I \rangle) = s^*$ is such that aRb? (where $R = P$ or $R = I$)?

NEC-UNDOM:

Input: A finite profile X of binary relations $\langle P, I \rangle$ on A and an alternative a from A.

Question: Is it the case that for every interval order $\langle P, I \rangle$ on A such that $score_X(\langle P, I \rangle) = s^*$, we have $a(P \cup I)b$ for every $b \in A$?

Those decision problems are "mildly" hard, since they belong to the complexity class Θ_2^p, consisting of all decision problems which can be solved in deterministic polynomial time using logarithmically many calls to an NP oracle. In order to prove the membership of NEC-COMP(R) and NEC-UNDOM to Θ_2^p, we consider the complementary problems and show them in Θ_2^p as well (this class is closed under complementation). We have already seen that SCORE is in NP. Now, the value of s^* can be computed by binary searching it within the bounds 0 and $m.n^2.max(p2p, p2i)$ which is a (rough) upper bound of s^*, and has a value linear in the size of X since $max(p2p, p2i)$ is a constant. Hence, s^* can be computed in deterministic polynomial time using logarithmically many calls to an NP oracle (used to solve the SCORE instances encountered during the search, associated to the successive values of k). Once this is done, it remains to guess a binary relation $\langle P, I \rangle$ on A using a last call to the NP oracle, check that it is an interval order such that $score_X(\langle P, I \rangle) = s^*$, and finally check that $a\bar{R}b$ (resp. that there exists a $b \in A$ such that bPa). We conjecture that NEC-COMP(R) and NEC-UNDOM are Θ_2^p-complete. Noticeably, when s^* is part of the input, the complexity of NEC-COMP(R) and NEC-UNDOM falls down to coNP. From the practical side, when several instances of NEC-COMP(R) or NEC-UNDOM sharing the same profile X are to be solved, it can prove useful to compute s^* once for all during a pre-processing phase, then to exploit it in order to solve those instances in a more efficient way.

4 Some Theoretical Results

We analyze here some expected properties for aggregation procedures such as respect of unanimity, independance, majority, etc., and our objective is to determine whether or not our approach satisfies some of them. We begin by the properties at work in Arrow's theorem:

Universality. An aggregation procedure is universal if it accepts all configurations for the input profile. Since the input of our procedure can be any finite set of interval orders, we can conclude that our procedure is universal.

Transitivity. Arrow's theorem imposes the transitivity of the preference and the indifference relation. Our procedure provides an interval order which has a transitive preference relation P but the indifference relation I is not necessarily transitive. However as we mentioned in the introduction, in order to find an undominated alternative, transitivity of P is enough.

Weak-unanimity. Our procedure satisfies the weak unanimity condition since unanimity is imposed by our formulation as a hard constraint to be respected.

Non-dictatorship. Our procedure obviously satisfies the non-dictatorship condition.

Independence. Our procedure does not satisfy the condition of independence of irrelevant alternatives: let us show it on a new example. The set of alternatives is $A = \{a, b, c, d\}$ and we have two different profiles X and X' which have the same votes on the subset $A' = \{c, d\}$ of A:

Example 3. Table 3 shows the compact matrix of each profile.

Table 4. The compact matrix of profile X and X' 4

X	a	b	c	d		X'	a	b	c	d
a	0	0	0	1		a	0	3	0	0
b	0	0	6	2		b	3	0	0	3
c	4	0	0	6		c	8	1	0	6
d	8	6	4	0		d	2	2	4	0

Our procedure concludes that for the profile X there are two optimum solutions, in the first one c is indifferent to d and in the second one d is preferred to c. However, even if the profile X' has the same votes for the comparison between c and d, our procedure concludes for X' that c is preferred to d.

We consider now some other properties that an aggregation procedure should preferably satisfy.

Anonymity. The result of the aggregation depends only on the preferences of voters (and for instance, not on the age, sex or seniority of candidates). Let $\mathcal{P}$ be the set of permutations of A, π one element of $\mathcal{P}$. We denote by $\pi(R)$, the binary relation such as $\pi(a)\pi(R)\pi(b) \iff aRb$. An aggregation procedure is anonymous if and only if $\forall \pi \in \mathcal{P}, f(R_1, R_2, \ldots, R_m) = \pi(f(\pi(R_1), \pi(R_2), \ldots, \pi(R_m)))$.
It is easy to see that our procedure is anonymous.

Loyalty. If there is just one voter the procedure must provide as a result the same preference as her: $m = 1 \implies f(R_1) = \{a \in A : aR_1b, \forall b \in A\}$. Again, it is easy to check that our procedure satisfies the loyalty condition

Majority condition. If there is a majority of voters who prefers a to b then the result of the aggregation procedure must agree with this comparison: f satisfies the *majority condition* if and only if $\forall (R_1, R_2, \ldots, R_m) \in X, \ \forall a, b \in A$

$$\#p(a,b) > \#p(b,a) \Longrightarrow aPb,$$
$$\#p(a,b) = \#p(b,a) \Longrightarrow aIb.$$

Our aggregation procedure does not satisfy the majority condition as the following example shows it.

Example 4. Table 5 gives the number of votes for pairwise comparisons between four alternatives given by 11 voters

Table 5. The number of voters agreeing for a strict preference

P	a	b	c	d
a	0	6	5	1
b	2	0	8	3
c	6	2	0	1
d	4	3	3	0

Even if the majority of voters prefer c to a, the result of our procedure concludes that a is preferred to c (the output is the interval order such that aPb, aPc, aId, aIa, bPc, bId, bIb, cId, cIc, dId and its distance to the profile is 39).

5 Conclusion

In this paper, we have presented an optimization-based approach to interval orders aggregation. In this approach, to every profile of interval orders, one associates an instance of a propositional optimization problem (namely the Binate Covering Problem); solving the latter gives in a straightforward way an interval order (the "closest" to the input profile in some sense), which is considered as the aggregation looked for. Among other things, we have computed an upper bound of the size of the BCP instance associated to every profile (showing that it is only logarithmic in the number of voters), identified some properties satisfied (or not) by the aggregation approach. An interesting feature of such an optimization-based approach to aggregation is that it can be easily tuned to fit with other preference structures (e.g. preorders, semiorders, etc.). Indeed, it is enough to point out the corresponding hard constraints. Investigating in more depth such extensions is a perspective for further research.

The Binate Covering Problem can be seen as a very specific case of an Integer Linear Program in which case efficient ILP frameworks exist (e.g. CPLEX). However, it looks that tools dedicated to Boolean reasoning are better suited to solve such problems: Weighted Partial MAX SAT [1] and Pseudo Boolean Optimization [11] engines are currently receiving a lot of attention since international evaluations are organized regularly and many systems are freely available for the research community.

We designed a proof of concept tool based on the SAT4J library[9], a library of Boolean search engines dedicated to solving SAT, MAX SAT and Pseudo Boolean problems. That tool can be downloaded from http://sat4j.ow2.org/.

In order to have an idea of the applicability of our approach on a real scenario, we used the publicly available results of the SAT RACE 2006[3]. It is a competitive event between 16 SAT solvers on a set of 100 benchmarks. Here each benchmark is a voter and each solver is an alternative. A given benchmark b prefers the SAT solver x to the SAT solver y iff x solved b faster than y. A given benchmark b is indifferent between the SAT solvers x and y iff none of x and y solved b or both of them solved b but with a roughly the same CPU time (the difference is less than 1 second). By definition, each vote is an interval order. Computing the aggregation of such votes means solving a binate covering problem with 376 variables and 47536 clauses. Our aggregator takes less than one second to generate the BCP from the compact matrix. SAT4J takes one second to find a solution, but fails to prove it is optimal even after running for several hours.

We plan to test several MAXSAT and Pseudo Boolean engines on aggregation of real interval orders instances (including LP based ones).

References

1. Argelich, J., Li, C.-M., Manyà, F., Planes, J.: The first and second max-sat evaluations. Journal on Satisfiability, Boolean Modeling and Computation (JSAT) 4, 251–278 (2008)
2. Arrow, K.J.: Social choice and individual values, 2nd edn. J. Wiley, New York (1951/1963)
3. de Condorcet, M.: Essai sur l'application de l'analyse à la probabilité des décisions rendues à la pluralité des voix. Imprimerie Royale, Paris (1785)
4. Coudert, O.: On solving covering problems. In: Design Automation Conference, pp. 197–202 (1996)
5. Fishburn, P.C.: Interval Orders and Interval Graphs. J. Wiley, New York (1985)
6. Gibbard, A.: Social choice and the arrow conditions (unpublished, 1969)
7. Hudry, O.: Np-hardness results for the aggregation of linear orders into median orders. Annals of Operations Research 163, 63–88 (2008)
8. Kemeny, J.G.: Mathematics without numbers. Daedalus 88, 575–591 (1959)
9. Le Berre, D., Parrain, A.: SAT4J, a SATisfiability library for java, http://www.sat4j.org
10. Luce, R.D.: Semi-orders and a theory of utility discrimination. Econometrica 24 (1956)
11. Manquinho, V., Roussel, O.: The first evaluation of pseudo-boolean solvers (pb 2005). Journal on Satisfiability, Boolean Modeling and Computation (JSAT) 2, 103–143 (2006)
12. Papadimitriou, C., Steiglitz, I.: Combinatorial Optimization: algorithms and complexity. Prentice-Hall, Englewood Cliffs (1982)
13. Pirlot, M., Vincke, P.: Semi Orders. Kluwer Academic, Dordrecht (1997)
14. Sen, A.K.: Collective Choice and Social Welfare. North Holland, Amsterdam (1970)
15. Stanley, R.: Hyperplanes arrangements, interval orders and trees. Proc. Nat. Acad. Sci. 93, 2620–2625 (1996)
16. Wiener, N.: A contribution to the theory of relative position. Proc. of Cambridge Philosophical Society 17, 441–449 (1914)

---

[3] http://fmv.jku.at/sat-race-2006/

Circular Representations of a Valued Preference Matrix

Karim Lidouh, Yves De Smet, and Minh Tuan Huynh

Université Libre de Bruxelles,
Department of Computer and Decision Engineering,
Brussels, Belgium

Abstract. In this paper we propose a model for the graphical representation of valued preference matrices. These are obtained from multicriteria outranking methods such as for instance ELECTRE or PROMETHEE. As a consequence, they are often known to be non-symmetric, making two-dimensional representations seldom possible. An optimization model is defined and a particle swarm optimization algorithm is used to solve it. Validation is based on artificial tests. Finally, an illustrative example is given.

1 Introduction

It is common knowledge that problems involving several conflicting criteria are ill-defined [1,2]. The ranking of alternatives is not an obvious task since a given action is seldom better than others on all the criteria. As a consequence, finding a solution that is simultaneously optimal for every considered point of view is utopian in practice.

Therefore, many authors insist on the fact that multicriteria methods do not allow to solve a problem but only support an individual or a group of individuals during the decision process. The main challenge is thus to provide adequate tools dedicated to this decision aid activity. Among them, graphical representations are especially appreciated by end users.

In this paper, we propose a particular graphical representation of valued preference matrices. These can be obtained for instance by applying the outranking method PROMETHEE [3,4] but are not restricted to it. In section 2, we describe the underlying model. Section 3 presents the algorithm that was used. Artificial tests are performed in section 4 and an illustrative example is presented in section 5.

2 The Model

Let $A = \{a_1, a_2, ..., a_n\}$ be a set of n alternatives that are evaluated on a set of q criteria denoted by $F = \{f_1, f_2, ..., f_q\}$. Without loss of generality, we may assume that these criteria have to be maximized. Let Π be a valued preference matrix built on the basis of all the pair-wise comparisons of elements belonging

F. Rossi and A. Tsoukis (Eds.): ADT 2009, LNAI 5783, pp. 261–271, 2009.
© Springer-Verlag Berlin Heidelberg 2009

to A. More precisely, $\Pi(a_i, a_j) = 1$ will denote that a_i is strictly preferred to a_j while $\Pi(a_i, a_j) = 0$ will denote that a_i is not strictly preferred to a_j. We assume that elements from the matrix Π are such that:

$$\begin{cases} \Pi(a_i, a_j) \geq 0 \\ \Pi(a_i, a_j) + \Pi(a_j, a_i) \leq 1. \end{cases} \tag{1}$$

One way to compute such a matrix is to use the PROMETHEE method [3,4]. At first, the differences between any pair of alternatives are computed for a given criterion f_k:

$$d_k(a_i, a_j) = f_k(a_i) - f_k(a_j). \tag{2}$$

Then, this difference is transformed into a preference degree denoted $\Pi_k(a_i, a_j)$ degree using a non-decreasing function $Pk : \Re \to [0,1]$:

$$\Pi_k(a_i, a_j) = P_k[d_k(a_i, a_j)]. \tag{3}$$

Finally, a global preference degree is computed by aggregating these uni-criteria degrees:

$$\Pi(a_i, a_j) = \sum_{k=1}^{q} w_k \Pi_k(a_i, a_j). \tag{4}$$

Where w_k represents the weight associated to criterion f_k. Without loss of generality, we assume that $w_k > 0$ and $\sum_{k=1}^{q} w_k = 1$.

Graphical tools are key elements to analyze problems involving multiple dimensions. They offer a convenient way to structure them and to lower their complexity. Traditional statistical approaches such as boxplots, piecharts, scatter plots, principal component analysis (PCA), ... are commonly used by practitioners during the decision process. However none of the methods really take into account the multicriteria nature of the problem. In other words, they do not incorporate the fact that the criteria have to be optimized and have to be interpreted with regard to the decision maker's preferences. To our knowledge, only a few approaches have been proposed to visualize multicriteria problems. For instance, in the PROMETHEE methodology, the GAIA method [5] is based on a PCA applied on the matrix of unicriterion net flow scores. It allows to bring information about conflicting and redundant criteria or alternatives and indicates regions where the best compromise solution is lying. We do think that this research area still needs to be explored.

As already stressed, the aim of this contribution is to propose a graphical representation of the matrix. When matrices are symmetric, the isotonic representation method can be used [6]. The underlying idea of this approach is to find the coordinates of the n points in a space of dimension l such that the distances between the pairs of elements are equal to the values of the matrix.

Due to the multicriteria context, π matrices are seldom symmetric. Moreover, the distance notion does not take into account the fact that criteria have to be optimized. Therefore, a new approach has to be developed.

First we start with the representation of two elements. The concept of circular representation is the following: the elements a_i and a_j will be placed on a circle the radius of which is equal to $[\Pi(a_i, a_j) + \Pi(a_j, a_i)]/(2\pi)$. This will lead to a perimeter of $[\Pi(a_i, a_j) + \Pi(a_j, a_i)]$. A clock-wise interpretation of the circle is the following: the arc from a_i to a_j is equal to $\Pi(a_i, a_j)$ and the arc from a_j and a_i is equal to $\Pi(a_j, a_i)$. As a consequence, two elements such that $\Pi(a_i, a_j) \simeq \Pi(a_j, a_i) \simeq 0$ are superposed. The graphical interpretation is thus consistent with the fact that they are indifferent. Figure 1 illustrates different preference settings.

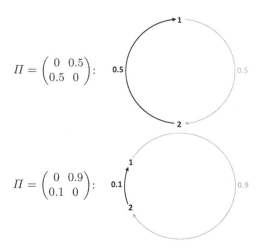

$$\Pi = \begin{pmatrix} 0 & 0.5 \\ 0.5 & 0 \end{pmatrix}:$$

$$\Pi = \begin{pmatrix} 0 & 0.9 \\ 0.1 & 0 \end{pmatrix}:$$

Fig. 1. Circular representation of two different preference settings

It is obvious to notice that such a representation is always possible when only two alternatives are considered. When more actions are taken into account, a two dimensional representation does not perfectly correspond to the Π matrix. Let $p(a_i, a_j)$ be the length of the arc going from a_i to a_j for a given circular representation. Our aim is to find the circular representation that will be as close as possible to the Π matrix. More formally, we want to find the circular representation such that:

$$min \sum_{i=1}^{n} \sum_{j=1}^{n} [\Pi(a_i, a_j) - p(a_i, a_j)]^2. \tag{5}$$

An empirical approach based on particle swarm optimization is used to solve this problem.

3   The Algorithm

Different approaches can be considered to solve the model presented in the previous section. As a first attempt, we have chosen to apply a Particle Swarm

Optimization (PSO) [7,8] algorithm which is generally well performing for continuous optimization problems. Of course other heuristics could be considered but this goes beyond the scope of this contribution. Our main focus remains to present an original view of a preference matrix.

To solve our problem using the PSO approach we will have to define the variables of our search space. There will be two sets of variables:

- Positioning variables (x_i, y_i): These variables define the position of an action i (where $i = 1, 2, ..., n$). There will be $2n$ variables in this set. As two actions will ever be separated by a maximum distance of $1/2\pi$, we can reduce the search space for these variables. To make the implementation easier, we will allow each variable to be in $[\frac{-1}{2\pi}, \frac{1}{2\pi}]$ even though we could reduce the space to a disc of diameter $1/2\pi$.
- Variables defining the circles k_{ij}: A single variable per pair of actions will be enough to define the preference circle once the coordinates of action i and j are chosen (where $i = 1, 2, ..., n - 1$ and $j = i + 1, ..., n$). The number of variables of this set will be C_2^n.

Figure 2 shows how the variable k_{ij} defines a circle for two actions i and j whose coordinates are known. We consider the chord between the two actions and the normal line that cuts it in the middle. On this line, k_{ij} will give us the position of the circle's center. Since this circle can only have a perimeter of 1 at most, as described in Section 2, k_{ij} will only have values in $[0, \frac{1}{2\pi}]$ if $\Pi(a_i, a_j) \geq \Pi(a_j, a_i)$ and in $[\frac{-1}{2\pi}, 0]$ if $\Pi(a_i, a_j) < \Pi(a_j, a_i)$.

Figure 3 gives an example of representation obtained using the previously described algorithm. The example involved 6 actions with randomly generated preferences. The color of the arcs indicates the difference between their length and the preference that they are meant to represent. Table 1 gives a legend of the colors and the range of error symbolized by them.

Studying the evolution of the mean error gave us an understanding of how effective particle swarm optimization can be for this kind of problems. We can see

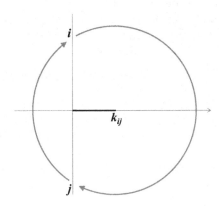

Fig. 2. Variable k_{ij} defining the circle for two actions

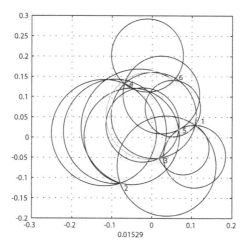

Fig. 3. Complete circular representation of a problem

Table 1. Error ranges on the representation

Color	Range of error
Red	$0.1 < \text{error}$
Blue	$0.01 < \text{error} < 0.1$
Green	$0.001 < \text{error} < 0.01$
Yellow	$\text{error} < 0.0001$

in Figure 4 that no matter where the algorithm starts, the error drops tremendously during the first iterations before stabilizing itself after about 20000 evaluations of the objective function. This behavior can be explained by three possible situations [9,10]:

- Convergence: all the particles have converged towards the same position and are stuck in the same local or global minimum.
- Divergence: the particles fail in finding a good solution. They keep moving yet the error of the best solution found does not change.
- Stagnation: the particles have assembled in two or more separated positions that act as attractive poles for the particles near them. The movements of the particles are influenced by the different local minima but the particles cannot escape from the current situation.

By observing the kinetic energy of the particle we can see if these are still in movement when the error stabilizes or if they are still looking for a better solution far from the best one discovered. We can also display the volume of the swarm of particles to see if these are dispersed or if they are concentrated around a single solution. Figure 5 shows the evolution of these two measures while the algorithm is solved.

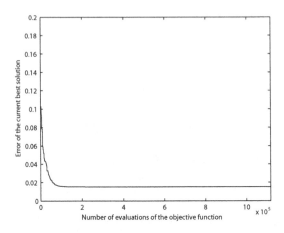

Fig. 4. Evolution of the mean error with each evaluation of the objective function

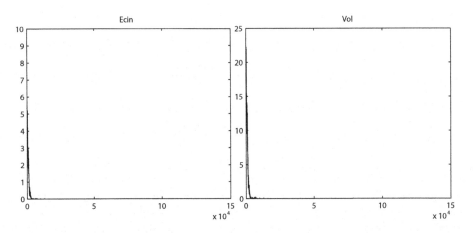

Fig. 5. Evolution of the kinetic energy and the volume of the swarm

It indicates that both the kinetic energy and the volume of the swarm reach zero, meaning that the algorithm has converged towards a single best solution.

4   Some Empirical Tests

Using empirical tests on artificial matrices we tried to optimize the parameters chosen for our algorithm. The parameters that we chose to modify for the tests were the following:

– The number of particles N: initially set at 20, we suspected the number had to be increased to broaden our coverage of the search space.

- The size of the neighborhood K: initially of 3 particles. A higher number would speed the propagation of good solutions to the other particles [11].
- A parameter γ that defines the movement equations of the particles.

We tried to find the values that allowed us to improve the algorithm in terms of speed and accuracy when searching for representations of attainable matrices (i.e. matrices that can have a perfect representation).

To evaluate the different parameter settings, several indicators were considered:

- The mean error on all arcs of the representation μ_{error}.
- The standard deviation of the mean error σ_{error}: giving the robustness of the performance for a high number of executions.
- The success rate of the algorithm τ: indicating the proportion of executions that yielded an acceptable representation.
- The speed of convergence μ_{conv}: in number of iterations before a stable representation is found.
- The standard deviation of the speed of convergence σ_{conv}.

After some tests where we appreciated the effects of modifying the parameters, we obtained a configuration that improved the results significantly. Table 2 summarizes the results obtained with an initial and the last configurations. We found out that increasing the number of particles improved the results until a number of 300 to 600. As for the other parameters, using a ring neighbourhood with 3 particles and a γ of 2.15 seemed to produce the best output.

Table 2. Quality of results with two configurations

Parameters	$N = 20$ $K = 3$ $\gamma = 2.01$	$N = 300$ $K = 3$ $\gamma = 2.15$
Indicators	$\mu_{error} = 0.1058$ $\sigma_{error} = 0.0109$ $\tau = 0\%$ $\mu_{conv} = 192,460$ $\sigma_{conv} = 55,182$	$\mu_{error} = 0.000858$ $\sigma_{error} = 0.0015$ $\tau = 59\%$ $\mu_{conv} = 68,877$ $\sigma_{conv} = 8,951$

5   An Illustrative Example

The example we will use to illustrate our model is taken from the multicriteria literature: a power plant localization problem described by Brans et al. [3]. In that example, 6 alternatives are evaluated using the PROMETHEE and GAIA methods. The preference matrix we obtain for it is the following:

$$\Pi = \begin{pmatrix} 0 & 0.296 & 0.25 & 0.268 & 0.1 & 0.185 \\ 0.463 & 0 & 0.689 & 0.334 & 0.296 & 0.5 \\ 0.235 & 0.18 & 0 & 0.334 & 0.055 & 0.43 \\ 0.398 & 0.506 & 0.305 & 0 & 0.224 & 0.212 \\ 0.444 & 0.515 & 0.486 & 0.379 & 0 & 0.448 \\ 0.286 & 0.399 & 0.25 & 0.431 & 0.133 & 0 \end{pmatrix} \tag{6}$$

When carefully studying this matrix, we can see that it cannot be exactly represented by a 2-dimensional arrangement. This can be easily demonstrated by showing that the triangle inequality is not verified for most sets of 3 actions from our problem:

$$|l(a,b) - l(a,c)| \le l(a,c) \le l(a,b) + l(b,c) \tag{7}$$

where $l(a,b)$ represents the length of the chord between actions a and b.

Using our swarm particle optimization algorithm we can obtain a complete circular representation as in Figure 3 or limit ourselves to the arcs that give us the most interesting information. The following matrix contains the net flow (or global score) of the 6 actions in our problem:

$$\phi = \begin{pmatrix} -0.1454 \\ 0.0772 \\ -0.1492 \\ -0.0202 \\ 0.2928 \\ -0.0552 \end{pmatrix}. \tag{8}$$

By comparing the net flows, we can see that the best alternative is action 5. Figure 6 shows a circular representation of the preferences for action 5 as well as the error on each arc. The color of the arcs indicates the error in length of the arc as already explained in Table 1. As the only colors present are yellow and blue, we are sure that this representation is reliable.

On the circular representation in Figure 6 we also have a representation of the net flow as small colored disks that are blue when the net flow is positive and red otherwise. The size of the disks indicates the absolute value of the net flow. This additional information allows us to quickly detect which action is the next best one in the ranking as well as compare its behavior with regard to the best one. However action 5's higher net flow makes it by far the most interesting option. This assumption is further reinforced by the length of the blue arcs leaving action 5 in comparison to the red ones.

When considering action 1 on Figure 7, we can remark that even though the action is globally considered as one of the worst, it is highly preferred to action 3 and is reasonably preferred to actions 2 and 4. Another interesting information is the big size of the circle between actions 1 and 2, indicating that the two actions are clearly differenciated (i.e. the sum of their respective preferences is close to 1) whereas actions 1 and 3 are more similar (i.e. the sum of their preferences has a lower value).

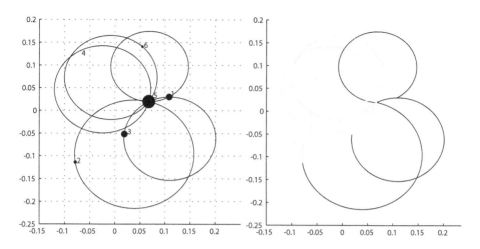

Fig. 6. Circular representation of the preferences for action 5

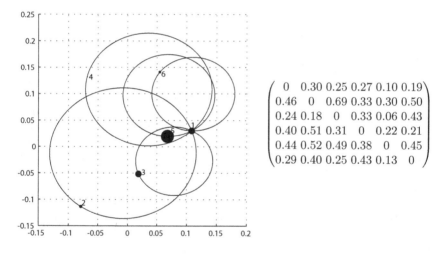

$$\begin{pmatrix} 0 & 0.30 & 0.25 & 0.27 & 0.10 & 0.19 \\ 0.46 & 0 & 0.69 & 0.33 & 0.30 & 0.50 \\ 0.24 & 0.18 & 0 & 0.33 & 0.06 & 0.43 \\ 0.40 & 0.51 & 0.31 & 0 & 0.22 & 0.21 \\ 0.44 & 0.52 & 0.49 & 0.38 & 0 & 0.45 \\ 0.29 & 0.40 & 0.25 & 0.43 & 0.13 & 0 \end{pmatrix}$$

Fig. 7. Circular representation of the preferences for action 1

Considering only one action at a time we are actually looking at a representation of the row and column of that action in the matrix Π. The elements considered are highlighted in the matrix of Figure 7 for action 1. This graphical representation allows us to quickly identify remarkable properties that would have taken us much more time to discover when simply looking at the corresponding numbers. As the cognitive effort of the decision maker has been reduced when analyzing a given preference matrix, they can afford to analyze more situations making it likely to take better decisions.

6 Conclusion

In this paper we described a new model to graphically represent valued preference matrices using circles. However this model cannot be used in all cases to accurately represent problems with more than two actions. Therefore we make use of particle swarm optimization to find a representation as close as possible to a given matrix. The developed algorithm is tested on several cases and its parameters are optimized to find better solutions. We also considered different variants of particle swarm optimization such as attractive-repulsive optimization [12] but the best results were obtained using the classical approach. Finally, an example was used to illustrate potential uses for the model such as the analysis of individual actions of a decision problem or the understanding of the relations between two of them.

Further work involves additional testing of the algorithm on different problems and the use of other optimization means to find more accurate representations. In particular, we could compute the theoretical minimum error of a given matrix and try to reach it using the algorithm. Or further analyze the search space to improve the way the algorithm explores it.

References

1. Vincke, P.: Multicriteria Decision-Aid. J. Wiley, Chichester (1992)
2. Roy, B.: Multicriteria Methodology for Decision Aiding. Kluwer Academic, Dordrecht (1996)
3. Brans, J., Mareschal, B.: PROMETHEE-GAIA. Une Méthodologie d'Aide à la Décision en Présence de Critères Multiples. Ellipses, Paris (2002)
4. Bouyssou, D., Marchant, T., Pirlot, M., Tsoukiàs, A., Vincke, P.: Evaluation and decision models with multiple criteria: Stepping stones for the analyst. In: International Series in Operations Research and Management Science, Boston, vol. 86 (2006)
5. Brans, J., Mareschal, B.: The PROMCALC and GAIA decision support system for multicriteria decision aid. Decision Support Systems 12, 297–310 (1994)
6. Dattorro, J.: Convex Optimization & Euclidean Distance Geometry. Lulu.com (July 2006)
7. Dioşan, L., Oltean, M.: Evolving the structure of the particle swarm optimization algorithms. In: Gottlieb, J., Raidl, G.R. (eds.) EvoCOP 2006. LNCS, vol. 3906, pp. 25–36. Springer, Heidelberg (2006)
8. Poli, R.: Analysis of the publications on the applications of particle swarm optimisation. Journal of Artificial Evolution and Applications 8(2), 1–10 (2008)
9. Clerc, M., Kennedy, J.: The particle swarm-explosion, stability, and convergence in a multi-dimensional complex space. IEEE Transactions on Evolutionary Computation 6(1), 58–73 (2000)
10. Clerc, M.: Stagnation analysis in particle swarm optimisation or what happens when nothing happens
11. Mohais, A.S., Mendes, R., Ward, C., Posthoff, C.: Neighborhood re-structuring in particle swarm optimization. In: Zhang, S., Jarvis, R.A. (eds.) AI 2005. LNCS (LNAI), vol. 3809, pp. 776–785. Springer, Heidelberg (2005)

12. Murphy, C.: Effects of swarm size on attractive-repulsive particle swarm optimisation (2004)
13. Huynh, M.T.: Application de l'optimisation par essaim particulaire à la représentation de matrices de préférence valuées. Master's thesis, Université Libre de Bruxelles, Belgium (May 2009)

The First Belief Dominance: A New Approach in Evidence Theory for Comparing Basic Belief Assignments

Mohamed Ayman Boujelben[1], Yves De Smet[2], Ahmed Frikha[3],
and Habib Chabchoub[1]

[1] Faculté des sciences économiques et de gestion de Sfax, Université de Sfax,
Sfax, Tunisie
[2] Faculté des sciences appliquées, Université Libre de Bruxelles,
Bruxelles, Belgique
[3] Institut supérieur de gestion industrielle de Sfax, Université de Sfax,
Sfax, Tunisie
ayman_boujelben@yahoo.fr, yves.de.smet@ulb.ac.be,
ahmed.frikha@isgis.rnu.tn, habib.chabchoub@fsegs.rnu.tn

Abstract. In this paper, we consider problems where the data is uncertain and/or imprecise and given by basic belief assignments (BBA's). In order to compare the different pairs of the BBA's, a new concept called the first belief dominance is proposed. This is naturally inspired by the concept of first stochastic dominance that allows comparing probability distributions. Finally, an application in multicriteria decision aid context is presented to illustrate the proposed technique.

Keywords: Evidence theory, First belief dominance.

1 Introduction

Evidence theory has been initially developed by Arthur Dempster in 1967 [5], formalized by Glenn Shafer in 1976 [16], and axiomatically justified by Philippe Smets in his transferable belief model [17]. This theory has been proposed as a generalisation of the Bayesian theory. It provides a convenient framework for modelling uncertainty and imprecision in situations where the available information is imperfect.

Basically, the imperfection in data within evidence theory is modelled using an evidential function called basic belief assignment (BBA). In practice, one is sometimes confronted with the necessity of comparing several BBA's. For instance, we can imagine a multicriteria decision problem where the evaluations of alternatives are given by BBA's. In order to apply a multicriteria procedure, the decision-maker has to compare these BBA's. However according to our knowledge, there are no procedures developed within evidence theory allowing the comparison of the BBA's.

In the probability theory, the comparison between two probability distributions is performed using a technique called the stochastic dominance. This approach has been introduced in statistics [13] [14] and has been addressed fundamentally in [3] [7] [8].

F. Rossi and A. Tsoukis (Eds.): ADT 2009, LNAI 5783, pp. 272–283, 2009.

It has been applied in several domains especially in the fields of finance, economics and multicriteria decision aid.

Initially, the stochastic dominance concept has been used in comparing probability distributions for example in [15]. Then, the use of stochastic dominance has been extended to compare imprecise probability distributions for instance in [12]. Recently, this concept has been employed to compare mixed functions, i.e., probability distributions, fuzzy membership functions, possibility measures and belief masses [4]. However, the use of this concept necessitates the transformation of these functions to others of which the proprieties are similar to those of probability functions. For instance, the pignistic transformation developed essentially to make decisions in evidence theory [18] is used to transform a BBA into a pignistic probability function.

In this paper, we propose a new concept in evidence theory called the first belief dominance allowing the comparison of the BBA's. This approach generalizes the first stochastic dominance concept which is the simplest case of the stochastic dominance approach. Let us note that similar extensions of stochastic ordering to belief functions, called credal orderings, have been developed by Thierry Denoeux and have been published recently in [6]. Some of these orderings have been introduced, without development, in [1] [2] in the context of novelty detection.

This paper is organized as follows: in section 2 we introduce some concepts of evidence theory. The notion of first belief dominance is presented in section 3. Finally, the proposed approach is illustrated by an application in multicriteria decision aid context in section 4.

2 Evidence Theory: Some Concepts

The aim of this section is to clarify the notation and terminology of evidence theory that will be used in the rest of the paper. We will only present the main functions that enable modelling the knowledge.

Let Θ be a finite set of mutually exclusive and exhaustive hypotheses called the frame of discernment and let 2^Θ be the set of all subsets of Θ. A BBA [16] is a function m from 2^Θ to $[0,1]$ verifying $m(\emptyset) = 0$ and $\sum_{A \in 2^\Theta} m(A) = 1$.

The quantity $m(A)$ represents the belief that is committed exactly to A which, in the case of a disjunction of hypotheses, hasn't been assigned to a subset of A because of insufficient information. When $m(A) \neq 0$, A is called focal element. The set of focal elements is called the focal set.

Two functions derived from the BBA, are the belief (or the credibility) function Bel and the plausibility function Pl [16]. These functions are defined respectively by:

$$Bel(A) = \sum_{B \subseteq A} m(B) \tag{1}$$

$$Pl(A) = \sum_{A \cap B \neq \emptyset} m(B) \tag{2}$$

The quantity $Bel(A)$ is interpreted as the total belief associated to A whereas $Pl(A)$ is viewed as the amount of belief that could potentially be placed in A. These two functions can be connected by the equation $Pl(A) = 1 - Bel(\overline{A})$ where $\overline{A}$ denotes the complement of A.

Obviously, if all the belief masses of a given BBA are associated to singletons, the induced belief function is nothing else than a probability function. In evidence theory, this special function is called a Bayesian belief function.

In his transferable belief model, Smets has proposed a technique called the pignistic transformation for translating the belief functions models to probability models in order to make decisions [18]. This transformation consists in distributing equally each belief mass $m(A)$ among the elements of A. This leads to the pignistic probability function $BetP$ defined as follows:

$$BetP(H_i) = \sum_{A \in 2^\Theta / H_i \in A} \frac{m(A)}{|A|}, \forall H_i \in A \tag{3}$$

where $|A|$ is the cardinal of the subset A.

3   First Belief Dominance

3.1   Notation

Let m_i and m_j denote the BBA's of a discrete variable taking the values x_h (with $h = 1, 2, ..., r$) and let Bel_i and Bel_j be their respective belief functions. Throughout this paper, we shall adopt the convention of labelling the values of the variable in accordance with their magnitudes, i.e., $x_1 \prec x_2 \prec ... \prec x_r$. The set of these values is denoted by X.

For all $k \in \{0, 1, ..., r\}$, let:

$$A_k = \begin{cases} \varnothing & \text{if } k = 0 \\ \{x_1, ..., x_k\} & \text{otherwise} \end{cases} \tag{4}$$

and let $\vec{S}(X)$ denotes the set $\{A_1, A_2, ..., A_r\}$. Similarly, for all $l \in \{0, 1, ..., r\}$ such as $l = r - k$, let:

$$B_l = \begin{cases} \varnothing & \text{if } l = 0 \\ \{x_{r-l+1}, ..., x_r\} & \text{otherwise} \end{cases} \tag{5}$$

and let $\overleftarrow{S}(X)$ denotes the set $\{B_1, B_2, ..., B_r\}$.

k and l represent respectively the number of elements of the sets A_k and B_l. Obviously, $|\vec{S}(X)| = |\overleftarrow{S}(X)| = r$, $\overline{A_k} = B_{r-k} = B_l$ for all $k \in \{0, 1, ..., r\}$ and $\overline{B_l} = A_{r-l} = A_k$ for all $l \in \{0, 1, ..., r\}$.

3.2 Definitions

Before introducing the first belief dominance concept, let us define the ascending and descending belief functions.

Definition 1. The ascending belief function denoted $\overrightarrow{Bel_i}$ and induced by m_i is a function $\overrightarrow{Bel_i} : S(X) \rightarrow [0,1]$ defined such as $\overrightarrow{Bel_i}(A_k) = \sum_{C \subseteq A_k} m_i(C)$ for all $A_k \in \vec{S}(X)$.

Definition 2. The descending belief function denoted $\overleftarrow{Bel_i}$ and induced by m_i is a function $\overleftarrow{Bel_i} : S(X) \rightarrow [0,1]$ defined such as $\overleftarrow{Bel_i}(B_l) = \sum_{C \subseteq B_l} m_i(C)$ for all $B_l \in \overleftarrow{S}(X)$.

The ascending belief function represents the beliefs of the nested sets $A_1, A_2, \ldots, A_r$, i.e., the sets $\{x_1\}, \{x_1, x_2\}, \ldots, \{x_1, \ldots, x_r\}$. Similarly, the descending belief function represents the beliefs of the nested sets $B_1, B_2, \ldots, B_r$, i.e., the sets $\{x_r\}, \{x_{r-1}, x_r\}, \ldots, \{x_1, \ldots, x_r\}$.

The third definition is that of the first belief dominance. This condition holds between two BBA's m_i and m_j whenever the two following conditions are verified simultaneously:

- The ascending belief function $\overrightarrow{Bel_i}$ lies, entirely or partly, below the ascending belief function $\overrightarrow{Bel_j}$,

- The descending belief function $\overleftarrow{Bel_i}$ lies, entirely or partly, above the descending belief function $\overleftarrow{Bel_j}$.

Definition 3. m_i is said to dominate m_j according the first belief dominance if and only if the following two conditions are verified simultaneously:

- $\overrightarrow{Bel_i}(A_k) \leq \overrightarrow{Bel_j}(A_k)$ for all $A_k \in \vec{S}(X)$,
- $\overleftarrow{Bel_i}(B_l) \geq \overleftarrow{Bel_j}(B_l)$ for all $B_l \in \overleftarrow{S}(X)$.

In the case where the two conditions are not verified simultaneously, then m_i does not dominate m_j according the first belief dominance concept. As a conclusion, the use of this approach leads to the following situations:

- FBD corresponds to the situation where m_i dominates m_j (we denote $m_i \, FBD \, m_j$),

- $\overline{FBD}$ corresponds to the situation where m_i does not dominate m_j (we denote $m_i \, \overline{FBD} \, m_j$).

3.3   The Preference Situations between the BBA's

The first belief dominance concept allows concluding if m_i (m_j, resp.) dominates m_j (m_i, resp.) or not. So, four preference situations can be established between m_i and m_j:

- If m_i FBD m_j and m_j FBD m_i, then m_i is indifferent to m_j;
- If m_i FBD m_j and m_j $\overline{\text{FBD}}$ m_i, then m_i is strictly preferred to m_j;
- If m_i $\overline{\text{FBD}}$ m_j and m_j FBD m_i, then m_j is strictly preferred to m_i;
- If m_i $\overline{\text{FBD}}$ m_j and m_j $\overline{\text{FBD}}$ m_i, then m_i and m_j are incomparable.

These preference situations have been also defined when the stochastic dominance rules are used to compare probability distributions [4]. The incomparability situation represents the fact that we cannot pronounce on a clear comparison between two BBA's.

Example 1. Let us consider the following three BBA's of a discrete variable taking the values x_1, x_2, and x_3 :

$$\begin{cases} m_1(x_1) = 0.2 \\ m_1(x_1, x_2) = 0.4 \\ m_1(x_3) = 0.4 \end{cases} \qquad \begin{cases} m_2(x_1) = 0.2 \\ m_2(x_2) = 0.4 \\ m_2(x_1, x_2, x_3) = 0.4 \end{cases} \qquad \{ m_3(x_2, x_3) = 1$$

At first, the ascending belief functions related to the BBA's are computed ($A_1 = \{x_1\}$, $A_2 = \{x_1, x_2\}$ and $A_3 = \{x_1, x_2, x_3\}$) :

$$\begin{cases} \overrightarrow{Bel_1}(A_1) = 0.2 \\ \overrightarrow{Bel_1}(A_2) = 0.6 \\ \overrightarrow{Bel_1}(A_3) = 1 \end{cases} \qquad \begin{cases} \overrightarrow{Bel_2}(A_1) = 0.2 \\ \overrightarrow{Bel_2}(A_2) = 0.6 \\ \overrightarrow{Bel_2}(A_3) = 1 \end{cases} \qquad \begin{cases} \overrightarrow{Bel_3}(A_1) = 0 \\ \overrightarrow{Bel_3}(A_2) = 0 \\ \overrightarrow{Bel_3}(A_3) = 1 \end{cases}$$

Similarly, the descending belief functions related to the BBA's are determined ($B_1 = \{x_3\}$, $B_2 = \{x_2, x_3\}$ and $B_3 = \{x_1, x_2, x_3\}$) :

$$\begin{cases} \overleftarrow{Bel_1}(B_1) = 0.4 \\ \overleftarrow{Bel_1}(B_2) = 0.4 \\ \overleftarrow{Bel_1}(B_3) = 1 \end{cases} \qquad \begin{cases} \overleftarrow{Bel_2}(B_1) = 0 \\ \overleftarrow{Bel_2}(B_2) = 0.4 \\ \overleftarrow{Bel_2}(B_3) = 1 \end{cases} \qquad \begin{cases} \overleftarrow{Bel_3}(B_1) = 0 \\ \overleftarrow{Bel_3}(B_2) = 1 \\ \overleftarrow{Bel_3}(B_3) = 1 \end{cases}$$

Then, the first belief dominance concept is applied. The observed belief dominances are illustrated on table 1.

Table 1. The observed belief dominances

	m_1	m_2	m_3
m_1	-	FBD	$\overline{\text{FBD}}$
m_2	$\overline{\text{FBD}}$	-	$\overline{\text{FBD}}$
m_3	$\overline{\text{FBD}}$	FBD	-

Based on the observed belief dominances, the preference situations between the BBA's are established:

- m_1 FBD m_2 and m_2 $\overline{\text{FBD}}$ m_1, then m_1 is strictly preferred to m_2;

- m_1 $\overline{\text{FBD}}$ m_3 and m_3 $\overline{\text{FBD}}$ m_1, then m_1 and m_3 are incomparable;

- m_2 $\overline{\text{FBD}}$ m_3 and m_3 FBD m_2, then m_3 is strictly preferred to m_2.

Finally, given two BBA's m_i and m_j and their related pignistic probability functions $BetP_i$ and $BetP_j$, it is possible to have m_i dominates (does not dominate, resp.) m_j according the first belief dominance approach whereas $BetP_i$ does not dominate (dominates, resp.) $BetP_j$ according the first stochastic one. Therefore, the dominance (the non dominance, resp.) according the first belief dominance concept does not imply necessarily the dominance (the non dominance, resp.) according the stochastic ordering applied on the pignistic probability functions.

Example 2. Let us consider again the data of example 1. The application of the first stochastic dominance (FSD) approach on the pignistic probability functions leads to the following stochastic dominances.

Table 2. The observed stochastic dominances

	$BetP_1$	$BetP_2$	$BetP_3$
$BetP_1$	-	$\overline{\text{FSD}}$	$\overline{\text{FSD}}$
$BetP_2$	FSD	-	$\overline{\text{FSD}}$
$BetP_3$	FSD	FSD	-

Based on the results of tables 1 and 2, we remark that the first belief and stochastic dominance approaches lead to similar results except in the following cases:

- m_1 FBD m_2 whereas $BetP_1$ $\overline{\text{FSD}}$ $BetP_2$;

- m_3 $\overline{\text{FBD}}$ m_1 whereas $BetP_3$ FSD $BetP_1$.

3.4 First Stochastic Dominance: A Particular Case of First Belief Dominance

Of course, it seems natural that the first belief dominance definition coincides with the first stochastic dominance definition in the context of Bayesian belief functions.

The first stochastic dominance is defined as follows. Let p_i and p_j be two probability functions and P_i and P_j be their respective cumulative distributions. According to [7], p_i is said to dominate p_j according the first stochastic dominance if and only if $P_i(x_h) \le P_j(x_h)$ for all $x_h \in X$.

Proposition 1. If Bel_i and Bel_j are two Bayesian belief functions over the frame X, then the two conditions of definition 3 are equivalent.

Proof. *If Bel_i is a Bayesian belief function over the frame X, then $\overrightarrow{Bel_i}(A_k) + \overrightarrow{Bel_i}(\overline{A_k}) = 1$ for all $k = 0, 1, ..., r$. Or, we have $\overline{A_k} = B_{r-k} = B_l$ for all $k = 0, 1, ..., r$, thus $\overrightarrow{Bel_i}(A_k) + \overleftarrow{Bel_i}(B_l) = 1$. The first condition of definition 3 can be written as $1 - \overrightarrow{Bel_i}(A_k) \geq 1 - \overrightarrow{Bel_j}(A_k)$ for all $k = 1, 2, ..., r$, then $\overleftarrow{Bel_i}(B_l) \geq \overleftarrow{Bel_j}(B_l)$ for all $l = 0, 1, ..., r-1$. As a result, the two conditions of definition 3 are equivalent when Bel_i and Bel_j are Bayesian belief functions.*

Proposition 2. If Bel_i and Bel_j are two Bayesian belief functions over the frame X, then the first stochastic dominance is a particular case of the first belief dominance.

Proof. *According to proposition 1, since the two conditions of definition 3 are equivalent, then m_i dominates m_j according the first belief dominance if and only if $\overrightarrow{Bel_i}(A_k) \leq \overrightarrow{Bel_j}(A_k)$ for all $A_k \in \vec{S}(X)$. Bel_i is a Bayesian belief function, then all the focal sets of m_i are singletons. As a result, $\overrightarrow{Bel_i}(A_k) = \sum_{h=1}^{k} m_i(x_h) = P_i(x_k)$ for all $A_k \in \vec{S}(X)$. Therefore, m_i dominates m_j according our approach if and only if $P_i(x_h) \leq P_j(x_h)$ for all $x_h \in X$. So, the first stochastic dominance is a particular case of the first belief dominance.*

4 Application in Multicriteria Decision Aid Context

In order to illustrate the first belief dominance concept, let us consider the following multicriteria decision problem.

A company wants to recruit a new collaborator for the marketing department. Five candidates are considered. A decision for selecting a candidate c_i (with $i = 1, 2, ..., 5$) has to be made based on four qualitative criteria to maximize: the learning capacities, the past experience, the decision-making capacities and the communication skills. The criteria weights are respectively equal to 0.3, 0.3, 0.3 and 0.1.

The candidates are evaluated by the director of human resources department. For each criterion, five assessment grades are considered: x_1 "very bad", x_2 "bad", x_3 "average", x_4 "good" and x_5 "excellent". The set of these grades is denoted by X.

The candidates' performances on each criterion are given by BBA's and are presented in table 3. For instance, the evaluation of candidate c_1 on criterion g_1 is established as follows: the director of human resources department hesitates between the third and the fourth assessment grades. He is sure that the candidate has either average or good learning capacities without being able to refine his judgment.

Table 3. The candidates' performances

	g_1	g_2	g_2	g_3
c_1	$m_1(x_3) = 0.94$ $m_1(x_3,x_4) = 0.06$	$m_1(x_4) = 1$	$m_1(x_1) = 0.04$ $m_1(x_3) = 0.6$ $m_1(x_1,x_2,x_3) = 0.36$	$m_1(x_2) = 1$
c_2	$m_2(x_1) = 0.6$ $m_2(x_1,x_2) = 0.4$	$m_2(x_3) = 1$	$m_2(x_4) = 0.6$ $m_2(x_4,x_5) = 0.4$	$m_2(x_3) = 1$
c_3	$m_3(x_2) = 1$	$m_3(x_3) = 0.28$ $m_3(x_4) = 0.44$ $m_3(x_3,x_4) = 0.28$	$m_3(x_2) = 0.33$ $m_3(x_3) = 0.67$	$m_3(x_3) = 0.84$ $m_3(x_3,x_4) = 0.16$
c_4	$m_4(x_1) = 0.67$ $m_4(X) = 0.33$	$m_4(x_4) = 0.67$ $m_4(x_3,x_4) = 0.33$	$m_4(x_1,x_2) = 0.44$ $m_4(x_2,x_3) = 0.19$ $m_4(x_3) = 0.37$	$m_4(x_3) = 0.9$ $m_4(x_4) = 0.1$
c_5	$m_5(x_2) = 1$	$m_5(x_4) = 1$	$m_5(x_3) = 0.6$ $m_5(x_3,x_4) = 0.4$	$m_5(x_2) = 0.9$ $m_5(x_3) = 0.05$ $m_5(x_2,x_3) = 0.05$

The first belief dominance approach is applied to compare the BBA's characterizing the candidates' evaluations on each criterion. The results are illustrated in tables 4 to 7.

Table 4. The observed belief dominances between candidates on criterion g_1

	c_1	c_2	c_3	c_4	c_5
c_1	-	FBD	FBD	FBD	FBD
c_2	FBD	-	FBD	FBD	FBD
c_3	FBD	FBD	-	FBD	FBD
c_4	FBD	FBD	FBD	-	FBD
c_5	FBD	FBD	FBD	FBD	-

Table 5. The observed belief dominances between candidates on criterion g_2

	c_1	c_2	c_3	c_4	c_5
c_1	-	FBD	FBD	FBD	FBD
c_2	FBD	-	FBD	FBD	FBD
c_3	FBD	FBD	-	FBD	FBD
c_4	FBD	FBD	FBD	-	FBD
c_5	FBD	FBD	FBD	FBD	-

Table 6. The observed belief dominances between candidates on criterion g_3

	c_1	c_2	c_3	c_4	c_5
c_1	-	FBD	FBD	FBD	FBD
c_2	FBD	-	FBD	FBD	FBD
c_3	FBD	FBD	-	FBD	FBD
c_4	FBD	FBD	FBD	-	FBD
c_5	FBD	FBD	FBD	FBD	-

Table 7. The observed belief dominances between candidates on criterion g_4

	c_1	c_2	c_3	c_4	c_5
c_1	-	FBD	FBD	FBD	FBD
c_2	FBD	-	FBD	FBD	FBD
c_3	FBD	FBD	-	FBD	FBD
c_4	FBD	FBD	FBD	-	FBD
c_5	FBD	FBD	FBD	FBD	-

Then, the preference situations between the candidates' evaluations on each criterion are determined using the rules presented in section 3.3. Figure 2 gives the preference graphs of criteria 1 to 4. The candidates are represented by nodes and the preference situations are represented by the graph conventions given in figure 1.

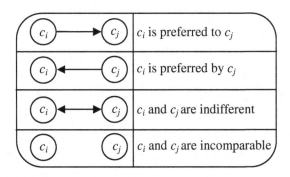

Fig. 1. Graphic representation of the preference situations

Finally, the preference graphs on each criterion are aggregated in order to obtain a global preference graph from which the best candidate is chosen. The aggregation is performed using the algorithm AL3 proposed by Jabeur and Martel [11]. This algorithm has been applied in the context of group decision making [9] [10]. It has also been used to aggregate the single-criterion preference relations between alternatives provided by the application of the stochastic dominance rules for comparing mixed evaluations (i.e. evaluations expressed by probability distributions, fuzzy membership functions, possibility measures and belief masses) [4].

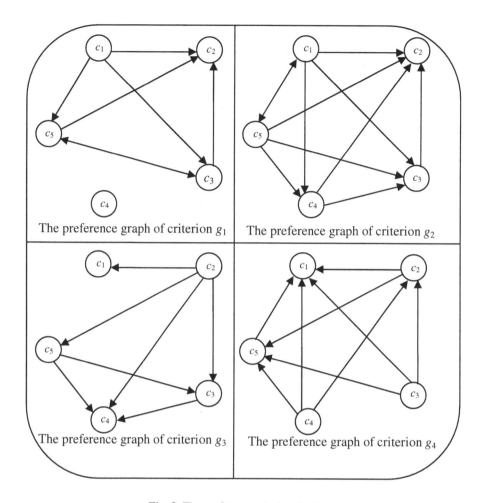

Fig. 2. The preference graphs of criteria

The algorithm AL3 allows determining, for each pair of candidates, the nearest global preference relation to the single-criterion ones. For this purpose, a divergence degree that measures the deviation between the global preference relation and each single-criterion one is calculated. This degree takes into account the criteria weights. The global preference relation is the one that minimizes the divergence degrees. More details concerning this algorithm can be found in [11]. Figure 3 gives the global preference graph.

As one may remark, since candidates c_2 and c_3 are preferred by candidates c_1 and c_5, they can not be chosen. c_1 and c_5 are indifferent. Moreover, c_4 is preferred by c_5. Then c_1 and c_5 are the set of best candidates.

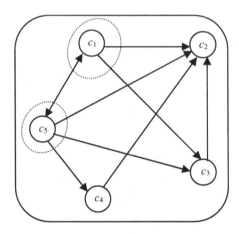

Fig. 3. The global preference graph

5   Conclusion

In this paper, we have introduced a new approach in evidence theory that allows to compare the BBA's called the first belief dominance. We have shown that it is a natural extension of the first stochastic dominance concept which is the simplest case of the stochastic dominance approach. In addition, we have illustrated the proposed concept by an application in multicriteria decision aid field. Of course there are still many directions for future research. Among others, we can mention the extension of the first belief dominance concept to second and third degrees.

References

1. Aregui, A., Denoeux, T.: Fusion of one-class classifiers in the belief function framework. In: Proceedings of the 10th International Conference on Information Fusion, Quebec, Canada (2007)
2. Aregui, A., Denoeux, T.: Novelty detection in the belief functions framework. In: Proceedings of IPMU 2006, Paris (2006)
3. Bawa, V.S.: Stochastic dominance: A research bibliography. Management Science 28, 698–712 (1982)
4. Ben Amor, S., Jabeur, K., Martel, J.M.: Multiple criteria aggregation procedure for mixed evaluations. European Journal of Operational Research 181, 1506–1515 (2007)
5. Dempster, A.P.: Upper and lower probabilities induced by a multi-valued mapping. Annual Mathematics and Statistics 38, 325–339 (1967)
6. Denoeux, T.: Extending stochastic ordering to belief functions on the real line. Information Sciences 179, 1362–1376 (2009)
7. Hadar, J., Russell, W.R.: Rules for Ordering Uncertain Prospects. The American Economic Review 59, 25–34 (1969)
8. Hanoch, G., Levy, H.: The efficiency analysis of choices involving risk. Review of Economic Studies 36, 335–346 (1969)

9. Jabeur, K., Martel, J.M.: A collective choice method based on individual preferences relational systems (p.r.s.). European Journal of Operational Research 177, 469–485 (2005)
10. Jabeur, K., Martel, J.M.: An ordinal sorting method for group decision-making. European Journal of Operational Research 177, 1549–1565 (2007)
11. Jabeur, K., Martel, J.M.: Détermination d'un (ou plusieurs) systèmes(s) relationnel(s) de préférence (s.r.p) collectif(s) à partir des s.r.p individuels. Document de travail # 011-2002, Faculté des Sciences de l'Administration (FSA), Université Laval (2002)
12. Langewish, A., Choobineh, F.: Stochastic dominance tests for ranking alternatives under ambiguity. European Journal of Operational Research 95, 139–154 (1996)
13. Lehmann, E.: Ordered families of distributions. Annals of Mathematical Statistics 26, 399–419 (1955)
14. Mann, H.B., Whitney, D.R.: On a test of whether one of two random variables is stochastically larger than the other. Annals of Mathematical Statistics 18, 50–60 (1947)
15. Martel, J.M., Zaras, K.: Stochastic dominance in multicriterion analysis under risk. Theory and Decision 39, 31–49 (1995)
16. Shafer, G.: A Mathematical Theory of Evidence. Princeton University Press, Princeton (1976)
17. Smets, P., Kennes, R.: The transferable belief model. Artificial Intelligence 66, 191–234 (1994)
18. Smets, P.: Decision Making in the TBM: the Necessity of the Pignistic Transformation. International Journal of Approximate Reasoning 38, 133–147 (2005)
19. Vincke, P.: L'Aide Multicritère à la Décision. Editions de l'Université de Bruxelles, Editions Ellipses, Bruxelles (1989)

Interpreting GUHA Data Mining Logic in Paraconsistent Fuzzy Logic Framework

Esko Turunen

Tampere University of Technology, Finland
esko.turunen@tut.fi

Abstract. A natural interpretation of GUHA style data mining logic in paraconsistent fuzzy logic framework is introduced. Significance of this interpretation is discussed.

Keywords: Data mining, fuzzy logic, paraconsistent logic.

1 Introduction

Classical Boolean logic is the logic of mathematics. In pure mathematical world things are binary: either a number is a prime number or is not, either a theorem is proved or not, *tertium non datur* is valid. Outside mathematics in the real world, in data analysis and decision making, however, applying Boolean logic causes anomalies: the law of the excluded middle is problematic, the use of classical quantifiers ∀ (for all) and ∃ (there exists) is clumsy and truth and falsehood need not to be each others complements. To overcome these problems several non–classical logics were born. In various many–valued logics such as mathematical fuzzy logic [16] the law of the excluded middle does not hold in general, in GUHA data mining logic [3] there are several non–classical quantifiers e.g. 'in most cases', 'above average' etc, and in paraconsistent logic [2], besides true or false, a statement can be unknown or contradictory, too. In this paper we explore the mutual relation of these non–classical logics. We show, in particular, how GUHA logic is related to paraconsistent fuzzy logic.

2 The GUHA Method in Data Mining

GUHA - General Unary Hypotheses Automaton - introduced in [3] and still developing, is a method of automatic generation of hypotheses based on empirical data, thus a method of data mining. GUHA is a kind of automated exploratory data analysis: it generates systematically hypotheses supported by the data.

The GUHA method is based on well–defined first order monadic logic containing generalized quantifiers on finite models. A GUHA procedure generates statements on association between complex Boolean attributes. These attributes are constructed from the predicates corresponding to the columns of the data matrix.

F. Rossi and A. Tsoukis (Eds.): ADT 2009, LNAI 5783, pp. 284–293, 2009.

GUHA is primary suitable for exploratory analysis of large data. The processed data forms a rectangle matrix, where rows correspond to objects belonging to the sample and each column corresponds to one investigated variable. A typical data matrix processed by GUHA has hundreds or thousands of rows and tens of columns. Exploratory analysis means that there is no single specific hypothesis that should be tested by our data; rather, the aim is to get orientation in the domain of investigation, analyze the behavior of chosen variables, interactions among them etc. Such inquiry is not blind but directed by some general direction of research. GUHA is not suitable for testing a single hypothesis: routine packages are good for this.

GUHA systematically creates all hypotheses interesting from the point of view of a given general problem and on the base of given data. This is the main principle: all interesting hypotheses. Clearly, this contains a dilemma: "all" means most possible, "only interesting" means "not too many". To cope with this dilemma, one may use different GUHA procedures and, having selected one, by fixing in various ways its numerous parameters.

GUHA procedures not only hypotheses relating one variable with another one, but expressing relations among single variables, pairs, triples, quadruples of variables etc. GUHA offers hypotheses. Exploratory character implies that the hypotheses produced by the computer (numerous in number: typically tens or hundreds of hypotheses) are just supported by the data, not verified. You are assumed to use this offer as inspiration, and possibly select some few hypotheses for further testing.

For a complete description of the GUHA method for data mining, see [3], [4]. A software implementation of GUHA called LispMiner is available freely from http://lispminer.vse.cz/

Since mathematical fuzzy logic (cf. [16]) and the GUHA method are both extension of classical Boolean logic and are related to vagueness and partial truth, it is not a surprising news that there are several approaches to connect mathematical fuzzy logic to the GUHA method. Here we mention Holeňa who has introduced a fuzzy version of the GUHA method in [6,7]. Novák et al. show in [8] that, by evaluating real–valued data by linguistic expressions and then using the GUHA method, one obtains data mining outcomes that are easily understandable as they are close to human way of thinking. – Such a target we, too, set when writing this paper. We aim to show that GUHA method has a natural interpretation in paraconsistent mathematical fuzzy logic which is our [17] recent extension of Belnap's paraconsistent logic [2].

3   Paraconsistent Fuzzy Logic

Quoting from Stanford Encyclopedia of Philosophy [13]: *The contemporary logical orthodoxy has it that, from contradictory premises, anything can be inferred. To be more precise, let $\models$ be a relation of logical consequence, defined either semantically or proof–theoretically. Call $\models$ explosive if it validates $\{\alpha, \neg\alpha\} \models \beta$ for every α and β (ex contradictione quodlibet). The contemporary orthodoxy,* *i.e., classical logic, is explosive, but also some non-classical logics such as intu-itionist logic and most other standard logics are explosive. The major motivation behind paraconsistent logic is to challenge this orthodoxy. A logical consequence relation,* $\models$*, is said to be* paraconsistent *if it is not explosive. Thus, if* $\models$ *is paraconsistent, then even if we are in certain circumstances where the available information is inconsistent, the inference relation does not explode into trivial-ity. Thus, paraconsistent logic accommodates inconsistency in a sensible manner that treats inconsistent information as informative.*

Four possible values associated with a statement (logic formula) Φ in Belnap's first order paraconsistent logic [2] are **true**, **false**, **contradictory** and **unknown**: if there is evidence for Φ and no evidence against Φ, then Φ obtains the value **true** and if there is no evidence for Φ and evidence against Φ, then Φ obtains the value **false**. A value **contradictory** corresponds to a situation where there is simultaneously evidence for Φ and against Φ and, finally, Φ is labeled by value **unknown** if there is no evidence for Φ nor evidence against Φ. More formally, the values are associated with ordered couples $\langle 1, 0 \rangle$, $\langle 0, 1 \rangle$, $\langle 1, 1 \rangle$ and $\langle 0, 0 \rangle$, respectively.

In [15] Tsoukias introduced an extension of Belnap's logic (named DDT) most importantly because the corresponding algebra of Belnap's original logic is not a Boolean algebra, while the extension is. Indeed, in that paper it was introduced and defined the missing connectives in order to obtain a Boolean algebra. Moreover, it was explained why we get such a structure. Among others it was shown that negation, which was reintroduced in [15] in order to recover some well known tautologies in reasoning, is not a complementation.

In [12] and [10], a continuous valued extension of DDT logic was studied. The authors imposed reasonable conditions this continuous valued extension should obey and, after a careful analysis, they came to the conclusion that the graded values are to be computed via

$$t(\Phi) = \min\{\alpha, 1 - \beta\}, \tag{1}$$
$$k(\Phi) = \max\{\alpha + \beta - 1, 0\}, \tag{2}$$
$$u(\Phi) = \max\{1 - \alpha - \beta, 0\}, \tag{3}$$
$$f(\Phi) = \min\{1 - \alpha, \beta\}. \tag{4}$$

where an ordered couple $\langle \alpha, \beta \rangle$, called *evidence couple*, is given. The intuitive meaning of α and β is the degree of evidence of a statement Φ and against Φ, respectively. Moreover, the set of 2×2 *evidence matrices* of a form

$$\begin{bmatrix} f(\Phi) & k(\Phi) \\ u(\Phi) & t(\Phi) \end{bmatrix}$$

is denoted by $\mathcal{M}$. The values $f(\Phi), k(\Phi), u(\Phi)$ and $t(\Phi)$ are values on the real unit interval $[0, 1]$ such that $f(\Phi) + k(\Phi) + u(\Phi) + t(\Phi) = 1$. Their intuitive meaning is *falsehood, inconsistency, unknown* and *truth*, respectively, of the statement Φ. One of the most important features of paraconsistent logic is that truth and falsehood are not each others complements.

In [12] it is shown how such a fuzzy version of Belnap's logic can be applied in preference modeling. However, several open problems related to this fuzzy extension of paraconsistent logic were posed in [10], most notably a need of a complete truth calculus and a more thorough investigation of the set $\mathcal{M}$ and its algebraic structure.

We continued Tsoukias' et al. work in [17]. To our understanding, the algebraic structure of $\mathcal{M}$ should *not* be a Boolean algebra. This opinion is based on our basic observation that the algebraic operations in (1) – (4) are expressible by the Lukasiewicz t–norm and the corresponding residuum, i.e. in the Lukasiewicz structure, which is an example of an injective MV–algebra and is not, in general, a Boolean algebra. It is known [16] that Lukasiewicz–Pavelka style fuzzy sentential logic is a complete logic in a sense that if the truth value set $\mathbf{L}$ forms an injective MV–algebra, then the set of a–tautologies and the set of a–provable formulae coincide for all $a \in \mathbf{L}$. We therefore considered the problem that, given a truth value set which is an injective MV–algebra, is it possible to transfer an injective MV–structure to the set $\mathcal{M}$, too. The answer turned out to be affirmative, consequently, the corresponding paraconsistent sentential logic is essentially Pavelka style fuzzy logic. This means that having any set of injective MV-algebra valued evidence couples $\langle \alpha, \beta \rangle$, the structure of the evidence matrices

$$\begin{bmatrix} \alpha^* \wedge \beta & \alpha \odot \beta \\ \alpha^* \odot \beta^* & \alpha \wedge \beta^* \end{bmatrix} \tag{5}$$

forms an injective MV–algebra, too. Here the operations $\odot, \wedge$ and * are the algebraic operations *product, meet* and *complement*, respectively, of the original injective MV–algebra. If, in particular, the original MV–algebra is the Lukasiewicz structure on the real unit interval, then $a \odot b = \max\{0, a+b-1\}$, $a \wedge b = \min\{a, b\}$, $a^* = 1 - a$ for all $a, b \in [0, 1]$. Moreover, in MV–algebras there is an additional operation $\oplus$, in the Lukasiewicz structure it is defined by $a \oplus b = \min\{1, a + b\}$, $a, b \in [0, 1]$.

Our result that continuous valued paraconsistent logic can be seen as a special case of Lukasiewicz–Pavelka style fuzzy logic has a consequence that a rich logical semantics and syntax is available. For example, all Lukasiewicz tautologies as well as Intuitionistic tautologies can be expressed in the framework of this logic. This follows by the fact that we have two sorts of logical connectives conjunction, disjunction, implication and negation interpreted either by the monoidal operations $\odot, \oplus, \longrightarrow, ^*$ or by the lattice operations $\wedge, \vee, \Rightarrow, ^\star$, respectively (however, neither * nor * is a lattice complementation). Besides, there are many other logical connectives available.

4   GUHA in Paraconsistent Logic Framework

Assume we have a data file composed of k columns and m rows, for example the following fancied allergy matrix:

Child	Tomato	Apple	Orange	Cheese	Milk
Anna	1	1	0	1	1
Aina	1	1	1	0	0
Naima	1	1	1	1	1
Rauha	0	1	1	0	1
Kai	0	1	0	1	1
Kille	1	1	0	0	1
Lempi	0	1	1	1	1
Ville	1	0	0	0	0
Ulle	1	1	0	1	1
Dulle	1	0	1	0	0
Dof	1	0	1	0	1
Kinge	0	1	1	0	1
Laade	0	1	0	1	1
Koff	1	1	0	1	1
Olavi	0	1	1	1	1

The meaning of '0's and '1's is obvious: Anna for example is allergic to tomato, apple, cheese and milk and is not allergic to orange. Now consider two Boolean attributes ϕ and ψ (in the above allergic matrix ϕ could mean 'child is allergic to tomato and apple' and ψ could mean 'child is allergic to milk'). A *four-fold contingency table* $\langle a, b, c, d \rangle$ related to these attributes is composed from numbers of objects in the data satisfying four different binary combinations of these attributes:

$$
\begin{array}{c|c|c}
 & \psi & \neg\psi \\
\hline
\phi & a & b \\
\hline
\neg\phi & c & d \\
\end{array}
$$

where
- a is the number of objects satisfying both ϕ and ψ,
- b is the number of objects satisfying ϕ but not ψ,
- c is the number of objects not satisfying ϕ but satisfying ψ,
- d is the number of objects not satisfying ϕ nor ψ,
- $m = a + b + c + d$.

Various relations between ϕ and ψ can be measured in the data by different *four-fold table quantifiers*, denoted by $\phi \sim \psi$, $\sim (a, b, c, d)$, $\sim (a, b, c)$ or $\sim (a, b)$ depending on context, which here are understood as functions with values in the real unit interval $[0, 1]$. Among the most well-known are the following two quantifiers:

1. Basic implicational quantifier. A statement connecting two attributes ϕ and ψ by *basic implicational quantifier* is defined to be true in a given data if

$$
a \geq n \text{ and } \frac{a}{a + b} \geq p,
$$

where $n \in \mathcal{N}$ and $p \in [0,1]$ are parameters given by user. Notice that basic implicational quantifier is corresponding to association rules obtained by apriori algorithm intoduced by Agrawal et al. in 1993 [1]. A fuzzy logic interpretation of this quantifier is the following

Given a data, the determining subset A is formed of cases that satisfy ϕ; there must be enough of them. The data supports a relation 'ϕ implies ψ' if there are few cases in A not satisfying ψ.

We recognize that a proposition *Cases in A not satisfying ψ* has a truth value $\frac{b}{a+b}$ which should be low, therefore, its complement – in terms of Lukasiewicz logic – should be large enough, i.e.

$$\left(\frac{b}{a+b} \right)^* = 1 - \frac{b}{a+b} = \frac{a+b-b}{a+b} = \frac{a}{a+b} \geq p.$$

2. Basic double implicational quantifier. A statement connecting two attributes ϕ and ψ by *basic double implicational quantifier* is defined to be true in a given data if

$$a \geq n \text{ and } \frac{a}{a+b+c} \geq p,$$

where $n \in \mathcal{N}$ and $p \in [0,1]$ are parameters given by user.

A fuzzy logic interpretation of this quantifier is now the following

Given a data, the determining subset A is formed of cases that satisfy ϕ or ψ; there must be enough cases satisfying both of them. The data supports a relation 'ϕ implies ψ and ψ implies ϕ' if there are few cases in A not satisfying ψ or few cases in A not satisfying ϕ.

It is easy to see that a proposition *Cases in A not satisfying ψ* has a truth value $\frac{b}{a+b+c}$ and a proposition *Cases in A not satisfying ϕ* has a truth value $\frac{c}{a+b+c}$, therefore a proposition *Cases in A not satisfying ψ or not satisfying ϕ* is related to a Lukasiewicz logic truth value $\frac{b}{a+b+c} \oplus \frac{c}{a+b+c}$ which should be low enough, therefore, its complement should be high enough, i.e.

$$\left(\frac{b}{a+b+c} \oplus \frac{c}{a+b+c} \right)^* = \left(\frac{b}{a+b+c} \right)^* \odot \left(\frac{c}{a+b+c} \right)^*$$

$$= \left(1 - \frac{b}{a+b+c} \right) \odot \left(1 - \frac{c}{a+b+c} \right)$$

$$= \max \left\{ \left(1 - \frac{b}{a+b+c} \right) + \left(1 - \frac{c}{a+b+c} \right) - 1, 0 \right\}$$

$$= \frac{a+b+c-b-c}{a+b+c}$$

$$= \frac{a}{a+b+c} \geq p.$$

These two examples introduced in [18] show that GUHA data mining logic is related to Lukasiewicz–Pavelka logic. However, there are several other quantifiers in LispMiner software implementation that have rather a statistic than logic character.

Our novel observation is that a value $\alpha = \frac{a}{m}$ can be seen as the degree that ϕ and ψ occur simultaneously, a value $\beta = \frac{b+c}{m}$ can be seen as the degree that ϕ and ψ do not occur simultaneously and a value $\frac{d}{m}$ the degree that ϕ and ψ do not occur at all – a kind of indifferent situation. Recalling $m = a + b + c + d$ and using the Lukasiewicz operations (equations (1) – (4)) it is easy to see that

$$\alpha^* \wedge \beta = \min\{1 - \tfrac{a}{m}, \tfrac{b+c}{m}\} = \tfrac{b+c}{m} = \beta,$$
$$\alpha \odot \beta = \max\{0, \tfrac{a}{m} + \tfrac{b+c}{m} - 1\} = 0,$$
$$\alpha^* \odot \beta^* = \max\{0, 1 - \tfrac{a}{m} + 1 - \tfrac{b+c}{m} - 1\} = \tfrac{d}{m},$$
$$\alpha \wedge \beta^* = \min\{\tfrac{a}{m}, 1 - \tfrac{b+c}{m}\} = \tfrac{a}{m} = \alpha.$$

Therefore $\langle \alpha, \beta \rangle = \langle \frac{a}{m}, \frac{b+c}{m} \rangle$ can be seen as an evidence couple for a statement Φ: 'ϕ and ψ occur simultaneously'. The correspondent evidence matrix is then

$$\begin{bmatrix} f(\Phi) & k(\Phi) \\ u(\Phi) & t(\Phi) \end{bmatrix} = \begin{bmatrix} \beta & 0 \\ \frac{d}{m} & \alpha \end{bmatrix}.$$

In practical data mining tasks run by LispMiner it often happens that 'indifferent cases rule over interesting cases', that is to say, value d in a four–fold contingency table is much bigger that values a, b, c. However, even in such cases it is useful to look for statements Φ such that the truth value of Φ is, say at least k times bigger than the falsehood of Φ, i.e. $\alpha \geq k\beta$, which is equivalent to $a \geq k(b + c)$. On the other hand such a statement Φ is stamped by label **true** if

$$\tfrac{a}{a+b+c} \geq p \text{ iff } a \geq p(a + b + c) \text{ iff } a(1 - p) \geq p(b + c) \text{ iff } a \geq \tfrac{p}{1-p}(b + c).$$

This means that $k = \frac{p}{1-p}$, $p \neq 1$, or equivalently $p = \frac{k}{k+1}$. We have proved

Theorem 1. *Given a data, all statements Φ such that the truth value of Φ is at least k times bigger than the falsehood of Φ in the sense of paraconsistent logic, can be found by the basic double implicational quantifier and setting $p = \frac{k}{k+1}$.*

Examples. Consider the above data about children's allergies.
(a) Let ϕ stand for 'child is allergic to tomato and apple' and ψ stand for 'child is allergic to milk'. Compute the corresponding contingency table, the evidence couple and the evidence matrix for a statement Φ: 'ϕ and ψ occur simultaneously'.

Solution. First write the corresponding table where the connective '&' is interpreted as a Boolean conjunction.

Child	Tomato & Apple	Milk
Anna	1	1
Aina	1	0
Naima	1	1
Rauha	0	1
Kai	0	1
Kille	1	1
Lempi	0	1
Ville	0	0
Ulle	1	1
Dulle	0	0
Dof	0	1
Kinge	0	1
Laade	0	1
Koff	1	1
Olavi	0	1

This leads to

	ψ	$\neg\psi$
ϕ	5	1
$\neg\phi$	7	2

Thus, the evidence couple is $\langle \frac{5}{15}, \frac{7+1}{15} \rangle$ and the correspondent evidence matrix is

$$\begin{bmatrix} f(\Phi) & k(\Phi) \\ u(\Phi) & t(\Phi) \end{bmatrix} = \begin{bmatrix} \frac{8}{15} & 0 \\ \frac{2}{15} & \frac{5}{15} \end{bmatrix}$$

Since $f(\Phi)$, the degree of falsehood of Φ, is larger that $t(\Phi)$, the degree of truth of Φ, we conclude that the given data does not support the statement that childen who are allergic to tomato and apple are simultaneously allergic to milk, too.

(b) Let ϕ stand for 'child is allergic to cheese' and ψ stand for 'child is allergic to milk'. Compute the corresponding contingency table, the evidence couple and the evidence matrix for the statement Φ: 'ϕ and ψ occur simultaneously'.

Solution. From the original data matrix we get the following contingency table

	Milk	$\neg$Milk
Cheese	8	0
$\neg$Cheese	4	3

Thus, the evidence couple is $\langle \frac{8}{15}, \frac{4+0}{15} \rangle$, and the correspondent evidence matrix is

$$\begin{bmatrix} f(\Phi) & k(\Phi) \\ u(\Phi) & t(\Phi) \end{bmatrix} = \begin{bmatrix} \frac{4}{15} & 0 \\ \frac{3}{15} & \frac{8}{15} \end{bmatrix}$$

We conclude that, based on the given data, the paraconsistent truth value of the statement Φ: 'cheese allergy and milk allergy occur simultaneously' is two times bigger than the paraconsistent falsehood of Φ and, thus, the data supports Φ.

5   Conclusion and Future Work

Various non–classical logics are useful in data analysis and algorithmic decision making. We have deepened the connection between fuzzy logic, paraconsistent logic and GUHA data mining logic. Basic implicational quantifier and basic double implicational quantifier, for example, have a natural interpretation in Lukasiewicz–Pavelka fuzzy logic. In this study we have shown that paraconsistent logic, too, has an interpretation and connection to GUHA data mining logic; indeed, first we proved that paraconsistent degrees of truth and falsehood, which are not mutually each other complements, of simultaneous occurrence of two GUHA attributes can be calculated by basic double implicational quantifier, and then we gave an example showing how to use this connection to express more sophisticated relationships between properties that could be possible by classical logic or even standard fuzzy logic. Lukasiewicz–Pavelka fuzzy logic as well as paraconsistent logic and GUHA logic are sound and theoretically well established, widely studied and acknowledged non–classical logics. Therefore all connections between these approaches clear the road for future investigation into better understanding of data analysis, knowledge extraction and decision theory.

References

1. Agrawal, R., Imielinski, T., Swami, A.N.: Mining Association Rules between Sets of Items in Large Databases. SIGMOD 22(2), 207–216 (1993)
2. Belnap, N.D.: A useful four–valued logic. In: Epstein, G., Dumme, J. (eds.) Modern uses of multiple valued logics, pp. 8–37. D. Reidel, Dordrecht (1977)
3. Hájek, P., Havel, I., Chytil, M.: The GUHA method of Automatic Hypotheses Determination. Computing 1, 293–308 (1966)
4. Hájek, P., Havránek, T.: Mechanizing Hypothesis Formation. Mathematical Foundations for a General Theory, p. 396. Springer, Berlin (1978)
5. Hájek, P.: Metamathematics of Fuzzy Logic. Kluwer, Dordrecht (1998)
6. Holeňa, M.: Fuzzy Hypotheses for GUHA Implications. Fuzzy Sets and Systems 98, 101–125 (1998)
7. Holeňa, M.: Fuzzy Hypotheses Testing in the Framework of Fuzzy Logic. Fuzzy Sets and Systems 145, 229–252 (2004)
8. Novák, V., Perfilieva, I., Dvořák, A., Chen, G., Wei, Q., Peng, Y.: Mining Pure Linguistic Associations from Numerical Data. Int. Journal of Approximate Reasoning (2007) (to appear)
9. Novák, V.: Fuzzy Logic Theory of Evaluating Expressions and Comparative Quantifiers. In: Proc. 11.th Int. Conf. IPMU, Éditions EDK, Les Cordeliers, Paris, July 2006, vol. 2, pp. 1572–1579 (2006)
10. Öztürk, M., Tsoukias, A.: Modeling uncertain positive and negative reasons in decision aiding. Decision Support Systems 43, 1512–1526 (2007)

11. Pavelka, J.: On fuzzy logic I, II, III. Zeitsch. f. Math. Logik 25, 45–52, 119–134, 447–464 (1979)
12. Perny, P., Tsoukias, A.: On the continuous extensions of four valued logic for preference modeling. In: Proceedings of the IPMU conference, pp. 302–309 (1998)
13. http://plato.stanford.edu/entries/logic-paraconsistent/
14. Rauch, J., Simunek, M.: Mining 4ft–rules. In: Arikawa, S., Morishita, S. (eds.) Discovery Science, pp. 268–272. Springer, Berlin (2000)
15. Tsoukias, A.: A first order, four valued, weakly paraconsistent logic and its relation to rough sets semantics. Foundations of Computing and Decision Sciences 12, 85–108 (2002)
16. Turunen, E.: Mathematics behind Fuzzy Logic. Springer, Heidelberg (1999)
17. Turunen, E.: A Para Consistent Fuzzy Logic. In: Ramanujam, R., Sarukkai, S. (eds.) Logic and Its Applications. LNCS (LNAI), vol. 5378, pp. 77–88. Springer, Heidelberg (2009)
18. Ylirinne, E., Turunen, E.: Interpreting Data Mining Quantifiers in Mathematical Fuzzy logic. In: Symposium on Fuzzy Systems in Computer Science, FSCS 2006, Magdeburg, Germany, September 27-28, pp. 33–41 (2006)

Insuring Risk-Averse Agents

Greg Hines and Kate Larson

Cheriton School of Computer Science
University of Waterloo
Waterloo, Canada
ggdhines@cs.uwaterloo.ca
klarson@cs.uwaterloo.ca

Abstract. In this paper we explicitly model risk aversion in multiagent interactions. We propose an *insurance mechanism* that be can used by risk-averse agents to mitigate against risky outcomes and to improve their expected utility. Given a game, we show how to derive Pareto-optimal insurance policies, and determine whether or not the proposed insurance policy will change the underlying dynamics of the game (*i.e.*, the equilibrium). Experimental results indicate that our approach is both feasible and effective at reducing risk for agents.

1 Introduction

In almost every decision people make, risk is a factor. When negotiating a business contract, there is the risk of either side being unable to fulfill its obligations. When bidding for multiple items in an auction, there is the risk of winning too many or too few items. Even when using the Internet, there is the risk of congestion depending on the routing policy used. In most of these cases people are risk averse. The importance of the influence of risk aversion on peoples' decisions is reflected in the size of the insurance industry, a multi-trillion dollar business, [4] and the amount of research in economics relating to risk [17].

There is considerable research in multiagent systems on helping people make better decisions in settings such as those mentioned above. However, this research generally assumes that people are risk neutral [5,15,16]. Given the prevalence of risk aversion in the real world, we believe it is important to study how to manage the effects of risk and risk aversion in multiagent systems.

In this paper, we study non-cooperative multiagent systems. Our main contribution is an insurance mechanism that can be used in games to reduce agents' risk and increase their utility. We present a characterization of our mechanism that allows us to easily determine if the mechanism can be applied to any given game. Experimental results show that our mechanism is usable in many different situations and is scalable. The experimental results also examine how much risk aversion matters in different settings. We conclude with a discussion of related work and some promising areas for future work.

F. Rossi and A. Tsoukis (Eds.): ADT 2009, LNAI 5783, pp. 294–305, 2009.

2   The Model and Background

In this section we introduce our model of risk aversion for multiagent systems, as well as define the key game-theoretic concepts used in this paper. For a more thorough introduction to game theory, we refer the reader to [8].

2.1   A Model of Risk Aversion

In this section we propose a model of risk aversion for a multiagent setting. The approach we take in modeling risk is motivated by models used in experimental economics [6]. If an agent is risk averse, then it dislikes uncertainty. For example, if given a choice between a lottery and a guaranteed payoff, a risk-averse agent will often prefer the guaranteed payoff, *even when the expected payoff from the lottery is higher.* In this paper, we model risk aversion by the concavity of an agent's utility function. Specifically, given an income I, if an agent's utility is of the form

$$U = I^r, \tag{1}$$

where $0 < r < 1$ is the *risk-attitude factor*, then the agent is risk averse. This utility function is depicted in Figure 1. For this paper we will assume that income is always greater than or equal to zero; we make this assumption since studies show that humans (and thus the agents designed to represent humans in different interactions) treat loss of income differently from equivalent gain of income [18].

Typically, a model of risk aversion requires a distinction between *income* and *utility* [11], therefore we generalize the notion of a game to reflect this distinction.

Definition 1. *Let N be a set of agents, $|N| = n$. An* income-based game *is defined as* $G^I = \langle N, A, I_1, \ldots, I_n, u_1, \ldots, u_n \rangle$ *where*

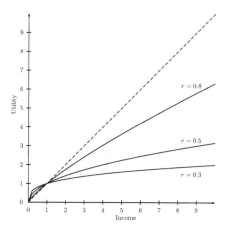

Fig. 1. A graphical depiction of income versus utility. The dashed line represents the utility of a risk-neutral agent while the solid lines represent the utilities of a risk averse agent for different values of r.

- A_i is the set of possible actions for agent i, and $A = A_1 \times \ldots \times A_n$,
- $I_i : A \to \mathbb{R}$ specifies the income to agent i given joint action $a \in A$, and
- $u_i : \mathbb{R} \to \mathbb{R}$ is the utility function of agent i.

We refer to the normal form (or matrix) representation of the income function as the *income matrix* and to the normal form representation of the utility function as the *utility matrix*.

Agents interact by following strategies, that is, by selecting actions to play according to some distribution. We are particularly interested in *correlated strategies*.

Definition 2. *A correlated strategy $\sigma_A = \{\sigma_A(a) | a \in A\}$ is any probability distribution over A. The conditional correlated strategy $\sigma_{A_{-i}}(a_{-i} | a_i)$ is the probability of the joint action (a_{-i}, a_i) according to σ_A given the action a_i where $a_{-i} = (a_1, \ldots, a_{i-1}, a_{i+1}, \ldots, a_n)$.*

In an income-based game, agents try to maximize utility, not income. That is, agents are trying to maximize their expected utility, given by

$$\sum_a \sigma_A(a) u_i(I_i(a)). \tag{2}$$

In this paper, we are interested in analyzing situations involving equilibrium play. Specifically, we are interested in studying situations where agents are playing a correlated equilibrium.

Definition 3. *A correlated strategy σ_A^* is a correlated equilibrium if for every agent i and every $a_i \in A_i$,*

$$\sum_{a_{-i} \in A_{-i}} \sigma_{A_{-i}}^*(a_{-i} | a_i) u_i(I_i(a_i, a_{-i})) \tag{3}$$
$$\geq \sum_{a_{-i} \in A_{-i}} \sigma_{A_{-i}}^*(a_{-i} | a_i) u_i(I_i(a_i', a_{-i})),$$

for all $a_i' \in A_i$.

We choose correlated equilibria as our solution concept for several reasons. First, it is a generalization of Nash equilibria, and so the techniques we propose in this paper can also be applied when a Nash equilibrium is chosen as the solution concept. Second, correlated equilibria can be "less risky" than Nash equilibria in that the correlation can help the agents in reducing the risk of mis-coordinating. Thus, if we can show that our approach is effective when risk is already being mitigated to some extent, then we believe that we can extrapolate our results and argue that our approach is broadly applicable and widely effective.

To illustrate the effects of risk, we consider an example of agents determining routing policies.

Example. *Consider two agents (Row and Column) on the Internet, each deciding on a routing policy to use. There is a public route available with a bandwidth of 200 kb/sec*

	L	R
T	75,75	75,200
B	200,75	100,100

	L	R
T	8.7,8.7	8.7,14.1
B	14.1, 8.7	10,10

Fig. 2. Income (left) and utility (right) matrices for a routing problem

that is shared evenly if both agents decide to use it. Each agent also has a private route available with a bandwidth of 75 kb/sec. Suppose that both agents' utilities are given by Equation 1 with r = 0.5. Figure 2 shows the income (measured in bandwidth) and utility matrices of the game. Note that the unit of income is in kb/sec while utility has no units.

If agents are playing the correlated strategy $(\sigma_A((B, L)), \sigma_A((T, R))) = (0.5, 0.5)$, the expected income for both agents is 137.5 kb/sec and the expected utility is 11.4. However, if the agents are guaranteed a bandwidth of 137.5 kb/sec, their utilities would increase to 11.7.

2.2 Characterization of the Set of Correlated Equilibria

Since correlated equilibria play a central role in our paper, we wish to derive a formal characterization of the set of correlated equilibria in an income-based game. (For simplicity, the rest of the paper will consider games with only two agents each with two actions. Our results can be generalized to an arbitrary number of agents and actions.) Given the income and utility matrices for a general income-based game shown in Figure 3, suppose Agent 1 is trying to determine whether or not to follow a given correlated strategy σ_A (assuming Agent 2 does as well).

The expected utility for Agent 1 from playing T when following σ_A is

$$a^r \sigma_{A_2}(L|T) + c^r \sigma_{A_2}(R|T), \tag{4}$$

and the expected utility for Agent 1 for instead playing B is

$$e^r \sigma_{A_2}(L|T) + g^r \sigma_{A_2}(R|T). \tag{5}$$

Therefore, for Agent 1 to be willing to follow σ_A, we require that

$$a^r \sigma_{A_2}(L|T) + c^r \sigma_{A_2}(R|T) \geq e^r \sigma_{A_2}(L|T) + g^r \sigma_{A_2}(R|T)$$

or

$$(a^r - e^r)\sigma_A(TL) + (c^r - g^r)\sigma_A(TR) \geq 0. \tag{6}$$

	L	R
T	a,b	c,d
B	e,f	g,h

	L	R
T	a^r,b^r	c^r,d^r
B	e^r,f^r	g^r,h^r

Fig. 3. Income matrix (left) and utility matrix (right) for a general game

Similar constraints can be created for every possible agent-action combination. The set of all such constraints completely defines the set of correlated equilibria for a game.

3   An Insurance Policy for Risk-Averse Agents

In this section we present an insurance policy mechanism for risk-averse multiagent systems. By buying insurance for certain outcomes, risk-averse agents are able to reduce their risk and increase their expected utility. We study the use of such a mechanism in a setting where agents are playing a correlated equilibrium.

The basic idea is for agents to be both buyers and sellers of insurance. For example, Agent i can buy coverage for the joint action a by selling coverage for another joint action, a'. If the outcome of the game is a, agent i *receives* income from another agent. If the outcome of the game is a', agent i *gives* some of its income to another agent. The unit cost of buying insurance (and the unit revenue from selling insurance) is set by a price vector $p = \{p(a)|a \in A\}$. The price vector is set by some third party; it is reasonable to assume that the correlating device of the correlated equilibrium also sets p.

3.1   Creating the Insurance Policy

Given a correlated equilibrium σ_A^* and a price vector p, each agent must determine how much insurance to buy and sell for each joint action a, *i.e.*, its demand $d_i(a)$. If $d_i(a) > 0$ then agent i wishes to buy insurance coverage for the joint action a and if $d_i(a) < 0$, agent i wishes to sell insurance coverage for the joint action a. Since agents are utility maximizers, the demand for each joint action can be computed by solving the constraint maximization problem:

$$\max_{d_i} \sum_{a \in A} \sigma_A^*(a) u_i(I_i(a) + d_i(a)), \tag{7}$$

$$\text{s.t.} \sum_{a \in A} p(a) \cdot (I_i(a) + d_i(a)) = \sum_{a \in A} p(a) \cdot (I_i(a)), \tag{8}$$

where $p(a)$ is the unit price for buying or selling insurance for the joint action a. The RHS of Equation 8 is agent i's budget given p, or the maximum amount of insurance it can possibly buy. Thus, agent i is simply trying to determine which insurance to buy ($d(a) > 0$) or to sell ($d(a) < 0$)), to maximize its expected utility while being constrained by its budget. For simplicity, let

$$x_i(a) = I_i(a) + d_i(a). \tag{9}$$

Let

$$R_i(a) = \frac{\partial \sigma_A(a) u_i(x_i(a))}{\partial x_i(a)} \cdot \frac{1}{p(a)}, \tag{10}$$

be the ratio of marginal utility compared to cost for buying insurance coverage for the joint action a. Agent i's utility is maximized when

$$R_i(a) = R_i(a'), \tag{11}$$

for all $a, a' \in A$. If Equation 11 is not satisfied, for example if $R_i(a) > R_i(a')$, then agent i would increase its utility by buying more coverage for the joint action a and buying less (or selling more) coverage for the joint action a'. For our utility function, this gives

$$x_i(a) = \sqrt[r-1]{\frac{\sigma(a')}{\sigma(a)}\frac{p(a)}{p(a')}}x_i(a'). \tag{12}$$

Equation 12 can be substituted into agent i's budget constraint (Equation 8) to determine its overall demand.

Example. *Continuing the example from Section 2, suppose the insurance price vector*

$$\{p((B,L)) = 1, p((T,R)) = 2\} \tag{13}$$

is announced and we wish to determine Row's optimal insurance coverage. In this case, Row's budget will be 350 kb/sec. Assuming $r = 0.5$, Equation 12 simplifies to

$$x_{Row}((B,L)) = \sqrt{2}x_{Row}((T,R)). \tag{14}$$

Substituting this into Equation 8, we get

$$\sqrt{2}x_{Row}((T,R)) + 2x_{Row}((T,R)) = 350, \tag{15}$$

$$x_{Row}((T,R)) = 102.5. \tag{16}$$

Similarly, we find $x_{Row}((B,L)) = 150.0$. Therefore, agent Row wishes to purchase insurance coverage of 27.5 kb/sec for the outcome (T,R) by selling 50 kb/sec of insurance coverage for the outcome (B,L).

Since, given p, each agent can determine its demand, the next challenge is to find an appropriate p. The insurance price vector should be set with several goals in mind. First, the resulting insurance policy should be budget balanced, *i.e.*, no external source of funding is required and the insurance policy does not make a profit. Second, the insurance policy should also be Pareto optimal, *i.e.*, no agent's expected utility can be increased without decreasing another's.

The insurance policy can be guaranteed to be budget balanced by choosing a p that results in supply equaling demand, *i.e.*, for all $a \in A$

$$\sum_i d_i(a) = 0. \tag{17}$$

To find which price vectors result in supply equaling demand, note that with our insurance mechanism, agents can only trade coverage and not create it. Therefore, our insurance mechanism is an example of an *exchange market* [8]. Arrow and Debreu proved that for every exchange economy, there exists a price vector which results in supply equaling demand. With respect to our insurance mechanism, Arrow and Debreu's theorem is as follows:

Theorem 1. *[2] For a given game G and correlated equilibrium σ_A^*, there exists some price vector p^* such that*

$$\sum_i d_i(a) = 0, \tag{18}$$

for every a ∈ A, assuming that u_i is continuous, strictly concave and strongly mono-tone.[1] *Such an d is known as a competitive equilibrium.*

It is straightforward to check that the utility function in Equation 1 satisfies all the required conditions in Theorem 1. Theorem 1 also requires that agents are *price-takers* – that is, each agent is unable to influence the price of the insurance policy. If an agent's demand (or supply) of insurance for a joint action is only a small fraction of the overall supply (or demand) then we can reasonably assume that the agent is a price-taker. Thus, our approach will work for games where there are many agents. However, if there are only a few agents, then they may be able to influence prices, and an alternative approach might be necessary. We propose that the price-setter also guarantees that the market will clear: that is, the third party promises to meet any extra demand and buy any extra supply.

Now that we have determined that p^* exists, we would like to know if it results in a Pareto optimal allocation. To do so, we rely on the following result [8].

Theorem 2 (First Fund. Thm. of Welfare Economics). *Any competitive equilibrium will always result in a Pareto optimal allocation.*

While for most exchange markets there are multiple p^*, with our utility function p^* is unique [8]. Since p^* is unique, this implies that p^* is also social-welfare maximizing.

At the same time, there is the complication that since the insurance policy will change agents' incomes and utilities, σ_A^*, the correlated equilibrium in the original game, may not remain an equilibrium once the insurance policy is in place. That is, since agents' utilities will have changed, $\sigma_{A_i}^*$ may no longer be a best response to $\sigma_{A_{-i}}^*$ and agents may wish to play other strategies. In this case, the insurance policy will no longer work since agents may now demand more coverage for certain joint actions or be willing to supply less coverage for other joint actions. Therefore, we are interested in finding a correlated equilibrium that will still be one after the insurance policy is in place. We call such an equilibrium an *insurable equilibrium*. Furthermore, for a given game G we would also like to provide a test to determine if any insurable equilibria exist.

To determine whether σ_A^* is an insurable equilibrium, we start by characterizing the set of all Pareto optimal allocations. The set of Pareto optimal allocations of income can be determined by solving the following constraint maximization problem:

$$\max_{x_i} u_i(x_i) \tag{19}$$

$$\text{s.t. } u_{-i}(I - x_i) = \bar{u} \tag{20}$$

for some fixed utility $\bar{u}$. This maximization problem can be solved using the Lagrangian method to find the constraint

$$\frac{x_i(a)}{x_i(a')} = \frac{I(a)}{I(a')}. \tag{21}$$

---

[1] A utility function is strongly monotone if $u(y) > u(x)$ if $y \geq x$ and $y \neq x$.

Table 1. The set of Pareto optimal allocation games where $0 \leq l \leq 1$

	L	R
T	$(a+b)l, (a+b)(1-l)$	$(c+d)l,(c+d)(1-l)$
B	$(e+f)l, (e+f)(1-l)$	$(g+h)l,(g+h)(1-l)$

For brevity, we omit the derivation.[2] Therefore, the set of Pareto optimal allocations is given by

$$x_1 = \{I_{a_1} \cdot l, I_{a_2} \cdot l, \dots\}, \tag{22}$$

$$x_2 = \{I_{a_1} \cdot (1-l), I_{a_2} \cdot (1-l), \dots\} \tag{23}$$

for $0 \leq l \leq 1$. The resulting set of *Pareto optimal allocation games* is shown in Table 1.

Using the same reasoning that we used to find the correlated equilibrium constraint in Equation 6, we can find the analogous constraint for the Pareto optimal allocation game:

$$\{[(a+b)l]^r - [(e+f)l]^r\}\sigma(TL)$$
$$+ \{[(c+d)l]^r - [(g+h)l]^r\}\sigma(TR) \geq 0. \tag{24}$$

Note that l^r cancels out in Equation 24 leaving the simplified constraint

$$[(a+b)^r - (e+f)^r]\sigma_A(TL)$$
$$+ [(c+d)^r - (g+h)^r]\sigma_A(TR) \geq 0. \tag{25}$$

Since Equation 25 does not depend on l, this constraint must hold for every Pareto optimal allocation, including the one created by p^*. Therefore, to determine if σ_A^* is an insurable equilibrium, we simply check whether it satisfies the general set of constraints for correlated equilibria in a Pareto optimal allocation game. By incorporating a convex programming method similar to that of Papadimitriou and Roughgarden [10], our approach can easily determine whether any insurable equilibria exist for a given income-based game.

4   Experimental Evaluation

In this section we describe the experimental evaluation of our insurance mechanism. In particular, we study the extent of its applicability and the amount that it improves the utility of the participating agents.

---

[2] Note that if the agents did not all have the same value for r, this characterization would not be possible.

4.1   Experimental Setup

We conducted our experiments using games with 2 agents with 2 actions per agent, 3 agents with 3 actions per agent, and 4 agents with 4 actions per agent. For each game, the income values were drawn randomly from one of two distributions: the first was the uniform distribution over $[0, 10]$, and the second distribution was a bi-modal Gaussian with $N_1(10, 3)$ and $N_2(100, 3)$. This second distribution was used to generate high-risk games. For all experiments we set the risk-attitude factor $r = 0.6$. Experimental evidence suggests that this value often captures humans' risk-attitudes [6], without being too extreme in either direction. We repeated each experiment 100 times, and all results reported are averaged over these repetitions. A linear program was used to determine if an equilibrium was insurable. We then used the tâtonnement process to find the resulting insurance prices [8].

In each of our experiments, there were three things we studied. First, we were interested in determining what percentage of games actually had insurable equilibria. This measurement allows us to determine the applicability of our approach. Second, for insurable equilibria, we were interested in understanding how effective our insurance mechanism was at reducing risk. For an agent to completely remove risk from a game, it must receive the same income for every possible outcome, so for a given game, we define u_i^{Orig} to be the expected utility to agent i in the original (non-insured) game, u_i^{Ins} to be the expected utility in the insured game, and u_i^{RF} to be the expected utility if the game was made to be risk-free. We define the *insurance effectiveness* for agent i (IE_i) as

$$IE_i = \frac{u_i^{Ins} - u_i^{Orig}}{u_i^{RF} - u_i^{Orig}} \times 100\%. \tag{26}$$

Finally, we were interested in understanding the underlying cost of risk aversion in terms of utility loss for an agent i. We define the *cost of risk aversion* of agent i ($CORA_i$) as

$$CORA_i = \frac{u_i^{RF} - u_i^{Orig}}{u_i^{RF}} \times 100\%. \tag{27}$$

4.2   Results

Table 2 presents our findings on how often insurable equilibria exist. We present our findings from games generated from both distributions. In general, we found that for games with more than two agents, insurable equilibria were very common, and over

Table 2. The percentage of insurable equilibria in randomly-generated games of different sizes, generated by two different distributions

Game Size	Distribution	
	Uniform	Bi-modal Gaussian
2 agents, $|A_i| = 2$	5%	5%
3 agents, $|A_i| = 3$	95%	94%
4 agents, $|A_i| = 4$	91%	100%

Table 3. Average insurance effectiveness (IE) and costs of risk aversion ($CORA$) in games generated using the bi-modal Gaussian distribution. For IE, values closer to 100% show that the insurance mechanism is improving the expected utility for an agent (if $IE = 100\%$ then the optimal utility is achieved). For $CORA$, a value of 4.5%, for example, indicates that risk aversion leads to a 4.5% decrease in utility, compared to a risk-neutral approach.

Game Size	IE	CORA
2 agents, $|A_i| = 2$	25%	4.5 %
3 agents, $|A_i| = 3$	72%	8.2 %
4 agents, $|A_i| = 4$	76.8%	12 %

90% of all games generated had insurable equilibria. For the randomly generated two-agent, two-action games, we found that insurable equilibria were quite rare, occurring in only 5% of games. While at first glance this was disappointing, upon further investigation of the two-agent, two-action games, we noticed that most of these games had a single pure-strategy Nash equilibrium. For such games there is no risk in miscoordinating, and thus no need for an insurance policy.

Table 3 presents our findings for the IE and $CORA$ measurements for different sizes of games, drawn from the bimodal distribution. The results presented are averaged over all games where there was an insurable equilibrium, and over all agents in those games. We make two important observations. First, as the game increases in size, the impact that risk has (as measured by $CORA$) also increases. Second, as CORA increases, so does the effectiveness of our insurance mechanism (as measured by IE). When games were generated using the uniform distribution (results not presented), we observed that there was less overall risk. In particular, the $CORA$ measurement was never greater than 3% on average, and thus, overall, the insurance effectiveness was also quite low. Given these results, we conclude that when risk is an important factor in a game, our insurance mechanism is highly effective. When there is little risk, however, it provides only minimal advantage.

5   Related Work

The standard insurance model assumes an initial level of wealth with some probability of an accident, *i.e.*, some loss of wealth, represented by a probability density function [1,12]. Someone interested in buying insurance decides on the type of coverage they want: the maximum coverage, the deductible, the level of coinsurance, *etc.* The insurer then decides on the premium to charge for that particular insurance policy. Research in insurance has examined questions such as determining the optimal policy to buy and the optimal premium to charge [1,12]. Other work has dealt with the effects of asymmetric information, moral hazards, and adverse selection [14].

Game theory has been used to a limited degree in the study of insurance; for example, in analyzing the actions of insurance companies in competitive markets [14]. However, this work ignores any strategic interaction between insurance buyers by assuming the insurance companies can supply any and all requested insurance policies with non-negative returns. Arrow and Raviv have both used decision theory in determining the optimal actions of both buyer and seller [1,12]. Since their work focuses on the actions of an isolated buyer and seller, this is more an application of decision theory than game theory.

The insurance research closest to our work is the study of *reciprocal reinsurance*: the exchange of risk between insurance companies [3]. Borch considered the problem as an n-person coalitional game and was able to solve it for $n = 2$. This work made several assumptions that our work does not, such as assuming that the probabilities of the different outcomes are independent and companies have additional outside money to use. The goal is for the two companies, A and B, to reach a deal where A pays B to cover a specific amount of A's risk (or vice-versa). Borch presented this as a bargaining problem, where companies had to find an amount to be paid and the risk to be covered, and suggested the Nash bargaining solution as the desired outcome.

There has been limited work on risk in multiagent systems. Exceptions include work by Lam *et al.*, which proposed an insurance scheme for agents trying to obtain resources [7]. In their work insurance premiums were paid to specific *insurance agents* who, in return, guaranteed that necessary resources were always available. The effects of risk aversion have been studied more often in auction design. Page, for example, studied the problem of optimal auction design with both risk-averse buyers and a risk-averse seller [9] while the effects of risk aversion in sequential auctions was studied by Robu and La Poutré [13].

6   Conclusion

In this paper we commenced a formal study of risk and risk-aversion in multiagent systems. We presented a mechanism that allows agents to buy and sell *insurance* in order to protect themselves against undesirable outcomes. We described how to derive Pareto-optimal insurance policies, and provided a characterization of *insurable equilibria* for two-player games. Experimental results indicated that when risk is prevalent in the agents' interactions, our insurance mechanism effectively mitigates the risk and improves the expected utility of all agents.

There are many interesting open challenges related to our insurance mechanism. First, we would like to study our mechanism in a 2-stage game model. This might allow for a more generalized equilibrium model, and also be useful in a repeated game model. The repeated game model could be used to study a non-equilibrium setting; a non-equilibrium model might involve relaxing the balanced-budget requirement and using a *targeted optimality* approach, where we would optimize the insurance mechanism for specific types of agents and games. Studying repeated games may also allow the use of credit and savings to reduce risk, and it would be interesting to compare the advantages of an insurance mechanism against a credit-and-savings mechanism. Secondly, we are interested in trying to apply our insurance mechanism to other models of multiagent systems such as cooperative games and collaborative multiagent systems. Thirdly, we would like to investigate whether other models of risk aversion are more useful. We are specifically interested in how cumulative prospect theory and loss aversion could be used in multiagent systems [18]. It would also be interesting to compare the advantages and disadvantages of the core and competitive equilibrium as different solution concepts. Finally, we would like to implement our insurance mechanism in real life settings. In such settings, agents may be unaware of their own degree of risk aversion or may choose to lie about it. As a result, there would be a need to use preference elicitation and mechanism design.

References

1. Arrow, K.: Essays in the Theory of Risk-Bearing. North-Holland, Amsterdam (1971)
2. Arrow, K., Debreu, G.: Existence of an equilibrium for a competitive economy. Econometrica 22(3), 265–290 (1954)
3. Borch, K.: The Mathematical Theory of Insurance. Lexington Books (1974)
4. Swiss Reinsurance Company. World insurance in 2006: Premiums came back to "life" (2006)
5. Conitzer, V., Sandholm, T.: AWESOME: A general multiagent learning algorithm that converges in self-play and learns a best response against stationary opponents. Machine Learning 67(1-2), 23–43 (2006)
6. Goeree, J.K., Holt, C.A., Palfrey, T.R.: Risk averse behavior in generalized matching pennies games. Games and Economic Behavior 45(1), 97–113 (2003)
7. Lam, Y.-H., Zhang, Z., Ong, K.-L.: Insurance Services in Multi-agent Systems. In: Zhang, S., Jarvis, R.A. (eds.) AI 2005. LNCS (LNAI), vol. 3809, pp. 664–673. Springer, Heidelberg (2005)
8. Mas-Colell, A., Whinston, M., Green, J.R.: Microeconomic Theory. Oxford University Press, Oxford (1995)
9. Page, F.H.: Optimal auction design with risk aversion and correlated information. Technical report, Tilburg University (1994)
10. Papadimitriou, C.H., Roughgarden, T.: Computing correlated equilibria in multi-player games. Journal of the ACM 55(3) (2005)
11. Rabin, M.: Risk aversion and expected-utility theory: A calibration theorem. Econometrica 68, 1281–1292 (2000)
12. Raviv, A.: The design of an optimal insurance policy. The American Economic Review 69, 84–96 (1979)
13. Robu, V., Poutré, H.L.: Designing bidding strategies in sequential auctions for risk averse agents: A theoretical and experimental investigation. In: Proceedings of the 9th Workshop on Agent Mediated Electronic Commerce, Honolulu, USA, pp. 76–89 (2007)
14. Rothschild, M., Stiglitz, J.: Equilibrium in competitive insurance markets: An essay on the economics of imperfect information. The Quarterly Journal of Economics 90, 629–649 (1976)
15. Rozenfeld, O., Tennenholtz, M.: Routing mediators. In: Proceedings of IJCAI 2007, Hyderabad, India, pp. 1488–1493 (2007)
16. Shoham, Y., Powers, R., Grenager, T.: If multiagent learning is the answer, what is the question? Artificial Intelligence 171(7), 365–377 (2007)
17. Stiglitz, J.: Nobel lecture (2001),
   http://nobelprize.org/nobel_prizes/economics/laureates/2001/stiglitz-lecture.pdf
18. Tversky, A., Kahneman, D.: Advances in prospect theory: Cumulative representation of uncertainty. Journal of Risk and Uncertainty 5(4), 297–323 (1992)

Adversarial Risk Analysis: Applications to Basic Counterterrorism Models

Jesus Rios[1] and David Rios Insua[2]

[1] Manchester Business School, UK
[2] Royal Academy of Sciences, Spain

Abstract. Interest in counterterrorism modelling has increased recently. A common theme in the approaches adopted is the need to develop methods to analyse decisions when there are intelligent opponents ready to increase our risks. Most of the approaches have a clear game theoretic flavour, although there have been some decision analytic based approaches. We have recently introduced a framework for adversarial risk analysis, aimed at dealing with problems with intelligent opponents and uncertain outcomes. In this paper, we shall explore how such framework may cope with two of the standard counterterrorism model formulations: sequential defend-attack and simultaneous defend-attack moves.

1 Introduction

Recent high-profile terrorist attacks have demanded significant investments in protective responses. This has stirred a great deal of interest in modelling issues to deal with counterterrorism decisions. Good accounts and introductions to this field may be seen in Parnell et al. (2008) and Bier and Azaiez (2008). Besides some reliability analysis studies based on tools such as fault trees, much of this literature has a distinct game theoretic flavour. Two examples include Zhuang and Bier (2007), who compute best responses and Nash equilibria as a basis for allocating resources against terrorism when the defender and attacker have different multiattribute utility functions, in situations of both simultaneous and sequential play; and Brown et al. (2006), who present max-min, min-max and min-max-min optimization models for defender-attacker, attacker-defender and defender-attacker-defender problems. Following Raiffa (2002), we remain skeptical about the relevance of such concepts in counterterrorism modelling, based on common knowledge assumptions that entail that parts have too much information about their counterparts, in a field in which secrecy tends to be an advantage.

The other mainstream literature in the field has a decision analytic flavour. Among others, we mention Von Winterfeldt and O'Sullivan (2006), who use decision trees to evaluate Man-Portable Air Defense Systems countermeasures; and Pinker (2007), who applies influence diagrams to assess the deployment of various short-term countermeasures. Their recurrent (critical and criticised) problem is the need to assess the probabilities of the actions of the others, which is a key

F. Rossi and A. Tsoukis (Eds.): ADT 2009, LNAI 5783, pp. 306–315, 2009.

issue of the Bayesian approach to games, see Kadane and Larkey (1982) or Raiffa (2002). Banks and Anderson (2006) provide a simple numerical comparison to both approaches to game theory within a smallpox threat problem. This tension between game theoretic and decision analytic approaches to decision making problems with adversaries is not exclusive of counterterrorism models but appears in other business and industrial areas, see e.g. van Bingsbergen and Marx (2007) or Rothkopf (2007).

In Rios Insua et al. (2009), we have introduced Adversarial Risk Analysis (ARA) a framework to cope with risk analysis situations in which we have one or more opponents ready to increase our risks. We applied it in simple auction contexts. ARA has a Bayesian game theoretic flavour. We use principled procedures which employ the game theoretical structure and other information available to assess probabilities on the opponents's actions. In this paper, we explore the application of such framework to simplified versions of two standard models used in counterterrorism contexts, see Zhuang and Bier (2007): sequential defend-attack and simultaneous defend-attack models. We aim at supporting the Defender in choosing her best defense against the attacker. We use coupled decision trees to illustrate our discussion. Our emphasis is on how we may coherently assess the probabilities of various attacks by the attacker. We first study the sequential model. We then consider the simultaneous one. We end up with some discussion on possible extensions and actual applications of our streamlined models. We shall assume all throughout the paper that both the defender and the attacker only have two actions available.

2   Defend-Attack Sequential Model

We start by considering a simple sequential Defend-Attack situation, in which the Defender starts by choosing a defense from a discrete set $\mathcal{D} = \{d_1, d_2\}$ and, then, the Attacker, having observed the defense, chooses an attack within the discrete set $\mathcal{A} = \{a_1, a_2\}$. The problem has a clear sequential game theoretic structure, as in Stackelberg games, see Aliprantis et al (2000). The only uncertainty deemed relevant is a binary outcome $S \in \{0, 1\}$ representing the success or failure of the Attack. For both players, the consequences depend on the success of this attack.

Fig. 1 shows a coupled decision tree representing this situation. The utility functions over the consequences for the Defender and the Attacker are, respectively, $u_D(a, d, S)$ and $u_A(a, d, S)$, depending on both actions and the eventual success S of the attack, which is probabilistically dependent on the actions of both the Attacker and the Defender: $S \mid d, a$. The sequence of nodes reflects that the Defender's choice is observed by the Attacker.

The standard game theoretic approach computes Nash equilibria under strong common knowledge assumptions: in this case, the Defender would know the beliefs and preferences of the Attacker, modeled, respectively, in (p_A, u_A), which is not realistic. Note that the attacker does not need to know the Defender's probabilities and utilities (p_D, u_D) as he observes the Defender's decision. We now weaken such common knowledge assumptions in this sequential decision

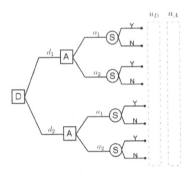

Fig. 1. The Defend-Attack sequential decision game

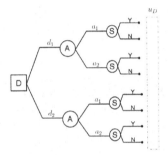

Fig. 2. The decision problem as seen by the Defender

game. Suppose, thus, that the Defender does not know (p_A, u_A). The Defender's decision tree denotes uncertainty about the Attacker's attack by replacing $\boxed{A}$ with $\textcircled{A}$. Note that, in order to solve this game, the Defender has already assessed $p_D(S|d, a)$ and $u_D(a, d, S)$, but she also needs $p_D(A = a|d)$, which is her assessment of the probability that the Attacker will choose attack a after observing that the Defender has chosen defense d. This assessment requires the Defender to analyze the problem from the Attacker's perspective, as follows.

For that, the Defender must place herself in the Attacker's shoes, and consider his decision problem. Fig. 3 represents the Attacker's problem, as seen by the Defender. We assume that the Defender will analyze the Attacker's problem considering that he is an expected utility maximizer. Thus, she will use all the information and judgment that she can about the Attacker's utilities and probabilities. Therefore, to find $p_D(A|d)$, she should first estimate the Attacker's utility function and his probabilities about success S, conditional on (d, a), and, consequently, compute the required probability. However, instead of using point estimates for p_A and u_A to find the Attacker's optimal decision $a^*(d)$, the Defender's uncertainty about the Attacker's decision should derive from her uncertainty about the Attacker's (p_A, u_A), which we describe through a distribution F. This will induce a distribution on the Attacker's expected utility $\psi_A(a, d)$.

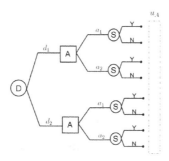

Fig. 3. The Defender's analysis of the Attacker's problem

Thus, assuming the Attacker is rational, the Defender's predictive distribution on the Attacker's attack choice given her defense choice d is

$$p_D(A = a|d) = \mathbb{P}_F[a = \mathrm{argmax}_{x \in \mathcal{A}} \Psi_A(d, x)], \quad \forall a \in \mathcal{A},$$

where

$$\Psi_A(d, a) = P_A(S = 0 \mid d, a)\, U_A(a, d, S = 0) + P_A(S = 1 \mid d, a)\, U_A(a, d, S = 1)$$

for $(P_A, U_A) \sim F$.

She may use Monte Carlo simulation to approximate $p_D(A|d)$ by drawing n samples $\{(p_A^i, u_A^i)\}_{i=1}^n$ from F, which produce $\{\psi_A^i\}_{i=1}^n \sim \Psi_A$, and approximating $p_D(A = a|d)$ by

$$\hat{p}_D(A = a|d) = \frac{\#\{a = \mathrm{argmax}_{x \in \mathcal{A}}\ \psi_A^i(d, x)\}}{n}.$$

Once the Defender has completed these assessments, she can solve her problem. Her expected utilities at node (S) in Figure 2 for each $(d, a) \in \mathcal{D} \times \mathcal{A}$ are:

$$\psi_D(d, a) = p_D(S = 0|d, a)\, u_D(a, d, S = 0) \ + \ p_D(S = 1|d, a)\, u_D(a, d, S = 1).$$

Working up the tree in Fig. 2, her estimated expected utilities at node (A) for each $d \in \mathcal{D}$ are:

$$\hat{\psi}_D(d) = \psi_D(d, a_1)\, \hat{p}_D(a_1|d) + \psi_D(d, a_2)\, \hat{p}_D(a_2|d).$$

Finally, her approximate optimal decision is $d^* = \mathrm{argmax}_{d \in \mathcal{D}}\ \hat{\psi}_D(d)$.

We summarize the previous discussion with the following procedure to find a recommendation for the Defender in the Defend-Attack sequential decision model.

1. Assess (p_D, u_D) from the Defender
2. Assess $F = (P_A, U_A)$, describing the Defender's uncertainty about (p_A, u_A)

3. For each d, simulate to approximate $p_D(A|d)$ as follows:
   (a) Generate $(p_A^i, u_A^i) \sim F$, $i = 1, \ldots, n$
   (b) Solve $a_i^*(d) = \text{argmax}_{a \in \mathcal{A}} \, \psi_A^i(d, a)$
   (c) Approximate $\hat{p}_D(A = a|d) = \#\{a = a_i^*(d)\}/n$
4. Solve the Defender's problem

$$d^* = \text{argmax}_{d \in \mathcal{D}} \, \psi_D(d, a_1) \, \hat{p}_D(a_1|d) \; + \; \psi_D(d, a_2) \, \hat{p}_D(a_2|d)$$

3   Defender-Attacker Simultaneous Action Model

We now present a simultaneous-move case in which the Defender and the Attacker must make their defense and attack decisions, without knowing the action chosen by each other. Again, each of them has just two options to choose from: $\mathcal{D} = \{d_1, d_2\}$ for the Defender and $\mathcal{A} = \{a_1, a_2\}$ for the Attacker. The only uncertainty is S: the success or failure of the Attack. This problem can be represented as a joint decision tree as in Fig. 4.

The standard game-theoretic approach computes Nash equilibria under the common knowledge assumption that both the Defender's and Attacker's preferences and beliefs are available. This is unrealistic in the counterterrorism context. When these are unknown, another standard game theoretic approach computes Bayes-Nash equilibria under the common knowledge assumption about priors over types of defenders and attackers, as in Harsanyi (1967). But again, such assumption is unrealistic in the counterterrorism setting. We now weaken this common prior knowledge assumption and solve the Defender-Attacker simultaneous action game for the Defender. In this model, the assessment of probabilities on the adversary's actions tend to be more elaborated than in the Defend-Attack sequential game from Section 2, as we shall see.

The Defender has to choose a defence action $d \in \mathcal{D}$, whose consequences depend on the success of the attack chosen simultaneously by the Attacker, as reflected in Fig. 5. By standard decision theory, see French and Rios Insua (2000),

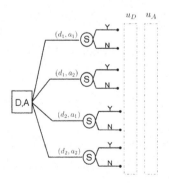

Fig. 4. The two players simultaneous decision game

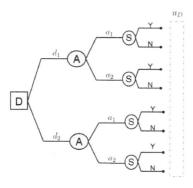

Fig. 5. The Defender's decision analysis

the defender should maximize her expected utility. Therefore, the decision problem she should solve is

$$d^* = \operatorname{argmax}_{d \in \mathcal{D}} \sum_{a \in \mathcal{A}} \left[\sum_{s \in \{0,1\}} u_D(a, d, s)\, p_D(S = s \mid d, a) \right] \pi_D(A = a),$$

where the Defender's utility function $u_D(a, d, s)$ and her personal probabilistic assessment p_D over S conditional on (d, a) are known to her. As the Defender does not know the Attacker's decision at node A, she expresses her uncertainty through a probability distribution $\pi_D(A = a)$, over all $a \in \mathcal{A}$.

The main difficulty for the Defender is assessing the probability distribution $\pi_D(A)$, which represents her uncertainty about what attack will the Attacker choose. To do so, she can think of the Attacker as an expected utility maximizer who tries to solve a decision problem as the one shown in Fig. 6, by finding the attack $a \in \mathcal{A}$ that provides him maximum expected utility:

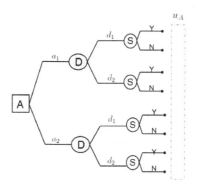

Fig. 6. The Attacker's decision analysis as seen by the Defender

$$a^* = \mathrm{argmax}_{a \in \mathcal{A}} \sum_{d \in \mathcal{D}} \left[\sum_{s \in \{0,1\}} u_A(a, d, s)\, p_A(S = s \mid d, a) \right] \pi_A(D = d).$$

However, the Defender cannot solve the Attacker's decision problem, and, thus, find his maximum expected utility attack a^*, since in general she will not know the Attacker's true probabilities and utility function (u_A, p_A, π_A). But, she may model all information available to her about (u_A, p_A, π_A) through a probability distribution $(U_A, P_A, \Pi_A) \sim F$. Thus, the Defender's assessment of $\pi_D(A)$ is reduced to the computation of the following probability distribution

$$A \mid D \sim \mathrm{argmax}_{a \in \mathcal{A}} \sum_{d \in \mathcal{D}} \left[\sum_{s \in \{0,1\}} U_A(a, d, s)\, P_A(S = s \mid d, a) \right] \Pi_A(D = d). \tag{1}$$

Note that Π_A represents the Defender's assessment of the probabilistic model π_A used by the Attacker to predict what defense the Defender will choose, whereas D represents the Attacker's beliefs about what choice the Defender will make. We are, thus, identifying two sources of uncertainty:

1. one stemming from the Attacker's uncertainty about the Defender's choice (D), and
2. one coming from the Defender's uncertainty about how the Attacker assesses D through Π_A.

Again, the Defender may presume that the Attacker thinks she is an expected utility maximizer trying to solve a decision problem analogous to that in Fig. 5. Therefore, in order for the Defender to solve Eq. (1), she will elicit $(U_A, P_A, \Pi_A) \sim F$ from her viewpoint, and assess D through the analysis of her decision problem as thought by the Attacker, reducing the assessment of D to the computation of the following distribution

$$D \mid A^1 \sim \mathrm{argmax}_{d \in \mathcal{D}} \left[\sum_{a \in \mathcal{A}} \sum_{s \in \{0,1\}} U_D(a, d, s)\, P_D(S = s \mid d, a) \right] \Pi_D(A^1 = a),$$

where the Defender needs to elicit $(U_D, P_D, \Pi_D) \sim G$ representing her assessment on the Attacker's estimation of her utility function $u_D(a, d, s)$ and her probability p_D over $S \mid d, a$, when she analyzes how the Attacker thinks about her decision problem, as well as the Defender's confidence in her assessment model leading to A^1, which represents the Attacker's decision when he is modeled by the Defender within her second level of recursive modeling. This may require again further recursive thinking from the Defender, which will lead to the following recursive computations, starting with $i = 0$.

Repeat

$$A^i \mid D^i \sim \operatorname{argmax}_{a \in \mathcal{A}} \sum_{d \in \mathcal{D}} \left[\sum_{s \in \{0,1\}} U_A^i(a,d,s) \, P_A^i(S = s \mid d, a) \right] \Pi_{A^i}(D^i = d)$$

$$\text{where } (U_A^i, P_A^i, \Pi_{A^i}) \sim F^i$$

$$D^i \mid A^{i+1} \sim \operatorname{argmax}_{d \in \mathcal{D}} \sum_{a \in \mathcal{A}} \left[\sum_{s \in \{0,1\}} U_D^i(a,d,s) \, P_D^i(S = s \mid d, a) \right] \Pi_{D^i}(A^{i+1} = a)$$

$$\text{where } (U_D^i, P_D^i, \Pi_{D^i}) \sim G^i$$

$i = i + 1$

To simplify the discussion, we have assumed that the decision models recursively used to assess A^i and D^i, from $i = 0$ and so forth, are a reflection of each other and have the same structure as those in Figs. 6 and 5, respectively. Moreover, the choice sets for the Defender and the Attacker are the same in all models in the recursive hierarchy of analysis: $\mathcal{D}$ and $\mathcal{A}$, respectively. This hierarchy of nested models would stop at the level in which the Defender lacks the kind of information necessary to assess the distribution F^i (or G^i). At this point, the Defender will holistically assign an unconditional probability distribution over A^i (or D^i), without going deeper in the hierarchy. Of course, if she feels that she still has no information available to do so, she can assign a noninformative or reference probability distribution, see French and Rios Insua (2000).

4   Discussion

We have provided here an account of how Bayesian decision analysis can support a Defender against an intelligent adversary (the Attacker), in contrast with the standard game-theoretic approach, in two stylised problems. In our framework, the Defender assesses the probabilities of the adversarial actions before computing her maximum expected utility defense strategy. We assume that the Attacker is an expected utility maximizer and the Defender's uncertainty about the Attacker' decision stems from her uncertainty about his decision analysis, specifically about his probabilities and utilities. The Bayesian approach weakens the unrealistic common knowledge assumptions necessary in the game-theoretic approach.

Of course, real problems in counterterrorism are extremely complex. For example, they involve hundreds of possible decisions, and there are large uncertainties associated with the goals and resources of the terrorists. For this reason, we have focused on illustrating with two simple examples a general methodology based on decision analysis principles to support decision making against an intelligent Attacker, making parallel comparisons with the standard solutions proposed by the game-theoretic approach. More complex dynamic interactions, say as in defend-attack-defend models or general coupled defend and attack influence diagrams, require more complex analysis but the methodology would stay essentially the same. Extensions of the methodology to the case in which there are more than one Attacker, or an uncertain number of Attackers, and more than one defender also need to be explored. This last one might require negotiations about risk sharing, as in Rios and Rios Insua (2010).

Acknowledgements

Research supported by grants from the Spanish Ministry of Science and Innovation, the E-DEMOCRACIA-CM program and the COST Algorithmic Decision Theory Action. We are grateful to discussions by the referees.

References

Aliprantis, C., Chakrabarti, S.: Games and Decision Making, Oxford, U.P (2000)

Banks, D., Anderson, S.: Game theory and risk analysis in the context of the smallpox threat. In: Wilson, A., Wilson, G., Olwell, D. (eds.) Statistical Methods in Counterterrorism, pp. 9–22 (2006)

Bier, V.S., Azaiez, N.: Game Theoretic Risk Analysis of Security Threats. Springer, Heidelberg (2008)

Bier, V., Oliveros, S., Samuelson, L.: Choosing what to protect. Journal of Economic Literature 9(4), 563–587 (2007)

Brown, G., Carlyle, M., Salmeron, J., Wood, K.: Defending critical infrastructure. Interfaces 36(6), 530–544 (2006)

Brown, G., Carlyle, W.M., Wood, R.: Optimizing Department of Homeland Security Defense Investments: Applying Defender-Attacker(-Defender) Optimization to Terror Risk Assessment and Mitigation, Appendix E. National Academies Press, Washington (2008)

French, S., Rios Insua, D.: Statistical Decision Theory, Arnold (2000)

Gibbons, R.: A Primer in Game Theory. Pearson Education Ltd., Harlow (1992)

Harsanyi, J.: Games with incomplete information played by "Bayesian" players, I-III. Part I. The basic model. Management Science 14(3), 159–182 (1967)

Harsanyi, J.: Subjective probability and the theory of games: Comments on Kadane and Larkey's paper. Management Science 28(2), 120–124 (1982)

Kadane, J.B., Larkey, P.D.: Subjective probability and the theory of games. Management Science 28(2), 113–120 (1982); reply: 124

Parnell, G., Banks, D., Borio, L., Brown, G., Cox, L.A., Gannon, J., Harvill, E., Kunreuther, H., Morse, S., Pappaioanou, M., Pollack, S., Singpurwalla, N., Wilson, A.: Report on Methodological Improvements to the Department of Homeland Security's Biological Agent Risk Analysis. National Academies Press, Washington (2008)

Pinker, E.J.: An analysis of short-term responses to threats of terrorism. Management Science 53(6), 865–880 (2007)

Raiffa, H.: Negotiation Analysis. Harvard University Press, Cambridge (2002)

Rios, J., Rios Insua, D.: Balanced increment and concession methods for negotiation support. Journal of the Spanish Royal Academy of Sciences, RACSAM (in press, 2010)

Rios Insua, D., Rios, J., Banks, D.: Adversarial risk analysis. Journal of the American Statistical Association 104(486), 841–854 (2009)

Rothkopf, M.: Decision Analysis: The right tool for auctions. Decision Analysis 4, 167–172 (2007)

van Bingsbergen, J.H., Marx, L.M.: Exploring relations between decision analysis and game theory. Decision Analysis 4, 32–40 (2007)

von Winterfeldt, D., O'Sullivan, T.M.: Should we protect commercial airplanes against surface-to-air missile attacks by terrorists? Decision Analysis 3(2), 63–75 (2006)

Zhuang, J., Bier, V.: Balancing Terrorism and Natural Disasters. Defensive Strategy with endogenous attack effort. Operations Research 55(5), 976–991 (2007)

Game Theory without Decision-Theoretic Paradoxes

Pierfrancesco La Mura

Department of Economics and Information Systems
HHL - Leipzig Graduate School of Management
Jahnallee 59, 04109 Leipzig, Germany
plamura@hhl.de

Abstract. Most work in game theory is conducted under the assumption that the players are expected utility maximizers. Expected utility is a very tractable decision model, but is prone to well-known paradoxes and empirical violations (Allais 1953, Ellsberg 1961), which may induce systematic biases in game-theoretic predictions. La Mura (2009) introduced a projective generalization of expected utility (PEU) which avoids the dominant paradoxes, while remaining quite tractable. We show that every finite game with PEU players has an equilibrium, and discuss several examples of PEU games.

Keywords: game theory, expected utility, paradoxes, Allais, Ellsberg, projective.

1 Introduction

The expected utility hypothesis is the *de facto* foundation of game theory. The von Neumann - Morgenstern axiomatization of expected utility, and later on the subjective formulations by Savage (1954) and Anscombe and Aumann (1963) were immediately greeted as simple and intuitively compelling. Yet, in the course of time, a number of empirical violations and paradoxes (Allais 1953, Ellsberg 1961) came to cast doubt on the validity of the hypothesis as a foundation for the theory of rational decisions in conditions of risk and subjective uncertainty. In economics and in the social sciences, the shortcomings of the expected utility hypothesis are generally well-known, but often tacitly accepted in view of the great tractability and usefulness of the corresponding mathematical framework. In fact, the hypothesis postulates that preferences can be represented by way of a utility functional which is linear in probabilities, and linearity makes expected utility representations particularly tractable in models and applications. In particular, in a game-theoretic context, linearity of expected utility ensures that the best response correspondence is also linear, and hence that any finite game has a Nash equilibrium.

In economics and the social sciences, the importance of accounting for violations of the expected utility hypothesis has long been recognized (Tversky 1975), but so far none of its numerous alternatives (e.g., Machina 1982, Schmeidler 1989, to quote only two particularly influential papers in a rich and constantly evolving literature) has fully succeeded in replacing expected utility as a standard foundation for decisions under uncertainty, partly due to the great mathematical tractability of expected utility relative to many of its proposed generalizations.

F. Rossi and A. Tsoukis (Eds.): ADT 2009, LNAI 5783, pp. 316–327, 2009.

La Mura (2009) introduced projective expected utility (PEU), a decision-theoretic framework which accommodates the dominant paradoxes while retaining significant simplicity and tractability. This is obtained by weakening the expected utility hypothesis to its projective counterpart, in analogy with the quantum-mechanical generalization of classical probability theory. We first review the EU and PEU frameworks, and show that the latter is sufficiently general to avoid both Allais' and Ellsberg's paradoxes. We then extend the notion of Nash equilibrium to games with PEU preferences, and prove that any finite game with PEU players has an equilibrium. Finally, we discuss several examples of games with PEU preferences and identify observable deviations from the theory of games with EU preferences.

2   Von Neumann - Morgenstern Expected Utility

Let Ω be a finite set of outcomes, and Δ be the set of probability functions defined on Ω, taken to represent risky prospects (or *lotteries*). Next, let $\succsim$ be a complete and transitive binary relation defined on $\Delta \times \Delta$, representing a decision-maker's preference ordering over lotteries. Indifference of $p, q \in \Delta$ is defined as $[p \succsim q$ and $q \succsim p]$ and denoted as $p \sim q$, while strict preference of p over q is defined as $[p \succsim q$ and not $q \succsim p]$, and denoted by $p \succ q$. The preference ordering is assumed to satisfy the following two conditions.

Axiom 1. *(Archimedean) For all $p, q, r \in \Delta$ with $p \succ q \succ r$, there exist $\alpha, \beta \in (0, 1)$ such that $\alpha p + (1 - \alpha)r \succ q \succ \beta p + (1 - \beta)r$.*

Axiom 2. *(Independence) For all $p, q, r \in \Delta$, $p \succsim q$ if, and only if, $\alpha p + (1 - \alpha)r \succsim \alpha q + (1 - \alpha)r$ for all $\alpha \in [0, 1]$.*

A functional $u : \Delta \to \mathbb{R}$ is said to represent $\succsim$ if, for all $p, q \in \Delta$, $p \succsim q$ if and only if $u(p) \geq u(q)$.

Theorem 1. *(von Neumann and Morgenstern) Axioms 1 and 2 are jointly equivalent to the existence of a functional $u : \Delta \to \mathbb{R}$ which represents $\succsim$ such that, for all $p \in \Delta$,*

$$u(p) = \sum_{\omega \in \Omega} u(\omega)p(\omega).$$

The von Neumann - Morgenstern setting is appropriate whenever the nature of the uncertainty is purely objective: all lotteries are associated with objective random devices, such as dice or roulette wheels, with well-defined and known frequencies for all outcomes, and the decision-maker only evaluates a lottery based on the frequencies of its outcomes.

3   Allais' Paradox

The following paradox is due to Allais (1953). First, please choose between

   A: A chance of winning 4000 dollars with probability 0.2
   B: A chance of winning 3000 dollars with probability 0.25.

Now suppose that, instead of A and B, your two alternatives are

C: A chance of winning 4000 dollars with probability 0.8
D: A chance of winning 3000 dollars with certainty.

If you chose A over B, and D over C, then you are in the modal class of respondents. The paradox lies in the observation that A and C are special cases of a two-stage lottery E which in the first stage either returns zero dollars with probably $(1 - \alpha)$ or, with probability α, leads to a second stage where one gets 4000 dollars with probability 0.8 and zero otherwise. In particular, if α is set to 1 then E reduces to C, and if α is set to 0.25 it reduces to A. Similarly, B and D are special cases of a two-stage lottery F which again with probability $(1 - \alpha)$ returns zero, and with probability α continues to a second stage where one wins 3000 dollars with probability 1. Again, if $\alpha = 1$ then F reduces to D, and if $\alpha = 0.25$ it reduces to B. Then it is easy to see that the $[A \succ B, D \succ C]$ pattern violates Axiom 2 (Independence), as E can be regarded as a lottery $\alpha p + (1 - \alpha)r$, and F as a lottery $\alpha q + (1 - \alpha)r$, where p and q represent lottery C and D, respectively, and r represents the lottery in which one gets zero dollars with certainty. When comparing E and F, why should it matter what is the value of α? Yet, experimentally one finds that it does.

4   Ellsberg's Paradox

Another disturbing violation of the expected utility hypothesis was pointed out by Ellsberg (1961). Suppose that an urn contains 300 balls of three possible colors: red, green, and blue. You know that the urn contains exactly 100 red balls, but are given no information on the proportion of green and blue.

You win if you guess which color will be drawn. Do you prefer to bet on red (R) or on green (G)? Many respondents choose R, on grounds that the probability of drawing a red ball is known to be $1/3$, while the only information on the probability of drawing a green ball is that it is between 0 and $2/3$. Now suppose that you win if you guess which color will *not* be drawn. Do you prefer to bet that red will not be drawn ($\overline{R}$) or that green will not be drawn ($\overline{G}$)? Again many respondents prefer to bet on $\overline{R}$, as the probability is known ($2/3$) while the probability of $\overline{G}$ is only known to be between 1 and $1/3$.

The pattern $[R \succ G, \overline{R} \succ \overline{G}]$ is incompatible with von Neumann - Morgenstern expected utility, which only deals with known probabilities, and is also incompatible with the Savage (1954) formulation of expected utility with subjective probability as it violates its Sure Thing axiom. Observe that, in Ellsberg's setting, the decision-maker ignores the actual composition of the urn, and hence operates in a state of subjective uncertainty about the true probabilistic state of affairs. In particular, the decision-maker is exposed to a combination of subjective uncertainty (on the actual composition of the urn) and objective risk (the probability of drawing a specific color from an urn of given composition). The paradox suggests that, in order to account for the choice pattern discussed above, subjective uncertainty and risk should be handled as distinct notions.

5 Projective Expected Utility

Let X be the positive orthant of the unit sphere in $\mathbb{R}^n$, where n is the cardinality of the set of relevant outcomes Ω. Then von Neumann - Morgenstern lotteries, regarded as elements of the unit simplex, are in one-to-one correspondence with elements of X, which can therefore be interpreted as risky prospects, for which the frequencies of the relevant outcomes are fully known. Observe that, while the projections of elements of the unit simplex (and hence, L^1 unit vectors) on the basis vectors can be naturally associated with probabilities, if we choose to model von Neumann - Morgenstern lotteries as elements of the unit sphere (and hence, as unit vectors in L^2) then probabilities are naturally associated with *squared* projections. The advantage of such move is that L^2 is the only L^p space which is also a Hilbert space, and Hilbert spaces have a very tractable projective structure which is exploited by the representation. In particular, it is unique to L^2 that the set of unit vectors is invariant with respect to projections.

Next, let $\langle .|. \rangle$ denote the usual inner product in $\mathbb{R}^n$. We denote the transpose of a vector or a matrix with a prime, *e.g.*, x' denotes the transpose of x. An orthonormal basis is a set of unit vectors $\{b^1, ..., b^n\}$ such that $\langle b^i|b^j \rangle = 0$ whenever $i \neq j$. In our context, orthogonality captures the idea that two events or outcomes are mutually exclusive (for one event to have probability one, the other must have probability zero). The natural basis corresponds to the set of degenerate lotteries returning each objective lottery outcome with certainty, and is conveniently identified with the set of objective lottery outcomes $\{\omega^1, \omega^2, ..., \omega^n\}$. Yet, in any realistic experimental setting, it is very unlikely that those objective outcomes will happen to coincide with the set of subjective consequences which are relevant from the perspective of the decision-maker. Moreover, even if the latter could be fully elicited, it would be generally problematic to relate a von Neumann - Morgenstern lottery, which only specifies the probabilities of the objective outcomes, with the probabilities induced on the subjective consequences, that is, the relevant dimensions of risk from the point of view of the decision-maker. The perspective of the observer or modeler is inexorably bound to objectively measurable entities, such as frequencies and prizes; by contrast, the decision-maker thinks and acts based on subjective preferences and subjective consequences, which in a revealed-preference context should be presumed to exist while at the same time assumed, as a methodological principle, to be unaccessible to direct measurement.

In this section we shall relax the assumption, implicit in the von Neumann - Morgenstern setting, that the objective outcomes and subjective consequences coincide, and replace it with the weaker requirement that there exists a set of mutually exclusive, jointly exhaustive subjective consequences with respect to which the decision-maker evaluates each uncertain prospect. As we shall see such weaker condition, together with the usual assumptions of completeness and transitivity of preferences, and the Archimedean and Independence conditions from the von Neumann-Morgenstern treatment, jointly characterize representability in terms of a projective generalization of the expected utility functional.

Let $B := \{b^1, ..., b^n\}$ represent a set of n mutually exclusive, jointly exhaustive subjective consequences, identified with an orthonormal basis in $\mathbb{R}^n$. For any lottery $x \in X$, its associated *risk profile* p_x with respect to B is defined by

$$p_x(b^i) = \langle x|b^i\rangle^2, i = 1, \ldots, n$$

The risk profile of a lottery x returns the probabilities induced on the subjective consequences by playing lottery x. Observe that the risk profile with respect to the natural basis simply returns the probabilities of the lottery outcomes. Hence, in the present setting
a lottery is identified both in terms of objective outcomes and subjective consequences.
Since the position of the subjective basis relative to the natural basis can be arbitrary, a
lottery can exhibit any combination of risk profiles on outcomes and consequences.

Axiom 3. *There exists an orthonormal basis* $Z := \{z^1, \ldots, z^n\}$ *such that any two lotteries* $x, y \in X$ *are indifferent whenever their risk profiles with respect to* Z *coincide.*

Axiom 3 requires that there exists a set of n mutually exclusive, jointly exhaustive
subjective consequences such that any two lotteries are only evaluated based on their
risk profiles, *i.e.*, on the probabilities they induce on those subjective consequences. In
the von Neumann - Morgenstern treatment, Axiom 3 is tacitly assumed to hold with
respect to the natural basis in $\mathbb{R}^n$. This implicit assumption amounts to the requirement
that lotteries are only evaluated based on the probabilities they induce on the objective
lottery outcomes.

While the probabilities with respect to the natural basis represent the relevant dimensions of risk as perceived by the modeler or an external observer (that is, the risk
associated with the occurrence of the objective outcomes), the *preferred basis* postu-
lated in Axiom 3 is allowed to vary across different decision-makers, capturing the idea
that the subjectively relevant dimensions of risk (that is, those pertaining to the actual
subjective consequences) may be perceived differently by different subjects. One case
in which the relevant dimensions of risk may differ across subjects is in the presence
of portfolio effects. Such effects are difficult to exclude or control for in experimental
settings, as the subject's portfolio is typically unaccessible to direct measurement.

Axiom 3 presumes that subjective consequences and objective outcomes have the
same cardinality; we relax this assumption later on, in the subjective formulation.

Once an orthonormal basis Z is given, each lottery x can be associated with a function
$p_x : Z \to [0, 1]$, such that $p_x(z^i) = \langle x|z^i\rangle^2$ for all $z^i \in Z$. Let B be the set of all such
risk profiles p_x, for $x \in X$, and let $\succsim$ be the complete and transitive preference ordering
induced on $B \times B$ by preferences on the underlying lotteries.

Note that a convex combination $\alpha p_x + (1 - \alpha)p_y$, where p_x and p_y are risk profiles,
is still a well-defined risk profile. We interpret this type of mixing as objective, while
subjective mixing will be later on captured by subjective probability over the underlying (Anscombe-Aumann) states. We postulate the following two axioms, which mirror
those in the von Neumann - Morgenstern treatment.

Axiom 4. *(Archimedean) For all* $x, y, t \in X$ *with* $p(x) \succ p(y) \succ p(t)$, *there exist*
$\alpha, \beta \in (0, 1)$ *such that* $\alpha p(x) + (1 - \alpha)p(t) \succ p(y) \succ \beta p(x) + (1 - \beta)p(t)$.

Axiom 5. *(Independence) For all* $x, y, t \in X$, $p_x \succsim p_y$ *if, and only if,* $\alpha p_x + (1-\alpha)p_t \succsim$
$a p_y + (1 - \alpha)p_t$ *for all* $\alpha \in [0, 1]$.

Some observations are in order at this point. First, note that the two axioms above impose conditions solely on risk profiles, and not on the underlying lotteries. This seems appropriate, as the decision-maker is not ultimately concerned with the risk associated to the objective outcomes, but only with the risk induced on the relevant subjective consequences. It is also worth noting that, while risk profiles are now defined with respect to subjective consequences, rather than objective outcomes as in the von Neumann - Morgenstern treatment, in our setting they are still interpreted as objective probability functions.

Theorem 2. *(La Mura 2009) Axioms 3-5 are jointly equivalent to the existence of a symmetric matrix U such that $u(x) := x'Ux$ for all $x \in X$ represents $\succsim$.*

6 Subjective Formulation

The following formulation extends the representation to situations of subjective uncertainty. First, we introduce the following setup and notation.

S is a finite set of states of Nature.

$\langle .|. \rangle$ denotes the usual inner product in Euclidean space.

Ω is the natural basis in $\mathbb{R}^n$, identified with a finite set $\{\omega^1, \ldots, \omega^n\}$ of lottery outcomes (prizes).

Z is an orthonormal basis in $\mathbb{R}^m$, with $m \geq n$, identified with a finite set of subjective consequences $\{z^1, \ldots, z^m\}$. V is an arbitrary $(m \times n)$ matrix chosen so that, for all ω^i in Ω, $V\omega^i$ is a unit vector in $\mathbb{R}^m$. Observe that V is always well defined as long as $m \geq n$. When $m = n$, we conventionally set $V \equiv I$, where I is the $n \times n$ identity matrix.

Lotteries correspond to L^2 unit vectors $x \in \mathbb{R}^n_+$; X is the set of all lotteries.

Since Ω is the natural basis, $\langle \omega^i | x \rangle^2 = x_i^2$; this quantity is interpreted as $p(\omega^i|x)$.

The quantity $\langle z^j | Vx \rangle^2$ is interpreted as $p(z^j|x)$, the conditional probability of subjective consequence z^j given lottery x. In particular, $\langle z^j | V\omega^i \rangle^2$ is interpreted as $p(z^j|\omega^i)$, the conditional probability of subjective consequence z^j given the degenerate lottery which returns objective outcome ω^i for sure.

Once the subjective consequences z^j are specified, for any lottery x one can readily compute $p(\omega^i|x) = x_i^2$ and $p(z^j|x) = \langle z^j | Vx \rangle^2$. Moreover, given the latter probabilistic constraints, one can readily identify a lottery x and an orthonormal basis Z which jointly satisfy them. Hence, in the above construction lotteries are identified with respect to two different frames of reference: objective lottery outcomes, and subjective consequences.

Observe that $p(z^j|x)$ generally differs from the probability of z^j given x computed according to the law of total probability, which is given by $\sum_i p(\omega^i|x)p(z^j|\omega^i) = \sum_i x_i^2 \langle z^j | V\omega^i \rangle^2$. To get a sense of why and how the law of total probability may fail, let us consider a decision-maker who really hates to lose whenever the probability of winning is high (more so than when the probability of winning is low), and loves to win when the probability of losing is high (even more so than when the latter probability is low). Clearly, in such case the probabilities of objective outcomes such as winning and losing are directly involved in the description of subjectively relevant consequences such as "I won (or lost) against all odds". Such dependency introduces an element of interference between lotteries and consequences that cannot be easily accounted for in the classical decision-theoretic setting, which presumes state independence.

An act is identified with a function $f : S \to X$. H is the set of all acts.

$\Delta(X)$ is the (nonempty, closed and convex) set of all probability functions on Z induced by lotteries in X.

M is the set of all vectors $(p_s)_{s \in S}$, with $p_s \in \Delta(X)$.

For each $f \in H$ a corresponding risk profile $p^f \in M$ is defined, for all $s \in S$ and all $z^j \in Z$, by $p_s^f(z^j) := \langle z^j | V f_s \rangle^2$.

As customary, we assume that the decision-maker's preferences are characterized by a rational (*i.e.*, complete and transitive) preference ordering $\succsim$ on acts. Next, we proceed with the following assumptions, which mirror those in Anscombe and Aumann (1963).

Axiom 6. *(Projective) There exists a finite orthonormal basis $Z := \{z_1, ..., z_m\}$, with $m \geq n$, such that any two acts $f, g \in H$ are indifferent if $p^f = p^g$.*

In Anscombe and Aumann's setting, the above axiom is implicitly assumed to hold with $Z \equiv \Omega$. Because of Axiom 6, preferences on acts can be equivalently expressed as preferences on risk profiles. For all $p^f, p^g \in M$, we stipulate that $p^f \succsim p^g$ if and only if $f \succsim g$.

Axiom 7. *(Archimedean) If $p^f, p^g, p^h \in M$ are such that $p^f \succ p^g \succ p^h$, then there exist $a, b \in (0, 1)$ such that $a p^f + (1 - a) p^h \succ p^g \succ b p^f + (1 - b) p^h$.*

Axiom 8. *(Independence) For all $p^f, p^g, p^h \in M$, and for all $a \in (0, 1]$, $p^f \succ p^g$ if and only if $a p^f + (1 - a) p^h \succ a p^g + (1 - a) p^h$.*

Axiom 9. *(Non-degeneracy) There exist $p^f, p^g \in M$ such that $p^f \succ p^g$.*

Axiom 10. *(State independence) Let $s, t \in S$ be non-null states, and let $p, q \in \Delta(X)$. Then, for any $p^f \in M$,*

$$(p_1^f, ..., p_{s-1}^f, p, p_{s+1}^f, ..., p_n^f) \succ (p_1^f, ..., p_{s-1}^f, q, p_{s+1}^f, ..., p_n^f)$$

if, and only if,

$$(p_1^f, ..., p_{t-1}^f, p, p_{t+1}^f, ..., p_n^f) \succ (p_1^f, ..., p_{t-1}^f, q, p_{t+1}^f, ..., p_n^f).$$

Theorem 3. *(Anscombe and Aumann) The preference relation $\succsim$ fulfills Axioms $6 - 10$ if and only if there is a unique probability measure π on S and a non-constant function $u : Z \to \mathbb{R}$ (unique up to positive affine rescaling) such that, for any $f, g \in H$, $f \succsim g$ if, and only if,*

$$\sum_{s \in S} \pi(s) \sum_{z^i \in Z} p_s^f(z^i) u(z^i) \geq \sum_{s \in S} \pi(s) \sum_{z^i \in Z} p_s^g(z^i) u(z^i).$$

La Mura (2009) provides a projective generalization of Theorem 3.

Theorem 4. *(La Mura) The preference relation $\succsim$ fulfills Axioms $6 - 10$ if and only if there is a unique probability measure π on S and a symmetric $(n \times n)$ matrix U with distinct eigenvalues such that, for any $f, g \in H$, $f \succsim g$ if, and only if,*

$$\sum_{s \in S} \pi(s) f_s' U f_s \geq \sum_{s \in S} \pi(s) g_s' U g_s.$$

7 Properties of the Representation

Our representation generalizes the Anscombe-Aumann expected utility framework in two directions. First, subjective uncertainty and risk are treated as distinct notions. Specifically, let us say that an act is *pure* or *certain* if it returns the same objective lottery in all states, and *mixed* or *uncertain* otherwise. While pure acts are naturally associated with risky decisions, in which the relevant frequencies are all known, mixed acts correspond to uncertain decisions, in which the decision-maker only has a subjective assessment of the true frequencies involved. Second, as we shall see, within this class of preferences both Allais' and Ellsberg's paradoxes are accommodated.

In the context of Theorem 2, for any two distinct outcomes ω^i and ω^j let $e_{i,j}$ be the objective lottery returning each of the two outcomes with equal frequency. Observe that

$$U_{ij} = u(e_{i,j}) - (\frac{1}{2}u(\omega^i) + \frac{1}{2}u(\omega^j)).$$

It follows that the off-diagonal entry U_{ij} in the payoff matrix can be interpreted as the discount, or premium, attached to a symmetric, objective lottery over the two outcomes with respect its expected utility base-line, and hence as a measure of preference for risk versus uncertainty along the specific dimension involving outcomes ω^i and ω^j. Let us say that a decision-maker is *averse to uncertainty* if she always weakly prefers $e_{i,j}$ to an equal subjective chance of ω^i or ω^j. Then a decision-maker is averse to uncertainty if and only if U is a Metzler matrix, *i.e.*, has non-negative off-diagonal elements.

Compared to existing generalizations of expected utility which avoid the Allais or Ellsberg paradoxes, such as the ones in Machina (1982), Schmeidler (1989), or Chew, Epstein and Segal (1991), among others, projective expected utility enjoys several advantages. Specifically, the representation is linear in the probabilities of states and consequences, hence remaining quite tractable, but can be nonlinear in the probabilities of the objective outcomes, hence allowing for portfolio effects. The axioms used to obtain the representation closely mirror those introduced by Anscombe and Aumann (1963), which are widely regarded as appealing. Finally, as shown in sections 8 and 9, a compatible specification of the payoff matrix avoids both Allais' and Ellsberg's paradox.

8 Example: Objective Uncertainty

Figure 1 below presents several examples of indifference maps for pure lotteries over three outcomes which can be obtained within our class of preferences for different choices of U.

The first pattern (parallel straight lines) characterizes von Neumann - Morgenstern expected utility. Within our class of representations, it corresponds to the special case of a diagonal payoff matrix U. All other patterns are impossible within von Neumann - Morgenstern expected utility. Observe that, even though the payoff matrix is an object of relatively limited algebraic complexity, the indifference curves can take a variety of different shapes: in particular, they do not need to be convex, or concave. Yet, since the indifference maps are generated by a limited number of parameters (the entries in the payoff matrix), the type and variety of preference patterns predicted by the model is also limited, and this in turn offers a basis for the empirical testability of the theory.

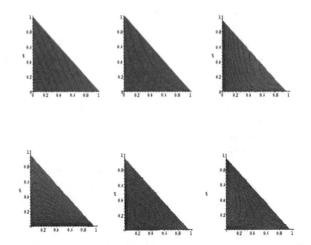

Fig. 1. Examples of indifference maps on the probability triangle

The representation is sufficiently general to accommodate Allais' paradox from section 3. In the context of the example in section 3, let $\{\omega^1, \omega^2, \omega^3\}$ be the outcomes in which 4000, 3000, and 0 dollars are won, respectively. To accommodate Allais' paradox, assume that the non-diagonal elements of the payoff matrix increase with the difference between the corresponding diagonal payoffs. In the example below, the non-diagonal elements are taken to be proportional to the fourth power of the difference.

$$U = \begin{array}{c} \\ \omega^1 \\ \omega^2 \\ \omega^3 \end{array} \begin{array}{ccc} \omega^1 & \omega^2 & \omega^3 \\ 1.1 & 0.00001 & 0.14641 \\ 0.00001 & 1 & 0.1 \\ 0.14641 & 0.1 & 0 \end{array}$$

The above formulation of the payoff matrix implies that, whenever the stakes involved are similar, the corresponding prospects are evaluated approximately at their (von Neumann-Morgenstern) expected utility values. By contrast, whenever the stakes are significantly different, the divergence from expected utility is also significant. Let the four lotteries A, B, C, D be defined, respectively, as the following unit vectors in $\mathbb{R}^3_+$: $a := (\sqrt{0.2}, 0, \sqrt{0.8})'$; $b := (0, \sqrt{0.25}, \sqrt{0.75})'$; $c := (\sqrt{0.8}, 0, \sqrt{0.2})'$; $d := (0, 1, 0)'$. Then lottery A is preferred to B, while D is preferred to C, as

$u(a) = a'Ua = 0.33713,$
$u(b) = b'Ub = 0.3366,$
$u(c) = c'Uc = 0.99713,$
$u(d) = d'Ud = 1.$

9   Example: Subjective Uncertainty

In the Ellsberg puzzle, suppose that either all the non-red balls are green (*i.e.*, 100 red, 200 green, 0 blue), or they are all blue (100 red, 0 green, 200 blue), with equal subjective probability. Further, suppose that there are just two objective outcomes, Win and Lose. Then the following specification of the payoff matrix accommodates the paradox.

$$U = \begin{array}{c} \\ \text{Win} \\ \text{Lose} \end{array} \begin{array}{cc} \text{Win} & \text{Lose} \\ 1 & \alpha \\ \alpha & 0 \end{array}$$

As we shall see, if $\alpha = 0$ we are in the expected utility case, where the decision-maker is indifferent between risk and uncertainty; when $\alpha > 0$, risk is preferred to uncertainty; and when $\alpha < 0$, the decision-maker prefers uncertainty to risk. In fact, let $\{Urn1, Urn2\}$ be the set of possible states of nature, with uniform subjective probability, and let $r := (\sqrt{1/3}, \sqrt{2/3})'$, $\bar{r} := (\sqrt{2/3}, \sqrt{1/3})'$ be the lotteries associated to pure acts R and $\bar{R}$, respectively. Furthermore, let $w := (1,0)'$, $l := (0,1)'$ be the lotteries corresponding to a sure win (W) and a sure loss (L), respectively. The mixed acts G and $\bar{G}$ have projective expected utilities given by

$$u(G) = p(Urn1)u(\bar{R})+p(Urn2)u(L),$$
$$u(\bar{G}) = p(Urn1)u(R)+p(Urn2)u(W).$$

One also has that $u(W) = w'Uw = 1$, $u(L) = l'Ul = 0$, $u(R) = r'Ur = 1/3 + \alpha\sqrt{8}/3$, $u(\bar{R}) = \bar{r}'U\bar{r} = 2/3 + \alpha\sqrt{8}/3$, and therefore

$$u(G) = 1/3 + \alpha\sqrt{2}/3$$
$$u(\bar{G}) = 2/3 + \alpha\sqrt{2}/3.$$

It follows that, whenever $\alpha > 0$, R is preferred to G and $\bar{R}$ to $\bar{G}$, so the paradox is accommodated. When $\alpha < 0$, the opposite pattern emerges: G is preferred to R and $\bar{G}$ to $\bar{R}$. Finally, when $\alpha = 0$ the decision-maker is indifferent between R and G, and between $\bar{R}$ and $\bar{G}$.

10   Games with PEU Preferences

Within the class of preferences characterized by Theorem 4, is it still true that every finite game has a Nash equilibrium? If the payoff matrix U is diagonal we are in the classical case, so we know that any finite game has an equilibrium, which moreover only involves objective risk (in our terms, this type of equilibrium should be referred to as "pure", as it involves no subjective uncertainty). For the general case, consider that $u(f)$ is still continuous and linear with respect to the subjective beliefs π, while possibly nonlinear (but still polynomial) with respect to risk. As we shall see, any finite game has an equilibrium even within this larger class of preferences, although the equilibrium may not be pure (in our sense): in general, an equilibrium will rest on a combination of objective randomization and subjective uncertainty about the players' decisions.

A finite, strategic-form game with PEU preferences is a n-tuple $G := (I, (A^i)_{i\in I}, (U^i)_{i\in I})$, where I is a finite set composed of k players, A^i represents the set of feasible actions for player i, and U^i is player i's payoff matrix on outcomes. An outcome is a complete assignment of actions $(a^1, ..., a^k)$, identified with the natural basis in $\mathbb{R}^d$, where $d := \prod_i |A^i|$. An act of player i is a function f^i from states S^i to lotteries on A^i. Let $\mathcal{H}^i$ be the set of feasible acts for player i. We assume that $\mathcal{H}^i$ is a compact set which is also convex with respect to both objective and subjective mixing, in the following sense.

Assumption (Convexity). For all $f^i, g^i \in \mathcal{H}^i$, and for all $a \in [0,1]$, there exist $h^i, l^i \in \mathcal{H}^i$ such that

(i) $p_s^{h^i} = a p_s^{f^i} + (1-a) p_s^{g^i}$ for all s (objective mixture)
(ii) $u(l^i) = a u(f^i) + (1-a) u(g^i)$ (subjective mixture).

A profile of acts is a unitary vector $f := (f^1, ..., f^k)$, and the utility of a profile f for player i is $u^i(f) = E_\pi[f'U^i f]$. An equilibrium is a profile f^* such that $u^i(f_i^*, f_{-i}^*) \geq u^i(f_i, f_{-i}^*)$ for all f^i.

Theorem 5. *Any finite game with PEU preferences has an equilibrium.*

Proof. First, let $b^i(f^{-i}) := \{f^i : [u^i(f^i, f^{-i}) \geq u^i(g^i, f^{-i})](\forall g^i \in \mathcal{H}^i)\}$. Next, let $\mathcal{H} := \times_i \mathcal{H}^i$, and let $b : \mathcal{H} \to \mathcal{H}$ be the best response correspondence, defined by $b(f) = \times_i b^i(f^i)$. Observe that b is a correspondence from a nonempty, convex, and compact set $\mathcal{H}$ to itself. In addition, b is a nonempty- and convex-valued, upper hemi-continuous correspondence. It follows that the conditions of Kakutani's fixed point theorem are satisfied, and hence the best response correspondence b has a fixed point: a profile of acts f^* such that $f^* \in b(f^*)$. The acts at this fixed point constitute an equilibrium since by construction $f^{*i} \in b^i(f^{*-i})$ for all i. ∎

11 Games with PEU Preferences: Examples

Consider the following game:

$Pl.1, Pl.2$	a^2	b^2
a^1	1.1,0	1,1.1
b^1	0,1.1	1.1,1

If the agents maximize vN-M expected utility, the unique mixed strategy equilibrium takes the following form: $(p,q) = (1/12, 1/12) \approx (0.08333, 0.08333)$, where p and q are the probabilities of playing a^1 and a^2, respectively. We shall call this equilibrium EU equilibrium.

If the players have the preferences we introduced to explain Allais paradox, what are the consequences on the equilibrium strategies of the players? In particular, does the equilibrium differ in the case of Allais agents? Consider the following game:

$Pl.1, Pl.2$	a^2	b^2
a^1	ω^1, ω^3	ω^2, ω^1
b^1	ω^3, ω^1	ω^1, ω^2

where the payoff matrix of both players is given by U in the Allais example. Observe that the payoff matrix U describes a decision-maker who is strictly uncertainty averse, and hence if all players have payoffs matrices given by U any equilibrium only involves pure acts. In this case, the unique equilibrium is given by $(p,q) = (0.17632, 0.17632)$.

Compared to the case of EU preferences, in the game with PEU preferences a^1 and a^2 are played more often. Starting from the EU equilibrium, observe that strategy a^2 becomes more attractive for P2. To re-establish equilibrium a^1 must be played more often, but then to keep P1 indifferent between her two strategies a^2 must also be played more often.

Next, consider the following very simple version of the Centipede game: P1 either quits (outcome; ω^2, ω^3) or passes, in which case P2 either quits (outcome; ω^3, ω^1), or passes (outcome; ω^1, ω^2). The payoff matrices of the two players are still given by U.

Let p denote the probability that P1 passes, and let q be the probability that P2 passes given that P1 passed. Then the PEU for P1 and P2 are given, respectively, by

$$((1-p)(1-q), p, (1-p)q)U((1-p)(1-q), p, (1-p)q)'$$
$$((1-p)q, (1-p)(1-q), p)U((1-p)q, (1-p)(1-q), p)'.$$

In a subgame-perfect equilibrium, player 2 chooses q so that it maximizes

$$(\sqrt{q}, \sqrt{(1-q)}, 0)U(\sqrt{q}, \sqrt{(1-q)}, 0)',$$

and hence will choose $q \approx 1$, in which case P1 will choose p to maximize

$$(0, \sqrt{p}, \sqrt{(1-p)})U(0, \sqrt{p}, \sqrt{(1-p)})',$$

and hence in equilibrium $p \approx 0.99029$. It follows that the unique subgame-perfect equilibrium in this game involves a small probability of continuation. Clearly, in a longer version of the centipede those continuation probabilities would be amplified.

References

Allais, M.: Le Comportement de l'Homme Rationnel devant le Risque: Critique des postulats et axiomes de l'École Americaine. Econometrica 21, 503–546 (1953)

Anscombe, F.J., Aumann, R.J.: A Definition of Subjective Probability. The Annals of Mathematical Statistics 34, 199–205 (1963)

Bernoulli, D.: Specimen theoriae novae de mensura sortis. Commentarii Academiae Scientiarum Imperialis Petropolitanae 5, 175–192 (1738); Translated as Exposition of a new theory of the measurement of risk. Econometrica 22, 123–136 (1954)

Chew, S.H., Epstein, L.G., Segal, U.: Mixture symmetry and quadratic utility. Econometrica 59, 139–164 (1991)

Deutsch, D.: Quantum Theory of Probability and Decisions. Proceedings of the Royal Society A455, 3129–3197 (1999)

Ellsberg, D.: Risk, Ambiguity and the Savage Axioms. Quarterly Journal of Economics 75, 643–669 (1961)

La Mura, P.: Projective expected utility. Journal of Mathematical Psychology (2009)

Machina, M.J.: Expected utility analysis without the independence axiom. Econometrica 50, 277–323 (1982)

von Neumann, J.: Mathematische Grundlagen der Quantenmechanic. Springer, Berlin (1932); Translated as Mathematical foundations of quantum mechanics. Princeton University Press, Princeton (1955)

von Neumann, J., Morgenstern, O.: Theory of Games and Economic Behavior. Princeton University Press, Princeton (1944)

Rabin, M., Thaler, R.H.: Anomalies: Risk Aversion. Journal of Economic Perspectives 15, 219–232 (2001)

Savage, L.J.: The Foundations of Statistics. Wiley, New York (1954)

Schmeidler, D.: Subjective probability and expected utility without additivity. Econometrica 57, 571–587 (1989)

Tversky, A.: A critique of expected utility theory: Descriptive and normative considerations. Erkenntnis 9, 163–173 (1975)

Ranking Methods Based on Dominance Measures Accounting for Imprecision

Alfonso Mateos, Antonio Jiménez, and José F. Blanco

Department of Artificial Intelligence, Technical University of Madrid (UPM),
28660 Boadilla del Monte, Madrid, Spain
{amateos,ajimenez}@fi.upm.es

Abstract. The additive multi-attribute utility model is widely used in multicriteria decision-making. However, it is often not easy to elicit precise values for the scaling weights representing the relative importance of criteria. In a group decision-making context a very widespread approach is to derive incomplete information, such as weight intervals or ordinal information rather than precise weights from a negotiation process. Different approaches have been proposed to deal with such situations. We advance two approaches based on dominance measures accounting for imprecise weights and compare them with other existing approaches using Monte Carlo simulation.

Keywords: Additive Multi-Attribute Utility Model, Imprecise Weights, Dominance Measure, Monte Carlo Simulation.

1 Introduction

In Multi-Attribute Utility Theory (MAUT) the functional form of the multi-attribute utility function differs subject to a variety of independence conditions, see [1]. The *additive model* is considered to be a valid approach in most practical situations for the reasons described in [2] and [3]. The functional form of this model is

$$u(A_l) = \sum_i w_i \, u_i(x_i^l),$$

where x_i^l is the performance of the attribute X_i for the alternative A_l, $u_i(x_i^l)$ is the utility associated with the above performance for $u_i(\cdot)$, the corresponding component utility function representing the decision maker's (DM) preferences over the possible attribute performances, and w_i are the weights representing the relative importance of each attribute. Note that $\sum_i w_i = 1$.

However, complex decision-making problems are usually plagued with uncertainty. Additionally, it is often not easy to elicit precise values for the scaling weights. They are often described within prescribed bounds or as just satisfying certain ordinal relations. Different authors refer to this situation as decision-making with *imprecise information*, with *incomplete information* or with *partial information*.

Several reasons are given in the literature to justify why a decision maker may wish to provide incomplete information ([4], [5]). Regarding weights, the DM may find it difficult to compare criteria or may not want to reveal his or her preferences in

F. Rossi and A. Tsoukis (Eds.): ADT 2009, LNAI 5783, pp. 328–339, 2009.

public. Moreover, the decision could be taken in a group decision-making situation, where incomplete information, such as weight intervals, is usually derived from a negotiation process ([6], [7]).

A lot of work on MAUT has dealt with incomplete information. Sage and White ([8]) proposed the model of imprecisely specified multi-attribute utility theory, where preference information about both weights and utilities is not assumed to be precise. Malakooti ([9]) suggested a new efficient algorithm for ranking alternatives when there exists incomplete information about the preferences and the value of the alternatives. This involves solving a single mathematical programming problem many times. Ahn ([10]) extends Malakooti's work.

Eum et al. ([11]) provided linear programming characterizations of dominance and potential optimality for decision alternatives when information about performances and/or weights is incomplete, extended the approach to hierarchical structures ([12]), and developed the concepts of potential weak potential optimality and strong potential optimality ([13]). More recently, Mateos et al. [14] considered the more general case where imprecision, described by means of fixed bounds, appears in alternative performances, as well as in weights and utilities.

Sarabando and Dias ([5]) give a brief overview of approaches proposed by different authors within the MAUT and MAVT (Multi-Attribute Value Theory) framework to deal with incomplete information.

A new approach is to use information about each alternative's intensity of dominance, known as *dominance measuring methods*. Ahn and Park ([15]) compute both dominating and dominated measures from a dominance matrix and then derive a *net dominance*. This is used as a measure of the strength of preference in the sense that a greater net value is better. They proposed and compared two alternative approaches with surrogate weighting methods and decision rules by means of a simulation study.

In this paper we propose to extend two dominance measuring methods proposed in [16]. The first one is based on dominating and dominated measures computed from the dominance matrix. They are combined into a net dominance value, but they are all computed differently than Ahn and Park's measures ([15]) to resolve deficiencies and improve in Ahn and Park's methods. In the second method, alternatives are ranked on the basis of a *dominance probability measure*. These dominance probability measures are based on the fact that the differences between utilities corresponding to alternatives A_k and A_j are always within the interval whose lower end-point is the element located at the kth row and jth column of the dominance matrix and whose upper end-point is located at the jth row and kth column of the dominance matrix with the sign changed. Moreover, alternative A_k dominates A_j for the positive values in the above interval, whereas it is dominated by A_j for the negative values. Both methods considered ordinal relations regarding attribute weights, i.e. DMs ranked attributes in descending order of importance.

In the proposed extensions we consider weight intervals rather than ordinal relations among attribute weights. Then, a simulation study is performed to compare the proposed methods with the measures reported in Ahn and Park ([15]) and with classical decision rules.

The paper is organized as follows. Section 2 introduces classical decision rules and Ahn and Park's methods, all of them compared with the proposed methodology in Section 4. Section 3 extends two dominance measuring methods for imprecise weights and compares them with other existing methods based on dominance measures. Section 4 evaluates and compares the methods in a simulation study and presents the study results. Finally, we outline our conclusions in section 5.

2 Ranking Methods

In this paper we consider a group decision-making problem with n attributes $(X_i, i=1,\ldots,n)$ and m alternatives $(A_j, j=1,\ldots,m)$, in which the DM preferences are represented by an additive multi-attribute utility function, with incomplete information about the weights. Specifically, the group of DMs provides a weight interval for each attribute. We denote it by

$$w \in W = \{w = (w_1,\ldots,w_n) \mid w_i \in [w_i^L, w_i^U], i=1,\ldots,n\},$$

where w_i^L and w_i^U are the lower and the upper end-points of the weight interval for the attribute X_i, $i=1,\ldots,n$.

On possible way of dealing with weight intervals described in the literature attempts to eliminate inferior alternatives based on the concept of *dominance*. Given two alternatives A_k and A_j, alternative A_k dominates A_j if $D_{kj} \geq 0$, D_{kj} being the optimum value of the optimization problem ([17]):

$$D_{kj} = min\{u(A_k) - u(A_l)\} = \sum_i w_i u_i(x_i^k) - \sum_i w_i u_i(x_i^l) \mid w \in W\}. \tag{1}$$

This can also be denoted by $D_{kj} = min\{wu_k - wu_j\} \mid w \in W\}$, where $u_k = (u_1(x_1^k),\ldots, u_n(x_n^k))$ and $u_j = (u_1(x_1^j),\ldots, u_n(x_n^j))$. This concept of dominance is called *pairwise dominance* and leads to the so-called *dominance matrix*:

$$D = \begin{pmatrix} - & D_{12} & \ldots & D_{1m-1} & D_{1m} \\ D_{21} & - & \ldots & D_{2m-1} & D_{2m} \\ D_{31} & D_{32} & \ldots & D_{3m-1} & D_{3m} \\ & & \ldots & & \\ D_{m1} & D_{m2} & \ldots & D_{mm-1} & - \end{pmatrix} \tag{2}$$

Another possibility is to use what is known as *absolute dominance* ([18]). Absolute dominance considers the following linear optimization problems:

$$U_k = max\{wu_k \mid w \in W\} \quad and \quad L_k = min\{wu_k \mid w \in W\}.$$

Alternative A_k absolutely dominates A_j if $L_k > U_j$, i.e. the lower bound of A_k exceeds the upper bound of A_j. Note that if A_k absolutely dominates A_j, then A_k dominates A_j, but the reverse does not hold.

Note that this dominance approach often results in almost no priorization of alternatives or too many non-dominated alternatives ([19]). However, pairwise and absolute dominance values can be used to further prioritize competitive alternatives, and hence recommend the best alternative and fully rank alternatives.

The following is an example of how these dominance values have been employed to modify three classical decision rules to operate in an imprecise decision-making context ([17], [18]):

- *maximax rule* or *optimist rule* (*OPT*): evaluating each alternative for its maximum guaranteed value, i.e. $max\{U_j, j=1,\ldots,m\}$.
- *maximin rule* or *pessimist rule* (*PES*): evaluating each alternative for its minimum guaranteed value, i.e. $max\{L_j, j=1,\ldots,m\}$.
- *minimax regret rule* (*REG*): evaluating each alternative for the maximum loss of value with respect to a better alternative, i.e. $min\{MR_k, k=1,\ldots,m\}$, where MR_k represents the maximum regret incurred when choosing alternative j, i.e. $MR_k = max\{max\{u(A_j) - u(A_k) \mid w \in W\} \; \forall j \neq k\}$.

Although none of these rules ensures that the best ranked alternative is the same as it would be if precise values were elicited for weights, simulations show that the selected alternative is generally one of the best ([20]).

A new approach is to use information about each alternative's intensity of dominance, known as dominance measuring methods. Ahn and Park [15] propose two approaches based on the dominance matrix D. In the first, denoted by *AP1*, alternatives are ranked according to a *dominating measure* $\alpha_k = \sum_{j=1, j \neq k}^{m} D_{kj}$. The higher this dominating measure is the more preferred the alternative will be, because the sum of the intensity of one alternative dominating the others will be also greater. In the second approach, denoted *AP2*, alternatives are ranked on the basis of the difference between the dominating measure α_k and a dominated measure $\beta_k = \sum_{j=1, j \neq k}^{m} D_{jk}$, i.e. on the basis of $\alpha_k - \beta_k$. A simulation study showed *AP1* to be better than *AP2*. Whereas *AP1* consists of just adding the paired dominance values in the kth row of D, *AP2* considers paired dominance values in both the kth row and the kth column of D. The reason why *AP1* is better than *AP2* is that *AP2* uses duplicated information (row and column values). On the other hand, *AP1* only takes into account the dominating measure, leading to a trade-off between positives and negatives. In the next section, we introduce two new methods aimed at overcoming these problems.

3 Dominance Measuring Extensions

The drawbacks associated with *AP1 and AP2* are that *AP1* only considers dominating measures (trade-off of positive and negative values), and *AP2* duplicates dominated measures. In this section, two new methods aimed at overcoming these problems are proposed. The first dominance measuring method (*DME1*) is based on the same idea as Ahn and Park suggested ([15]). First, we compute both *dominating* and *dominated measures* from the paired dominance values D_{kj} and then we derive a *net dominance*. This is used as a measure of the strength of preference in the sense that a greater net value is better. However, we compute the positives and negative dominating measures (step 2) and positive and negative dominated measures (step 4). They are used to compute first a proportion representing the strength of one alternative dominating the others (step 3) and second a proportion representing the intensity one alternative being dominated by the others (step 5). Finally, we subtract both proportions (step 6) to compute the intensity of the dominating over the dominated proportions. The ranking of alternatives will be based on this intensity value (step 7).

DME1 can be implemented in the following seven steps:

1. Obtain the paired dominance values D_{kj} as in (1) and the dominance matrix D as in (2).

2. Compute the *dominating measures* α_k, α_k^+ and α_k^- for each alternative A_k:
    $\alpha_k = \sum_{j=1, j \neq k}^{m} D_{kj}$, $\alpha_k^+ = \sum_{j=1, j \neq k, D_{kj} > 0}^{m} D_{kj}$ and $\alpha_k^- = \sum_{j=1, j \neq k, D_{kj} < 0}^{m} D_{kj}$, $\forall k$.

    In other words, α_k is computed by adding the paired dominance values in the kth row of D, whereas α_k^+ and α_k^- are computed in the same way considering positive and negative values, respectively, in the corresponding row only. Note that $\alpha_k = \alpha_k^+ + \alpha_k^-$.

3. Compute the proportion
    $$P_k^\alpha = \frac{\alpha_k^+}{\alpha_k^+ - \alpha_k^-}.$$

    Note that we can assume that $0 < \alpha_k^+ - \alpha_k^- \leq \infty$, avoiding division by 0, as demonstrated at the end of the algorithm. Note also that $0 \leq P_k^\alpha \leq 1$.

4. Compute the *dominated measures* β_k, β_k^+ and β_k^- for each alternative A_k:
    $\beta_k = \sum_{j=1, j \neq k}^{m} D_{jk}$, $\beta_k^+ = \sum_{j=1, j \neq k, D_{jk} > 0}^{m} D_{jk}$ and $\beta_k^- = \sum_{j=1, j \neq k, D_{jk} < 0}^{m} D_{jk}$, $\forall k$.

    β_k is computed by adding the paired dominance values in the kth column of D, whereas β_k^+ and β_k^- are computed in the same way considering the positive and negative values in the respective column only. Note that $\beta_k = \beta_k^+ + \beta_k^-$.

5. Compute the proportion
    $$P_k^\beta = \frac{\beta_k^+}{\beta_k^+ - \beta_k^-}.$$

    Note that, as in the case of P_k^α, we can assume that $0 < \beta_k^+ - \beta_k^- \leq \infty$, and $0 \leq P_k^\beta \leq 1$.

6. Calculate the *net dominance value* P_k for each alternative A_k:
    $$P_k = P_k^\alpha - P_k^\beta, \quad k = 1, \dots, m.$$

    Note that $-1 \leq P_k \leq 1$, where $-1 = P_k$ ($\Leftrightarrow P_k^\alpha = 0$ and $P_k^\beta = 1$) when all alternatives dominate the alternative A_k, and $P_k = 1$ ($\Leftrightarrow P_k^\alpha = 1$ and $P_k^\beta = 0$) when all alternatives are dominated by alternative A_k.

7. Rank alternatives according to the P_k values, where the best (rank 1) is the alternative for which P_k is a maximum and the worst (rank m) is the alternative for which P_k is the minimum.

Let us demonstrate that $\alpha_k^+ - \alpha_k^- > 0$, as stated in the third step of the above algorithm, because $\alpha_k^+ - \alpha_k^- = 0 \Leftrightarrow \alpha_k^+ = 0$ and $\alpha_k^- = 0 \Leftrightarrow D_{kj} = 0, \forall j \neq k \Leftrightarrow \sum_i w_i u_i(x_i^k) - \sum_i w_i u_i(x_i^j) \geq 0, \forall j \neq k \Leftrightarrow \sum_i w_i u_i(x_i^j) - \sum_i w_i u_i(x_i^k) \leq 0, \forall j \neq k \Leftrightarrow D_{jk} \leq 0, \forall j \neq k$.

Therefore, we have two possibilities:

a. $D_{kj} = 0$ and $D_{jk} < 0 \Rightarrow$ alternative A_k dominates A_j.

b. $D_{kj} = 0$ and $D_{jk} = 0 \Leftrightarrow w(u_k - u_j) = 0, \forall w \in W \Leftrightarrow u_k = u_j$.

This demonstrates that $D_{kj} = 0$ and $D_{jk} = 0, \Leftrightarrow u_i(x_i^k) = u_i(x_i^j), \forall i$, i.e. if utility functions $u_i(\cdot)$ are strictly monotone $\forall i$, then A_k and A_j are the same alternative, else ($u_i(\cdot)$) are not strictly monotone) both alternatives A_k and A_j are indifferent. In both cases, we can discard alternative A_j and keep A_k.

In conclusion, if we assume that there are no two alternatives A_k and A_j with $u_i(x_i^k) = u_i(x_i^j), \forall i$ (in this case, alternative A_j would be discarded because they are indifferent) and $\alpha_k^+ - \alpha_k^- = 0$, then alternative A_k dominates $A_j, \forall j$, i.e. alternative A_k is preferred.

The difference between the utilities corresponding to A_k and A_j, $w(u_k - u_j)$ with $w \in W$, is always within $[D_{kj}, -D_{jk}]$ as demonstrated below.

$D_{kj} = min\{wu_k - wu_j \mid w \in W\} = min\{w(u_k - u_j) \mid w \in W\} \leq \{w(u_k - u_j) \mid w \in W\}$, and $\{w(u_k - u_j) \mid w \in W\} \leq max\{w(u_k - u_j) \mid w \in W\} = -min\{-w(u_k - u_j) \mid w \in W\} = -min\{w(u_j - u_k) \mid w \in W\} = -D_{jk}$.

It has been already demonstrated that $w(u_k - u_j) \in [D_{kj}, -D_{jk}], \forall w \in W$. Thus,

- If $-D_{jk} \leq 0 \Leftrightarrow D_{kj} < 0$ and $D_{jk} \geq 0 \Leftrightarrow$ alternative A_j dominates $A_k \Rightarrow$ the probability of alternative A_k dominating A_j is 0.

- If $D_{kj} \geq 0 \Leftrightarrow D_{kj} \geq 0$ and $D_{jk} < 0 \Leftrightarrow$ alternative A_k dominates $A_j \Rightarrow$ the probability of alternative A_k dominating A_j is 1.

- If $D_{kj} < 0$ and $D_{jk} < 0$ then interval $[D_{kj}, -D_{jk}]$ will consist of a positive subinterval with positive values in which alternative A_k dominates A_j and a negative subinterval in which alternative A_j dominates A_k. Thus, the probability of A_k dominating A_j is the proportion of the positive subinterval over the whole $[D_{kj}, -D_{jk}]$.

On the basis of this idea, we now propose a second method, denoted *DME2*. In *DME2*, paired dominance values D_{kj} are first transformed into dominance probabilities DP_{kj} (step 2) depending on the dominance among alternatives A_k and A_j. Then a dominance probability measure (DPM_k) is derived for each alternative A_k (step 3) as the sum of the dominance probabilities of alternative A_k regarding the others alternatives. This is used as a measure of the strength of preference in the sense that a greater dominance probability measure is better.

DME2 can be implemented as the following four steps:

1. Compute dominance matrix D (2) from the paired dominance values D_{kj} (1).
2. If $D_{kj} \geq 0$, then alternative A_k dominates alternative A_j, i.e. the probability of A_k dominating A_j is 1, $DP_{kj} = 1$.
   Else ($D_{kj} < 0$):

- If $D_{jk} \geq 0$, then alternative A_j dominates alternative A_k, therefore, the probability of A_k dominating A_j is 0, i.e. $DP_{kj}=0$.
- Else note that alternative A_j is preferred to alternative A_k for those values in w that satisfy $D_{kj} \leq w(u_k - u_j) \leq 0$, and A_k is preferred to A_j for those values in w that satisfy $0 \leq w(u_k - u_j) \leq -D_{jk} \Rightarrow$ the probability of A_k dominating A_j is

$$DP_{kj} = \frac{-D_{jk}}{-D_{jk}-D_{kj}}.$$

3. Compute a dominance probability measure (DPM) for each alternative A_k

$$DPM_k = \sum_{j=1, j \neq k}^{m} DP_{kj}.$$

Rank alternatives according to the DPM values, where the best (rank 1) is the alternative with greatest DPM and the worst is the alternative with the least DPM.

4 Computational Study

Having described the proposed dominance measuring extensions (DME1 and DME2), let us compare these methods with Ahn and Park's approach ([15]) and with decision rules modified to operate in an imprecise decision context.

We propose to carry out a simulation study of the above methods to analyze their performance. For a decision-making problem with m alternatives and n attributes, the process would be as follows:

1. Randomly generate component utilities for each alternative in each attribute from a uniform distribution in (0,1), leading to an $m \times n$ matrix. Normalize the columns in this matrix to make the smallest value 0 and the largest 1, and remove dominated alternatives.
2. Generate attribute weights representing their relative importance. Note that these weights are the TRUE weights and the derived ranking of alternatives will be denoted as the TRUE ranking. To generate the TRUE weights, we first select n-1 independent random numbers from a uniform distribution on (0,1), and then rank these numbers. Suppose the ranked numbers are $1 \geq r_{n-1} \geq ..., \geq r_2 \geq r_1 > 0$. The differences between adjacently ranked numbers are then used as the desired weights: $w_n^T = 1 - r_{n-1}$, $w_{n-1}^T = r_{n-1} - r_{n-2}$, ..., $w_1^T = r_1$. The resulting weights will sum 1 and be uniformly distributed in the weight space.
3. To derive the corresponding weight intervals, add and subtract the same quantity to precise values, leading to the lower and upper end-points of the weight intervals. We used the quantities, q, of 0.025/2, 0.05/2, 0.075/2 and 0.1/2 that represent 2.5%, 5%, 7.5% and 10% imprecision, respectively. In other words, $[w_i^L, w_i^U] = [w_i^T-q, w_i^T+q]$. If $w_i^T-q<0$ then $w_i^T-q=0$ and if $w_i^T+q>1$ then $w_i^T+q=1$ is considered. Throughout the simulation process weights will be randomly generated from these weight intervals, $[w_i^T-q, w_i^T+q]$.

4. Compute the ranking of alternatives for each method according to their procedures and compare with the TRUE ranking, computed in step 2. We use two measures of efficacy, *hit ratio* and *rank-order correlation* ([15], [21]). The hit ratio is the proportion of all cases in which the method selects the same best alternative as in the TRUE ranking. Rank-order correlation represents how similar the overall structures ranking alternatives are in the TRUE ranking and in the ranking derived from the method. It is calculated using Kendall's τ ([22]): $\tau = 1 - 2 \times$ (number of pairwise preference violations) / (total number of pair preferences).

Following ([15], [16]), four different levels of alternatives ($m = 3,5,7,10$) and five different levels of attributes ($n = 3,5,7,10,15$) were considered in order to validate the results output. Also, 10 replications of 10,000 trials were performed for each of the 20 design elements (alternatives × attributes).

Table 1 exhibits the average hit ratio for each of the 20 design elements when the interval length is 0.025, i.e. the average values of 10 replications of 10,000 trials, whereas the last row in this table is the mean of each column. The highest average hit values in each design element are highlighted in bold.

Looking at the modified decision rules in Table 1, the *REG* method appears to be better than the *PES* method. *PES* outperforms the *OPT* method. For the dominance measuring methods, the mean value for the *DME1* method is the greatest (0.886), its hit ratio being the highest for 12 out of the 20 design elements. The *DME1* method is

Table 1. Average hit ratios

Alternatives - Attributes		Modified decision rules			Dominance measuring		
		OPT	PES	REG	AP1	AP2	DME1	DME2
3	3	0.972	0.97	0.975	0.974	0.975	**0.975**	0.975
	5	0.942	**0.955**	0.952	0.954	0.952	0.954	0.952
	7	0.916	0.921	0.923	0.923	0.923	**0.926**	0.923
	10	0.854	0.855	0.859	0.86	0.86	**0.874**	0.859
	15	0.842	0.844	0.846	0.846	0.846	**0.854**	0.846
5	3	0.961	0.964	0.968	0.967	0.968	0.967	**0.968**
	5	0.933	0.935	0.94	0.938	**0.94**	0.939	0.939
	7	0.926	0.932	0.934	0.933	0.933	**0.934**	0.934
	10	0.844	0.85	0.856	0.856	0.857	**0.857**	0.856
	15	0.769	0.772	0.776	0.776	0.776	**0.776**	0.776
7	3	0.945	0.945	0.953	0.95	0.953	**0.953**	0.953
	5	0.938	0.944	0.944	0.943	**0.945**	0.944	0.944
	7	0.901	0.901	0.908	0.909	**0.909**	0.908	0.908
	10	0.821	0.829	0.836	0.837	0.837	**0.837**	0.836
	15	0.717	0.72	0.727	0.727	0.727	**0.727**	0.727
10	3	0.934	0.935	0.947	0.944	0.947	**0.947**	0.947
	5	0.908	0.908	0.915	0.913	**0.917**	0.916	0.916
	7	0.912	0.921	0.922	0.923	0.923	0.922	**0.923**
	10	0.8	0.809	0.817	0.813	0.818	0.816	**0.818**
	15	0.691	0696	0.702	0.702	0.702	**0.702**	0.702
Mean		0.876	0.88	0.885	0.884	0.885	**0.886**	0.885

followed by *AP2* and *DME2*, which are better than *AP1*. On the whole, *DME1* is superior to the others in terms of the hit ratio. Thus, the proposed *DME1* method performs better than the dominance measuring methods in [15].

The average hit ratio decreases the more attributes there are for all the methods under consideration and any given number of alternatives. This decrease is more pronounced than if the number of alternatives is increased for the same number of attributes.

Again, if we consider rank-order correlation, see Table 2, the extensions proposed in this paper (*DME1* and *DME2*), perform better than the approaches suggested by Ahn and Park and the modified decision rules. The highest rank-order correlation values in each design element are highlighted in bold.

Again, the rank-order correlation decreases the more attributes there are for any given number of alternatives. However, the number of alternatives does not affect the correlation.

Table 2. Rank-order correlation (Kendall's τ)

Alternatives - Attributes		Modified decision rules			Dominance measuring		
		OPT	PES	REG	AP1	AP2	DME1	DME2
3	3	0.963	0.96	0.961	0.964	0.966	**0.967**	0.966
	5	0.938	0.949	0.946	0.949	0.947	**0.949**	0.947
	7	0.894	0.896	0.894	0.899	0.9	**0.904**	0.9
	10	0.808	0.809	0.81	0.816	0.816	**0.836**	0.816
	15	0.771	0.771	0.772	0.775	0.775	**0.787**	0.775
5	3	0.96	0.959	0.956	0.963	0.965	**0.965**	0.965
	5	0.941	0.94	0.936	0.943	0.945	**0.945**	0.945
	7	0.91	0.912	0.91	0.915	0.916	**0.916**	0.916
	10	0.86	0.859	0.861	0.866	0.868	**0.868**	0.867
	15	0.786	0.786	0.788	0.792	0.792	**0.792**	0.792
7	3	0.961	0.961	0.961	0.964	0.966	0.966	0.966
	5	0.938	0.939	0.936	0.942	0.943	**0.943**	0.943
	7	0.908	0.907	0.905	0.912	0.914	0.913	**0.914**
	10	0.847	0.845	0.843	0.854	0.856	**0.856**	0.856
	15	0.775	0.776	0.779	0.782	0.782	**0.782**	0.782
10	3	0.957	0.959	0.958	0.962	0.963	**0.963**	0.963
	5	0.93	0.93	0.928	0.934	0.936	**0.936**	0.936
	7	0.915	0.916	0.914	0.919	0.921	**0.921**	0.921
	10	0.864	0.864	0.863	0.871	**0.874**	0.873	0.873
	15	0.79	0.79	0.792	0.795	0.796	**0.796**	0.796
		0.885	0.886	0.885	0.89	0.892	**0.894**	0.892

The simulation was also carried out for the other percentages of imprecision (5%, 7.5% and 10%), as pointed out before, and the results were similar, i.e. the proposed extensions (*DME1* and *DME2*) performed better than the other approaches.

Figure 1 and Figure 2 show the superiority of the *DME1* for the different percentages of imprecision under consideration. Moreover, the greater the imprecision the better *DME1* performed compared with the other methods under consideration. The worst performances are for modified decision rules, specifically for the *maximax* or *optimist rule* (*OPT*).

These results match up with outcomes reported in [16], where the same computational study was carried out, considering in that case the ordinal relations among attribute weights.

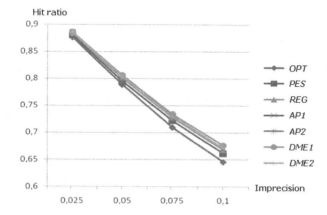

Fig. 1. Hit ratio means for the different percentages of imprecision

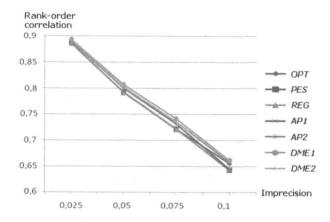

Fig. 2. Rank-order correlation for the different percentages of imprecision

5 Conclusions

In real complex decision-making problems it is not easy to elicit precise values for the weights representing the relative importance of criteria, which are often described within prescribed bounds or as just satisfying certain ordinal relations.

Two possible approaches for dealing with weight intervals are to rank alternatives based on the concept of dominance (*dominance measuring methods*) and so-called *decision rules*.

In this paper we have proposed the extension of two dominance measuring methods, *DME1* and *DME2*. The first one is based on dominating and dominated measures, which are combined into a *net dominance value* as a measure of the strength of preference. The second ranks alternatives on the basis of a *dominance probability measure*. Both extensions consider weight intervals rather than ordinal relations among attribute weights.

A simulation study was performed to compare the proposed extensions with the method suggested by Ahn and Park and with modified decision rules. The results show that *DME1* outperforms the other methods in terms of the identification of the best alternative and the overall ranking of alternatives. Different situations of imprecision were analyzed (2.5%, 5%, 7.5% and 10% imprecision), leading to the same conclusion. Furthermore, the greater the imprecision, the better *DME1* will perform compared with the other methods.

Acknowledgments. The paper was supported by Madrid Regional Government project S-0505/TIC/0230 and the Spanish Ministry of Education and Science project TIN 2008-06796-C04-02.

References

1. Keeney, R.L., Raiffa, H.: Decision with Multiple Objectives: Preferences and Value-Tradeoffs. Wiley, New York (1976)
2. Raiffa, H.: The Art and Science of Negotiation. Harvard University Press, Cambridge (1982)
3. Stewart, T.J.: Robustness of Additive Value Function Method in MCDM. Journal of Multi-Criteria Decision Analysis 5, 301–309 (1996)
4. Weber, M.: Decision Making with Incomplete Information. European Journal of Operational Research 28, 44–57 (1987)
5. Sarabando, P., Dias, L.C.: Simple Procedures of Choice in Multicriteria Problems without Precise Information about the Alternatives Values', C & OR (revision), INESC-Coimbra, Technical report, pp. 1–24 (2009)
6. Jiménez, A., Mateos, A., Ríos-Insua, S.: Monte Carlo Simulation Techniques in a Decision Support System for Group Decision-Making. Group Decision and Negotiation 14(2), 109–130 (2005)
7. Mateos, A., Jiménez, A., Ríos-Insua, S.: Monte Carlo Simulation Techniques for Group Decision-Making with Incomplete Information. European Journal of Operations Research 174(3), 1842–1864 (2006)
8. Sage, A., White, C.C.: Ariadne: a Knowledge-Based Interactive System for Planning and Decision Support. IEEE Transactions on Systems, Management and Cybernetics, Part A: Systems and Humans 14, 35–47 (1984)
9. Malakooti, B.: Ranking and Screening Multiple Criteria Alternatives with Partial Information and Use of Ordinal and Cardinal Strength of Preferences. IEEE Transactions on Systems, Management and Cybernetics: Part A 30(3), 787–801 (2000)
10. Ahn, B.S.: Extending Malakooti's Model for Ranking Multicriteria Alternatives with Preference Strength and Partial Information. IEEE Transactions on Systems, Management and Cybernetics, Part A: Systems and Humans 33(3), 281–287 (2003)

11. Eum, Y., Park, K.S., Kim, H.: Establishing Dominance and Potential Optimality in Multi-criteria Analysis with Imprecise Weights and Values. Computers and Operations Research 28(5), 397–409 (2001)

12. Lee, K., Park, K.S., Kim, H.: Dominance, Potential Optimality, Imprecise Information, and Hierarchical Structure in Multi-criteria Analysis'. Computers and Operations Research 29, 1267–1281 (2002)

13. Park, K.: Mathematical Programming Models for Characterizing Dominance and Potential Optimality when Multicriteria Alternative Values and Weights are Simultaneously Incomplete. IEEE Transactions on Systems, Management and Cybernetics, Part A: Systems and Humans 34, 601–614 (2004)

14. Mateos, A., Ríos-Insua, S., Jiménez, A.: Dominance, Potential Optimality and Alternative Ranking in Imprecise Decision Making. Journal of Operational Research Society 58(3), 326–336 (2007)

15. Ahn, B.S., Park, K.S.: Comparing Methods for Multiattribute Decision Making with Ordinal Weights'. Computers and Operations Research 35, 1660–1670 (2008)

16. Mateos, A., Jiménez, A., Blanco, J.F.: A MCDM Ranking Method Based on a Dominance Measure: Computational Study. Group Decision and Negotiation (2009) (in revision)

17. Puerto, J., Marmol, A.M., Monroy, L., Fernández, F.R.: Decision Criteria with Partial Information. International Transactions in Operational Research 7, 51–65 (2000)

18. Salo, A., Hämäläinen, R.P.: Preference Ratio in Multiattribute Evaluation (PRIME) - Elicitation and Decision Procedures under Incomplete Information. IEEE Transactions on Systems, Management and Cybernetics: Part A 31(6), 533–545 (2001)

19. Kirkwood, C.W., Corner, J.L.: The Effectiveness of Partial Information about Attribute Weights for Ranking Alternatives in Multiattribute Decision Making. Organization Behavior and Human Decision Processes 54, 456–476 (1993)

20. Sarabando, P., Dias, L.C.: Multi-attribute Choice with Ordinal Information: a Comparison of Different Decision Rules. IEEE Transactions on Systems, Management and Cybernetics, Part A (to appear, 2009)

21. Barron, F., Barrett, B.: Decision Quality Using Ranked Attribute Weights'. Management Science 42(11), 1515–1523 (1996)

22. Winkler, R.L., Hays, W.L.: Statistics: Probability, Inference and Decision. Holt, Rinehart & Winston, New York (1985)

Optimizing the Hurwicz Criterion
in Decision Trees with Imprecise Probabilities

Gildas Jeantet and Olivier Spanjaard

LIP6 - UPMC
104 avenue du Président Kennedy 75016 Paris, France
{gildas.jeantet,olivier.spanjaard}@lip6.fr

Abstract. This paper is devoted to sequential decision problems with imprecise probabilities. We study the problem of determining an optimal strategy according to the Hurwicz criterion in decision trees. More precisely, we investigate this problem from the computational viewpoint. When the decision tree is *separable* (to be defined in the paper), we provide an operational approach to compute an optimal strategy, based on a bicriteria dynamic programming procedure. The results of numerical tests are presented. When the decision tree is *non-separable*, we prove the NP-hardness of the problem.

Keywords: Sequential decision making, Imprecise probabilities, Hurwicz's criterion, Computational complexity, Exact algorithms.

1   Introduction

Decision under uncertainty is one of the main field of research in decision theory, due to its numerous applications (e.g. medical diagnosis, robot control, strategic decision, games...). Decision under uncertainty means that the consequences of a decision depends on uncertain events. In decision under risk, it is customary to assume that a precise probability is known for each event appearing in the decision problem. A decision can thus be characterized by a *lottery* over possible consequences. A popular criterion to compare lotteries (and therefore decisions) is the expected utility (EU) model proposed by von Neumann and Morgenstern [11]. In this model, a *utility function u* (specific to each decision maker) assigns a numerical value to every outcome. The evaluation of a lottery is then performed via the computation of its utility expectation (the greater the better). However, when several experts have divergent viewpoints or when empirical data are missing, it is not obvious to elicit sharp numerical probabilities for each event [1,2,12]. A natural way to take into account this difficulty is to use intervals of probabilities rather than scalar probabilities. This is known as decision making under imprecise probabilities.

Comparing decisions amounts then to comparing imprecise lotteries, i.e. lotteries where several possible probability distributions are taken into account.

F. Rossi and A. Tsoukis (Eds.): ADT 2009, LNAI 5783, pp. 340–352, 2009.

A pessimistic agent will make the decision that maximizes the worst possible expected utility. This is known as the Γ-*maximin* decision criterion. Conversely, an optimistic agent will make the decision that maximizes the best possible expected utility. This is known as the Γ-*maximax* decision criterion. Including these two extremes, Jaffray and Jeleva recently proposed to use the Hurwicz criterion, that enables to model intermediate attitudes by performing a linear combination of both previous criteria [4]. Note that Hurwicz introduced this criterion in the context of decision under complete ignorance (i.e., when absolutely no information is known about the probabilities), but the authors preserved its denomination of "Hurwicz's criterion" since it extends naturally to the case of imprecise probabilities.

To our knowledge, the algorithmic issues related to the use of Hurwicz's criterion in a sequential decision problem with imprecise probabilities have not been studied until now. It is indeed frequent to encounter *sequential decision problems* where one does not make a simple decision but one follows a *strategy* (i.e. a sequence of decisions conditioned by events) resulting in a non deterministic outcome. Several representation formalisms can be used for sequential decision problems, such as decision trees (e.g., [9]), influence diagrams (e.g., [10]) or Markov decision processes (e.g., [8]). A decision tree is an explicit representation of a sequential decision problem, while influence diagrams or Markov decision processes are compact representations and make it possible to deal with decision problems of greater size. It is important to note that, in all these formalisms, the set of potential strategies is combinatorial (i.e., its size increases exponentially with the size of the instance). The computation of an optimal strategy for a given representation and a given decision criterion is then an algorithmic issue in itself. It is well-know that an optimal strategy for EU in a decision tree endowed with scalar probabilities can be determined in linear time by backward induction. This is no more the case when dealing with imprecise probabilities and Hurwicz's criterion. In the particular case of Γ-maximin and Γ-maximax criteria, Kikuti *et al.* [2] have presented algorithms that employ *dynamic feasibility*, that is, one declares infeasible any strategy that includes a suboptimal substrategy (a substrategy is a strategy in a subtree). In the present paper, on the contrary, we consider that all strategies are feasible (i.e., even the ones that include a suboptimal substrategy), and we study the computational complexity of determining an optimal strategy according to Hurwicz's criterion in a decision tree endowed with imprecise probabilities. Furthermore, we propose algorithmic procedures to tackle the problem.

The remainder of the paper is organized as follows. We first give some preliminaries on imprecise probabilities and decision criteria used in such a setting (Section 2). Then, we present the difficulties raised by the use of imprecise probabilities in sequential decision problems, and we distinguish a separable case and a non-separable case (Section 3). The next two sections are devoted to the description of our results in these two cases (Section 4 and 5). Finally, we conclude by giving some avenues for future research (Section 6).

2   Single Stage Decision Making with Imprecise Probabilities

Several mathematical models of imprecise probabilities have been proposed in the literature [12,13]. A common point between these models is that they often define a probability interval $[P^-(E), P^+(E)]$ for each event E. Following Jaffray and Jeleva [4], we assume that there exists a *real* probability P_0 such that $P_0(E) \in [P^-(E), P^+(E)]$ for all events E. To compare imprecise lotteries (i.e., lotteries with imprecise probabilities), one must therefore consider a set $\mathcal{P}$ of possible probability distributions. This is close to the approach adopted to compare feasible solutions in discrete optimization with interval data [5], with the difference that the set of possible probability distributions is not the cartesian product of the probability intervals of the events. A probability distribution should indeed satisfy the Kolmogorov axioms ($P(E) \geq 0$, $P(\Omega) = 1$, $P(E_1 \cup E_2 \cup \ldots) = P(E_1) + P(E_2) + \ldots$ for pairwise disjoint events E_i).

Let us present popular decision criteria in such a setting. For instance, consider two lotteries f, g involving three pairwise disjoint events E_1, E_2, E_3. If E_1 (resp. E_2, E_3) occurs, f yields -50 (resp. 0,100). If E_1 (resp. E_2, E_3) occurs, g yields 130(resp. -30,-50). In the EU model with sharp probabilities, a lottery is evaluated by its expected utility, namely $E(f) = P(E_1)u(-50) + P(E_2)u(0) + P(E_3)u(100)$ for f. Assume now that probabilities are imprecise, e.g. $P_0(E_1) \in [0.2, 0.4]$, $P_0(E_2) \in [0.4, 0.6]$ and $P_0(E_3) \in [0.2, 0.3]$. The set $\mathcal{P}$ of possible probability distributions is therefore defined by $\mathcal{P} = \{P : P(E_i) \in [P^-(E_i), P^+(E_i)] \ \forall i, \text{ and } \sum_i P(E_i) = 1\}$. If the decision maker wants to hedge against the worst possible expected utility, a lottery f is evaluated by $\underline{E}(f) = \min\{E(f, P) : P \in \mathcal{P}\}$ where $E(f, P)$ denotes the expected utility of lottery f according to probability P. This is the so-called Γ-*maximin* decision criterion. The value of the Γ-maximin criterion can be computed by using the following simple result:

Proposition 1. *Consider a lottery f yielding utility u_i if event E_i occurs ($i = 1, \ldots, n$), with $u_1 \leq \ldots \leq u_n$ and $P(E_i) \in [P^-(E_i), P^+(E_i)]$. The probability distribution $\underline{P}_f$ in $\mathcal{P}$ recursively defined by*

$$\begin{cases} \underline{P}_f(E_1) = \min\{1 - \sum_{j=2}^n P^-(E_j), P^+(E_j)\} \\ \underline{P}_f(E_i) = \min\{1 - \sum_{j=1}^{i-1} \underline{P}_f(E_j) - \sum_{j=i+1}^n P^-(E_j), P^+(E_i)\} \quad \forall i \end{cases}$$

yields expected utility $\underline{E}(f)$.

Proof. Consider a probability distribution $P \neq \underline{P}_f$. Let us show that $E(f, \underline{P}_f) \leq E(f, P)$. We denote by i_0 the index such that $P(E_i) = \underline{P}_f(E_i)$ for $i < i_0$ and $P(E_{i_0}) < \underline{P}_f(E_{i_0})$ ($P(E_{i_0}) > \underline{P}_f(E_{i_0})$ is impossible). One should have $\sum_{i=i_0}^n P(E_i) = 1 - \sum_{i=1}^{i_0-1} \underline{P}_f(E_i)$. Consequently, $P(E_{i_0}) < \underline{P}_f(E_{i_0})$ implies that $P(E_i) > P^-(E_i)$ for some $i > i_0$. Let us set $i_1 = \min\{i : i > i_0 \text{ and } P(E_i) > P^-(E_i)\}$ and $\varepsilon = \min\{\underline{P}_f(E_{i_0}) - P(E_{i_0}), P(E_{i_1}) - P^-(E_{i_1})\} > 0$. We denote by P_1 the probability distribution defined by $P_1(E_{i_0}) = P(E_{i_0}) + \varepsilon$, $P_1(E_{i_1}) = P(E_{i_1}) - \varepsilon$ and $P_1(E_i) = P(E_i)$ for $i \neq i_0, i_1$. We have $E(f, P_1) \leq E(f, P)$ since

$E(f, P_1) - E(f, P) = \varepsilon(u_{i_0} - u_{i_1}) \le 0$. If $P_1 \ne \underline{P}_f$, by the same reasoning one can construct a probability distribution P_2 such that $E(f, P_2) \le E(f, P_1)$. In this way, one generates a sequence $P_1, \ldots, P_k$ of probability distributions such that $E(f, P_{i+1}) \le E(f, P_i)$ and $P_k = \underline{P}_f$. Therefore $E(f, \underline{P}_f) \le E(f, P)$.   ∎

For instance, let us come back to lotteries f, g previously mentioned. We have $\underline{P}_f(E_1) = \min\{1 - 0.4 - 0.2, 0.4\} = 0.4$, $\underline{P}_f(E_2) = \min\{1 - 0.4 - 0.2, 0.6\} = 0.4$ and $\underline{P}_f(E_3) = \min\{1 - 0.4 - 0.4, 0.3\} = 0.2$. Consequently, for $u(x) = x$, we have $\underline{E}(f) = 0.4 \times (-50) + 0.2 \times 100 = 0$. Similarly, one computes $\underline{P}_g(E_1) = 0.2$, $\underline{P}_g(E_2) = 0.5$, $\underline{P}_g(E_3) = 0.3$ and $\underline{E}(g) = -4$. Therefore lottery f is preferred to g for the Γ-maximin criterion.

Conversely, if the decision maker wants to maximize the best possible expected utility, a lottery f is evaluated by $\bar{E}(f) = \max\{E(f, P) : P \in \mathcal{P}\}$. This is the so-called Γ-maximax decision criterion. The probability distribution $\bar{P}_f$ yielding $\bar{E}(f)$ is defined by:

$$\begin{cases} \bar{P}_f(E_1) = \max\{1 - \sum_{j=2}^n P^+(E_j), P^-(E_j)\} \\ \bar{P}_f(E_i) = \max\{1 - \sum_{j=1}^{i-1} \bar{P}_f(E_j) - \sum_{j=i+1}^n P^+(E_j), P^-(E_i)\} \quad \forall i \end{cases}$$

Coming back again to lotteries f, g previously mentioned, we have $\bar{P}_f(E_1) = 0.2$, $\bar{P}_f(E_2) = 0.5$, $\bar{P}_f(E_3) = 0.3$, $\bar{E}(f) = 20$ on the one hand, and $\bar{P}_g(E_1) = 0.4$, $\bar{P}_g(E_2) = 0.4$, $\bar{P}_g(E_3) = 0.2$, $\bar{E}(g) = 30$ on the other hand. Therefore lottery g is preferred to f for the Γ-maximax criterion. This shows that the preferences are of course very dependent on the degree of pessimism of the decision maker.

For this reason, Jaffray and Jeleva [4] propose to extend the Hurwicz criterion for decision under complete ignorance to the case of imprecise probabilities. According to the Hurwicz criterion, a lottery f is evaluated by $\alpha \underline{E}(f) + (1 - \alpha)\bar{E}(f)$. In other words, the decision maker will look at the worse and best possible expected utilities and, according to its degree of pessimism, will put more or less weight on the former or the later. It reduces to Γ-maximin for $\alpha = 1$, and to Γ-maximax for $\alpha = 0$. When comparing lotteries f, g previously mentioned according to the Hurwicz criterion, we have f preferred to g for $\alpha > 5/7$, and g preferred to f for $\alpha < 5/7$. Note that the Hurwicz criterion is compatible with dominance, i.e. if a lottery has a greater expected utility than another one for all possible probability distributions, then its evaluation will be better [4]. This property is indeed desirable to guarantee a rational behavior.

3   Multistage Decision Making with Imprecise Probabilities

In multistage decision making, one studies problems where one has to take a sequence of decisions conditionally to events. The formalism of decision trees provides a simple and explicit representation of a sequential decision problem under risk. It is a tree with three kinds of nodes: decision nodes (represented by squares), chance nodes (represented by circles) and utility nodes (leaves of the tree). A decision node (resp. chance node) can be seen as a decision variable (resp. random variable), the domain of which corresponds to the labels of the branches starting from that node. When probabilities are imprecise, the sharp probability that a given random variable takes a given value is unknown: one only knows an interval of probabilities in which it is included. The values indicated at the leaves correspond to the *utilities* of the consequences. For the sake of illustration, we now give an example of a well-kown multistage decision problem, and its representation with a decision tree. Note that one omits the orientation of the edges when representing decision trees.

Example 1 (oil wildcatter's problem [9]). *An oil wildcatter has to decide whether to drill or not at a given site. For that purpose, he first has to decide whether to sound or not the geological structure of the site (decision D_1), which costs 10000$ and gives a better estimation of the quantity of oil to be found. The result of the sounding can be seen as a random variable T that can take three possible values: no if there is no hope of oil, open if some oil is expected, or closed if much oil is expected. Next, he decides whether to drill or not (decision D_2), which costs 70000$. Finally, if he decides to drill, the result of the drilling can be seen as a random variable S that can take three possible values: the hole is dry (the outcome is 0$), wet (120000$) or soaking (270000$). This problem can be represented by the decision tree on the left side of Figure 1. Note that decision D_2 is duplicated in several nodes (nodes D_2^1, D_2^2, D_2^3 and D_2^4) since it can be taken in several different contexts (a sounding has been performed or not, the result of the sounding is encouraging or not...).*

When sharp probabilities are known, each branch starting from a chance node representing random variable X is endowed with probability $P(X = x|past(X))$, where $past(X)$ denotes all the value assignments to random and decision variables on the path from the root to X. Furthermore, in this paper, we assume that $P(X = x|past(X))$ only depends on the random variables in $past(X)$. For instance, in the decision tree for the oil wildcatter problem, $P(S = soak|D_1 = sounding, T = no) = P(S = soak|T = no)$. When probabilities are imprecise, we assume that a conditional probability table is indicated for each chance node in the decision tree. In each cell of the table, an interval of probabilities is given. For the oil wildcatter problem, the conditional probability tables are presented besides the decision tree in Figure 1. So as to have complete conditional probability tables, we make an assumption of symmetry: the structures of subtrees of a same chance node are identical. Note that this assumption does not imply symmetric decision trees (as those obtained by unfolding an influence diagram [3]). For instance, the decision tree in Figure 1 is not symmetric but the condition holds: the three subtrees of node T have the same structure (the subtrees of nodes S are all leaves).

A *strategy* consists in setting a value to every decision variable conditionally to its past. The decision tree in Figure 1 includes 10 feasible strategies, among which for instance strategy $s = (D_1 = sounding, D_2^2 = not\,drill, D_2^3 = drill, D_2^4 = drill)$ (note that node D_2^1 cannot be reached when $D_1 = sounding$). In our setting, a strategy can be associated to a compound lottery over the utilities, where the probabilities of the involved events are imprecise. For instance,

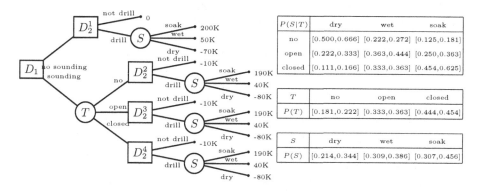

Fig. 1. Decision tree for the oil wildcatter problem

strategy s corresponds to the compound lottery yielding $-10K$ if $T = no$, $190K$ (resp. $40K, -80K$) if $T = open$ or $T = closed$ and then $S = soak$ (resp. wet, dry). Comparing strategies amounts therefore to compare compound lotteries. Given a decision tree T, the evaluation of a strategy (more precisely, of the corresponding compound lottery) according to the Hurwicz criterion depends on the set $\mathcal{P}_T$ of possible probability distributions *on decision tree T* (i.e., the set of assignments of sharp probabilities to the tables coming with T). This evaluation is a combinatorial problem in itself due to the combinatorial nature of $\mathcal{P}_T$. We distinguish two cases:

Non-separable decision trees. We say that a decision tree T is *non-separable* when $\mathcal{P}_T$ is a subset of the cartesian product of possible probability distributions at each chance node. In other words, the fact that the probabilities sum up to 1 at each chance node is not sufficient to ensure the global consistency of the probability distribution on the decision tree. This is the case for the decision tree of Figure 1. Consider for instance the following partial probability distribution on the tree: $P(S = dry|T = no) = 0.55$, $P(S = dry|T = open) = 0.33$, $P(S = dry|T = closed) = 0.12$, $P(T = no) = 0.20$, $P(T = open) = 0.35$, $P(T = closed) = 0.45$, $P(S = dry) = 0.22$. This partial probability distribution can be completed so that the probabilities sum up to 1 at each chance node, but is globally inconsistent since the total probability theorem does not hold: $P(S = dry|T = no)P(T = no) + P(S = dry|T = open)P(T = open) + P(S = dry|T = closed)P(T = closed) = 0.2795 \neq 0.22 = P(S = dry)$.

Separable decision trees. We say that a decision tree T is *separable* when $\mathcal{P}_T$ is equal to the cartesian product of possible probability distributions at each chance node. In other words, the only requirement to ensure that a probability distribution is globally consistent is that the probabilities sum up to 1 at each chance node. This is for instance the case for the decision tree of Figure 2 as soon as random variables A, B, C, D, E are mutually independent.

Solving a decision tree means finding an optimal strategy according to a given decision criterion (here, Hurwicz and its particular cases). Note that the number of potential strategies grows exponentially with the size of the decision tree, i.e. the number of decision nodes (this number has indeed the same order of magnitude as the number of nodes in $\mathcal{T}$). Indeed, one easily shows that there are $\Theta(2^{\sqrt{n}})$ strategies in a complete binary decision tree $\mathcal{T}$, where n denotes the number of decision nodes. This prohibitive number of potential strategies makes it impossible to resort to an exhaustive enumeration of the strategies when the size of the decision tree increases. For this reason, it is necessary to develop an optimization algorithm to determine the optimal strategy. It is well-known that the rolling back method makes it possible to compute in linear time an optimal strategy w.r.t. EU. Indeed, such a strategy satisfies the optimality principle: any substrategy of an optimal strategy is itself optimal. Starting from the leaves, one computes recursively for each node the expected utility of an optimal substrategy: the optimal expected utility for a chance node equals the expectation of the optimal utilities of its successors; the optimal expected utility for a decision node equals the maximum expected utility of its successors. This is however more difficult to optimize the Hurwicz criterion in decision trees with imprecise probabilities. In Section 5, we will show that this is actually an NP-hard problem in non-separable decision trees. Before that, in the next section, we will study the case of separable decision trees.

4   Optimizing the Hurwicz Criterion in Separable Decision Trees

When trying to optimize the Hurwicz criterion in a decision tree, it is important to note that the optimality principle does not hold. For instance, consider Figure 2 and assume complete ignorance about probabilities (i.e., all intervals of probabilities are $[0, 1]$). Let us set $\alpha = 0.5$ and perform backward induction on the decision tree with $u(x) = x$. In D_2, the decision maker prefers decision *up* to *down* (the Hurwicz criterion is equal to 15 for $D_2 = up$, compared to 12.5 for $D_2 = down$) and in D_3 he also prefers decision *up* to *down* (a sure utility of 10, compared to 9.5). In D_1, the decision maker has then the choice between a first lottery offering a minimum utility of 0 and a maximum utility of 20 if he decides *up*, and a second lottery offering a minimum of 5 and a maximum of 10 if he decides *down*. The best decision according to the Hurwicz criterion

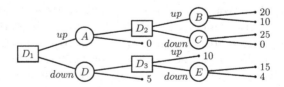

Fig. 2. A separable decision tree

Table 1. Strategies and their evaluations

D_1	D_2	D_3	$\alpha = 0$	$\alpha = 0.5$	$\alpha = 1$
up	up	–	20	10	0
up	down	–	25	12.5	0
down	–	up	10	7.5	5
down	–	down	15	9.5	4

is *up* (10 compared to 7.5). The strategy returned by dynamic programming is therefore $(D_1 = up, D_2 = up)$ with a value of 10. Table 1 indicates the value of every strategy with respect to α. For $\alpha = 0.5$, strategy $(D_1 = up, D_2 = down)$ is optimal with a value of 12.5. In this case, one thus observes that the strategy returned by dynamic programming is suboptimal. For this reason, a decision maker using the Hurwicz criterion should adopt a *resolute choice* behavior [7], i.e. he initially chooses a strategy and never deviates from it later. We focus here on determining an optimal strategy from the root.

Before showing how to compute an optimal strategy according to the Hurwicz criterion in a separable decision tree, we first show how to compute an optimal strategy according to Γ-maximin and Γ-maximax. It is well-known that the validity of the rolling back method on decision trees relies on the fulfillment of the independence axiom [6]. The *independence* axiom [11] states that the mixture of two lotteries f and g with a third one h should not reverse preferences (induced by the decision criterion used): if f is strictly preferred to g, then $\lambda f + (1 - \lambda)h$ (i.e., the compound lottery that yields lottery f (resp. h) with probability λ (resp. $1 - \lambda$)) should be strictly preferred to $\lambda g + (1 - \lambda)h$. The following result states that the independence axiom holds for Γ-maximin and Γ-maximax *under a separability condition*:

Proposition 2. *Let f, g, h denote lotteries with sets $\mathcal{P}_f, \mathcal{P}_g, \mathcal{P}_h$ of possible probability distributions. If the set $\mathcal{P}_{\lambda f + (1-\lambda)h}$ (resp. $\mathcal{P}_{\lambda g + (1-\lambda)h}$) of possible probability distributions on the compound lottery $\lambda f + (1-\lambda)h$ is the cartesian product of $\mathcal{P}_f$ (resp. $\mathcal{P}_g$) and $\mathcal{P}_h$ (separability condition), then the following properties hold:*

$$\underline{E}(f) \geq \underline{E}(g) \Rightarrow \underline{E}(\lambda f + (1 - \lambda)h) \geq \underline{E}(\lambda g + (1 - \lambda)h)$$
$$\bar{E}(f) \geq \bar{E}(g) \Rightarrow \bar{E}(\lambda f + (1 - \lambda)h) \geq \bar{E}(\lambda g + (1 - \lambda)h)$$

Proof. We show that $\underline{E}(\lambda f + (1 - \lambda)h) = \lambda \underline{E}(f) + (1 - \lambda)\underline{E}(h)$ under the assumptions of the proposition. We have indeed $\underline{E}(\lambda f + (1 - \lambda)h) = \min\{E(\lambda f + (1 - \lambda)h, P) : P \in \mathcal{P}_{\lambda g + (1 - \lambda)h}\}$. By linearity of expectation, it equals $\min\{\lambda E(f, P) + (1 - \lambda)E(h, P) : P \in \mathcal{P}_{\lambda g + (1-\lambda)h}\}$. By separability assumption, it equals $\min\{\lambda E(f, P_f) + (1 - \lambda)E(h, P_h) : P_f \in \mathcal{P}_f, P_h \in \mathcal{P}_h\} = \lambda \min\{E(f, P_f) : P_f \in \mathcal{P}_f\} + (1 - \lambda)\min\{E(h, P_h) : P_h \in \mathcal{P}_h\}$. By definition of $\underline{E}(\cdot)$, it equals $\lambda \underline{E}(f) + (1 - \lambda)\underline{E}(h)$. This implies the validity of the first property. The proof is similar for the second property.   ∎

In a separable decision tree, the separability condition of Proposition 2 holds at every chance node. For this reason, the rolling back method returns an optimal strategy when used with Γ-maximin or Γ-maximax in a separable decision tree. The computational complexity of this procedure is linear in the number of decision nodes.

Let us now explain our approach for computing an optimal strategy according to the Hurwicz criterion. We recall that the rolling back method does not work when operating directly with the Hurwicz criterion for $\alpha \neq 0, 1$. However, one can use the following simple property: if a substrategy is dominated by another one at the same node for both the Γ-maximin and Γ-maximax criteria (i.e., its value is smaller or equal for both criteria, and strictly smaller for at least one), then it cannot yield an optimal strategy for the Hurwicz criterion. The idea is to compute the set of non-dominated strategies (more precisely, one strategy for each non-dominated vector) by a bicriteria rolling back procedure from the leaves. At the root, one computes then the value of every non-dominated strategy according to the Hurwicz criterion, and one returns the best one. We describe here, for each node X of a decision tree (for simplicity, we assume here that the decision tree is binary), how the set $ND(X)$ of non-dominated vectors (the first (resp. second) component represents the minimum (resp. maximum) expected utility of a feasible strategy) are inferred from the non-dominated vectors of its successors X_1 and X_2:

- At a leaf X labelled by utility u, one sets $ND(X) = \{(u, u)\}$.
- At a chance node X, let us denote by $P(X_1) \in [P^-(X_1), P^+(X_1)]$ the probability variable assigned to edge (X, X_1). Then $ND(X)$ is the set of non-dominated vectors computed by the formula:

$$(\min_{P(X_1)} \{P(X_1)\underline{E_1} + (1 - P(X_1))\underline{E_2}\}, \max_{P(X_1)} \{P(X_1)\overline{E_1} + (1 - P(X_1))\overline{E_2}\})$$

  for all $(\underline{E_1}, \overline{E_1}) \in ND(X_1)$, $(\underline{E_2}, \overline{E_2}) \in ND(X_2)$.
- At a decision node X, $ND(X)$ is the set of non-dominated vectors in set $ND(X_1) \cup ND(X_2)$.

We now give an example to illustrate the operation of the procedure.

Example 2. *Let us come back to the decision tree of Figure 2 and assume again complete ignorance about probabilities, $\alpha = 0.5$ and $u(x) = x$ (for simplicity in the calculation). The trace of the algorithm is the following:*

- at leaf 20 (resp. 10, etc.) $ND(20) = \{(20, 20)\}$ (resp. $\{(10, 10)\}$, etc.);
- $ND(B) = \{(10, 20)\}$ since combining $(10, 10)$ and $(20, 20)$ yields $(10, 20)$;
- $ND(C) = \{(0, 25)\}$ since combining $(0, 0)$ and $(25, 25)$ yields $(0, 25)$;
- $ND(D_2) = \{(10, 20), (0, 25)\}$ since both vectors are non-dominated;
- $ND(A) = \{(0, 25)\}$ since combining $(10, 20)$ (resp. $(0, 25)$) and $(0, 0)$ yields $(0, 20)$ (resp. $(0, 25)$), and $(0, 25)$ dominates $(0, 20)$;
By proceeding similarly, one obtains $ND(D) = \{(4, 15), (5, 10)\}$. At the root, one obtains finally $ND(D_1) = \{(0, 25), (4, 15), (5, 10)\}$. By evaluating every vector according to the Hurwicz criterion, one finds that $(0, 25)$ is an optimal vector (corresponding to optimal strategy $(D_1 = up, D_2 = down)$).

The algorithm has been implemented in C++, and we have carried out numerical tests on a PC with a Pentium IV CPU 2.13Ghz processor and 3.5GB of RAM.

Table 2. Numerical results

Algorithms		Imprecision		Ignorance	
Depth (nodes)		Avg	Max	Avg	Max
13 (16, 383)	card.	144	600	24	39
	time	0	0	0	0
15 (65, 535)	card.	940	3, 290	47	68
	time	2.35	44	0.02	1
17 (262, 143)	card.	7, 182	40, 930	90	115
	time	1, 189.97	3, 595	0.14	1
19 (1, 048, 575)	card.	−	−	174	216
	time	−	−	0.58	1
21 (4, 194, 303)	card.	−	−	348	570
	time	−	−	2.09	3
23 (16, 777, 215)	card.	−	−	714	1, 164
	time	−	−	8.31	9
25 (67, 108, 863)	card.	−	−	−	−
	time	−	−	−	−

Our tests were performed on complete binary decision trees of even depth. The depth of these decision trees varies from 4 to 14 (5 to 5461 decision nodes), with an alternation of decision nodes and chance nodes. Utilities are real numbers randomly drawn within interval $[1, 500]$. The imprecise probabilities were generated by randomly drawning a sharp probability distribution for each chance node, and then randomly generating an interval of probabilities around each probability. The numerical results are summarized in Table 2. Column "Imprecision" (resp. "Ignorance") details results obtained in the case of imprecise probabilities (resp. complete ignorance). Note that some tuning of the bicriteria rolling back method is possible in the case of complete ignorance, that considerably speeds up the procedure. Furthermore, the number of non-dominated vectors at each node is upper bounded by n in this case (where n denotes the number of decision nodes), and therefore the whole procedure performs in $O(n^2)$. For each depth, 500 instances were randomly generated and one indicates the average (Avg) and maximum (Max) computation times (in sec.), as well as the cardinality of the set of non-dominated vectors at the root. Symbol "−" appears when the memory size was not sufficient to execute the algorithm. One can observe that the smaller memory space requirements make it possible to solve larger instances (up to 16 millions of nodes) in the case of complete ignorance.

5   Optimizing the Hurwicz Criterion in Non-separable Decision Trees

We now prove that the determination of an optimal strategy according to the Hurwicz criterion in a non-separable decision tree is an NP-hard problem, where the size of an instance is the number of involved decision nodes. Actually, we show a stronger result:

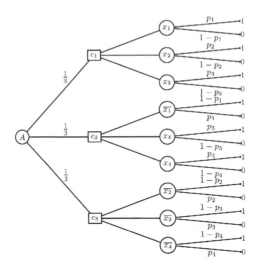

Fig. 3. An example of reduction

Proposition 3. *The determination of an optimal strategy according to the Γ-maximax criterion in a non-separable decision tree is an NP-hard problem.*

Proof. The proof relies on a polynomial reduction from problem 3-SAT, which can be stated as follows:

INSTANCE: a set X of boolean variables, a collection C of clauses on X such that $|c| = 3$ for every clause $c \in C$.
QUESTION: does there exist an assignment of truth values to the boolean variables of X that satisfies simultaneously all the clauses of C ?

Let $X = \{x_1, \ldots, x_n\}$ and $C = \{c_1, \ldots, c_m\}$. The polynomial generation of a decision tree from an instance of 3-SAT is performed as follows. One defines a decision node for each clause of C. Given c_i a clause in C, the corresponding decision node in the decision tree, also denoted by c_i, has three children (chance nodes), one for each literal in the clause. These chance nodes are denoted by the name of the corresponding literal. Every chance node x_i (resp. $\bar{x}_i$) has two children: a leaf of utility 1 with probability $p_i \in [0, 1]$ (resp. $1 - p_i$), and a leaf of utility 0 with probability $1 - p_i \in [0, 1]$ (resp. p_i). Finally, one adds a chance node A as root, predecessor of all decision nodes c_i, with probability $1/m$ on every branch. The obtained decision tree includes m decision nodes, $3m + 1$ chance nodes and $6m$ leaves. Furthermore, n probability variables are involved. This guarantees the polynomiality of the reduction. For the sake of illustration, on Figure 3, we represent the decision tree obtained for the following instance of 3-SAT: $(x_1 \vee x_2 \vee x_3) \wedge (\bar{x}_1 \vee x_3 \vee x_4) \wedge (\bar{x}_2 \vee \bar{x}_3 \vee \bar{x}_4)$.

Note that, in this kind of decision trees, the Γ-maximax value of any strategy is upper bounded by 1. Furthermore, given an assignment of truth values that makes satisfiable the 3-SAT expression, one can construct a strategy whose Γ-maximax value is 1. There exists indeed at least a literal whose truth value is "true" for every clause c_i. Let us denote by k_i the index of such a literal in c_i. At every node c_i, one makes decision leading to literal whose index is k_i. By setting $p_{k_i} = 1$ (resp. 0) if it is a positive (resp. negative) literal, the expected utility of the corresponding strategy is 1. Conversely, given a strategy whose Γ-maximax value is 1, one can construct an assignment of truth values that makes satisfiable the 3-SAT expression. Indeed, at every decision node c_i the chosen decision necessarily leads to a chance node returning a utility of 1 with a probability set to 1. Let us denote by k_i the index of the chance node chosen at c_i. One obtains a partial assignment by setting x_{k_i} to "true" (resp. "false") if $p_{k_i} = 1$ (resp. 0). Any completion of this partial assignment makes satisfiable the 3-SAT expression. This concludes the proof.   ∎

6 Conclusion

In this paper, we have proposed an operational procedure to determine an optimal strategy according to the Hurwicz criterion in a separable decision tree. Furthermore, we have proved that the problem becomes NP-hard in non-separable decision trees. For future research, it would be interesting to propose an algorithm for optimizing the Hurwicz criterion in a non-separable decision tree. In this purpose, a branch and bound is worth investigating. An upper bound easily computable would consist for instance in computing an upper bound of the value of a Γ-maximin strategy by determining the maximum expected utility for a feasible sharp probability distribution (i.e., consistent with the intervals of probabilities), and an upper bound of a Γ-maximax strategy by relaxing the non-separability constraints (and therefore using the procedure for Γ-maximax detailled in Section 4). Combining both upper bounds with α and $1 - \alpha$ would provide an upper bound for the Hurwicz criterion.

References

1. Bielza, C., Rios Insua, D., Rios-Insua, S.: Influence diagrams under partial information. In: Bayesian Statistics, Oxford, U.P, vol. 5, pp. 491–497 (1996)
2. de Campos, C.P., Kikuti, D., Cozman, F.G.: Partially ordered preferences in decision trees: computing strategies with imprecision in probabilities. In: IJCAI Workshop on Advances in Preference Handling (2005)
3. Howard, R., Matheson, J.: Influence Diagrams. Strategic Decisions Group, Menlo Park (1984)
4. Jaffray, J.-Y., Jeleva, M.: Information processing under imprecise risk with the hurwicz criterion. In: 5th International Symposium on Imprecise Probability: Theories and Applications, pp. 233–242 (2007)
5. Kasperski, A.: Discrete Optimization with Interval Data: Minmax Regret and Fuzzy Approach. Studies in Fuzziness and Soft Computing. Springer, Heidelberg (2008)

6. LaValle, I.H., Wapman, K.R.: Rolling back decision trees requires the independence axiom. Management Science 32(3), 382–385 (1986)
7. McClennen, E.F.: Rationality and Dynamic choice: Foundational Explorations. Cambridge University Press, Cambridge (1990)
8. Puterman, M.L.: Markov Decision Processes - Discrete Stochastic Dynamic Programming. Wiley & Sons, Chichester (1994)
9. Raiffa, H.: Decision Analysis: Introductory Lectures on Choices under Uncertainty. Addison-Wesley, Reading (1968)
10. Shachter, R.: Evaluating influence diagrams. Operations Research 34, 871–882 (1986)
11. von Neuman, J., Morgenstern, O.: Theory of games and economic behaviour. Princeton University Press, Princeton (1947)
12. Walley, P.: Statistical reasoning with imprecise probabilities. Monographs on statistics and applied probability, vol. 91. Chapman and Hall, Boca Raton (1991)
13. Weichselberger, K.: The theory of interval-probability as a unifying concept for uncertainty. In: 1st International Symposium on Imprecise Probabilities: Theories and Applications (ISIPTA), pp. 387–396 (1999)

Axioms for a Class of Algorithms of Sequential Decision Making

Murali Agastya[1] and Arkadii Slinko[2]

[1] Economics Discipline, University of Sydney NSW 2006, Australia
m.agastya@econ.usyd.edu.au
[2] Department of Mathematics, The University of Auckland, Private Bag 92019, Auckland, New Zealand
a.slinko@auckland.ac.nz

Abstract. We axiomatically characterise a class of algorithms for making sequential decisions in situations of complete ignorance. These algorithms assume that a decision maker (DM) (human or or a software agent) has exogenously defined utilities for prizes and she uses the empirical distribution of prizes to calculate the "expected utility" of each action maximising this expected utility at each stage of the decision making process. We show that this class of algorithms is defined by three simple axioms that highlight the independence of the given actions, the bounded rationality of the agent, and the principle of insufficient reason at margin.

Keywords: sequential decision making, ex-post rationality, fictitious play, multiset.

1 Introduction

Consider a Decision Maker (DM) who has to repeatedly choose from a finite set of actions. Each action results in a random reward, also drawn from a finite set. The environment is complex in the sense that the DM is either unable to offer a complete description of the states of the world or is unable to construct a meaningful prior probability distribution. Naturally, the well established Bayesian methods of say [12] or [1] would then be inapplicable.

Our approach is to postulate that the DM has a preference relation defined directly over the set of actions which is updated over time in response to the sequences of observed rewards. Thus, if $\mathcal{A}$ denotes the set of all actions and H the set of all histories, the DM is completely described by the family $D := (\succeq_{h_t})_{h_t \in H}$, where $\succeq_{h_t} \subseteq \mathcal{A} \times \mathcal{A}$ is a well defined preference relation on the actions following a history h_t at date t. A history consists of the sequences of rewards, drawn from a finite set $\mathcal{R}$, that are obtained over time to each of the actions. Later we will impose axioms on D of procedural rationality type.

There is a considerable literature in economics and psychology on a variety of "stimulus-response" models of individual choice behavior. In these models, the DM does not attempt to learn the environment, instead she looks at the past experiences and takes her decisions on the basis of her observations. Most of this

F. Rossi and A. Tsoukis (Eds.): ADT 2009, LNAI 5783, pp. 353–364, 2009.

literature prescribes some boundedly rational rule(s) for updating and the focus is on analysis of implied adaptive dynamics. These imputed rules of updating vary widely. They range from modifications of fictitious play and reinforcement learning to imitation of peers etc. See for example [2], [13], [7] and the references therein.

Our approach outlined above is different. We do not consider any particular updating rules but impose axioms on the updating procedure. These axioms impose some structural restrictions and postulate certain independence and we derive an ex-post utility representation for such a DM. This approach may be found in [4] where they axiomatically characterised replicator dynamics which makes [4] the closest relative of this paper. We note that the Case Based Decision Theory of Gilboa and Schmeidler [8], [9] is not applicable due to the assumption of infinitude of cases and the Archimedean axiom that they impose.

Chapter 2 introduces the model, Chapter 3 defines the ex-post utility representation, Chapter 4 introduces the axioms, Chapter 5 formulates the main theorem and outlines its proof, and finally Section 6 fills all the gaps and completes the proof of the main theorem.

2   The Model

A Decision Maker must choose from a finite set of m actions $\mathcal{A} = \{a_1, \ldots, a_m\}$, at each moment $t = 0, 1, 2, \ldots$. Every action results in a reward, drawn from a finite set $\mathcal{R} = \{1, \ldots, n\}$. The rewards are governed by a stochastic process unknown to the DM. Following her choice at date t, the vector of realised rewards, $\mathbf{r}_t = (r_1^{(t)}, \ldots, r_m^{(t)})$, where $r_i^{(t)}$ is the reward to action a_i at moment t, is revealed to the DM. Thus the DM observes the rewards for *all* actions and not only for the one she has chosen. A *history* at date t is a sequence of vectors of rewards $h_t = (\mathbf{r}_0, \ldots, \mathbf{r}_{t-1})$.

The sequential decisions of the DM are guided by the following principle. Following any history h_t, the DM works out a preference relation[1] $\succeq_{h_t}$ on the set of actions $\mathcal{A}$. At date t she chooses one of the maximal actions with respect to $\succeq_{h_t}$, observes the set of outcomes $\mathbf{r}_t$ and calculates a new preference relation $\succeq_{h_{t+1}}$ where $h_{t+1} = (h_t, \mathbf{r}_t)$. At the outset the DM is indifferent between all the actions so she chooses a random one.

Let H_t denote the set of all histories at date t and $H = \bigcup_{t \geq 1} H_t$. Thus, the family of preference relations $D := (\succeq_h)_{h \in H}$ completely describes the DM. Our objective is to discuss the behavior of this learning agent through the imposition of certain axioms that encapsulate the DM's procedural rationality. For a DM satisfying these axioms we will derive an ex-post utility representation theorem that is based on the empirical distribution of rewards in any history.

Before proceeding any further with the analysis, it is important to point out two salient features of the above formulation of the DM.

---

[1] Throughout, by a *preference relation* on any set, we mean a binary relation that is a complete, transitive and reflexive ordering of the elements.

First, as in [4], a history describes the rewards to all the actions in each period, including those that the DM did not choose. This implicitly assumes that decisions are taken in a social context where other people are taking other actions and the rewards for each action are publicly announced. Examples of such situations are numerous and include investing in a share market and betting on horses. Relaxing this assumption of learning in a social context is a topic of future research.

Second, note that the description requires a preference on actions to be specified after every conceivable history. This is much in the spirit of the theoretical developments in virtually all decision theory. The presumption underlying such an abstraction is that any subset of these acts may be presented to the DM and that a necessary aspect of a theory is that it is applicable with sufficient generality. Given the temporal nature of the problem at hand this assumption may be quite natural. For, all conceivable histories may appear by assuming that the underlying random process generates every $\mathbf{r} \in \mathcal{R}^m$ with a positive probability.

We make a non-triviality assumption on D for the rest of this paper. We assume that the DM is not indifferent between all actions following all histories.

3 Multisets and Ex-post Utility Maximisation

Here we will introduce the rule (a class of algorithms) that we will eventually axiomatise. For this rule, the number of times different rewards accrue to given action during a history is important. To progress further, we will need to introduce the idea of a *multiset*. A multiset over an underlying set may contain several copies of any given element of the latter. The number of copies of an element is called its *multiplicity*. Our interest is in multisets over $\mathcal{R}$. Therefore, multiset μ is identified with a vector $\mu = (\mu(1), \dots, \mu(n)) \in \mathbb{Z}_+^n$, where $\mu(i)$ is the multiplicity of the ith prize and the *cardinality* of this multiset is $\sum_{i=1}^{n} \mu(i)$.

Let $\mathcal{P}_t[n]$ denote the subset of all such multisets of cardinality t whereupon

$$\mathcal{P}[n] = \bigcup_{t=1}^{\infty} \mathcal{P}_t[n] \tag{1}$$

denotes the set of all non-empty multisets over $\mathcal{R}$. Mostly, we will write $\mathcal{P}_t$ instead of $\mathcal{P}_t[n]$ when the number of prizes is clear. The *union* of $\mu, \nu \in \mathcal{P}$ is defined as the multiset $\mu \cup \nu$ for which $(\mu \cup \nu)(i) = \mu(i) + \nu(i)$ for any $i \in \mathcal{R}$. Observe that whenever $\mu \in \mathcal{P}_t$ and $\nu \in \mathcal{P}_s$, then $\mu \cup \nu \in \mathcal{P}_{t+s}$.

Given any history $h \in H_t$, let $\mu_i(a, h)$ denote the number of times the reward i has occured in the history of rewards $h(a)$ corresponding to action a and $\mu(a, h) = (\mu_1(a, h), \dots, \mu_n(a, h))$.

For any two vectors $\mathbf{x} = (x_1, \dots, x_n)$, $\mathbf{y} = (y_1, \dots, y_n)$ of $\mathbb{R}^n$, we let $x \cdot y$ denote their dot product, i.e. $x \cdot y = \sum_{i=1}^{n} x_i y_i$.

Here comes the rule. A DM applying this rule must have exogenously defined utilities of the prizes. Let $\mathbf{u} = (\mathbf{u}_1, \dots, \mathbf{u}_n)$ be the vector of her utilities, where

u_i is the utility of the ith prize. At any moment t the DM calculates the total utility of the prices for each given action in the past and chooses the action which performed best in the past and for which the total utility of prizes is at least as high as for any other action. In other words she chooses any action belonging to $\mathrm{argmax}_i(\mu(a_i, h) \cdot \mathbf{u})$.

The problem of the DM is that she does not know the probabilities. In the absence of any knowledge about the environment the most reasonable thing to do is to assume that the process of generating rewards is stationary and to replace the probabilities of the rewards with their empirical frequencies. Due to the assumed stationarity of the process she expects that these frequencies approximate probabilities well (at least in the limit), so in a way the DM acts as an expected utility maximiser relative to the empirical distribution of rewards. This rule is very much in the spirit of the so-called fictitious play[2].

There is a good reason to allow the DM to use different vectors of utilities at different moments. This will allow the DM, at each moment, to refine her utilities from the previous period to reflect her preferences on larger multisets and longer histories. An obvious consistency condition must however be imposed: we require that the vector of utilities the DM uses at time t must be also suitable to evaluate actions in all previous moments.

Definition 1 (Ex-Post Utility Representation). *A sequence $(\mathbf{u}_t)_{t \geq 1}$ of vectors of $\mathbb{R}_+^n$ is said to be an* ex-post utility representation *of $D = (\succeq_h)_{h \in H}$ if, for all $t \geq 1$,*

$$a \succeq_h b \iff \mu(a, h) \cdot \mathbf{u}_t \geq \mu(b, h) \cdot \mathbf{u}_t \quad \forall a, b \in \mathcal{A}, \ \forall h \in H_s, \tag{2}$$

for all $s \leq t$. The representation is said to be global *if $\mathbf{u}_t \equiv \mathbf{u}$ for some $\mathbf{u} \in \mathbb{R}_+^n$.*

In what follows, we shall say that the DM is *ex-post rational* if she admits an ex-post utility representation.

We emphasise that the object that is of ultimate interest is the ranking of the actions following a history. The utility representation of a DM involves assigning non-negative weights to the rewards. However this assignment is not unique. A sequence $(\mathbf{u}_t')_{t \geq 1}$ obtained by applying some positive affine transformations $\mathbf{u}_t' = \alpha_t \mathbf{u}_t + \beta_t$ (with $\alpha_t > 0$) to a given utility representation $(\mathbf{u}_t)_{t \geq 1}$ is also a utility representation.

Therefore, we should adopt a certain normalisation. By $\Delta \subseteq \mathbb{R}^m$ we denote the $m - 1$ dimensional unit simplex consisting of all non-negative vectors $\mathbf{x} = (x_1, \ldots, x_n)$ such that $x_1 + \ldots + x_n = 1$. Due to the non-triviality assumption, for any $\mathbf{u}_t$, not all utilities are equal. Hence we may assume that at any $\mathbf{u}_t = (u_1, \ldots, u_n)$ in a representation, $\min\{u_i\} = 0$. We may further normalise the coordinates to sum to one so that every $\mathbf{u}_t$ may be assumed to lie in one of the following subsets of the unit simplex:

$$\Delta^i = \{\mathbf{u} = (u_1, \ldots, u_n) \in \Delta \mid u_i = 0\}, \tag{3}$$

which is one of the facets[3] of Δ.

---

[2] Ficitious play was introduced by [3]. See [5] for variations of fictitious play.

[3] Facet of a polytope is a face of the maximal dimension.

4   Axioms

Next, we turn to the axioms that are necessary and sufficient for D to admit an ex-post utility representation. The first axiom says that in comparing a pair of actions, the information regarding the other actions is irrelevant. Intuitively, this amounts to asserting that the agent believes that she is facing an environment in which consequences of actions are statistically uncorrelated.

Axiom 1. *Consider h_t, h'_t and actions $a, b \in \mathcal{A}$ such that $h_t(a) = h'_t(a)$ and $h_t(b) = h'_t(b)$. Then $a \succeq_{h_t} b$ if and only if $a \succeq_{h'_t} b$.*

Although the agent has the entire history at her disposal, we postulate that for any action, the algorithm only tracks the number of times different rewards were realised. This means that the agent believes that she is facing an environment generated by a stationary stochastic process.

Axiom 2. *Consider a history h_t at which for two actions a and b the multisets of prizes are the same, i.e. $\mu(a, h_t) = \mu(b, h_t)$. Then $a \sim_{h_t} b$.*

The next axiom describes how the DM learns to revise her preferences in response to new information.

Axiom 3. *For any history h_t and any $r \in \mathcal{R}$, if $h_{t+1} = (h_t, \mathbf{r}_t)$ where $\mathbf{r}_t = (r, \dots, r)$, then $\succeq_{h_{t+1}} = \succeq_{h_t}$.*

Due to Axiom 1, it implies that if at some history h_t the DM (weakly) prefers an action a to b and in the current period both these actions yield the same reward, according to the next axiom, the DM continues to prefer a to b. We view Axiom 3 as loosely capturing the "principle of insufficient reason at the margin".

5   The Main Theorem

In this section we will formulate and give an outline of the proof of the main theorem. Recall that $ri(C)$ denotes the relative interior of a convex set C.

Theorem 1 (Representation Theorem). *Suppose $m \geq 3$. The following are equivalent:*

1. *$D = (\succeq_h)_{h \in H}$ satisfies Axioms 1–3.*
2. *D has an ex-post utility representation. There exists a unique sequence of non-empty convex polytopes $(U_t)_{t \geq 0}$ such that $U_t \subseteq \Delta^i$ for some i and*
   (a) *$U_{t+1} \subseteq U_t$ for all $t \geq 1$.*
   (b) *$\bigcap_{t=1}^{\infty} U_t$ consists of a single vector.*
   (c) *A sequence $(\mathbf{u}_t)_{t \geq 1}$ of vectors of $\mathbb{R}_n^+$ is a utility representation of D if and only if $\mathbf{u}_t$ is a positive affine transformation of some $\mathbf{u}'_t \in ri(U_t)$. In particular, any sequence $(\mathbf{u}_t)_{t \geq 1}$ such that $\mathbf{u}_t \in ri(U_t)$ is a utility representation of D.*
   (d) *If $\bigcap_{t=1}^{\infty} U_t$ is in the interior of every U_t, then the representation is global.*

Remark 1. We note that despite an expected-utility-like calculation that is implicitly involved in Theorem 1, it is important to note that there is no connection with the expected utility hypothesis. Our DM is only ex-post rational.

Proof. It is easy to show that any DM with an ex-post utility representation satisfies the axioms. Let us show the non-trivial part of the theorem, which is, $1 \Rightarrow 2$. We begin by defining, for each $t \geq 1$, a binary relation $\succeq_t^*$ on $\mathcal{P}_t = \mathcal{P}_t[n]$ as follows: for any $\mu, \nu \in \mathcal{P}_t$,

$$
\begin{aligned}
\mu \succeq_t^* \nu \Longleftrightarrow \ & \text{there exists } a, b \in \mathcal{A} \text{ and a history } h_t \in H_t \\
& \text{such that } \mu = \mu(a, h_t) \text{ and } \nu = \mu(b, h_t) \text{ and} \qquad (4) \\
& a \succeq_{h_t} b
\end{aligned}
$$

Analogously we define also a strict version $\succ_t^*$ of $\succeq_t^*$. The latter needs to be proved to be antisymmetric. For, for a certain pair of multisets $\mu, \nu \in \mathcal{P}_t$, different choices of histories and actions can result in both $\mu \succeq_t^* \nu$ and $\nu \succ_t^* \mu$ at once. However, we claim that:

Claim 1. *For any $a, b, c, d \in \mathcal{A}$ and any two histories $h_t, h_t' \in H_t$ such that $\mu(a, h_t) = \mu(c, h_t')$ and $\mu(b, h_t) = \mu(d, h_t')$,*

$$
a \succeq_{h_t} b \Longleftrightarrow c \succeq_{h_{t'}} d.
$$

The above claim ensures that $\succ_t^*$ is antisymmetric since $\succ_h$ is antisymmetric. It is now also clear that the sequence $\succeq^* = (\succeq_t^*)_{t \geq 1}$ inherits the non-triviality assumption in the sense that for some t the relation $\succeq_t^*$ is not a complete indifference. Next we claim that

Claim 2. $\succeq_t^*$ *is a preference ordering on $\mathcal{P}_t$.*

Both of the above claims only rely on Axiom 1 and Axiom 2. The proofs of Claim 1 and Claim 2 are straightforward but nevertheless relegated to the Appendix. By a repeated application of Axiom 3, we see at once that

Claim 3. *The sequence $\succeq^* = (\succeq_t^*)_{t \geq 1}$ satisfies the following property: for any $\mu, \nu \in \mathcal{P}_t$ and any $\xi \in \mathcal{P}_s$,*

$$
\mu \succeq_t^* \nu \Longleftrightarrow \mu \cup \xi \succeq_{t+s}^* \nu \cup \xi \qquad (5)
$$

for all $t, s \in \mathbb{Z}_+$.

The remainder of the proof will follow from Theorem 2 proved in the next section and further considerations.

The requirement in Theorem 1 that there are at least three actions for the agent to choose from cannot be dropped. To see this we have the following counter-example with $m = 2$.

Example 1. Pick any utility vector $\mathbf{u} = (u_1, \ldots, u_n)$ for the rewards and define D as follows:

Following a history $h_t \in H_t$,

1. If $\mu(a_i, h_t) \cdot \mathbf{u} > \mu(a_j, h_t) \cdot \mathbf{u}$, the DM strictly prefers a_i to a_j, where $i \neq j$ and $i, j = 1, 2$.
2. If $\mu(a_1, h_t) \cdot \mathbf{u} = \mu(a_2, h_t) \cdot \mathbf{u}$, then
   (a) If the corresponding multisets of rewards are the same, i.e. $\mu(a_1, h_t) = \mu(a_2, h_t)$, then the actions are indifferent.
   (b) Otherwise a_1 is strictly preferred.

It may be readily verified that $\mathcal{D}$ described above satisfies Axioms 1-3 but does not admit an ex-post utility representation.

6   Orders on Multisets and Their Utility Representation

This section completes the proof of the main theorem.

As we know from Section 2, multisets of cardinality t are important for a DM as they are closely related to histories at date t. The DM has to be able to compare them for all t. At the same time in the context of this paper it does not make much sense to compare multisets of cardinalities of different sizes (it would if we had missing observations). Due to this, our main object in this subsection is a family of orders $(\succeq_t)_{t \geq 1}$, where $\succeq_t$ is an order on $\mathcal{P}_t$. In this case we denote by $\succeq$ the partial (but reflexive and transitive) binary relation on $\mathcal{P}$ whereby for any $\mu, \nu \in \mathcal{P}$, where $\mu \succeq \nu$ if both μ and ν are of the same cardinality, say t, and $\mu \succeq_t \nu$ and $\mu \succeq \nu$ is undefined otherwise.

To complete the proof of the main theorem we must study orders on $\mathcal{P}$ with the property (5). Due to their importance we will give them a special name.

Definition 2 (Consistency). *An order $\succeq = (\succeq_t)_{t \geq 1}$ on $\mathcal{P}$ is said to be* consistent *if it satisfies the condition (5) from Claim 3, that is, for any $\mu, \nu \in \mathcal{P}_t$ and any $\xi \in \mathcal{P}_s$,*

$$\mu \succeq_t \nu \iff \mu \cup \xi \succeq_{t+s} \nu \cup \xi. \tag{6}$$

We note that, due to the twosidedness of the arrow in (6), we have also

$$\mu \succ_t \nu \iff \mu \cup \xi \succ_{t+s} \nu \cup \xi. \tag{7}$$

One consistent linear order that immediately comes to our mind is the lexicographic order which is an extension of a linear order on $\mathcal{R}$. But, of course, this is not the only consistent order. Now we will define a large class of consistent orders on $\mathcal{P}$ to which the lexicographic order belongs.

Definition 3 (Local Representability). *An order $\succeq := (\succeq_t)_{t \geq 1}$ on $\mathcal{P}$ is locally representable if, for every $t \geq 1$, there exist $\mathbf{u}_t \in \mathbb{R}^n$ such that*

$$\mu \succeq_s \nu \iff \mu \cdot \mathbf{u}_t \geq \nu \cdot \mathbf{u}_t \qquad \forall \mu, \nu \in \mathcal{P}_s, \quad \forall s \leq t. \tag{8}$$

A sequence $(\mathbf{u}_t)_{t\geq 1}$ is said to locally represent $\succeq$ if (8) holds. The order $\succeq$ is said to be globally representable if there exist $\mathbf{u} \in \mathbb{R}^n$ such that (8) is satisfied for $\mathbf{u}_t = \mathbf{u}$ for all t.

The lexicographic order is locally representable but not globally.

Theorem 2. An order $\succeq = (\succeq_t)_{t\geq 1}$ on $\mathcal{P}$ is consistent if and only if it is locally representable.

Proof. If the order is locally representable it is straightforward to verify that it is consistent. Suppose the sequence of vectors $(\mathbf{u}_t)_{t\geq 1}$ represents $\succeq = (\succeq_t)_{t\geq 1}$. Let $\mu, \nu \in \mathcal{P}_s$ with $\mu \succeq_s \nu$ and $\eta \in \mathcal{P}_t$. Then $\mu \cdot \mathbf{u}_{s+t} \geq \nu \cdot \mathbf{u}_{s+t}$ since $\mathbf{u}_{s+t}$ can be used to compare multisets of cardinality t as $t < t + s$. But now

$$(\mu + \eta) \cdot \mathbf{u}_{s+t} - (\nu + \eta) \cdot \mathbf{u}_{s+t} = \mu \cdot \mathbf{u}_{s+t} - \nu \cdot \mathbf{u}_{s+t} \geq 0$$

which means $\mu + \eta \succeq_{s+t} \nu + \eta$.

To see the converse, let $\succeq = (\succeq_t)_{t\geq 1}$ be consistent. An immediate implication of consistency is that for any $\mu_1, \nu_1 \in \mathcal{P}_t$ and $\mu_2, \nu_2 \in \mathcal{P}_s$,

$$\mu_1 \succeq_t \nu_1 \text{ and } \mu_2 \succeq_s \nu_2 \implies \mu_1 \cup \mu_2 \succeq_{t+s} \nu_1 \cup \nu_2, \tag{9}$$

where we have $\mu_1 \cup \mu_2 \succ_{t+s} \nu_1 \cup \nu_2$ if and only if either $\mu_1 \succ_t \nu_1$ or $\mu_2 \succ_s \nu_2$. Indeed by consistency, we have

$$\mu_1 \cup \mu_2 \succeq_{t+s} \nu_1 \cup \mu_2 \succeq_{t+s} \nu_1 \cup \nu_2.$$

Now suppose, by way of contradiction, that local representability fails at some t which means that $\mathbf{u}_t$ is the first vector that cannot be found. Note that there are $N = \binom{n+t-1}{t}$ multisets of cardinality t in total. Let us enumerate all the multisets in $\mathcal{P}_t$ so that

$$\mu_1 \succeq_t \mu_2 \succeq_t \cdots \succeq_t \mu_{N-1} \succeq_t \mu_N. \tag{10}$$

Some of these relations may be equivalencies, the others will be strict inequalities. Let $I = \{i \mid \mu_i \sim_t \mu_{i+1}\}$ and $J = \{j \mid \mu_j \succ_t \mu_{j+1}\}$. If $\succeq_t$ is complete indifference, i.e. all inequalities in (10) are equalities, then it is representable and can be obtained by assigning 1 to all of the utilities. Hence at least one ranking in (10) is strict or $J \neq \emptyset$.

The non-representability of $\succeq_t$ is equivalent to the assertion that the system of linear equalities $(\mu_i - \mu_{i+1}) \cdot \mathbf{x} = 0$, $i \in I$, and linear inequalities $(\mu_j - \mu_{j+1}) \cdot \mathbf{x} > 0$, $j \in J$, has no semi-positive solution.

A standard linear-algebraic argument tells us that inconsistency of the system above is equivalent to the existence of a nontrivial linear combination

$$\sum_{i=1}^{N-1} c_i(\mu_i - \mu_{i+1}) = 0 \tag{11}$$

with non-negative coefficients c_j for $j \in J$ of which at least one is non-zero (see, for example, Theorem 2.9 of [6], page 48). Coefficients c_i, for $i \in I$, can be replaced by their negatives since the equation $(\mu_i - \mu_{i+1}) \cdot \mathbf{x} = 0$ can be replaced with $(\mu_{i+1} - \mu_i) \cdot \mathbf{x} = 0$. Thus we may assume that all coefficients of (11) are non-negative with at least one positive coefficient c_j for $j \in J$. Since the coefficients of vectors $\mu_i - \mu_{i+1}$ are integers, we may choose $c_1, \ldots, c_n$ to be non-negative rational numbers and ultimately non-negative integers.

The equation (11) can be rewritten as

$$\sum_{i=1}^{N-1} c_i \mu_i = \sum_{i=1}^{N-1} c_i \mu_{i+1}, \tag{12}$$

which can be rewritten as the equality of two unions of multisets:

$$\bigcup_{i=1}^{N-1} \underbrace{\mu_i \cup \ldots \cup \mu_i}_{c_i} = \bigcup_{i=1}^{N-1} \underbrace{\mu_{i+1} \cup \ldots \cup \mu_{i+1}}_{c_i} \tag{13}$$

which contradicts to $c_j > 0$, $\mu_j \succ \mu_{j+1}$ and (9). This contradiction proves the theorem.

The above equivalence lies at the heart of proof Theorem 1. Indeed, it already implies, via Claims 1-3 given in the previous section, that Axioms 1-3 imply the existence of an ex-post representation for $\mathcal{D}$. What remains to be shown is the characterization of all such representations.

Consistent orders on $\mathcal{P}_t$ can be represented geometrically [14]. Every point $\mathbf{u} = (u_1, \ldots, u_n) \in \mathbb{R}^n$ defines an order $\succeq_\mathbf{u}$ on $\mathcal{P}_t$, which obtains when we allocate utilities $u_1, \ldots, u_n$ to prizes $i = 1, 2, \ldots, n$, that is

$$\mu \succeq_\mathbf{u} \nu \iff \sum_{i=1}^{n} \mu(i) u_i \geq \sum_{i=1}^{n} \nu(i) u_i. \tag{14}$$

Any order on $\mathcal{P}_t$ that can be expressed as $\succeq_\mathbf{u}$ for some $\mathbf{u} \in \mathbb{R}^n$ is said to be *representable*. We will now argue that the representable linear orders on $\mathcal{P}_t$ are in one-to-one correspondence with the regions of the following hyperplane arrangment.

For any pair of multisets $\mu, \nu \in \mathcal{P}_t[n]$, we define the hyperplane

$$L(\mu, \nu) = \left\{ \mathbf{x} \in \mathbb{R}^n \mid \sum_{i=1}^{n} \mu(i) x_i - \sum_{i=1}^{n} \nu(i) x_i = 0 \right\}$$

and consider the hyperplane arrangement

$$A(t, n) = \{ L(\mu, \nu) \mid \mu, \nu \in \mathcal{P}_t[n] \}. \tag{15}$$

The set of representable linear orders on $\mathcal{P}_t[n]$ is in one-to-one correspondence with the regions of $A = A(t, n)$. In fact, then the linear orders $\succeq_\mathbf{u}$ and $\succeq_\mathbf{v}$ on

$\mathcal{P}_t$ will coincide if and only if **u** and **v** are in the same region of the hyperplane arrangement A. This immediately follows from the fact that the order $\mu \succ_{\mathbf{x}} \nu$ changes to $\mu \prec_{\mathbf{x}} \nu$ (or the other way around) when **x** crosses the hyperplane $L(\mu, \nu)$. The closure of every such region is a convex polytope.

Example 2. The 12 regions on the figure below represent all 12 representable orders on $\mathcal{P}_2[3]$.

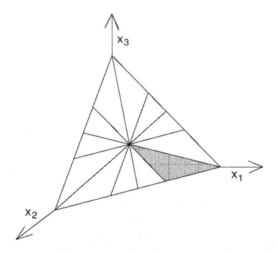

with the shaded region corresponding to the lexicographic order $1^2 \succ 12 \succ 13 \succ 2^2 \succ 23 \succ 3^2$.

Let us note that in (14) we can divide all utilities by $u_1 + \ldots + u_n$ and the inequality will still hold. Hence we could from the very beginning consider that all vectors of utilities are in the hyperplane J given by $x_1 + \ldots + x_n = 1$ and even in the simplex Δ given by $x_i \geq 0$ for $i = 1, 2, \ldots, n$.

Thus, every representable linear order on $\mathcal{P}_t$ is associated with one of the regions of the induced hyperplane arrangement $A^J = \{L \cap J \mid L \in A\}$.

Let us note that due to our non-triviality assumption the vector $\left(\frac{1}{n}, \ldots, \frac{1}{n}\right)$ does not correspond to any order. Consider a utility vector $\mathbf{u} \in \Delta$ different from $\left(\frac{1}{n}, \ldots, \frac{1}{n}\right)$ lying in one of the regions of A^J whose closure is V. We then can normalise **u** applying a positive affine linear transformation which makes its lowest utility zero. Indeed, suppose that without loss of generality $u_1 \geq u_2 \geq \ldots \geq u_n \neq \frac{1}{n}$. Then we can solve for α and β the system of linear equations $\alpha + n\beta = 1$ and $\alpha u_n + \beta = 0$ and since the determinant of this system is $1 - nu_n \neq 0$ its solution is unique. Then the vector of utilities $\mathbf{u}' = \alpha \mathbf{u} + \beta \cdot \mathbf{1}$ will lie on the facet Δ^n of Δ and we will have $\succeq_{\mathbf{u}'} = \succeq_{\mathbf{u}}$. Hence the polytope V has one face on the boundary of Δ. We denote it U. So if the order $\succeq$ on $\mathcal{P}_t$ is linear the dimension of U will be $n - 2$.

In general, when the order on $\mathcal{P}_t$ is not linear, the utility vector **u** that represents this order must be a solution to the finite system of equations and strict inequalities:

$$(\mu - \nu) \cdot \mathbf{u} = 0 \quad \text{whenever } \mu \sim_\mathbf{u} \nu,$$
$$(\mu - \nu) \cdot \mathbf{u} > 0 \quad \text{whenever } \mu \succ_\mathbf{u} \nu, \qquad \forall \, \mu, \nu \in \mathcal{P}_t. \qquad (16)$$

Then $\mathbf{u}$ will lie in one (or several) of the hyperplanes of $A(k, n)$. In that hyperplane an arrangement of hyperplanes of smaller dimension will be induced by $A(k, n)$ and $\mathbf{u}$ will belong to a relative interior of a polytope U of dimension smaller than $n - 2$.

Let now $\succeq = (\succeq_t)_{t \geq 1}$ be a consistent order on $\mathcal{P}$. By Theorem 2 it is locally representable. We have just seen that in such case, for any t, there is a convex polytope U_t such that any vector $\mathbf{u}_t \in ri(U_t)$ represents $\succeq_t$. Due to consistency any vector $\mathbf{u}_s \in ri(U_s)$, for $s > t$ will also represent $\succeq_t$ so $U_t \supseteq U_s$. Thus we see that our polytopes are nested. Note that only points in the relative interior of U_t are suitable points of utilities to rationalise $\succeq_t$. We have almost proved our main theorem. The only thing which is left to note is that the intersection $\bigcap_{t=1}^\infty U_t$ has exactly one element. This is immediately implied by the following

Proposition 1. *Let $\mathbf{u} \neq \mathbf{v}$ be two distinct vectors of normalised non-negative utilities. Then there exist a positive integer t and two multisets $\mu, \nu \in \mathcal{P}_t$ such that $(\mu - \nu) \cdot \mathbf{u} > 0$ but $(\mu - \nu) \cdot \mathbf{v} < 0$.*

Proof. Since $\mathbf{u}$ and $\mathbf{v}$ are normalised we have, in particular, $u_n = v_n = 0$. Since $\mathbf{u} \neq \mathbf{v}$, there will be a point $\mathbf{x} = (x_1, \ldots, x_n) \in \mathbb{R}^n$ such that $\mathbf{x} \cdot \mathbf{u} > 0$ but $\mathbf{x} \cdot \mathbf{v} < 0$. As rational points are everywhere dense in $\mathbb{R}^n$ we may assume that $\mathbf{x}$ has rational coordinates. Then multiplying by their common denominator we may assume all coefficients are integers. After that we may change the last coordinate x_n of $\mathbf{x}$ to x'_n so that to achieve $x_1 + x_2 + \ldots + x'_n = 0$. Now since $u_n = v_n = 0$, we will still have $\mathbf{x}' \cdot \mathbf{u} > 0$ and $\mathbf{x}' \cdot \mathbf{v} < 0$ for $\mathbf{x}' = (x_1, x_2, \ldots, x'_n)$. Now $\mathbf{x}'$ is uniquely represented as $\mathbf{x}' = \mu - \nu$ for two multisets μ and ν. Since the sum of coefficients of $\mathbf{x}'$ was zero, the cardinality of μ will be equal to the cardinality of ν. Let this common cardinality be t. Then $\mu, \nu \in \mathcal{P}_t$ and they are separated by a hyperplane from $A(t, n)$. The proposition is proved.

References

1. Anscombe, F., Aumann, R.: A definition of subjective probabiliy. Annals of Mathematical Statistics 34, 199–205 (1963)
2. Börgers, T., Morales, A.J., Sarin, R.: Expedient and monotone learning rules. Econometrica 72(2), 383–405 (2004)
3. Brown, G.W.: Iterative solutions of games by fictitious play. In: Koopmans, T.C. (ed.) Activity Analysis of Production and Allocation, pp. 374–376. Wiley, New York (1951)
4. Easley, D., Rustichini, A.: Choice without beliefs. Econometrica 67(5), 1157–1184 (1999)
5. Fudenberg, D., Levine, D.K.: The Theory of Learning in Games. MIT Press, Cambridge (1998)
6. Gale, D.: The Theory of Linear Economic Models. McGraw-Hill, New-York (1960)
7. Gigerenzer, G., Selten, R.: Bounded Rationality: The Adaptive Toolbox. MIT Press, Cambridge (2002)

8. Gilboa, I., Schmeidler, D.: A theory of case-based decisions. Cambridge University Press, Cambridge (2001)
9. Gilboa, I., Schmeidler, D.: Inductive Inference: An Axiomatic Approach. Econometrica 71(1), 1–26 (2003)
10. Lettau, M., Uhlig, H.: Rules of thumb and dynamic programming. Tilburg University Discussion Paper (1995)
11. Milnor, J.H.: Games against nature. In: Davis, C.H., Coombs, R.M., Thrall, R.L. (eds.) Decision Processes, pp. 49–60. John Wiley & Sons, Inc., Chichester (1954)
12. Savage, L.J.: The Foundations of Statistics. Harvard University Press, Cambridge (1954)
13. Schlag, K.H.: Why imitate, and if so, how? A boundedly rational approach to multi-armed bandits. Journal of Economic Theory 78(1), 130–156 (1998)
14. Sertel, M.R., Slinko, A.: Ranking Committees, Income Streams or Multisets. Economic Theory 30(2), 265–287 (2007)

Appendix

Proof of Claim 1. Take the hypothesis as given. If the actions $a, b, c, d \in \mathcal{A}$ are distinct, consider a history $g_t \in H_t$ such that $g_t(a) = h_t(a)$, $g_t(b) = h_t(b)$, $g_t(c) = h'_t(a)$ and $g_t(d) = h'_t(b)$. Applying Axiom 2, $a \sim_{g_t} c$ and $b \sim_{g_t} d$ and therefore, $a \succeq_{g_t} b \Leftrightarrow c \succeq_{g_t} d$. Apply Axiom 1 to complete the claim.

Suppose now that a, b, c, d are not all distinct. We will prove that if $\mu(a, h) = \mu(c, h')$ and $\mu(b, h) = \mu(b, h')$, then

$$a \succeq_{h_t} b \Longleftrightarrow c \succeq_{h'_t} b,$$

which is the main case. Let us consider five histories presented in the following table:

	h	h^1	h^2	h^3	h'
a	$h(a)$	$h(a)$	$h'(b)$	$h'(b)$	$h'(a)$
b	$h(b)$	$h(b)$	$h(b)$	$h'(b)$	$h'(b)$
c	$h(c)$	$h'(c)$	$h'(c)$	$h'(c)$	$h'(c)$

In what follows we repeatedly use Axiom 1 and Axiom 2 and transitivity of $\succeq_{h^i}$, $i = 1, 2, 3$. Comparing the first two histories, we deduce that $c \sim_{h^1} a \succeq_{h^1} b$ and $c \succeq_{h^1} b$. Now comparing h^1 and h^2 we have $c \succeq_{h^2} b \sim_{h^2} a$ and $c \succeq_{h^2} a$. Next, we compare h^2 and h^3 and it follows that $c \succeq_{h^3} a \sim_{h^2} b$, whence $c \succeq_{h^3} b$. Now comparing the last two histories we obtain $c \succeq_{h'} b$, as required.

Proof of Claim 2. Given the fact that actions must be ranked for all conceivable histories, $\succeq_t^*$ is a complete ordering of $\mathcal{P}_t$. From its construction, $\succeq_t^*$ is also is reflexive. Again, through appealing to Axiom 1 and Axiom 2 repeatedly, it may be verified that it is also transitive. Indeed, choose $\mu, \nu, \xi \in \mathcal{P}_t$ such that $\mu \succeq_t^* \nu$ and $\nu \succeq_t^* \xi$. Pick three distinct actions $a, b, c \in \mathcal{A}$ and consider a history $h_t \in H_t$ such that $\mu(a, h_t) = \mu$, $\mu(b, h_t) = \nu$ and $\mu(c, h_t) = \xi$. By definition, $a \succeq_{h_t} b$ and $b \succeq_{h_t} c$ while transitivity of $\succeq_{h_t}$ shows that $a \succeq_{h_t} c$. Hence $\mu \succeq_t^* \xi$.

Algorithmic Aspects of Scenario-Based Multi-stage Decision Process Optimization

Ronald Hochreiter

Department of Statistics and Decision Support Systems, University of Vienna
ronald.hochreiter@compmath.net

Abstract. Multi-stage decision optimization under uncertainty depends on a careful numerical approximation of the underlying stochastic process, which describes the future uncertain values on which the decision will depend on. The quality of the scenario model severely affects the quality of the solution of the optimization model. Various approaches towards an optimal generation of discrete-state approximations (represented as scenario trees) have been suggested in the literature. Direct scenario tree sampling based on historical data or econometric models, as well as scenario path simulation and optimal tree approximation methods are discussed from an algorithmic perspective. A multi-stage financial asset management decision optimization model is presented to outline strategies to analyze the impact of various algorithmic scenario generation methodologies.

Keywords: Optimization under uncertainty, scenario generation, scenario optimization, financial decision theory, risk management.

1 Introduction

In this paper, we consider algorithmic issues of the computation of multi-stage scenario-based stochastic decision processes for decision optimization models under uncertainty. We consider that given a specific discrete-time stochastic process on the decision horizon $t = 1, \ldots, T$, a decision maker observes the realization of this random process ξ_t at each decision stage t, and takes a decision x_t based on all observed values up to t $(\xi_1, \ldots, \xi_t)$. Let there now be a sequence of decisions $x_1, \ldots, x_T$. At the terminal stage T we observe a sequence of decisions $x = (x_1, \ldots, x_T)$ with realizations $\xi = (\xi_1, \ldots, \xi_T)$, which lead to cost $f(x, \xi)$ (or likewise profit). The stochastic optimization task is to find the sequence of decisions $x(\xi)$, which minimizes some probability functional (most commonly the expectation, a risk measure, or a combination of these two) of the respective cost function $f(x(\xi), \xi)$ - see e.g. [1] for a classification of risk measures in this context.

We consider the multi-stage case in such a way that there is at least one intermediary stage between root and terminal stage, i.e. $T > 2$. See [2] for a recent overview of the area of stochastic programming, and especially [3] for stochastic programming languages, environments, and applications. Unfortunately, the

F. Rossi and A. Tsoukis (Eds.): ADT 2009, LNAI 5783, pp. 365–376, 2009.
© Springer-Verlag Berlin Heidelberg 2009

topic of scenario generation is not sufficiently treated in most text books on stochastic programming. Formally speaking, we consider multi-stage stochastic programming problems as defined in Eq. 1.

$$\begin{aligned}
\text{minimize } x : \ & \mathbb{F}\big(f(x(\xi), \xi)\big) \\
\text{subject to} \quad & (x(\xi), \xi) \in \mathcal{X} \\
& x \in \mathcal{N}
\end{aligned} \tag{1}$$

The multi-variate, multi-stage stochastic process ξ describes the future uncertainty, i.e. the subjective part of the stochastic program, and the constraint set $\mathcal{X}$ defines feasible combinations of x and ξ. This constraint set is used to model the underlying real-world decision problem. Furthermore, a set of non-anticipativity constraints $\mathcal{N}$, consisting of functions $\xi \mapsto x$ which make sure that x_t is only based on realizations up to stage t $(\xi_1, \ldots, \xi_t)$, is necessary. To solve multi-stage programs with numerical optimization solvers, the underlying stochastic process has to be discretized into a scenario tree, and this scenario tree approximation will inherently fulfill these non-anticipativity constraints.

Two issues negatively affect the application of multi-stage stochastic programs for real-world decision problems: First, modeling the underlying decision problem is a non-trivial task. Multi-stage optimization models and stochastic scenario models require stable scenario tree handling procedures, which are considered to be too cumbersome to be applied to real-world applications, and the communication of tree-based models to non-experts is complicated. Secondly, modeling the underlying uncertainty is complex and messy. As discussed above, a good discrete-time, discrete-space scenario tree approximation of the underlying stochastic process has to be generated in order to numerically compute a solution, and the quality of the scenario model severely affects the quality of the solution (garbage in $\rightarrow$ garbage out).

In this paper, we focus on the second problem. While different tree generation methods have been proposed, there are hardly any comparisons of different methodologies, see e.g. [4], [5], or [6] for an overview of various methodologies proposed so far. Numerical comparisons as shown by [7] and [8] are often focused on a small set of methods and on specific numerical questions. In this paper, a summary of various approaches towards an optimal generation of decision processes is shown to serve as an overview and an outline for various ways to extend and improve the currently applied approaches.

This paper is organized as follows: In Section 2, a straight-forward multi-stage financial asset management problem is presented, before any scenario generation methodologies are discussed in order to introduce the application area of stochastic programming to non-experts. This example will be used for numerical studies. Section 3 outlines important issues regarding scenario tree generation, while Section 4 summarizes five different approaches towards tree generation, both direct tree creation techniques as well as optimal tree approximations of pre-sampled scenario paths. Section 5 presents selected numerical results, while Section 6 concludes the paper.

2   Multi-stage Financial Asset Management

To analyze the impact of different scenario generation methodologies we consider a multi-stage asset management decision optimization model. This model is an extension of the classical risk-return optimization approach which was introduced by Markowitz in the early 1950s [9]. In [10] this approach has been generalized to single-stage stochastic portfolio optimization using a wide range of risk measures (e.g. all coherent risk measures as shown by [11] among others) to overcome the problem of using the Variance as the only risk measure, and to use scenarios instead of using a pre-estimated correlation matrix, which can be problematic if the number of assets under consideration is large. Furthermore, a detailed multi-stage stochastic formulation of the classical Markowitz Mean-Variance optimization has been reported by [12].

Combining the latter two approaches, our decision taker faces a discrete-time decision horizon $t = 1, \ldots, T$, and a set of investment assets $\mathcal{A}$ with uncertain future returns V_a, i.e. a stochastic process represented as a multi-variate, multi-stage scenario tree. There is some investment budget B available at each stage up to $T - 1$. This amount is deterministically determined in advance. The objective function consists of the aforementioned risk-return bi-criteria functional, whereby the aim is to maximize the expected terminal wealth and to maximize[1] the Conditional Value-at-Risk[2] (CVaR at confidence level α) of the wealth at the terminal stage T. Both criteria are weighted using a risk-aversion parameter κ, which can be adapted to the needs of the investor and to the current market situation. CVaR can be reformulated as a linear programming model conveniently as shown by [13], which simplifies the numerical solution procedure. The main decision is concerned with the amount of budget b_a to be invested into each asset a, as the portfolio is rebalanced at the stages $T = 2, \ldots, T - 1$. There is no rebalancing at terminal stage T. One important constraint is that the amount of purchases p in each stage cannot exceed the sum of the amount of sales s plus the additional budget available at the respective stage.

Given the above problem specification, we may formulate our multi-stage stochastic programming model as shown in Eq. 2. The objective function as well as the constraints are shown in stage-based formulation, i.e. the numbers in square brackets represent the stage at which the respective constraint is active.

$$
\begin{aligned}
\text{maximize } & \mathbb{E}(\textstyle\sum_a b_a, T) + \kappa \text{CVaR}_\alpha(\textstyle\sum_a b_a, T) & & \\
\text{subject to } & \textstyle\sum_a b_a = B & & [1] \\
& b_a \le V_a b_a^{(-1)} + p_a - s_a & & \forall \mathcal{A} \ [2, \ldots, T-1] \\
& \textstyle\sum_a p_a \le \textstyle\sum_a s_a + B & & [2, \ldots, T-1] \\
& b_a \le V_a b_a^{(-1)} & & \forall \mathcal{A} \ [T] \\
& b_a, p_a, s_a \ge 0 & & \forall \mathcal{A} \ [1, \ldots, T]
\end{aligned}
\tag{2}
$$

---

[1] Maximizing CVaR is equivalent to minimizing the anticipated financial risk of the stochastic future wealth.

[2] CVaR is also called Expected Shortfall or Tail Value-at-Risk.

The multi-stage recourse can be observed in the second and the fourth constraint, as (the scenario tree) V_a represents the future asset return of asset a in the respective stage and is multiplied by the invested budget $b_a^{(-1)}$ of the previous stage, which is notated using the (-1) superscript.

The parameters which have to be specified by the decision taker are the asset returns V_a, which are stochastic and need to be approximated, as well as the budget B, which is deterministic in this model. The stochastic decision variables, which will be calculated by a numerical optimization solver are the amounts of current investment budget b_a, purchases p_a, as well as sales s_a of each asset a out of the given investment universe $\mathcal{A}$ at each stage t.

3   Multi-stage Scenario Tree Generation

The quality of the scenario tree severely affects the quality of the solution of the multi-stage stochastic decision model, such that any approximation should be done in consideration of optimality criteria, i.e. before a stochastic optimization model is solved, a scenario optimization problem has to be solved independently of the optimization model. It should be noted, that there are also scenario generation approaches where the generation is not decoupled from the optimization procedure, see e.g. [14]. A major drawback is that these methods are often limited to a certain restricted set of models, and cannot be generalized easily.

In the context of separated scenario optimization, optimality can be defined as the minimization of the distance between the original (continuous or highly discrete) stochastic process and the approximated scenario tree. Choosing an appropriate distance may be based on subjective taste, e.g. Moment Matching as proposed by [15], selected due to theoretical stability considerations (see [16] and [17]), which leads to probability metric minimization problems as shown by [18] and [19], or it may be predetermined by chosen approximation method, e.g. by using different sampling schemes like QMC in [20] or RQMC in [21], see also [22]. It is important to remark that once the appropriate distance has been selected, an appropriate heuristic to approximate the chosen distance has to be applied, which affects the result significantly.

Single-stage scenario generation, i.e. an optimal approximation of a multivariate probability distribution without any tree structure can be done via various sampling as well as clustering techniques. The real algorithmic challenge of multi-stage scenario generation is caring about the tree structure while still minimizing the distance. Only in rare cases, this problem can be solved without heuristics.

Different methods for two main strategies to design multi-stage scenario tree generation heuristics will be presented - direct scenario tree sampling, and optimal tree approximation of pre-simulated scenario paths.

Direct scenario tree sampling. Based on historical data or some econometric model, a sampling procedure or a node-wise approximation method is applied to build the scenario tree iteratively from the root node to the terminal stage.

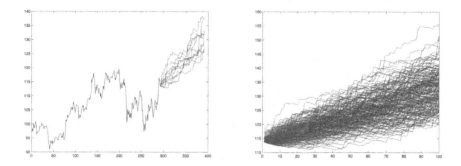

Fig. 1. GARCH(1,1) estimation of financial stock returns, sampling 200 paths

Scenario path simulation and optimal tree approximation. It may be more convenient to use pre-sampled scenario paths and build an optimal approximation out of this data. The time-dependency of the underlying process is preserved, see Fig. 1, and even more importantly any simulation model can be used, which is convenient for real-world applications - either plain econometric time-series models ((V)AR(I)MA, (G)ARCH, ...), specialized models (e.g. Wilkie model for actuarial use), or custom (company-specific) scenario generators may be used as input scenario path generator. The optimization task is to find the optimal tree node and arc links.

One issue, which is often disregarded due to the fact that most research papers on multi-stage scenario tree generation focus on one specific methodology is that different scenario tree methods are restricted to one of the following scenario tree structures, which leads to problems for numerical comparisons: Either the the scenario structure is represented nodes-by-node with a vector denoting the successors of each node in the respective stage, e.g. $N(5, 3, 2)$ results in a scenario tree with 30 scenarios and $1+5+15+30 = 51$ nodes, or it is represented nodes-at-stage-wise with a vector containing the total number of nodes in each stage, e.g. $S(10, 20, 30)$ results in a scenario tree with 30 scenarios and $1 + 10 + 20 + 30 = 61$ nodes. The first type is common for techniques, which build the tree from the root to the terminal stage node per node, while the second is inherently used for scenario generation methods, which try to find the optimal per-node branching structure in addition to approximating tree values.

Another issue regarding multi-stage scenario tree generation is the absence of a common tree format, i.e. no common standard for representing discretized stochastic processes is available, mainly because of the lack of commercial interest. To accommodate all different tree structures, and the implicit formulation of non-anticipativity constraints, a node-based vector/matrix data format of scenario trees is proposed in this paper, which is shown in Table 1. Every scenario generation method presented below produces exactly this representation.

Table 1. Vector/matrix data format for multi-stage scenario trees

$V(n,d)$ d-dimensional value of node n
$A(n)$ ancestor node of node n
$T(n)$ stage of node n
$P(n)$ probability to reach node n from its ancestor

4 Multi-stage Scenario Tree Generation Methods

4.1 Direct Scenario Tree Sampling

Iterative node-by-node data resampling. If we aim at building a scenario tree with a node-based tree structure $N(n_1, \ldots, n_T)$ given some (multi-variate) times-series, the most straightforward approach is to start at the root node, which represents the current (i.e. deterministic) value, and then to proceed as follows: for each stage $t = 1, \ldots, T$, and for each node in stage t sample n_t random values from the time series using time-series data as well as data from already sampled ancestor nodes.

Example. If a time-series $v_{a,t}$ consists of three years of daily asset return data, and we want to sample a scenario tree reflecting possible developments for the next three years in three stages, we may sample n_1 nodes for the first stage. For each of this n_1 nodes, we sample n_2 values using a shorter time series, e.g. only the last two years and include each previously sampled value from the first stage into the initial data set for each second stage node. For the third stage we use only one year of data and both ancestor nodes from stage 1 and 2.

While the advantage of this approach is that it is simple, and historical dependence of even highly multi-variate structures can be kept cheaply, it bears the disadvantage that if small trees are needed, two runs might generate completely different trees, and it may be used for a huge amount of stages, if the time-series are not long enough.

Iterative node-by-node data approximation. To improve the quality of the solution of the previous method, random sampling of n_t node values can be replaced with an optimal single-stage approximation, e.g. Principal Component Analysis (see e.g. [23]), K-Means clustering (see e.g. [24]), or moment matching. The result is affected by the chosen approximation method - both a distance and a heuristic needs to be specified.

Distribution fitting and direct node sampling. Sometimes the decision taker wants to use some specific stochastic process for the future development of uncertain values, and is aware of how to estimate the parameters of the respective process. Given a node-based tree structure $N(n_1, \ldots, n_T)$, the tree can be generated from the root node iteratively.

Example. A univariate process ξ might be normally distributed and follows an additive recursion, i.e.

$$\xi_0 = \mu_0, \quad \xi_1 \sim N(\mu_0, \sigma_0^2), \quad \xi_t = b\xi_{t-1} + \epsilon_t, \quad \epsilon_t \sim N(\mu, \sigma^2), \quad (3)$$

where

$$\mu = \mu_0(1 - b), \sigma = \sigma_0(1 - b^2),$$

and ξ_t and ϵ_t are independent. ξ is a stationary Gaussian Markov process. This process is used for numerical results shown below.

In general, this approach is simple and extremely flexible, as any stochastic process which can be estimated and simulated can be used. One disadvantage is the loss of time any dependency, which leads to bad convergence behavior.

4.2 Optimal Tree Approximations of Scenario Paths

Scenario merging - forward and backward. Given a stage-based tree structure $S(s_1, \ldots, s_T)$ and one specific distance function, a straightforward approach to build an optimal tree given a set of pre-sampled scenarios is to start by calculating the distance of each scenario path to each other with the chosen distance. Then proceed in a forward-sweep $t = 1, \ldots, T$ or in a backward-sweep $t = T, \ldots, 1$, and as long as there are more nodes than s_t within the current stage t merge the nodes of the two scenarios, which are closest to each other.

The advantage is that this process is rather straightforward, quite efficient and flexible as the distance calculation is done only once at the beginning, but forward and backward sweeps may lead to different results. Consider the simple simple example shown in Figure 2. We like to approximate 4 sampled scenarios to build a $S(2, 3, 4)$ tree. In this case there are already differences between using forward and backward-based algorithms, notably not in the terminal stage, but within intermediary stages, which are crucial in any multi-stage setting.

Forward-based scenario clustering. Another method which can be computed efficiently, and can be used for scenario trees with a large amount of stages (e.g. more than 100), is to resemble the node-per-node iterative approximation, but to use the future paths of approximated nodes for subsequent stages.

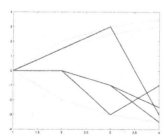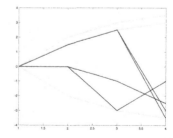

Fig. 2. Scenario merging example - forward sweep (left), and backward sweep (right)

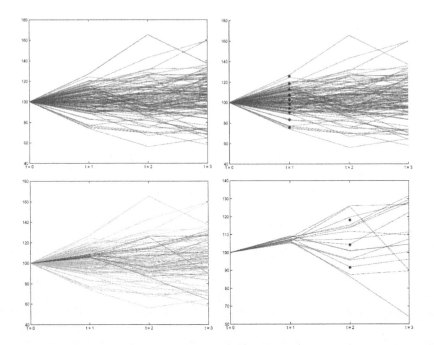

Fig. 3. Forward-based scenario clustering: given pre-sampled paths (top, left), approximate the first stage (top, right), and then use all future paths for each node (bottom, left) to continue the approximations iteratively node-by-node (bottom, right)

The implementation is quite tricky, and in the case of a high number of stages non-branching scenarios might occur. The process is schematically shown in Figure 3.

5   Numerical Results and Comparison

To obtain numerical results using the multi-stage stochastic financial asset management problem shown in Section 2, we use two different financial assets. A fixed-income instrument with a deterministic fixed return r at each stage, and a stochastic stock, which will be used for comparing scenario generation techniques. Historical data of the IBM stock has been used, using daily data from January 3rd, 2007 to February 29th, 2008, which is summarized in Figure 4.

The following additional parameters have been used for solving the asset management model: the initial budget is $B = 10000$ in the first stage, and 0 for all following stages ($t=2,\ldots,T$), i.e. budget is only available at the root stage, which does not have economic implications. The risk quantile CVaR $\alpha = 0.9$, and the risk aversion parameter $\kappa = 1$. The fixed-income return is $r = 1.03$ p.a. The return will be replicated for each node of the stock tree (adjusted given the respective stage of the node).

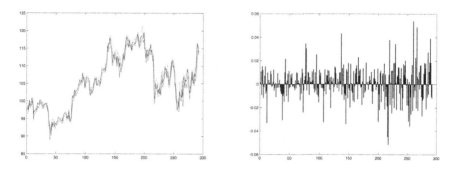

Fig. 4. IBM: historical closing prices (left) and daily returns (right)

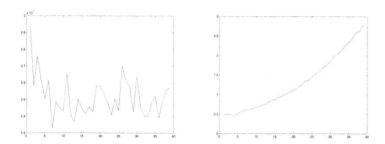

Fig. 5. Distribution fitting and direct node sampling. $N(k, k)$ $\forall k = 2, \ldots, 50$.

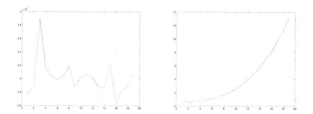

Fig. 6. Distribution fitting and direct node sampling. $N(k, k, k)$ $\forall k = 2, \ldots, 20$.

The node-based optimization model has been implemented using AMPL (implicit formulation of non-anticipativity constraints). All scenario generation methods have been implemented in MatLab 7.6. The optimization models have been solved using Mosek 5 on a MacBook Pro, Mac OS X 10.5.1, Intel, with Intel Core 2 Duo (2.4 GHz), i.e. 1 CPU with 2 Cores, and 2 GB 667 GHz DDR2 SDRAM.

Different aspects have to be analyzed: the run-time of the optimization problem given the specific tree, the convergence of the objective function, and more subtle, the convergence of the optimal decision, i.e. in our specific example the decision on how much budget should be invested into either the stock or the

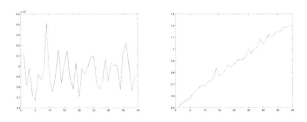

Fig. 7. Distribution fitting and direct node sampling. $N(k, 3, 3) \; \forall k = 2, \ldots, 40$.

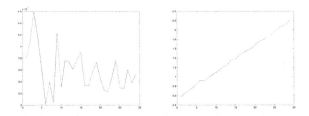

Fig. 8. Distribution fitting and direct node sampling. $N(k, 3, 3, 3) \; \forall k = 2, \ldots, 30$.

fixed-income product. The visualization of the convergence cannot be done satisfyingly in most cases due to the high dimensionality, such that we will focus on the first two aspects using the stochastic process from Eq. (3) above. The financial asset management problem is solved and both the convergence of the objective function (left) as well as the solution time of the optimization solver (right) is shown. The tree structure severely affects the required solution time and the stability of the underlying optimization problem solution. A trade-off has to be taken, which has to be analyzed for each chosen combination of an optimization problem and a scenario generation technique.

The results are shown for two stages in Fig. 5, for three stages using different node structures in Fig. 6 and Fig. 7, and for four stages in Fig. 8.

6   Conclusion

In this paper, we presented a unified comparison of different decision process optimization techniques for optimization under uncertainty models. Different generation techniques are described, some of which have not been explicitly described in the literature yet. This description and comparison should create an awareness of choosing some method, and may be used as a guideline for creating and improving new methodologies towards scenario generation, which is the crucial part for solving multi-stage decision problems.

References

1. Eichhorn, A., Römisch, W.: Polyhedral risk measures in stochastic programming. SIAM Journal on Optimization 16(1), 69–95 (2005)
2. Ruszczyński, A., Shapiro, A. (eds.): Stochastic programming. Handbooks in Operations Research and Management Science, vol. 10. Elsevier Science B.V., Amsterdam (2003)
3. Wallace, S.W., Ziemba, W.T. (eds.): Applications of stochastic programming. MPS/SIAM Series on Optimization, vol. 5. Society for Industrial and Applied Mathematics, SIAM (2005)
4. Dupačová, J., Consigli, G., Wallace, S.W.: Scenarios for multistage stochastic programs. Annals of Operations Research 100, 25–53 (2000)
5. Pennanen, T.: Epi-convergent discretizations of multistage stochastic programs. Mathematics of Operations Research 30(1), 245–256 (2005)
6. Kuhn, D.: Aggregation and discretization in multistage stochastic programming. Mathematical Programming 113(1, Ser. A), 61–94 (2008)
7. Hilli, P., Pennanen, T.: Numerical study of discretizations of multistage stochastic programs. Kybernetika 44(2), 185–204 (2008)
8. Kaut, M., Wallace, S.W.: Evaluation of scenario generation methods for stochastic programming. Pacific Journal of Optimization 3(2), 257–271 (2007)
9. Markowitz, H.M.: Portfolio selection. The Journal of Finance 7(1), 77–91 (1952)
10. Hochreiter, R.: Evolutionary stochastic portfolio optimization. In: Brabazon, A., ONeill, M. (eds.) Natural Computing in Computational Finance. Studies in Computational Intelligence, vol. 100, pp. 67–87. Springer, Heidelberg (2008)
11. Artzner, P., Delbaen, F., Eber, J.M., Heath, D.: Coherent measures of risk. Mathematical Finance 9(3), 203–228 (1999)
12. Steinbach, M.C.: Markowitz revisited: mean-variance models in financial portfolio analysis. SIAM Review 43(1), 31–85 (2001)
13. Rockafellar, R., Uryasev, S.: Optimization of Conditional Value-at-Risk. The Journal of Risk 2(3), 21–41 (2000)
14. Casey, M.S., Sen, S.: The scenario generation algorithm for multistage stochastic linear programming. Mathematics of Operations Research 30(3), 615–631 (2005)
15. Høyland, K., Wallace, S.W.: Generating scenario trees for multistage decision problems. Management Science 47(2), 295–307 (2001)
16. Rachev, S.T., Römisch, W.: Quantitative stability in stochastic programming: the method of probability metrics. Mathematics of Operations Research 27(4), 792–818 (2002)
17. Heitsch, H., Römisch, W., Strugarek, C.: Stability of multistage stochastic programs. SIAM Journal on Optimization 17(2), 511–525 (2006)
18. Pflug, G.C.: Scenario tree generation for multiperiod financial optimization by optimal discretization. Mathematical Programming 89(2, Ser. B), 251–271 (2001)
19. Dupačová, J., Gröwe-Kuska, N., Römisch, W.: Scenario reduction in stochastic programming. An approach using probability metrics. Mathematical Programming 95(3, Ser. A), 493–511 (2003)
20. Pennanen, T., Koivu, M.: Epi-convergent discretizations of stochastic programs via integration quadratures. Numerische Mathematik 100(1), 141–163 (2005)
21. Koivu, M.: Variance reduction in sample approximations of stochastic programs. Mathematical Programming 103(3, Ser. A), 463–485 (2005)
22. Pennanen, T.: Epi-convergent discretizations of multistage stochastic programs via integration quadratures. Mathematical Programming 116(1-2, Ser. B), 461–479 (2009)

23. Topaloglou, N., Vladimirou, H., Zenios, S.: Cvar models with selective hedging for international asset allocation. Journal of Banking and Finance 26(7), 1535–1561 (2002)
24. Hochreiter, R., Pflug, G.C.: Financial scenario generation for stochastic multi-stage decision processes as facility location problems. Annals of Operations Research 152(1), 257–272 (2007)

Choquet Optimization Using GAI Networks for Multiagent/Multicriteria Decision-Making

Jean-Philippe Dubus, Christophe Gonzales, and Patrice Perny

LIP6 - UPMC
104 avenue du président Kennedy, F-75016 Paris
firstname.lastname@lip6.fr

Abstract. This paper is devoted to preference-based recommendation or configuration in the context of multiagent (or multicriteria) decision making. More precisely, we study the use of decomposable utility functions in the search for Choquet-optimal solutions on combinatorial domains. We consider problems where the alternatives (feasible solutions) are represented as elements of a product set of finite domains and evaluated according to different points of view (agents or criteria) leading to different objectives. Assuming that objectives take the form of GAI-utility functions over attributes, we investigate the use of GAI networks to determine efficiently an element maximizing an overall utility function defined by a Choquet integral.

Keywords: GAI-nets, Choquet Integral, Multiobjective Combinatorial Optimization, Multiagent Decision-Making, Preference-based Configuration.

1 Introduction

The multiplication of preference-based configuration problems has stressed the need for compact preference representation languages and for preference-based optimization algorithms. In this area, graphical models are omnipresent. One can distinguish non-numerical models like CP-nets [1,2] and their extension to the multiagent case mCP-nets [3] on the one hand, and numerical models based on decomposable utility functions like UCP-nets [4] and GAI-nets [5,6,7] on the other hand. In this paper, we investigate the potential of GAI-networks to represent and solve decision making problems where the performance of a solution is evaluated according to different points of view. This type of problem occurs when several criteria, possibly conflicting, must be considered in the decision analysis, or when several agents are involved in the decision process. In both cases, any feasible solution is represented by a vector of utilities. Assuming that each of these utilities is defined by a GAI-decomposable model, we study the use of GAI-nets to determine efficiently a solution having a utility vector maximizing an overall utility function defined by a Choquet integral.

The paper is organized as follows: in Section 2 we explain how GAI-networks are used to represent preferences in multiobjective problems. Then, after recalling

F. Rossi and A. Tsoukis (Eds.): ADT 2009, LNAI 5783, pp. 377–389, 2009.

basic notions linked to capacities and Choquet integrals (Section 3), we introduce a vector-passing algorithm for Choquet-optimization (Section 4). Under the assumption of convex capacity, we propose a refinement of the previous algorithm and a second algorithm based on a ranking procedure (Section 5). Both algorithms have been implemented and tested on randomly drawn instances. The solutions times obtained are given for the sake of comparison (Section 6).

2   GAI Models for Individual and Collective Preferences

In configuration problems, alternatives (feasible solutions) are characterized by n variables (or attributes) $x_1, \ldots, x_n$ taking their values in finite domains $X_1, \ldots,$ X_n respectively. They can thus be seen as elements of the product set of these domains $\mathcal{X} = X_1 \times \cdots \times X_n$. Throughout this paper, by abuse of notation, for any set $\mathbf{Y} \subseteq \{1, ..., n\}$, $X_{\mathbf{Y}}$ refers to $\prod_{i \in \mathbf{Y}} X_i$ and $x_{\mathbf{Y}}$ to the projection of $x \in \mathcal{X}$ on $X_{\mathbf{Y}}$. We also consider preference relations over $\mathcal{X}$ representable by utility functions, i.e., by functions $u : \mathcal{X} \mapsto \mathbb{R}$ such that, for all $x, y \in \mathcal{X}$, $u(x) \geq u(y)$ if and only if x is preferred to y or x and y are judged equivalent. Such functions u are used within solvers to determine the best elements in $\mathcal{X}$ [8].

One major difficulty in using utilities lies in their elicitation: each agent has her own preferences and, hence, her own utility u that needs be constructed prior to being used for optimization tasks. However, on combinatorial domains such as $\mathcal{X}$, elicitation may be impossible as it may involve asking unreasonably large amounts of questions to the agent. Fortunately, it is often the case that subsets of attributes are considered independent by the agent. For instance, the brand of a car may be irrelevant to preferences over its colors, hence inducing an independence between color and brand. In such cases, these independences can be exploited to drastically reduce the elicitation burden. In the literature, different types of independence have been studied such as *preferential independence* or *utility independence* [9,8,10], that induce different decompositions of utility u as a function of subutilities, say u_i's, defined over small sets of attributes.

The most widely used decomposition is the additive one: $u(x) = \sum_{i=1}^n u_i(x_i)$ for any $x = (x_1, \ldots, x_n) \in \mathcal{X}$. Note that this model only requires eliciting and storing $u_i(x_i)$ for any $x_i \in X_i$, $i = 1, \ldots, n$. However, such a decomposition is not always appropriate as it inevitably rules out any interaction between attributes, which is far from being realistic. Some generalizations of additive utilities have thus been investigated. In particular, GAI (generalized additive independence) decompositions introduced by [11] are especially attractive as they allow quite general interactions between attributes while preserving some decomposability. Actually, GAI decomposition is a generalization of the additive decomposition in which subutilities u_i's are allowed to be defined over overlapping factors.

Definition 1. *Let $\mathbf{C}_1, \ldots, \mathbf{C}_k$ be subsets of $\mathbf{N} = \{1, \ldots, n\}$ such that $\mathbf{N} = \bigcup_{i=1}^k \mathbf{C}_i$. A utility function $u(\cdot)$ over $\mathcal{X}$ is GAI-decomposable w.r.t. the $X_{\mathbf{C}_i}$'s if and only if there exist functions $u_i : X_{\mathbf{C}_i} \mapsto \mathbb{R}$ such that:*

$$u(x_1, \ldots, x_n) = \sum_{i=1}^k u_i(x_{\mathbf{C}_i}), \quad \text{for all } x = (x_1, \ldots, x_n) \in \mathcal{X} \ .$$

For instance, $u(a, b, c, d, e, f) = u_1(a, b) + u_2(c, d) + u_3(a, c, e) + u_4(e, f)$ defined on $A \times B \times C \times D \times E \times F$ is a GAI-decomposable utility, with $X_{\mathbf{C}_1} = A \times B$, $X_{\mathbf{C}_2} = C \times D$, $X_{\mathbf{C}_3} = A \times C \times E$ and $X_{\mathbf{C}_4} = E \times F$. GAI decompositions can be represented by graphical structures called *GAI networks* [5]:

Definition 2. *Let* $u(x) = \sum_{i=1}^{k} u_i(x_{\mathbf{C}_i})$ *be a GAI utility. A GAI net representing* u *is an undirected graph* $\mathcal{G} = (\mathcal{C}, \mathcal{E})$ *satisfying the following properties:*

Prop 1: $\mathcal{C} = \{X_{\mathbf{C}_1}, \ldots, X_{\mathbf{C}_k}\}$. *Vertices* $X_{\mathbf{C}_i}$*'s are called* cliques. *To each vertex* $X_{\mathbf{C}_i}$ *is associated the corresponding factor* u_i *from the utility function* u*;*

Prop. 2: $(X_{\mathbf{C}_i}, X_{\mathbf{C}_j}) \in \mathcal{E} \Rightarrow \mathbf{C}_i \cap \mathbf{C}_j \neq \emptyset$. *Edges* $(X_{\mathbf{C}_i}, X_{\mathbf{C}_j})$*'s are labeled by* $X_{\mathbf{S}_{ij}}$, *where* $\mathbf{S}_{ij} = \mathbf{C}_i \cap \mathbf{C}_j$. $X_{\mathbf{S}_{ij}}$ *is called a* separator*;*

Prop. 3: *for all* $X_{\mathbf{C}_i}, X_{\mathbf{C}_j}$ *such that* $\mathbf{C}_i \cap \mathbf{C}_j = \mathbf{S}_{ij} \neq \emptyset$, *there exists a path between* $X_{\mathbf{C}_i}$ *and* $X_{\mathbf{C}_j}$ *in* $\mathcal{G}$ *such that for every clique* $X_{\mathbf{C}_h}$ *in this path* $\mathbf{S}_{ij} \subseteq \mathbf{C}_h$ *(running intersection property).*

Cliques are drawn as ellipses and separators as rectangles. For any GAI decomposition, by Definition 2, cliques should be the sets of variables of the subutilities. The edges in the network represent the intersections between subsets of attributes. Fig. 1 shows the GAI net's structure for the example given just below Definition 1. In this paper, we shall only be interested in GAI trees as it is not restrictive [5]. For the elicitation of GAI networks, refer to [5,12,6].

Consider now a finite set of objectives, criteria or agents, $M = \{1, \ldots, m\}$ and assume that any solution $x \in \mathcal{X}$ is characterized by a utility vector $(u^1(x), \ldots, u^m(x)) \in \mathbb{R}^m$ where $u^i : \mathcal{X} \to \mathbb{R}$ is the ith utility. It measures the relative utility of alternatives with respect to the ith point of view (criterion or agent) considered in the problem. Hence, the comparison of alternatives, say x and y, now reduces to that of their utility vectors $(u^1(x), \ldots, u^m(x))$ and $(u^1(y), \ldots, u^m(y))$.

Each u^i is actually a single utility and, as such, can be GAI decomposable. Assume that all u^i's have the same GAI structure, that is, the u^i's are decomposable as sums of functions u_j^i's whose domains are the same for all i's (but their values differ from one j to another). Then, a GAI net compactly encoding vectors $(u^1, \ldots, u^m)$ can easily be constructed: its graphical structure is that of the GAI net of any u^i (since they are all identical), and each clique $X_{\mathbf{C}_j}$ contains utility vectors $(u_j^1, \ldots, u_j^m)$. Fig. 1 shows how vectors of utilities u^i's decomposable as $u_1^i(a, b) + u_2^i(c, d) + u_3^i(a, c, e) + u_4^i(e, f)$ can be represented by a GAI net. In this figure, tables contain values of utility vectors $(u_j^1, \ldots, u_j^m)$, for fixed j's.

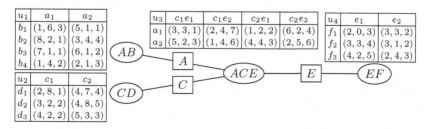

u_1	a_1	a_2
b_1	$(1, 6, 3)$	$(5, 1, 1)$
b_2	$(8, 2, 1)$	$(3, 4, 4)$
b_3	$(7, 1, 1)$	$(6, 1, 2)$
b_4	$(1, 4, 2)$	$(2, 1, 3)$

u_2	c_1	c_2
d_1	$(2, 8, 1)$	$(4, 7, 4)$
d_2	$(3, 2, 2)$	$(4, 8, 5)$
d_3	$(4, 2, 2)$	$(5, 3, 3)$

u_3	$c_1 e_1$	$c_1 e_2$	$c_2 e_1$	$c_2 e_2$
a_1	$(3, 3, 1)$	$(2, 4, 7)$	$(1, 2, 2)$	$(6, 2, 4)$
a_2	$(5, 2, 3)$	$(1, 4, 6)$	$(4, 4, 3)$	$(2, 5, 6)$

u_4	e_1	e_2
f_1	$(2, 0, 3)$	$(3, 3, 2)$
f_2	$(3, 3, 4)$	$(3, 1, 2)$
f_3	$(4, 2, 5)$	$(2, 4, 3)$

Fig. 1. Example of a GAI network with three criteria

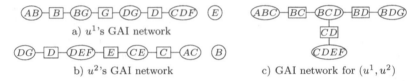

a) u^1's GAI network

b) u^2's GAI network

c) GAI network for (u^1, u^2)

Fig. 2. The GAI trees representing u^1 and u^2 and that for (u^1, u^2)

Of course, in practice, u^i's are seldom decomposable w.r.t. the same GAI structure. For instance, if $\mathcal{X}$ is a set of cars, then, in a family, the utility over $\mathcal{X}$ of the father may be decomposable as $u^1(\text{car}) = u_1^1(\text{price,brand}) + u_2^1(\text{power,speed,consumption}) + u_3^1(\text{speed,security})$ whereas that of the mother may be $u^2(\text{car}) = u_1^2(\text{price,consumption}) + u_2^2(\text{color,brand})$, and the utility of their son $u^3(\text{car}) = u_1^3(\text{brand}) + u_2^3(\text{color}) + u_3^3(\text{power,speed})$. In such a case, we need to find a GAI net with "bigger" cliques that can contain all the u_j^i's functions while encompassing as much as possible the decompositions of the u^i's. For instance, utilities u^1 and u^2 decomposable w.r.t. the GAI nets on the left of Fig. 2 can both be represented (less compactly) by that of Fig. 2.c. Hence vectors (u^1, u^2) are GAI decomposable according to Fig. 2.c. This graph can be constructed by triangulation of the union of the u^i's Markov graphs [13,5].

3   Preference Aggregation with the Choquet Integral

In a multiagent/multicriteria problem, comparing elements of $\mathcal{X}$ amounts to comparing their respective utility profiles. The basic preference model to compare solutions is Pareto dominance defined, for any pair $x, y \in \mathcal{X}$ by: $x \succ_P y \Leftrightarrow u^i(x) \geq u^i(y)$ for all $i \in \{1, \ldots, m\}$ and $u^j(x) > u^j(y)$ for some j. This naturally leads to a primary optimality concept known as Pareto-optimality. Pareto optimal elements in a set $X \subseteq \mathcal{X}$ are those that are Pareto-dominated by no other element in X, i.e., they have a utility profile that cannot be improved on one component without downgrading another one. As shown in [14] Pareto-optimal elements in $\mathcal{X}$ can be computed using vector-valued GAI-networks (see Fig. 1). However, Pareto-dominance is only a partial weak order that leaves many pairs of solutions uncompared. Hence, the Pareto set can be huge due to the combinatorial nature of the problem, and its exact determination requires, for some instances, prohibitive computation times. Fortunately, decision theory provides various preference models refining Pareto dominance and modeling various attitudes in preference aggregation. Among them, the Choquet integral [15] is one of the most expressive decision criteria. It is an aggregation function that generalizes weighted averages when weights are not only attached to each component (criteria or agent) but also possibly to any subset of components. These weights are possibly non additive and are represented by a *capacity* on $M = \{1, \ldots, m\}$.

Definition 3. *A capacity on M is a set function $v : 2^M \rightarrow [0,1]$ such that:*
$$v(\emptyset) = 0; \quad v(M) = 1; \quad \forall A, B \in 2^M \text{ such that } A \subseteq B, \ v(A) \leq v(B).$$

For any subset $A \subseteq M$, $v(A)$ represents the importance of coalition A. Let us first recall some definitions about capacities.

Definition 4. *A capacity v is said to be* convex *(or supermodular) when $v(A \cup B) + v(A \cap B) \geq v(A) + v(B)$ for all $A, B \subseteq M$, and it is said to be* concave *(or submodular) when $v(A \cup B) + v(A \cap B) \leq v(A) + v(B)$ for all $A, B \subseteq M$.*

Definition 5. *To any capacity v, we can associate a* dual *capacity $\bar{v}$ defined by $\bar{v}(A) = 1 - v(M \setminus A)$ for all $A \subseteq M$.*

It is well known that $\bar{v}$ is concave if and only if v is convex and vice-versa. Remark that when v is convex, we have $v(A) + v(M \setminus A) \leq 1$, hence $v(A) \leq \bar{v}(A)$. As we shall see later, a useful concept in this case is the core of v defined by:
$$core(v) = \{\lambda \in \mathcal{L} : v(A) \leq \lambda(A) \leq \bar{v}(A)\},$$
where $\mathcal{L}$ is the set of probability distributions on M and $\lambda(A) = \sum_{i \in A} \lambda_i$ represents the probability of A. The core is known to be non-empty as soon as v is convex [16]. This result will be used in Section 5. The Choquet integral of a utility vector $u(x) = (u^1(x), \ldots, u^m(x))$ w.r.t. a capacity v is defined by:
$$C_v(u(x)) = \sum_{i=1}^{m} \left[v(X^{(i)}) - v(X^{(i+1)})\right] u^{(i)}(x) = \sum_{i=1}^{m} \left[u^{(i)}(x) - u^{(i-1)}(x)\right] v(X^{(i)}) \quad (1)$$

where $(.)$ is a permutation on $\{1, \ldots, m\}$ such that $0 = u^{(0)}(x) \leq u^{(1)}(x) \leq \ldots \leq u^{(m)}(x)$, $X^{(i)} = \{j \in M, u^j(x) \geq u^{(i)}(x)\} = \{(i),(i+1), \ldots, (m)\}$ for $i \leq m$ and $X^{(m+1)} = \emptyset$. Note that $X^{(i+1)} \subset X^{(i)}$, hence $v(X^{(i)}) \geq v(X^{(i+1)})$ for all i. The Choquet integral generalizes averages with the following interpretation based on Eq. (1): for a given utility vector $(u^1(x), \ldots, u^m(x))$, the outcome is at least $u^{(1)}(x)$ with weight $v(X^{(1)}) = 1$, then it increases from $u^{(1)}(x)$ to $u^{(2)}(x)$ with weight $v(X^{(2)})$, then from $u^{(2)}(x)$ to $u^{(3)}(x)$ with weight $v(X^{(3)})$, and so on... The overall integral thus results from aggregation of marginal utility increments $[u^{(i)}(x) - u^{(i-1)}(x)]$ weighted by $v(X^{(i)})$. Note that Choquet Integral includes weighted averages as particular cases. Indeed, when v is additively decomposable, $v(A) = \sum_{i \in A} v_i$ for all $A \subseteq M$, where $v_i = v(\{i\})$. Hence $v(X^{(i)}) - v(X^{(i+1)}) = v^{(i)}$ for all i and $C_v(u(x)) = \sum_{i=1}^{m} v_{(i)} u^{(i)}(x) = \sum_{i=1}^{m} v_i u^i(x)$. When used with a non-additive capacity, it offers enhanced descriptive possibilities.

Example 1. Consider a case with 3 criteria or agents ($M = \{1,2,3\}$) and 3 solutions x, y, z with utility vectors $u(x) = (15,5,10)$, $u(y) = (10,10,10)$ and $u(z) = (5,15,10)$ respectively, and the convex capacity v defined in Table 1 (we also give its dual $\bar{v}$ and an additive capacity $p \in core(v)$). $C_v(u(x)) = 5 \times 1 + (10 - 5) \times 0.5 + (15 - 10) \times 0.2 = 8.5$, $C_v(u(y)) = 10 \times 1 = 10$ and $C_v(u(z)) = 5 \times 1 + (10 - 5) \times 0.5 + (15 - 10) \times 0.1 = 8$. Hence according to the model, we get: $y \succ x \succ z$. If we use the dual capacity $\bar{v}$, which is concave, we get $C_{\bar{v}}(u(x)) = 5 \times 1 + (10 - 5) \times 0.9 + (15 - 10) \times 0.5 = 12$, $C_{\bar{v}}(u(y)) = 10 \times 1 = 10$ and $C_{\bar{v}}(u(z)) = 5 \times 1 + (10 - 5) \times 0.8 + (15 - 10) \times 0.5 = 11.5$. Hence with $\bar{v}$

we get: $x \succ z \succ y$. Note that none of the two orders obtained are representable with a weighted sum because $x \succ y$ would imply $y \succ z$, $y \succ x$ would imply $z \succ y$ and conversely. When v is convex we can see that solution y with a flat utility profile is better ranked. On the contrary, with a concave capacity, it seems that solutions x and z with contrasted profiles are preferred to y. As we shall see in Section 5, this is a general feature of Choquet integral: we have to use a convex capacity to exhibit preference for well-balanced solutions and conversely.

In the next section, we investigate the determination of the optimal tuple in $\mathcal{X}$ w.r.t. the Choquet integral. Note that, when choosing $v(A) = 1$ for all non-empty $A \subseteq M$, then $C_v(u(x)) = u^{(m)}(x) = \max_{i \in M} u^i(x)$. Hence the determination of a Choquet-optimal solution reduces to a min-max optimization problem which is known to be NP-hard even when every function u^i is an additive utility [17].

4   A Vector-Passing Algorithm for Choquet Optimization

All GAI message-passing algorithms rely on the same principle: a clique called *root* is chosen to concentrate during a *collect* phase all the information relevant to compute some quantity to be optimized. This phase is processed recursively: *root* asks its neighbor cliques to send it messages containing the aforementioned relevant information; in turn, these neighbors ask their other neighbors to send relevant information, and so on. Once a clique has received all the information it requested, it computes and sends the message it was asked for. Once *root* has received all the information it requested, a *distribute* phase is applied that propagates recursively optimal attributes instantiations from *root* toward the outside of the GAI net. The result of this phase is an optimal instantiation tuple of all the attributes of the GAI network w.r.t. the quantity to be optimized.

As an illustration in the scalar case, consider a utility decomposable according to the graph of Fig. 1: $u(a, b, c, d, e, f) = u_1(a, b) + u_2(c, d) + u_3(a, c, e) + u_4(e, f)$, where each u_i's codomain is $\mathbb{R}$. Assume we wish to find a tuple maximizing u. Let clique EF act as *root*. Collect consists in EF asking ACE to send a message, which in turn asks both AB and CD to send messages. Clique AB sends message $\phi_1(A) = \{\max_b u_1(a, b) : a \in A\}$ and clique CD sends $\phi_2(C) = \{\max_d u_2(c, d) : c \in C\}$, that is, messages $\phi_1(A)$ and $\phi_2(C)$ contain the optimal values of u_1 and u_2 for each value of separators A and C respectively. Then clique ACE sends message $\phi_3(E) = \{\max_{a,c}[u_3(a, c, e) + \phi_1(a) + \phi_2(c)] : e \in E\}$. Finally, *root* EF computes $\max_{e,f}[u_4(e, f) + \phi_3(e)]$, which is the optimal value for u. Actually, as described more formally in [7], we just computed:

$$\max_{e,f}[u_4(e, f) + \max_{a,c}((\max_b u_1(a, b)) + (\max_c u_2(c, d)) + u_3(a, c, e))] = \max_{a,b,c,d,e,f} u .$$

The distribute phase just traces back the Argmax's to find the optimal tuple.

In a multiagent/multicriteria setting, where the overall criterion to optimize is a Choquet integral, there is an additional source of complexity: the Choquet integral is not a GAI decomposable function even when u^i's are GAI decomposable. As a consequence, optimality of the solutions cannot be guaranteed by

Table 1. A capacity v for three criteria (named 1,2,3)

A	$\emptyset$	$\{1\}$	$\{2\}$	$\{3\}$	$\{1,2\}$	$\{1,3\}$	$\{2,3\}$	$\{1,2,3\}$
$v(A)$	0	0.2	0.1	0.1	0.4	0.5	0.5	1
$p(A)$	0	0.5	0.4	0.1	0.9	0.6	0.5	1
$\overline{v}(A)$	0	0.5	0.5	0.6	0.9	0.9	0.8	1

passing messages containing only one locally optimal scalar per value of sep-arator (such as $\phi_1(a)$ above). Messages have to carry multiple utility vectors. Indeed, assume we wish finding an optimal tuple w.r.t. a Choquet integral with capacity v defined by Table 1 and utility u over $A \times B \times C$ decomposable as $u_1(a,b) + u_2(b,c)$. For a given value b of B, assume that the message sent by clique AB to BC could be $u_1(a,b) = (3,2,2)$ or $u_1(a',b) = (2,2,4)$. Both utility vectors yield the same Choquet integral: $C_v(3,2,2) = C_v(2,2,4) = 2.2$, hence it is tempting to send only one vector to BC since both vectors seem *a priori* equiv-alent. However, if the vector received by BC is added to $u_2(b,c) = (1,2,3)$, then $C_v(u_1(a,b) + u_2(b,c)) = 4.1 > C_v(u_1(a',b) + u_2(b,c)) = 3.7$, and if it is added to $u_2(b,c') = (3,2,1)$, then $C_v(u_1(a,b) + u_2(b,c')) = 3.9 < C_v(u_1(a',b) + u_2(b,c')) = 4.4$. As a consequence, locally in clique AB, it is not possible to determine which of $u_1(a,b)$ and $u_1(a',b)$ should be sent to clique BC to determine the optimal solution and we thus need to send both utility vectors on the separator.

Fortunately, not all utility vectors need be sent on separators: for any fixed value of a separator, only Pareto-nondominated vectors need be. As Choquet integral increases with each component, if $x \succsim_P y$, then $C_v(x) \geq C_v(y)$ for any capacity v. Now, once the value of a separator is fixed in a GAI net, it breaks its underlying utility into an additive utility. For instance, in Fig. 1, fixing the value of E to e' decomposes $u(a,b,c,d,e',f)$ into $w_1(a,b,c,d) + w_2(f)$ where $w_1(a,b,c,d) = u_1(a,b) + u_2(c,d) + u_3(a,c,e')$ and $w_2(f) = u_3(e',f)$. Hence, if $w_1(a,b,c,d) \succsim_P w_1(a',b',c',d')$, then adding to both vectors $w_2(f)$ results in $u(a,b,c,d,e',f)$ and $u(a',b',c',d',e',f)$ respectively, the former Pareto dominat-ing the latter. Hence, $C_v(u(a,b,c,d,e',f)) \geq C_v(u(a',b',c',d',e',f))$, and tuple (a',b',c',d',e',f) need not be considered as the optimal tuple. Applying this re-sult on utility u of Fig. 1, if EF is chosen as *root*, only nondominated utilities of message $\mathcal{M}_A$ of Fig. 3 need be sent from clique AB to ACE. Similarly, only non dominated vectors of $\mathcal{M}_C$ need be sent from CD. ACE now needs only send vectors of $\mathcal{M}_E$ to clique EF. Thus, considering messages containing only non-dominated utility vectors, we need not examine all the 288 instantiation tuples of $\mathcal{X}$, but we just examine the 8 vectors in u_1 and propagate 4 of them in $\mathcal{M}_A$; we just propagate 4 vectors out of 6 in $\mathcal{M}_C$; clique ACE combines these messages with u_3, thus creating 32 new vectors, of which only 11 are transmitted to clique EF. Finally, EF combines $\mathcal{M}_E$ with u_4, thus creating 33 vectors, and selects that which optimizes c_v, thus highlighting the efficient optimization process.

An additional (global) pruning can be used in conjunction with the above lo-cal pruning to speed-up the search: assume that, during our search, we exhibited a complete instantiation x having utility vector $u(x)$. For a new instantiation y to be optimal, $u(y)$ must not be Pareto dominated by $u(x)$. As in the preceding

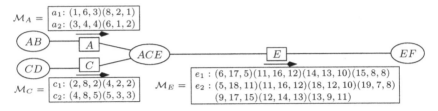

Fig. 3. Sending nondominated messages toward *root EF*

paragraph, assume that the value of separator E is fixed to e', hence decomposing u into $w_1 + w_2$ as described above and assume that we know for sure that, for any $f \in F$, $u_3(e', f) \precsim_P h$ for a given vector h. Then, if $u(x) \succsim_P w_1(a, b, c, d) + h$, no instantiation f is such that $w_1(a, b, c, d) + w_2(f) \succ_P u(x)$ and, thus, $w_1(a, b, c, d)$ needs not be sent on separator E. In this paper, we considered the following heuristic h: given a set of vectors $Z = \{(z_i^1, \ldots, z_i^m), i \in \{1, \ldots, r\}\}$, h is defined as $h = \nabla Z = (z^1, \ldots, z^m)$ where $z^j = \max\{z_1^j, \ldots, z_r^j\}$ for all $j \in \{1, \ldots, m\}$. For clique AB of Fig. 1, $h(a) = \nabla\{u_3(a, c, e) + u_2(c, d) + u_4(e, f)\}$ for all $a \in A$. However, to speed up h's computation, we approximate it by $h'(a) = \nabla\{u_3(a', c, e) + \nabla\{u_2(c', d) : c' = c\} + \nabla\{u_4(e', f) : e' = e\} : a' = a\}$ as follows: first compute $\mathcal{H}_E^D = \nabla\{u_4(e, f)\}$ for all $e \in E$, then $\mathcal{H}_C^C = \nabla\{u_3(c, d)\}$ for all $c \in C$. Finally, compute $u_3 + \mathcal{H}_C^C + \mathcal{H}_E^D$ and apply operator ∇ on it, resulting in $\mathcal{H}_A^D$ (see Fig. 4). The same applies to clique CD: $h'(c) = \nabla\{u_3(a, c', e) + \nabla\{u_4(e', f) : e' = e\} : c' = c\}$. Here again, there just needs to compute $\mathcal{H}_A^C = \nabla\{u_1(a', b) : a' = a\}$, then $u_3 + \mathcal{H}_A^C + \mathcal{H}_E^D$ and apply operator ∇.

To avoid redundant computations, we can use the following message-passing scheme: let EF act as root. Collect: EF asks ACE to send a message, which in turn asks AB and CD to send messages. AC sends message $\mathcal{H}_A^C = \{\nabla\{u_1(a', b) : a' = a\} : a \in A\}$. Similarly, clique CD sends message $\mathcal{H}_C^C = \{\nabla\{u_3(c', d) : c' = c\} : c \in C\}$. Finally, ACE sends on separator E message $\nabla\{u_3 + \mathcal{H}_A^C + \mathcal{H}_C^C\}$ for each value $e \in E$ (see Fig. 4.a). Distribute phase: EF sends on E message $\mathcal{H}_E^D = \{\nabla\{u_4(e', f) : e' = e\} : e \in E\}$. Clique ACE now computes $\nabla\{u_3 + \mathcal{H}_A^C + \mathcal{H}_E^D\}$ and sends it to CD, and $\nabla\{u_3 + \mathcal{H}_C^C + \mathcal{H}_E^D\}$ and send it to AB. In other words, before sending a message to a neighbor, a clique combines the messages it received from all its other neighbors and, then, applies operator ∇. At the end of the distribute phase, each message $\mathcal{H}^D$ corresponds to heuristic h' (Fig. 4.b).

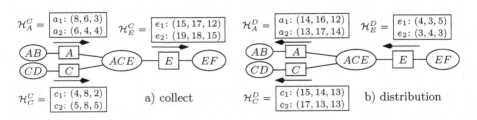

Fig. 4. Propagation of heuristics information

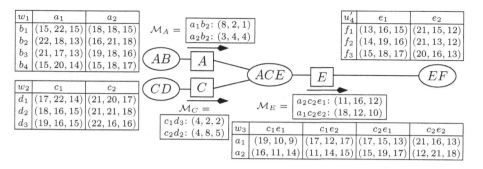

Fig. 5. The vector message-passing algorithm

We can now propose an algorithm in the spirit of MOA* [18] pruning dominated utility vectors, thus reducing the number of vectors sent on separators. First, using h' and a collect, we propagate toward *root* (here EF) on each separator the most promising utility vectors, i.e., those that, given h', have the highest Choquet integrals: clique AB thus sends toward ACE message $\mathcal{M}_A$ containing, for each $a \in A$, the vector $u_1(a, b)$ maximizing $C_v(w_1(a, b))$, where $w_1(a, b) = u_1(a, b) + \mathcal{H}_A^D(a)$. Fig. 5 shows the values of vectors w_1's. In addition, vectors u_1's that are not inserted into $\mathcal{M}_A$ and that are not Pareto dominated by other u_1's are stored into a set of "open vectors" denoted by $\mathcal{L}$ (see Table 2). Set $\mathcal{L}$ thus corresponds to *a priori* less promising vectors that may yet be optimal and, as such, that will need to be sent later on on separators to guarantee the correctness of the algorithm. Similarly, clique CD sends message $\mathcal{M}_C$ containing the u_2's maximizing, for each $c \in C$, $C_v(w_2(c, d))$ where $w_2(c, d) = u_2(c, d) + \mathcal{H}_C^D(c)$. In addition, non dominated u_2's not belonging to $\mathcal{M}_C$ are added to $\mathcal{L}$. Clique ACE now computes vectors $u_3' = u_3 + \mathcal{M}_A + \mathcal{M}_C$, which correspond to utility of instantiations of attributes A, B, C, D, E. Clique ACE sends in $\mathcal{M}_E$ those u_3''s that maximize, for each value of separator E, $C_v(w_3)$, where $w_3(a, c, e) = u_3'(a, c, e) + \mathcal{H}_E^D(e)$, i.e., the most promising (so far) utility vectors. The other u_3' vectors are stored into $\mathcal{L}$. Now EF can compute vectors $u_4'(e, f) = u_4(e, f) + \mathcal{M}_E(e)$, which correspond to utilities of complete tuples, and select that which maximizes C_v. Let us call u^* this utility vector. By Fig. 5, $u^* = (15, 18, 17)$ and $C_v(u^*) = 16.1$.

To ensure correctness, we need to send messages of $\mathcal{L}$ toward *root* and check whether they yield better C_v's. But before, we can prune from $\mathcal{L}$ all vectors u_i's such that $C_v(w_i) \leq C_v(u^*)$, since $w_i = u_i + h'$ is an upper bound on the utility of complete tuples compatible with u_i. The set $\mathcal{L}$ of Table 2 thus reduces to the utility vectors of instantiations a_2b_3 and c_2d_3. Note the efficiency of this

Table 2. The set of open vectors $\mathcal{L}$

tuple	u_i, u_i'	$C_v(w_i)$	tuple	u_i, u_i'	$C_v(w_i)$	tuple	u_i, u_i'	$C_v(w_i)$
a_1b_1	$(1, 6, 3)$	15.7	a_2b_3	$(6, 1, 2)$	17.2	c_1d_1	$(2, 8, 1)$	16.0
c_2d_3	$(5, 3, 3)$	17.2	$a_1c_1e_1$	$(15, 7, 4)$	11.3	$a_1c_2e_1$	$(13, 12, 8)$	14.4
$a_2c_1e_1$	$(12, 8, 9)$	12.9	$a_1c_1e_2$	$(14, 8, 10)$	13.3	$a_2c_2e_2$	$(9, 17, 15)$	15.3

pruning rule. We now add the most promising vector of $\mathcal{L}$ to its appropriate separator (here, we can add $u_1(a_2, b_3)$ to $\mathcal{M}_A$) and remove it from $\mathcal{L}$, then clique ACE updates its u'_3 table by computing all the new combinations $u'_3(a_2, c, e) = u_3(a_2, c, e) + u_1(a_2, b_3) + \mathcal{M}_C(c)$. These new combinations are added $\mathcal{L}$, provided that their $C_v(w_3) > C_v(u^*)$ and that they are not Pareto dominated by other u'_3's (for fixed values of separator E). The same process is applied until $\mathcal{L}$ is empty. When a vector from $\mathcal{L}$ is added to the separator adjacent to $root$, the latter updates the value of u^* and prunes $\mathcal{L}$. When $\mathcal{L}$ is empty, all possible optimal combinations have been tested and u^* is an optimal utility vector.

This process is formalized in function Choquet below. In this function, we do not use u_i's but rather labels, i.e., triples $\langle v, X_{\mathbf{C}_i}, x_{\mathbf{D}} \rangle$, where v is a utility vector, $X_{\mathbf{C}_i}$ denotes the clique that created the label and $x_{\mathbf{D}}$ is the partial instantiation yielding v. For a given clique $X_{\mathbf{C}_i}$, $Labels(u_i)$ is table u_i in which all vectors are substituted by their label. For a given set of labels $\mathcal{M}$, $\mathcal{M}[x_{\mathbf{E}}]$ denotes the subset of labels $\langle v, X_{\mathbf{C}_i}, y_{\mathbf{D}} \rangle \in \mathcal{M}$ such that $x_{\mathbf{D} \cap \mathbf{E}} = y_{\mathbf{D} \cap \mathbf{E}}$, and $\mathcal{M} \oplus \mathcal{N} = \{ \langle v + w, X_{\mathbf{E}}, x_{\mathbf{C} \cup \mathbf{D}} \rangle : \langle v, X_{\mathbf{C}_i}, x_{\mathbf{C}} \rangle \in \mathcal{M}$ and $\langle w, X_{\mathbf{C}_j}, x_{\mathbf{D}} \rangle \in \mathcal{N} \}$. Finally, for any $X_{\mathbf{C}_i}$, we denote $X_{\mathbf{C}_{p(i)}}$ the clique adjacent to $X_{\mathbf{C}_i}$ on the path between $root$ and $X_{\mathbf{C}_i}$, and we denote $\mathrm{Adj}(X_{\mathbf{C}_i})$ the set of cliques adjacent to $X_{\mathbf{C}_i}$ except $X_{\mathbf{C}_{p(i)}}$.

Function Choquet ()
01 set of open labels $\mathcal{L} \leftarrow \emptyset$; let $root$ be any clique
02 **for all** cliques $X_{\mathbf{C}_i}$ from the leaves to the $root$ **do**
03      $\mathcal{U} \leftarrow Labels(u_i) \oplus_{X_{\mathbf{C}_k} \in \mathrm{Adj}(X_{\mathbf{C}_i})} \mathcal{M}_k$
04      $\mathcal{M}_i \leftarrow \{ \mathcal{U}$'s most promising label for each value of separator $X_{\mathbf{S}_{ip(i)}} \}$
05      $\mathcal{L} \leftarrow \mathcal{L} \cup \{ ParetoNonDom(\mathcal{U}[x_{\mathbf{S}_{ip(i)}}] \backslash \mathcal{M}_i[x_{\mathbf{S}_{ip(i)}}])$ for all $x_{\mathbf{S}_{ip(i)}} \in X_{\mathbf{S}_{ip(i)}} \}$
06   send message $\mathcal{M}_i$ on separator $X_{\mathbf{S}_{ip(i)}}$
07 **done**
08 $C_v^{best} \leftarrow \max_{L \in \mathcal{M}_{\mathbf{root}}} C_v(L)$
09 **while** $\mathcal{L} \neq \emptyset$ **do**
10   let $L = \langle v, X_{\mathbf{C}_i}, x_{\mathbf{D}} \rangle$ be the most promising label in $\mathcal{L}$; remove L from $\mathcal{L}$
11      $\mathcal{M}_i \leftarrow \mathcal{M}_i \cup \{L\}$; $\mathcal{U}'_{p(i)} \leftarrow \{L\} \oplus Labels(u_{p(i)}) \oplus_{X_{\mathbf{C}_k} \in \mathrm{Adj}(X_{\mathbf{C}_{p(i)}}) \backslash \{X_{\mathbf{C}_i}\}} \mathcal{M}_k$
12   **if** $p(i) = root$ **then**
13          $C_v^{best} \leftarrow \max\{C_v^{best}, \max_{L \in \mathcal{U}'_{p(i)}} C_v(L)\}$
14   remove from $\mathcal{L}$ labels L's such that $C_v(L \oplus \mathcal{H}_{X_{\mathbf{C}_i}}^{\mathcal{D}}) \leq C_v^{best}$
15   **else**
16          $\mathcal{F} \leftarrow \{L \in \mathcal{L}$ whose clique is $X_{\mathbf{C}_{p(i)}}\}$; $\mathcal{L} \leftarrow (\mathcal{L} \backslash \mathcal{F}) \cup ParetoNonDom(\mathcal{F} \cup \mathcal{U}'_{p(i)})$
17 **done**
18 **return** C_v^{best}

5   The Case of Convex Capacities

Convex capacities are of special interest for Choquet integrals due to their interpretation in terms of preference aggregation: they convey an idea of compromise or fairness named hereafter *preference for well balanced solutions*, meaning intuitively that smoothing or averaging a cost vector improves the alternative. A useful formalization of this idea has been introduced in [19] through an axiom named *"preference for diversification"* due to its interpretation in the context of portfolio management. This axiom can be reformulated in our framework as:

Definition 6 (Preference for well-balanced solutions). *Preference for well-balanced solutions holds for a relation $\succsim$ on $\mathcal{X}$ if, for any n utility vectors $u_1, \ldots,$ $u_n \in \mathbb{R}$, and for all real numbers $\alpha_1, \ldots, \alpha_n \geq 0$ such that $\sum_{i=1}^{n} \alpha_i = 1$:*

$$[u_1 \sim u_2 \sim \ldots \sim u_n] \Longrightarrow \sum_{i=1}^{n} \alpha_i u_i \succsim u_k, \quad k = 1, \ldots, n .$$

When using a Choquet Integral, the above axiom is equivalent to choosing a convex capacity v as shown in [19]. Coming back to Example 1, we can imagine a fourth solution w with utility vectors $u(w) = (12.5, 5, 10)$ such that $C_v(u(w)) = 5 \times 1 + (10 - 5) \times 0.5 + (12.5 - 10) \times 0.2 = 8$. Now, remarking that $C_v(u(z)) = C_v(u(w)) = 8$, preference for well-balanced solutions induced by the convexity of v implies that vector $0.5u(z) + 0.5u(w) = (9.75, 10, 10)$ would be preferred to $u(z)$ and $u(w)$. Observing that $u(y)$ Pareto-dominates $u(w)$, we deduce that $u(y)$ is also preferred to $u(z)$ and $u(w)$ by monotonicity of the Choquet integral. Hence resorting to a convex capacity might be natural in many decision situations where a compromise is sought among conflicting points of view. Let us show now how the convexity of v can be exploited on the algorithmic side.

Let $C_v(u(x))$ be the value of the Choquet integral for any x, and let $\bar{u} : \mathcal{X} \mapsto \mathbb{R}$ be a convex combination of marginal utilities defined by: $\bar{u}(x) = \sum_{i=1}^{m} p_i u^i(x)$ where p is a probability distribution in $core(v)$. Such a distribution can be determined using a greedy algorithm [20] (such a p is given in Table 1). Then the following property holds: $\bar{u}(x) \geq C_v(u(x))$ for all $x \in \mathcal{X}$ [21,22]. Hence $\bar{u}(x)$ is an upper bound for the Choquet integral. This can be exploited in the algorithm of Section 4 in conjunction with the pruning rule of heuristic h': after scalarizing the tables of utility vectors of Fig. 1 by $\bar{u}_i(x) = \sum_{j=1}^{m} p_j u_i^j(x)$, applying the same process that enabled us to compute $\mathcal{H}^D$ with these new tables yields a scalar heuristic $\mathcal{H}^P$ stored on each separator. $\mathcal{H}^P$ can be used to prune any utility vector u'_j to be sent by clique X_{C_j} on a separator whenever $\bar{u}_j + \mathcal{H}_j^P \leq C_v(u^*)$.

Utility $\bar{u}$ can also be used directly for computing a Choquet-optimal element. As a linear combination of GAI utilities, $\bar{u}$ is also a GAI function. Hence, we can rank efficiently elements of $\mathcal{X}$ by decreasing value of $\bar{u}(x)$ [17]. Now, assume that $x^1, \ldots, x^k$, the k-best elements on $\mathcal{X}$ w.r.t. $\bar{u}$, have been computed. Let $\hat{x}^k = \arg\max_{i=1,\ldots,k} C_v(u(x^i))$. If $C_v(u(\hat{x}^k)) \geq \bar{u}(x^k)$, then $\hat{x}^k$ is the optimal choice for C_v. Indeed, as we rank elements w.r.t. $\bar{u}$, for any $k' > k$, $\bar{u}(x^{k'}) \leq \bar{u}(x^k)$, and since $\bar{u}(x) \geq C_v(u(x))$ for all $x \in \mathcal{X}$, $C_v(u(x^{k'})) \leq \bar{u}(x^{k'}) \leq \bar{u}(x^k) \leq C_v(u(\hat{x}^k))$. Consequently, the optimal choice for C_v can be obtained by ranking elements x^i w.r.t. $\bar{u}$ until the maximum value of the $C_v(u(x^j))$'s for all the x^j's found so far, exceeds $\bar{u}(x^i)$. This is another way of generating C_v optimal elements in $\mathcal{X}$ but, contrary to algorithm presented in Section 4, this algorithm is only feasible for a capacity with a non-empty core. This is obviously the case when v is convex.

Table 3. Response times for Choquet optimization

	$m = 2$	$m = 5$		$m = 2$, v convex				$m = 5$, v convex		
n	VPA*	VPA*	n	VPA*	VPA*+S	Rank	n	VPA*	VPA*+S	Rank
5	0.004	0.015	5	0.004	0.003	0.007	5	0.012	0.008	0.009
10	0.06	34.27	10	0.06	0.05	0.47	10	33.34	0.28	9.62
15	13.62	467.35	15	11.30	6.55	>1200	15	451.84	443.76	>1200
	Table 3.a			Table 3.b				Table 3.c		

6   Experimentations

In order to evaluate in practice the performance of our algorithms, we compared the vector-passing algorithm of Section 4 (hereafter denoted VPA*), VPA* with heuristic $\mathcal{H}^P$ (denoted VPA*+S) and the ranking algorithm mentioned above (Rank). In each experimentation, X_i's domains were all of size 5. GAI networks were randomly generated with an average of 3.5 attributes per clique. Utility tables were filled with numbers drawn randomly between 0 and 100. All experiments were performed on a 2.13GHz PC with 3GB of RAM. The tables below report average response times in seconds over 2000 experiments. For Table 3.a, we generated random capacities (not necessarily convex). VPA* reveals efficient for instances where the ranking approach does not apply (non-convex cases). Tables 3.b and 3.c are given for the sake of comparison of the two variants of VPA* with Rank in the convex case. Rank is known to be very efficient when criteria are positively correlated. We deliberately generated instances with conflicting (negatively correlated) criteria to check whether VPA* was able to outperform Rank in this case. The answer is clearly positive considering Tables 3.b and 3.c.

References

1. Domshlak, C., Brafman, R.: CP-nets: Reasoning and consistency testing. In: Proc. of KR (2002)
2. Boutilier, C., Brafman, R., Domshlak, C., Hoos, H., Poole, D.: CP-nets: A tool for representing and reasoning with conditional ceteris paribus preference statements. Journal of Artificial Intelligence Research 21, 135–191 (2004)
3. Rossi, F., Venable, K.B., Walsh, T.: mCP nets: Representing and reasoning with preferences of multiple agents. In: Proc. of AAAI, pp. 729–734 (2004)
4. Boutilier, C., Bacchus, F., Brafman, R.: UCP-networks; a directed graphical representation of conditional utilities. In: Proc. of UAI (2001)
5. Gonzales, C., Perny, P.: GAI networks for utility elicitation. In: KR (2004)
6. Braziunas, D., Boutilier, C.: Local utility elicitation in GAI models. In: Proc. of UAI (2005)
7. Gonzales, C., Perny, P., Queiroz, S.: GAI networks: Optimization, ranking and collective choice in combinatorial domains. Foundations of computing and decision sciences 32(4), 3–24 (2008)
8. Keeney, R.L., Raiffa, H.: Decisions with Multiple Objectives - Preferences and Value Tradeoffs. Cambridge University Press, Cambridge (1993)
9. Krantz, D., Luce, R.D., Suppes, P., Tversky, A.: Foundations of Measurement (Additive and Polynomial Representations), vol. 1. Academic Press, London (1971)

10. Bacchus, F., Grove, A.: Graphical models for preference and utility. In: Proc. of UAI (1995)
11. Fishburn, P.C.: Interdependence and additivity in multivariate, unidimensional expected utility theory. International Economic Review 8, 335–342 (1967)
12. Gonzales, C., Perny, P.: GAI networks for decision making under certainty. In: IJCAI 2005 – Workshop on Advances in Preference Handling (2005)
13. Cowell, R., Dawid, A., Lauritzen, S., Spiegelhalter, D.: Probabilistic Networks and Expert Systems. Stats for Engineering and Information Science. Springer, Heidelberg (1999)
14. Dubus, J.P., Gonzales, C., Perny, P.: Multiobjective optimization using GAI models. In: Proc. of IJCAI (2009)
15. Choquet, G.: Theory of capacities. Annales de l'Institut Fourier 5, 131–295 (1953)
16. Shapley, L.: Cores of convex games. Int. J. of Game Theory 1, 11–22 (1971)
17. Gonzales, C., Perny, P., Queiroz, S.: Preference aggregation with graphical utility models. In: Proc. of AAAI, pp. 1037–1042 (2008)
18. Stewart, B.S., White III, C.C.: Multiobjective A*. J. ACM 38(4), 775–814 (1991)
19. Chateauneuf, A., Tallon, J.M.: Diversification, convex preferences and non-empty core in the choquet expected utlity model. Economic Theory 19, 509–523 (2002)
20. Jaffray, J.Y.: On the maximum probability which is consistent with a convex capacity. Int. J. of Uncertainty, Fuzziness and KB Systems 3(1), 27–34 (1995)
21. Galand, L., Perny, P.: Search for choquet-optimal paths under uncertainty. In: Proc. of UAI, pp. 125–132 (2007)
22. Queiroz, S.: Multiperson Choquet-compromise search on large combinatorial domains. In: 2nd IEEE Int. Workshop on Soft Comp. Applications, pp. 187–192 (2007)

Compact Preference Representation in Stable Marriage Problems

Enrico Pilotto[1], Francesca Rossi[1], Kristen Brent Venable[1], and Toby Walsh[2]

[1] Department of Pure and Applied Mathematics, University of Padova, Italy
{epilotto, frossi, kvenable}@math.unipd.it
[2] NICTA and UNSW Sydney, Australia
Toby.Walsh@nicta.com.au

Abstract. The stable marriage problem has many practical applications in two-sided markets like those that assign doctors to hospitals, students to schools, or buyers to vendors. Most algorithms to find stable marriages assume that the participants explicitly expresses a preference ordering. This can be problematic when the number of options is large or has a combinatorial structure. We consider therefore using CP-nets, a compact preference formalism in stable marriage problems. We study the impact of this formalism on the computational complexity of stable marriage procedures, as well as on the properties of the solutions computed by these procedures. We show that it is possible to model preferences compactly without significantly increasing the complexity of stable marriage procedures and whilst maintaining the desirable properties of the matching returned.

1 Introduction

The stable marriage problem is a well-known problem with many practical applications. It is usually defined as the problem of matching men to women so that no man and woman, who are not married to each other, both prefer each other to their current partner [6]. Problems of this kind arise in many real-life situations, such as assigning residents to hospitals, students to schools, as well as in two-sided market trading. A specific application is a web-based stable marriage system for matching sailors to ships in the US Navy.

Surprisingly, a stable matching always exists whatever preferences are held by the men and women. The Gale-Shapley algorithm finds a stable matching in polynomial time [4]. The matching computed is male-optimal since the men have the best possible partners. Since this might be considered unfair to the women, many other stable marriage algorithms have been developed. For example, Gusfield gives a polynomial algorithm to compute the stable matching where the regret of the most unsatisfied person is minimal [5]. We will focus on these two algorithms since they contain the main features of many other stable marriage algorithms.

Both algorithms assume that agents express their preferences (over the members of the other gender) explicitly as a totally ordered list of members of the other gender. In some applications, the number of men and women can be large. It may therefore be unreasonable to assume that each man and woman provides a strict ordering of the other

F. Rossi and A. Tsoukis (Eds.): ADT 2009, LNAI 5783, pp. 390–401, 2009.

gender. In addition, eliciting their preferences may be a costly and time-consuming process. The sets of men and women may have a combinatorial structure. It could therefore be costly to give preference over all options.

For instance, consider a large set of hospitals offering residencies. Doctors might not want to rank explicitly all the hospitals, but might wish to express preferences over features. For example, they might say "I prefer a position close to my home town", or "If the hospital is far away from my home town, then I want a better salary". Based on this information, we can rank the hospitals.

Our challenge is to adapt algorithms to find stable marriages to work with such preference statements. We will investigate whether this changes the computational complexity of stable marriage algorithms like Gale-Shapley's and Gusfield's as well as properties of the matchings computed.

To model preferences compactly, we will use (acyclic) CP-nets [1]. These let agents state their preferences simply and naturally by means of qualitative conditional statements. We will show how to use the preferences orderings induced by CP-nets within GS and Gusfield's algorithms with little additional computational cost. This claim is supported by both theoretical and experimental studies.

2 Background

2.1 CP-nets

CP-nets [1] are a graphical model for compactly representing conditional and qualitative preference relations. CP-nets are sets of *ceteris paribus (cp)* preference statements. For instance, the statement *"I prefer red wine to white wine if meat is served"* asserts that, given two meals that differ *only* in the kind of wine served *and* both containing meat, the meal with red wine is preferable to one with white wine.

A CP-net has a set of features (also called variables) $F = \{x_1, \ldots, x_n\}$ with finite domains $\mathcal{D}(x_1), \ldots, \mathcal{D}(x_n)$. For each feature x_i, we are given a set of *parent* features $Pa(x_i)$ that can affect the preferences over the values of x_i. This defines a *dependency graph* in which each node x_i has $Pa(x_i)$ as its immediate predecessors. Given this structural information, the agent explicitly specifies her preference over the values of x_i for *each complete assignment* on $Pa(x_i)$. This is by means of a total order over $\mathcal{D}(x_i)$. An *acyclic* CP-net is one in which the dependency graph is acyclic.

Consider a CP-net whose features are A, B, C, and D, with binary domains containing f and $\overline{f}$ if F is the name of the feature, and with the following preference statements: $a \succ \overline{a}$, $b \succ \overline{b}$, $(a \wedge b) \vee (\overline{a} \wedge \overline{b}) : c \succ \overline{c}$, $(a \wedge \overline{b}) \vee (\overline{a} \wedge b) : \overline{c} \succ c$, $c : d \succ \overline{d}$, $\overline{c} : \overline{d} \succ d$. Here, $a \succ \overline{a}$ represents the unconditional preference for $A = a$ over $A = \overline{a}$, while $c : d \succ \overline{d}$ states that $D = d$ is preferred to $D = \overline{d}$ given that $C = c$.

The semantics of CP-nets depends on the notion of a *worsening flip*. This is a change in the value of a feature to a less preferred value according to the preference statement for that feature. For example, in the CP-net above, passing from $abcd$ to $ab\overline{c}d$ is a worsening flip since c is better than $\overline{c}$ given a and b.

A solution (also called outcome) of a CP-net is an assignment to all its variables of values from their domains. One solution α is *better* than another solution β (written $\alpha \succ \beta$) iff there is a chain of worsening flips from α to β. This definition induces in general a preorder over the solutions. If the CP-net is acyclic, the solution ordering is a partial order with only one top element.

In general, finding the optimal solution of a CP-net is NP-hard. However, in acyclic CP-nets, the unique optimal solution can be found in linear time. We simply sweep through the dependency graph assigning each variable to the its most preferred value. For instance, in the CP-net above, we would choose $A = a$ and $B = b$, then $C = c$, and then $D = d$.

Determining if one solution is better than another (called a dominance query) is NP-hard even for acyclic CP-nets. Whilst tractable special cases exist, there are also acyclic CP-nets in which there are exponentially long chains of worsening flips between two solutions.

2.2 Stable Marriage Problems

The *stable marriage problem* (SMP) is the problem of finding a matching between the elements of two sets. Usually, the members of the two sets are called men and women. More precisely, given n men and n women, where each person strictly orders all members of the opposite sex, we wish to marry the men to the women such that there is not a man and woman who would both rather be married to each other than to their current partners. If there is no such couple, the matching is called *stable*. We will write $pref(x)$ for the preference ordering of man or woman x.

The *Gale-Shapley algorithm* (GS) [4] is a well-known algorithm to solve the SMP problem:

Algorithm 1. GS

Set all men and women as free
while *there is a free man m* **do**
  $w \leftarrow$ the first woman in $pref(m)$ to which he has not yet proposed
  if *w is free* **then**
  └ match m with w
  if $m >_{pref(w)} z$, *where z is w's current partner* **then**
  └ match m with w and set z free
  else
  └ w rejects m and m remains free

This algorithm consists of a number of rounds in which each un-engaged man proposes to the most preferred woman to whom he has not yet proposed. Each woman receiving a proposal becomes "engaged", provisionally accepting the proposal from her most preferred man. In subsequent rounds, an already engaged woman can "trade up", becoming engaged to a more preferred man and rejecting a previous proposal, or if she prefers him, she can stick with her current partner. The algorithm takes $O(n^2)$ steps and construct a matching that is *male-optimal*, since every man is paired with his highest ranked feasible partner, and *female-pessimal*, since each woman is paired with her lowest ranked feasible partner.

Consider $n = 3$. Let $W = \{w_1, w_2, w_3\}$ and $M = \{m_1, m_2, m_3\}$ be respectively the set of women and men. The following sequence of strict total orders defines an SMP:

- $m_1 : w_1 > w_2 > w_3$ (i.e., man m_1 prefers woman w_1 to w_2 to w_3); $m_2 : w_2 > w_1 > w_3$; $m_3 : w_3 > w_2 > w_1$
- $w_1 : m_1 > m_2 > m_3$; $w_2 : m_3 > m_1 > m_2$; $w_3 : m_2 > m_1 > m_3$

For this SMP, the Gale-Shapley algorithm returns the male-optimal marriage $\{(m_1, w_1), (m_2, w_2), (m_3, w_3)\}$. On the other hand, the female-optimal marriage is $\{(w_1, m_1), (w_2, m_3), (w_3, m_2)\}$.

Male-optimality might be considered unfair to the women. Other proposal-based algorithms to compute stable matchings have been proposed that might be considered fairer. For example, Gusfield gives an algorithm to compute the minimum-regret stable matching [5]. This is the best stable matching as measured by the person who has the largest regret in it. The regret of a man in a matching is the number of women that are more preferred than its current partner. The regret of a woman is defined analogously. Gusfield's algorithm passes from one matching to another. In each step, the person with the maximum regret is identified, and their current marriage is broken to pass to another matching with a smaller maximum regret.

3   Operations Opt, Next, and Compare in the SMP Algorithms

In algorithms such as GS and Gusfield's, men make proposals, starting from their most preferred woman and going down in their ordering, whilst women receive proposals and compare these against the men to whom they are currently engaged. Moreover, in both algorithms, proposals are made in increasing order of regret. This is especially exploited by Gusfield's algorithm, where the notion of regret is also used to decide how to modify the current matching in order to obtain one with a smaller regret. Three operations are thus needed by both algorithms:

- $Opt(pref(m))$: Given a man m, we compute his optimal woman. This is needed the first time a man makes a proposal.
- $Next(pref(m), w)$: Given a man m and a woman w, we compute the next best woman for m. This is needed when a man makes a new proposal.
- $Compare(pref(w), m_1, m_2)$: Given a woman w and two men m_1 and m_2, we decide if m_2 is preferred to m_1 for w. This is needed when a woman compares two proposals to decide whether to remain with the current man (m_1) or to leave him for a new man who is proposing (m_2).

Operations Opt and $Next$ return a woman, while $Compare$ returns a Boolean value.

If preferences are given explicitly as strict total orders, as in the traditional SMP setting, these operations all take constant time. However, if preferences are represented with a compact representation language such as CP-nets, then this is not the case. Thus, to understand the impact of using a compact preference formalisms within algorithms like GS, we consider the computational complexity of these operations on CP-nets.

4   Opt, Next, and Compare on CP-nets

We consider a stable marriage problem with n men and women, where each man and each woman specifies their preferences over the other sex via a CP-net. We call this a

Compact SMP (CSMP). $pref(m)$ is now the solution ordering induced by a CP-net, which can be a partial ordering.

For simplicity, we will consider CP-nets with two values in each domain. However, the results can be easily generalized to non-binary domains. Each man and woman is described by a set of Boolean features, and n is the size of the Cartesian product of such domains. Thus the number of features f of each CP-net is $log(n)$. Conversely, if we are given a set of f features, we assume that each assignment to such features corresponds to a man (resp. a woman). We could, however, relax this assumption by using a constrained CP-net to rule out infeasible combinations.

Let us consider the three operations used within the SMP algorithms. In general, finding the optimal solution of a CP-net is a computationally difficult problem, as is dominance testing (the problem of comparing two solutions in the CP-net ordering) [1]. We are not aware of any study of the complexity of finding the next best solution. However, as two of the three operations are computationally intractable in general, and as we wish to find settings where such operations take just polynomial time, we turn our attention to acyclic CP-nets. These are more restrictive but may be sufficiently expressive in many contexts [1].

As mentioned before, in acyclic CP-nets there is always one optimal solution, and it can be found in linear time in the number of features f by a simple forward sweep algorithm. Operation Opt thus takes $O(f)$ time.

While the solution ordering of an acyclic CP-net may be partial, operations $Next$ and $Compare$ need a total order, since $Next$ returns one new proposal to be made, and $Compare$ chooses between two proposals. Therefore, we will consider linearizations of the CP-net solution ordering.

This does not contradict a user's preference statements, since a linearization only orders pairs of elements that were incomparable. Notice also that, even if we could work with partial orders, dominance testing (and thus $Compare(pref(w), m_1, m_2)$) is intractable in general for an acyclic CP-net. Here, on the other hand, we aim to find linearizations where all three operations are tractable.

We will focus on those linearizations where the regret is larger as we descend the order. As noted before, our stable marriage algorithms make proposals in this order.

5 A Linearization of the CP-net Solution Ordering

Linearizations of the solution ordering of acyclic CP-nets have been considered in [2,3]. A consequence of these results is that, given a feature order which is compatible with their topological order in the dependency graph, any lexicographical ordering over the solutions is a linearization of the original partial ordering. Computing Next in such a linearization is polynomial since it simply requires the next tuple of feature values in the lexicographic ordering. Also the Compare operation is polynomial since it reduces to a comparison of tuples over the lexicographical relation. Unfortunately, such linearizations do not in general satisfy the regret condition. We therefore consider a different linearization, where the Next and Compare operations are polynomial, and where solutions closer to the top of the partial order (that is, with a smaller regret) come first.

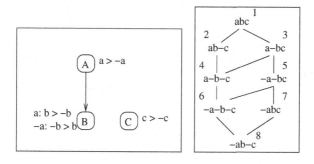

Fig. 1. A CP-net and its induced solution ordering

We recall that the regret of a man is the distance between his partner in the current marriage and his most preferred woman. This notion has been originally defined over total orders [6]. However, it can be generalized to be used on partial orders. More precisely, the regret of a man m when married to a woman w is the longest path, in the preference ordering of m, between w and the top element of his ordering.

For example, let us consider the CP-net, as well as its induced solution ordering, shown in Figure 1 (where $\bar{x}$ is written as $-x$ for all values x). This CP-net has three features A, B, and C, where B depends on A. The regret of $\bar{a}bc$ is 3 since there are at most 2 solutions between the top and this one.

While $pref(m)$ is the preference ordering induced by the CP-net of m, we will call *lex-pref(m)* our linearization of $pref(m)$.

This linearization is based on a lexicographical order over feature levels. Given an acyclic CP-net, we divide its features into levels, each containing all the features that have the same longest path length to a feature without outgoing edges in the dependency graph. For example, in the CP-net of Figure 1, we have two levels: level 2, containing only A and corresponding to a longest path of length 1, and level 1, containing B and C, corresponding to a longest path of length 0.

Given a solution, we then associate to it a vector v of length equal to the number of levels, say k, whose elements $v_1, \ldots, v_k$, corresponding to levels k to 1, are Boolean vectors of length equal to the number of features in each level. Features are ordered within each level in some fixed order.

For the previous example, we have a vector with two elements (since we have two feature levels), where the first one has one Boolean value (corresponding to the value for A), and the second one is a two-element Boolean vector (corresponding to the values for B and C). The Boolean values in the vectors are set according to the values of the features: given the values of the parents, if the value of the considered variable is the most preferred, we put 0, otherwise 1.

Consider again the CP-net in Figure 1. The solution $a\bar{b}\bar{c}$ gives the vector $[0, 11]$, since a is the most preferred value in the CP-table of feature A, while, given $A = a$, both $\bar{b}$ and $\bar{c}$ are the least preferred values in the CP-tables of features B and C.

Given any two such vectors, say $v = [v_1, \ldots, v_k]$ and $v' = [v'_1, \ldots, v'_k]$, we now define how to order them. Let us denote by $sum(x)$ the sum of all elements in vector

x. Then $v <_{lex-pref} v'$ (that is, v precedes v' in the ordering and is more preferred) iff $[sum(v_1), \ldots, sum(v_k)]$ is lexicographically smaller than $[sum(v'_1), \ldots, sum(v'_k)]$, or $[sum(v_1), \ldots, sum(v_k)] = [sum(v'_1), \ldots, sum(v'_k)]$ and $[v_1, \ldots, v_k]$ is lexicographically smaller than $[v'_1, \ldots, v'_k]$.

For example, vector $[00, 01]$ is preferred to vector $[00, 10]$, since the vectors of the sums are equal ($[0, 1]$) and $[00, 01]$ is lexicographically smaller than $[00, 10]$. Also, $[00, 01]$ is preferred to $[00, 11]$, and $[10, 00]$ is preferred to $[01, 01]$.

In Figure 1, the linearization of the ordering is shown via the numbers above each solution. As the following theorem shows, this is a linearization of the solution ordering induced by the CP-net.

Theorem 1. *Given a CP-net and two solutions s and s', with associated vectors v and v', if $s \succ s'$, then $v <_{lex-pref} v'$.*

Proof. The vectors of the sums represents the number of violations in each level of the CP-net. If s' has a larger number of violations in higher levels of the CP-net with respect to s (thus s' is less preferred than s), the vector of the sums of v' will be larger than the one of v. Notice also that solutions which are incomparable in the CP-net ordering are strictly ordered in our linearization.   □

6   Complexity of Opt, Next, and Compare, and Regret Condition

Since we have linearized a partial order with one top element, $Opt(pref(m)) = Opt(lex-pref(m))$. Therefore, finding $Opt(lex-pref(m))$ is polynomial since we can use the sweep forward algorithm to find $Opt(pref(m))$ in acyclic CP-nets.

We will now consider the complexity of the operations $Next$ and $Compare$ on this linearization.

For operation $Compare(lex-pref(w), m_1, m_2)$, we can directly use the definition of $<_{lex-pref}$ given above: $Compare(lex-pref(w), m_1, m_2) = true$ iff $v(m_2) <_{lex-pref} v(m_2)$, where $v(m_i)$ is the vector associated to m_i in the CP-net $pref(w)$.

Theorem 2. *Given a CSMP of size n, a woman w and two men m_1 and m_2, $Compare(lex-pref(w), m_1, m_2)$ can be computed in $O(f)$, where f is the number of features of the CP-net of w.*

Proof. The vectors associated to each of the two men can be computed in linear time in the number of features of the CP-net: for each feature, we check the position of the feature value in a row of the appropriate CP-table. Given the two vectors, we need to compute the vectors of their sums, which is linear in the number of features, and to compare them (and possibly also the original vectors) lexicographically. This takes time linear in the number of features.   □

For operation $Next(lex-pref(m), w)$, given a vector, we need to find the next vector in the $<_{lex-pref}$ ordering. This can be done by following a procedure similar to incrementing a Boolean counter.

Given a vector $v = [v_1, \ldots, v_k]$, we build a new vector v' as follows. For j from k to 1, we compute the first j such that $sum(v_j) = sum(next-lex(v_j))$, where $next-lex$ is just the next element in a standard lexicographical order. If such a j exists, then $v' = [v_1, \ldots, next\text{-}lex(v_j), v'_{j+1}, \ldots, v'_k]$, where v'_h with $h > j$ is the minimum vector with sum equal to $sum(v_h)$. Otherwise, we compute $next\text{-}lex([sum(v_1), \ldots, sum(v_k)])$, and set v' as the smallest vector with these sums.

For example, consider solution $ab\bar{c}$ for the CP-net in Figure 1. The corresponding vector is $[0, 11]$. There is no way to increase lexicographically either 11 or 0 while maintaining the same sums, so we must consider the sum vector $[0, 2]$, increment it to $[1, 0]$, and then build vector $[1, 00]$, which corresponds to solution $\bar{a}bc$. If instead we consider solution $ab\bar{c}$, whose vector is $[0, 01]$, we can modify 01 into 10 while maintaining the same sums, so we get vector $[0, 10]$, corresponding to solution $a\bar{b}c$.

Algorithm 2. $Next$

Input: acyclic CP-net N of man m, vector $v = [v_1, \ldots, v_k]$,
Output: vector v', successor of v in $<_{lex\text{-}pref(m)}$
$v' \leftarrow v$
$i \leftarrow k$
while $i > 1$ *and* $v'_i = LEX\_NEXT(v'_i)$ **do**
$\quad \llcorner \; i \leftarrow i - 1$
if $i \geq 1$ **then**
$\quad \mid \; v'[i] \leftarrow LEX\_NEXT(v'_i)$
$\quad \mid \; RESET\_ALL\_SUCC(v', i)$
$\quad \llcorner \;$ return v'
else
$\quad \mid \; P \leftarrow SUM\_NEXT([sum(v'_1), \ldots, sum(v'_k)])$
$\quad \mid \;$ **if** $P = nil$ **then**
$\quad \mid \quad \llcorner \;$ return "No more solutions"
$\quad \llcorner \;$ return $SUM\_MIN(P)$

In Algorithm 2, procedure LEX_NEXT takes as input a vector v'_i and returns the lexicographical successor of v'_i that has the same sum, if it exists, and v'_i itself otherwise. Procedure RESET_ALL_SUCC, given a vector v' and an index i, resets the sub-vector with components v'_j, with index $j \geq i$, to the minimal lexicographic Boolean vector with sum equal to $sum(v'_j), j \geq i$. Procedure SUM_NEXT computes the lexicographic successor of a vector of sums taking into account the maximum cost sum of each level. If no such successor exists, it returns nil. Finally, procedure SUM_MIN computes the lexicographical minimal vector having the sum vector given in input.

Theorem 3. *Given a CSMP of size n, a woman w and a man m with a CP-net with f features, Algorithm 2 computes $Next(lex\text{-}pref(m), w)$ in $O(f)$ time.*

Proof. Computing the next vector in a lexicographical order, computing the sum of the vector, and computing the minimum vector with the same sum are all tasks that can be done in time linear in the size of the vector. Thus the above algorithms can run in time linear in the size of the vector associated to woman w, which is the number of features of the CP-net of man m. $\qquad\square$

We have shown that all the three operations (Opt, Next, and Compare) can be computed in polynomial time on our linearization of the CP-net ordering. We now show that this linearization also has the property that later elements have a larger regret.

Theorem 4. *Given a CP-net of a man m, and two women w_1 and w_2, if $s' <_{lex-pref} s$, then the regret of m when married with w_1 is smaller than or equal to his regret when married to w_2.*

7 Properties of the Generated Stable Marriage: Stability, Male-Optimality, and Minimum Regret

7.1 Stability

If we run the GS algorithm on the linearization just defined, by definition we obtain a matching which is stable w.r.t. this linearization. However, one may wonder if the generated matching is stable w.r.t. the partial order of the CP-net.

As is standard in SMPs with ties [6], also in our case, where the orders induced by the CP-nets may be partial, we define a matching to be stable when there is no man and woman who *strictly prefer* each other to their partner in the matching.

Since our linearization orders more pairs than the partial order of the CP-net, it is easy to see that any matching which is stable for the linearization is also stable for the partial order. In fact, if there is a blocking pair (that is, a man and a woman who would prefer to be together rather than with their current partners) in the partial order, such a pair will be blocking also according to the linearization. Therefore stability is assured, both w.r.t. the linearization and w.r.t. the original CP-net solution ordering.

7.2 Male-Optimality

The matching found by the GS algorithm on the linearization will of course be male-optimal w.r.t. the linearization. However, it may be not male-optimal w.r.t. the original partial order.

In the presence of partial orders, the definition of male-optimality is the same as for total orders: a stable marriage is male-optimal if, for each man m and partner w, there is no stable marriage in which m is married to another woman w' which he strictly prefers to w. Notice that, while a male-optimal matching always exists when we work with totally ordered preferences, this is not true when we have partial orders.

Even if a male-optimal matching exists, running the GS algorithm on our linearization can return a matching which is not male-optimal. Consider the following SMP with 2 men and 2 women, where $\bowtie$ means incomparability: $m_1 : w_1 \bowtie w_2$; $m_2 : w_1 \prec w_2$; $w_1 : m_1 \bowtie m_2$; $w_2 : m_1 \bowtie m_2$. Given the linearization where we order incomparable element in incresing index order, the stable matching obtained by GS on this linearization is $((m_1, w_1), (m_2, w_2))$. However, the only male-optimal matching in the original problem is $((m_1, w_2), (m_2, w_1))$.

Notice that, when preferences are totally ordered, the output of the GS algorithm is not affected by the order in which men propose to women. With partial orders, this order may affect the result, since women leave the current partner only if the new proposal is strictly better. For example, in ther SMP above, if man m_2 proposes first (to w_1), then the resulting marriage is the male-optimal one $(((m_1, w_2), (m_2, w_1)))$.

There is a way to "do our best" w.r.t. male-optimality, by following the policy that the next men making a proposal is one of those, if any, with a single next best woman. If no such man exists, then we can choose any man. If we do this, then it is possible to show that the generated marriage is never worse w.r.t. male-optimality than the one obtained by any other policy.

To (partially) implement this policy, we can exploit the fact that only incomparable women can have the same vectors of sums (see Theorem 1). Therefore, we can identify a man with a single next best woman by executing twice the Next operation and by comparing the vectors of the sums of the two women obtained. If they are the same, then the two women are incomparable (thus the man does not have a single next best woman). On the other hand, if they are different, the two women may be ordered or incomparable in the partial order. Thus we should choose a man with two different vectors of the sums. Notice that this is an approximation of the desired policy, since some incomparable women may have different vectors of the sums. Notice also that the implementation of the proposal policy does not add to the worst-case cost of the GS algorithm, since the result of the second Next operation can be saved for future use.

7.3   Minimum Regret

Computing a stable matching which minimizes the regret of the person who is worst-off may be perceived to be fairer than computing the male-optimal matching. As mentioned in Section 2, such a stable matching is found by Gusfield's algorithm. As with GS, our linearization allows us to use Gusfield's algorithm with compact preference formalisms.

Theorem 5. *Running Gusfield's algorithm on a CSMP where each CP-net has f features, with Opt to find the first proposal, Next to find the next proposal, Compare to compare two proposals, and r to compute the regret of a matching, we obtain a stable matching with minimum regret in $O(n^2 f)$ time.*

Proof. Our linearization respects the ordering induced by the regret and allows to compute the regret efficiently. Notice that this is important in Gusfield's algorithm, since regret is not only used to establish a proposal order but also to identify the person who is worst-off in the current matching.

Given that Gusfield's algorithm runs in $O(n^2)$ time, and considering the complexity of Opt, $Next$, and $Compare$ and that of computing r, it is easy to see that our version of the algorithm runs in $O(n^2 f)$ time.   □

8   Experimental Analysis

Given the linearization described above, we can either pre-compute it and then run GS as usual over a strict linear order, or we can just compute the part of the linearization that GS needs during the execution of the algorithm. In the first scenario, we need to compute the $Next$ operation n^2 times (n times for each man and woman), and then GS can run in the usual $O(n^2)$ time. In terms of space, however, we need to store all the n linearizations, which takes $O(n^2)$ space. In the second scenario, there is no pre-computation burden, but each step of GS requires additional time to perform a $Next$ (and possibly a $Compare$) operation. Given the theoretical complexity results above, GS will run in $O(n^2 log(n))$ time. The space needed is just to store the CP-nets, and not the linearization, which is now $O(nlog(n))$.

We ran some experiments to see which of these two scenarios is more effective. Given a number of features f, we randomly generate acyclic CP-nets with f Boolean features where each feature has at most two parents. For each feature, we then generate a CP-table by making sure that the dependency graph is respected. To generate a whole CSMP with $n = 2^f$ men and women, we generate $2n$ CP-nets with f features each. The experiments have been performed on an Intel Core Duo 3GHz processor with 4GB RAM, and show the average over 100 instances.

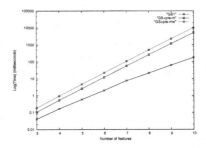

Fig. 2. Execution time for three versions of GS

Figure 2 gives the log-scale time needed to run three different versions of the GS algorithm: the one with the Next and Compare operation executed on demand (GS1), the one with the pre-computation of the linearization for the men, and the Compare operation executed on demand (GS+pre-m), and the one with the pre-computation of all the linearizations (GS+pre-mw). It is easy to see that it is inefficient to pre-compute the linearizations, even for just the men.

This is perhaps not too surprising, since computing the linearizations needs to run the Next operation exactly n^2 (or $2n^2$) times, while the GS algorithm needs $O(n^2)$ time in the worst case but may in practice require only a much smaller number of proposals. This is confirmed by Figure 3, where we plot the number of proposals made by the GS1 algorithm as a function of the number of features in the CP-nets. The GS algorithm usually makes only a small number of proposals. For example, for 10 features, we have $n = 2^{10} = 1024$, thus $n^2 = 1,048,576$, but the GS algorithm makes less than 16,000 proposals on average.

Notice that algorithm GS1 takes very little time even for CP-nets with 10 features (less than 1 second). This scenario is realistic, since it models problems with about a thousand members of each sex. In this setting, it might be impractical to ask each agent to rank all members of the other sex, whilst it is more practical to specify a CP-net over just 10 features.

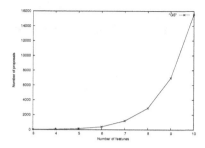

Fig. 3. Number of proposals made by the GS1 algorithm

9 Conclusions and Future Work

We have considered using a qualitative compact preference representation, namely CP-nets, in the context of stable marriage problems. We have shown that the benefits brought by compactness do not impact greatly on the complexity of computing stable matchings nor on the properties of the returned matching.

The significance of our study on the complexity of the Next operation on CP-nets goes beyond its use in SMPs. In fact, such an operation is also needed when computing the top k solutions, such as in web search, or when an additional solution is looked for.

In the future, we plan to investigate the use of other compact approaches to preferences such as soft constraints, as well as to consider other versions of the stable marriage problem (such as with ties). We also plan to see whether tractable cases exists for operations like Next on classes of soft constraints or constrained CP-nets.

References

1. Boutilier, C., Brafman, R.I., Domshlak, C., Hoos, H.H., Poole, D.: CP-nets: A tool for representing and reasoning with conditional ceteris paribus preference statements. J. Artif. Intell. Res. (JAIR) 21, 135–191 (2004)
2. Brafman, R.I., Chernyavsky, Y.: Planning with goal preferences and constraints. In: ICAPS, pp. 182–191 (2005)
3. Brafman, R.I., Domshlak, C., Kogan, T.: Compact value-function representations for qualitative preferences. In: UAI, pp. 51–59. AUAI Press (2004)
4. Gale, D., Shapley, L.S.: College admissions and the stability of marriage. Amer. Math. Monthly 69 (1962)
5. Gusfield, D.: Three fast algorithms for four problems in stable marriage. SIAM Journal of Computing 16(1) (1987)
6. Gusfield, D., Irving, R.W.: The Stable Marriage Problem: Structure and Algorithms. MIT Press, Cambridge (1989)

Neuroevolutionary Inventory Control
in Multi-Echelon Systems*

Steve D. Prestwich[1], S. Armagan Tarim[2], Roberto Rossi[3], and Brahim Hnich[4]

[1] Cork Constraint Computation Centre, Ireland
[2] Operations Management Division, Nottingham University Business School, Nottingham, UK
[3] Logistics, Decision and Information Sciences Group, Wageningen UR, The Netherlands
[4] Faculty of Computer Science, Izmir University of Economics, Turkey
`s.prestwich@cs.ucc.ie`, `armtar@yahoo.com.tr`,
`roberto.rossi@wur.nl`, `brahim.hnich@ieu.edu.tr`

Abstract. Stochastic inventory control in multi-echelon systems poses hard problems in optimisation under uncertainty. Stochastic programming can solve small instances optimally, and approximately solve large instances via scenario reduction techniques, but it cannot handle arbitrary nonlinear constraints or other non-standard features. Simulation optimisation is an alternative approach that has recently been applied to such problems, using policies that require only a few decision variables to be determined. However, to find optimal or near-optimal solutions we must consider exponentially large scenario trees with a corresponding number of decision variables. We propose a neuroevolutionary approach: using an artificial neural network to approximate the scenario tree, and training the network by a simulation-based evolutionary algorithm. We show experimentally that this method can quickly find good plans.

1 Introduction

In the area of optimisation under uncertainty, one of the most mature fields is inventory control. This field has achieved excellent theoretical and practical results using techniques such as dynamic programming, but some problems are too large or complex to be solved by classical methods. Particularly hard are those involving *multi-echelon systems*, in which multiple stocking points form a supply chain. In such cases we may resort to simulation-based methods. Simulation alone can only evaluate a plan, but when combined with an optimisation algorithm it can be used to find near-optimal solutions (or plans). This approach is called *simulation optimisation* (SO) and has a growing literature in many fields including production scheduling, network design, financial planning, hospital administration, manufacturing design, waste management and distribution. It is a practical approach to optimisation under uncertainty that can handle problems containing features that make them difficult to model and solve by other methods: for example non-linear constraints and objective function, and demands that are correlated or have unusual probability distributions.

---

* B. Hnich is supported by the Scientific and Technological Research Council of Turkey (TUBITAK) under Grant No. SOBAG-108K027. This material is based in part upon works supported by the Science Foundation Ireland under Grant No. 05/IN/I886.

F. Rossi and A. Tsoukis (Eds.): ADT 2009, LNAI 5783, pp. 402–413, 2009.
© Springer-Verlag Berlin Heidelberg 2009

SO approaches to inventory control are typically based on policies known to be optimal in certain situations, involving a small number of reorder points and reorder quantities. For example in (s, S) policies whenever a stock level falls below s it is replenished up to S, while in (R, S) policies the stock level is checked at times specified by R, and if it falls below S then it is replenished up to S. SO can apply standard optimisation techniques such as genetic algorithms to these policies by assigning genes to reorder points and replenishment levels. In more complex situations involving constraints, multiple stocking points, etc, these policies may be suboptimal in terms of expected cost, though they can have other desirable properties such as improved planning stability. But a cost-optimal plan for a multi-stage problem with recourse must specify an order quantity in every possible scenario, so the plan must be represented via a *scenario tree*. The number of scenarios might be very large, or infinite in the case of continuous probability distributions, making the use of SO problematic. Scenario reduction techniques may be applied to approximate the scenario tree, but it might not always be possible to find a small representative set of scenarios.

An alternative form of approximation is to use an artificial neural network (ANN) to represent the policy. For example, the inputs to the ANN could be the current stock levels and time, and the outputs could be the recommended actions (whether or not to replenish and by how much). We must then train the ANN so that its recommendations correspond to a good plan. No training data is available for such a problem so the usual ANN backpropagation training algorithm cannot be applied. Instead we may use an evolutionary algorithm to train the network to minimise costs. This *neuroevolutionary* approach has been applied to control problems [8,9,21] and to playing strategies for games such as Backgammon [16] and Go [14], but it has not been extensively applied to inventory control. In this paper we apply neuroevolution to stochastic inventory control in multi-echelon systems. Section 2 presents our method, Section 3 evaluates the method experimentally, Section 4 surveys related work, and Section 5 concludes the paper.

2   A Neuroevolutionary Approach

To approximate the scenario tree, we construct a function whose input is a vector containing the time period and current inventory levels, and whose output is a vector of order quantities (which might be zero). We design the function automatically by simulation optimisation.

2.1   Scenario Tree Compression by Neural Network

An obvious choice for this function is an artificial neural network (ANN), which can approximate any function with arbitrary accuracy given a sufficient number of units. ANNs also come with a ready-made algorithm for optimisation: the well-known backpropagation algorithm. However, there is a problem with this approach: we do not have training data available (this also precludes the use of Support Vector Machines). To obtain training data we would have to solve a set of instances, and there is no known method for solving the harder instances to optimality. Instead we must use an ANN to solve a problem in *reinforcement learning*, in which we must choose its weights in order

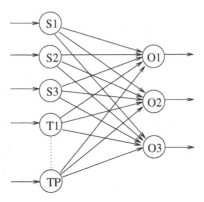

Fig. 1. The feedforward ANN used

to maximise reward (in this case to minimise expected cost). Backpropagation cannot be used for this task, but we can instead use an evolutionary algorithm (EA) whose genes are the weights and whose fitness function is minus the cost. This neuroevolutionary approach has been applied to control problems and game learning.

In our experiments we began with a standard three-layer feedforward ANN, which is a universal function approximator: it can approximate any function to arbitrary accuracy given a sufficient number of hidden units. We tried different numbers of hidden units, including multiple hidden layers, with different transfer functions in all the units (including sigmoids, limiter functions and polynomial expressions), and with two alternative representations of time period t: as an integer $t = 1 \ldots P$ and using the well-known *unary encoding* which is often used to represent symbolic ANN inputs and gave better results here. In the unary encoding we associate a binary variable with each period, and period t is represented by a vector $(0_1, \ldots, 0_{t-1}, 1, 0_{t+1}, \ldots, 0_P)$. Surprisingly, we obtained best results using an extremely simple network, with no hidden layer and the identity transfer function $f(x) = x$. No bias term is needed because the unary encoding already provides a time-dependent bias.

The ANN corresponding to three stocking points is shown in Figure 1, where Si denotes the ith stock level, Oi the ith order level, and Tj the jth binary variable in the unary time encoding. All units use the identity transfer function. Each arrowed line connecting two units in the diagram has an associated weight, so the ANN has $K(P + K)$ weights, where K is the number of stocking points. This ANN represents a simple set of affine relationships

$$O_j = \sum_i S_i w_{ij} + w_{tj}$$

where w_{ij} is the ANN weight between stock level S_i and order level O_j, and w_{tj} is the ANN weight between time t and order level O_j. (An affine transformation is a linear transformation followed by a translation.) One would not expect this to yield an efficient or even a sensible policy, but our policy is not yet complete as we have not handled the problem constraints.

2.2 Constraint Handling

The ANN forms only part of the policy. We also need a way of handling the constraints of the problem, which forbid (i) negative orders (corresponding to selling unused stock back to the supplier), and (ii) negative stock levels. We will train the ANN by an EA and there are several ways of handling constraints in EAs. We use a *decoder* which transforms the (possibly infeasible) ANN solution into one that violates no constraints. Decoders are a way of finding feasible solutions from chromosomes that represent infeasible states. They are problem-specific and ours works as follows. Suppose at period t we have stock levels S_i and the ANN suggests ordering quantities O_i. We modify each quantity O_i by

$$O_i \leftarrow \max(O_i, 0)$$

to avoid violating constraints of type (i). Then for any stocking point i that supplies a set of stocking points X_i we modify its order level O_i by

$$O_i \leftarrow \max\left(O_i, \left(\sum_{j \in X_i} O_j\right) - S_i\right)$$

This ensures that each supplier orders sufficient stock to fulfil its deliveries, and avoids violating constraints of type (ii). The policy is now the composition of the ANN and the decoder, which transforms the affine function of the ANN into a continuous piecewise affine function.

Note that we must modify the order levels of the stocking points earlier in the supply chain first. This is always possible if the supply chain is in the form of a directed acyclic graph. If lateral transshipments are allowed (orders between stocking points at the same level) or if there are constraints on order sizes or storage capacities then the decoder must be modified; we leave this issue for future work.

2.3 The Evolutionary Algorithm

To train the ANN we use an EA. There are many such algorithms in the literature, and we now describe our choice and the design decisions behind it. Firstly, we decided not to use genetic recombination. When training an ANN by EA one can encounter the well-known *competing conventions* problem (see [20] for example). This is caused by two forms of symmetry: an ANN's hidden units can be permuted without changing its output, and a hidden unit's weights can all be multiplied by -1 without changing its output. Thus if there are h hidden units then there are $2^h h!$ equivalent ANNs. Crossover is unlikely to give good results unless the parent chromosomes are aiming for symmetrically similar representations, though it is possible to design crossover operators that handle the symmetries [23]. This problem does not apply to our simple ANN because it has no hidden units, but in experiments crossover did not improve results so we do not use it.

We decided to use a $(\mu + 1)$-Evolution Strategy (ES) because it is almost exactly a steady-state genetic algorithm without crossover, and an efficient method for handling noise in the fitness function is known for a steady-state genetic algorithm (see below).

However, we adapted it to a *cellular* ES, in which each chromosome is notionally placed in an artificial space and nearby chromosomes form its neighbourhood. Cellular algorithms can reduce premature convergence, which we found to be a problem with our initial standard ES. In our ES the population size is μ, at each iteration a new chromosome c' is created by mutating a randomly selected chromosome c, and if c' is fitter than the least-fit chromosome c^* in the neighbourhood of c then it replaces c^*, otherwise c' is discarded. We used a ring topology and define the neighbourhood of a chromosome to be its two adjacent chromosomes.

A common form of mutation adds normally distributed noise to each gene, but we use a method that gave better results in experiments. For each chromosome we generate two uniformly distributed random numbers, p in the range $(0, 1)$ and q in the range $(0, 0.5)$. Then for each allele in the chromosome, with probability p we change it, otherwise with probability $1 - p$ we leave it unchanged. If we do change it then with probability q we set it to 0, otherwise with probability $1 - q$ we add a random number with Cauchy distribution to it. *Cauchy mutation* has been shown to speed up EAs [24]. It can be computed as $s \tan(u)$ where u is a uniformly distributed random variable in the range $(-\pi, \pi)$ and s is a scale factor. For each chromosome we compute a random scale factor, itself with Cauchy distribution and fixed scale factor 100. Finally, if no allele was modified (which is possible for small p) then we modify one randomly selected allele as described. This rather complex mutation operator is designed to generate a variety of random moves, with different numbers of modified alleles and different scale factors. All chromosomes initially contain alleles generated randomly using the same Cauchy distribution. We do not use the well-known technique of self-adapting step sizes, because in a $(\mu + 1)$-ES offspring with reduced mutation variances are always preferred [2].

2.4 Handling Uncertainty

When demand is probabilistic the fitness function of the EA is noisy. In such cases we must average costs over a number of simulations. In some previous SO approaches to inventory control, this problem was tackled by averaging costs over a small number of simulations because the simulations were computationally expensive: for example [13] use 3 samples. The standard deviation of the sample mean of a random variable with standard deviation σ is $\sigma/\sqrt{n}$ where n is the number of samples, so a large number of samples may be needed for very noisy fitness functions. Here we use smaller problems than those in [13] so we can afford to use a much larger number of simulations and obtain reliable cost estimates. To do this for every chromosome would be expensive but there are more efficient methods, and we use the *greedy averaged sampling* resampling scheme of [17]. This requires two parameters to be tuned by the user: U and S. On generating a new chromosome c it takes S samples to estimate its fitness before placing it into the population. It then selects another chromosome c' (which may be c) for *resampling*: another S samples are taken for c' and used to refine its fitness estimate. c' is the chromosome with highest fitness among those with fewer than U samples, so the function of U is to prevent any chromosome from being sampled more times than necessary. If all chromosomes in the population have been sampled U times then no resampling is performed. The algorithm is summarised in Figure 2.

```
train(μ, S, U)
   create ANN population of size μ
   evaluate population using S samples
   while not(termination condition)
      select a parent
      breed an offspring O by mutation
      evaluate O using S samples
      if O fitter than locally least-fit chromosome L
         replace L by O
      select globally fittest chromosome F with #samples < U
      if F exists
         re-evaluate F using S more samples
   return best chromosome found with #samples ≥ U
```

Fig. 2. Cellular evolution strategy with resampling

The aim of this resampling method is to obtain chromosomes with good fitness averaged over many samples, while expending a smaller number of samples on less-promising chromosomes. In our experiments we set $U = 10000$ so that cost estimates are obtained over 10000 samples, but by setting $S = 1$ we only expend approximately 200 samples per chromosome on average (this number was found by experiment). As the population size is 50, and $50 \times 200 = 10000$, this implies that a typical chromosome uses little more than one sample before being rejected as unfit. Using small S also has an effect beyond reducing the average number of samples per chromosome: it encourages exploration by preserving less-fit chromosomes for longer. We found this to be a very beneficial effect.

Some points are glossed over in Figure 2 for the sake of readability. Firstly, if S is not a divisor of U then fewer than S samples are needed in the final resampling of any chromosome to bring its total to U. Secondly, the termination condition is unspecified, and we simply use a timeout. Thirdly, if no chromosome has U samples on termination then we must choose another chromosome to return. To avoid this, S should be assigned a sufficiently large value so that in experiments there is always a chromosome with U samples on termination. This value must be chosen by experimentation.

2.5 Discussion of the Method

We refer to our method as NEMUE[1] (Neuro-Evolution for MUlti-Echelon systems). NEMUE is the result of many experiments with alternative versions. We experimented with an array of ANNs, one for each time period. This model has $12P$ weights and clearly subsumes the model above: any plan that can be represented by that model can also be represented by this one. The results should therefore be at least as good, but in experiments they were significantly worse. We believe that the ANN array is simply harder to train than a single ANN.

---

[1] The "lady of the lake" in Arthurian legend.

We also tried a non-unary encoding of time, in which order levels are linear functions of stock levels and polynomial functions of time. Fixing the polynomial degree makes the size of the ANN independent of the number of time periods. Using a cubic function of time gave reasonable results but was inferior to the unary encoding.

We used a decoder to handle the problem constraints, but there are other ways of handling constraints in EAs. The simplest is to use a *penalty function* which adds a large artificial cost for each violated constraint. In our problem this forces the ANN to learn to order sufficient stock in order to avoid stockout. We tried a penalty function but it gave inferior results to the decoder.

3   Experiments

Ultimately we are interested in solving large, realistic inventory problems with multiple stocking points, stochastic lead times, correlated demands and other features that make classical approaches impractical. Unfortunately there are no known methods for solving such problems to optimality, so there is no way of evaluating our method. Instead we consider more modest problems to test the ability of NEMUE to find good plans.

Our benchmark problems have two multi-echelon topologies: *arborescent* and *serial*. In the arborescent case we have three stocking points A, B and C, with C supplying A and B, while in the serial case C supplies B which supplies A. In both cases we have linear holding costs, linear penalty costs, fixed ordering costs, and stationary probabilistic demands. The closing inventory levels for period t are $I_t^A = I_{t-1}^A + Q_t^A - d_t^A$, $I_t^B = I_{t-1}^B + Q_t^B - d_t^B$ and $I_t^C = I_{t-1}^C + Q_t^C - Q_t^A - Q_t^B$ where Q_t is the order placed in period t and d_t is the demand in period t. If $I_t < 0$ then the incurred cost is $-I_t.\pi$, otherwise it is $I_t.h$, where π is the penalty cost and h the holding cost. Suppliers are not allowed to run out of stock. We prepared 28 instances of both the arborescent and serial types, with various costs and 2–9 time periods, giving a total of 56 instances with a range of characteristics as follows. The holding costs for A, B and C are 4, 5 and 1 respectively for arborescent instances 1–14; 3, 2 and 1 for arborescent instances 15–28; and 3, 2 and 1 for all serial instances. For the arborescent instances the penalty costs for A and B are 12 and 25 respectively for instances 1–14; and 3 and 6 for instances 15–28. For all the serial instances the penalty cost for instance A is 12. The ordering costs for A, B and C are 150, 130 and 170 respectively for arborescent instances 1–14; 80, 75 and 100 for arborescent instances 15–28; and 75, 80 and 100 for all serial instances. For space reasons we do not specify the demands in detail, but we used 10 patterns for arborescent instances and 4 patterns for serial instances. In each period we specify a deterministic demand which is then multiplied by either $\frac{2}{3}$ with probability 0.25, 1 with probability 0.5, or $\frac{4}{3}$ with probability 0.25. Thus the number of possible scenarios is 3^P, giving 59,049 scenarios for the largest problems ($P = 10$).

We solved these problems in two ways: using Stochastic Programming (SP) [3] and NEMUE. SP is a field of Operations Research designed to solve optimisation problems under uncertainty via scenario reduction techniques: a representative subset of all possible scenarios is selected and used to generate a deterministic equivalent optimisation problem, which is then typically solved using integer linear programming. We use the

SP results to evaluate the quality of plans found by NEMUE. The optimal replenishment plans are obtained using the following Stochastic Integer Programming model:

$$\min \mathsf{E}[C] = \sum_{t=1}^{N} \sum_{p \in P} \left(a_p \delta_{pt} + h_p I_{pt}^{+} + \pi_p I_{pt}^{-} \right)$$
$$\text{s.t. } t = 1, \ldots, N \text{ and } p \in P$$
$$I_{pt} = I_{p,t-1} + Q_{pt} - Q_{P_p,t} - d_{pt}$$
$$I_{pt} = I_{p,t}^{+} - I_{p,t}^{-}$$
$$Q_{pt} \le M \delta_{pt}$$
$$\delta_{pt} \in \{0, 1\} \quad Q_{pt} \ge 0$$

where

C : total holding and ordering/set-up cost of the system over N periods;
a : fixed ordering/set-up cost;
h : proportional inventory holding cost per period;
P : the set of all stocking points;
P_p : the set of stocking points supplied directly by the stocking point p;
d_{pt} : random demand at stocking point p, in period t;
δ_{pt} : a binary variable that takes the value of 1 if a replenishment occurs
    : at stocking point p in period t and 0 otherwise;
I_{pt} : the inventory level at the end of period t at stocking point p;
Q_{pt} : the order quantity at the beginning of period t at stocking point p;

and I^{+} and I^{-} denote positive and negative closing inventory levels. Except for the lowest echelon stocking points, I^{-} is zero. M is some large positive number. In this stochastic model a *here-and-now* policy is adapted: all decision variables are set before observing the realisation of the random variables. The certainty equivalent model is obtained using the compiler described in [22] and solved with CPLEX 11.2.

Results comparing SP and NEMUE are shown in Table 1. All SP runs were terminated after one hour and all NEMUE results after 30 minutes on a 2.8 GHz Pentium (R) 4 with 512 RAM, each figure being the best of six five-minute runs. The NEMUE parameters used were $S = 1$, $U = 10000$ and $\mu = 50$. SP runs that were aborted because of memory problems are denoted by "—". (In the few cases that SP found and proved optimality, this sometimes took much less than one hour.) The columns marked "%opt" denote the optimality gap: a reported cost c and gap g means that SP proved that the optimal solution cannot have cost lower than $c' = c(100 - g)/100$ (this does not imply the existence of a solution with cost c'). In several cases NEMUE finds superior plans to those found by SP, showing that on larger instances SP fails to find optimal plans. In a few cases NEMUE appears to find plans that are slightly better than optimal: this is of course impossible, and is a consequence of the empirical nature of the data. In such cases we assume that NEMUE found an optimal plan.

SP was unable to find provably optimal plans for all but the smallest instances. We believe that for the medium-sized instances SP finds optimal plans but does not prove optimality before timeout. For the largest instances SP ran out of memory, though we use the state-of-the-art CPLEX solver and a powerful machine (an Intel Core 2 Duo CPU E7200 with 2.53 GHz and 3GB of RAM). On the largest instances for which SP did not run out of memory, it was unable to prove optimality even within several days. Thus our benchmark problems straddle the borderline of solvability by classical methods.

Table 1. Experimental results

arborescent					serial				
	SP		NEMUE			SP		NEMUE	
# periods	cost	%opt	cost	%opt	# periods	cost	%opt	cost	%opt
1	4 2507	0	2573	2.6	1	4 995	0	993	0
2	5 3124	1.4	3180	3.1	2	5 1269	0.7	1298	2.9
3	6 3657	2.7	3775	5.7	3	6 1493	1.8	1491	1.7
4	7 4214	5.6	4250	6.4	4	7 1794	7.4	1797	7.6
5	8 4654	8.2	4722	9.5	5	8 2087	12.0	1987	7.6
6	9 5472	16.9	5164	11.9	6	9 2741	25.7	2295	11.3
7	10 —		? 5590	?	7	10 —		? 2603	?
8	4 2100	0	2169	3.2	8	4 1311	0.2	1306	0.0
9	5 2626	0.6	2722	4.1	9	5 1598	2.2	1594	2.0
10	6 3311	1.8	3409	4.6	10	6 1833	4.3	1832	4.2
11	7 4065	2.5	4153	4.6	11	7 2024	6.7	2024	6.7
12	8 4454	3.4	4542	5.3	12	8 2160	9.3	2142	8.5
13	9 5158	10.3	5115	9.5	13	9 2678	25.1	2264	11.4
14	10 —		? 5432	?	14	10 —		? 2407	?
15	4 1342	0.2	1340	0.1	15	4 1104	0	1104	0
16	5 1657	1.8	1671	2.6	16	5 1417	2.1	1423	2.5
17	6 1930	2.2	1938	2.6	17	6 1759	4.1	1763	4.3
18	7 2180	4.5	2192	5.0	18	7 2057	5.4	2055	5.3
19	8 2428	6.1	2393	4.7	19	8 2266	6.6	2258	6.3
20	9 2853	13.9	2617	6.1	20	9 2706	17.7	2479	10.2
21	10 —		? 2864	?	21	10 —		? 2627	?
22	4 1086	0	1096	0.9	22	4 828	0	828	0
23	5 1334	0.2	1330	0.0	23	5 931	0	934	0.3
24	6 1680	0.6	1677	0.4	24	6 1259	1.3	1265	1.8
25	7 2055	0.7	2051	0.5	25	7 1633	2.4	1639	2.8
26	8 2219	1.1	2220	1.1	26	8 1757	2.7	1766	3.2
27	9 2479	2.0	2531	4.0	27	9 1983	3.9	2000	4.7
28	10 —		? 2665	?	28	10 —		? 2150	?

Despite the simplicity of its policy and the large number of scenarios (at least on the larger instances) the NEMUE results are remarkably good. On 13 of the 28 arborescent instances and 19 of the 28 serial instances, NEMUE found plans that were at least as good as those found by SP. On the three serial instances for which SP found provably optimal plans, NEMUE found equally good plans. On most of the larger instances NEMUE found better plans than SP. These results show that: (i) a relatively simple, continuous, piecewise affine function can closely approximate a large policy tree for multi-echelon systems; (ii) such a function can be effectively represented by an affine function followed by a decoder function; (iii) the affine function can be learned in a reasonable time by evolutionary search; (iv) that our approach is more scalable than SP.

It is tempting to speculate that with improved heuristics and longer runtimes we might find optimal strategies for *all* instances. But there is no guarantee that all scenario trees can be well-approximated in this way, and in more extensive experiments on arborescent instance 1 (for example) we have been unable to find an optimal plan. Nevertheless, the results are very promising.

4   Related Work

Though simulation was originally used only to evaluate solutions found by other means, the field of SO has recently become more popular — see the survey of [5]. SO may be *recursive* or *non-recursive*. In the non-recursive approach an approximate cost function is learned during a simulation phase, then this function is minimised using an optimisation algorithm during a second phase. NEMUE is an example of recursive SO in which simulation and optimisation alternate and inform each other.

A tutorial and survey of the application of SO to inventory control is given in a recent paper [12]. Relatively little work has been done on applying SO to multi-echelon systems, and we have been unable to find any other work on inventory control via neuroevolution, though several papers use EAs for inventory control (for example [1,13,15,18]). Another difference of NEMUE is that it aims to approximate optimal policy trees, whereas most SO methods aim to find parameters for special policies such as (s, S). A different way of using ANNs for inventory control is to solve a set of training instances by some other method, then train an ANN to learn how to find good solutions from new instances (for example [6]). But we then need another algorithm to solve the problems, which is the aim of NEMUE. A related approach to neuroevolution is *genetic programming*, in which an EA is used to evolve an algorithm for solving the problem, instead of an ANN. This approach has also been applied to inventory control [11].

Another interesting approach to sequential decision problems such as those in inventory control is the field variously referred to as *neuro-dynamic programming*, *temporal difference learning* and *approximate dynamic programming*. This blend of dynamic programming and simulation has been applied to many problems including inventory control: see for example [4,7,10,19]. A drawback is that special techniques are needed to cope with the well-known "curse of dimensionality": the vast number of states that result from a simple discretisation of the continuum of states in these problems. In contrast, neuroevolution can directly handle a continuum of states.

5   Conclusion

We have proposed what seems to be the first neuroevolutionary method for approximating optimal plans in multi-echelon stochastic inventory control problems. Large or infinite scenario trees are approximated by a neural network, which is trained by an evolutionary algorithm with resampling, while problem constraints are handled by a decoder. Because the method is simulation-based and uses general-purpose techniques such as evolutionary algorithms and neural networks, it does not rely on special properties of the problem and can be applied to inventory problems with non-standard features. We showed experimentally that the method can find near-optimal solutions. In future work we will extend the method to handle problem features such as capacity constraints.

References

1. Arnold, J., Köchel, P.: Evolutionary Optimisation of a Multi-Location Inventory Model With Lateral Transshipments. In: 9th International Working Seminar on Production Economics, Igls, vol. 2, pp. 401–412 (1996); Preprints
2. Bäck, T., Hoffmeister, F., Schwefel, H.-P.: A Survey of Evolution Strategies. In: 4th International Conference on Genetic Algorithms (1991)
3. Birge, J.R., Louveaux, F.: Introduction to Stochastic Programming. Springer, New York (1997)
4. Chaharsooghi, S.K., Heydari, J., Zegordi, S.H.: A Reinforcement Learning Model for Supply Chain Ordering Management: an Application to the Beer Game. Decision Support Systems 45(4), 949–959 (2008)
5. Fu, M.C.: Optimization for Simulation: Theory vs Practice. INFORMS Journal of Computing 14, 192–215 (2002)
6. Gaafar, L.K., Choueiki, M.H.: A Neural Network Model for Solving the Lot-Sizing Problem. Omega 28(2), 175–184 (2000)
7. Giannoccaro, I., Pontrandolfo, P.: Inventory Management in Supply Chains: a Reinforcement Learning Approach. International Journal of Production Economics 78(2), 153–161 (2002)
8. Gomez, F., Schmidhuber, J., Miikkulainen, R.: Efficient Non-Linear Control Through Neuroevolution. Journal of Machine Learning Research 9, 937–965 (2008)
9. Hewahi, N.M.: Engineering Industry Controllers Using Neuroevolution. Artificial Intelligence for Engineering Design, Analysis and Manufacturing 19(1), 49–57 (2005)
10. Jiang, C., Shenga, Z.: Case-Based Reinforcement Learning for Dynamic Inventory Control in a Multi-Agent Supply-Chain System. Expert Systems with Applications 36(3 part 2), 6520–6526 (2009)
11. Kleinau, P., Thonemann, U.W.: Deriving Inventory-Control Policies With Genetic Programming. OR Spectrum 26(4), 521–546 (2004)
12. Köchel, P.: Simulation (Optimisation) in Inventory Theory. Tutorial, 8th ISIR Summer School on New and Classical Streams in Inventory Management: Advances in Research and Opening Frontiers (2007)
13. Köchel, P., Nieländer, U.: Simulation-Based Optimisation of Multi-Echelon Inventory Systems. International Journal of Production Economics 1, 503–513 (2005)
14. Lubberts, A., Miikkulainen, R.: Co-Evolving a Go-Playing Neural Network. In: Genetic and Evolutionary Computation Conference, pp. 14–19. Kaufmann, San Francisco (2001)
15. Olsen, A.L.: An Evolutionary Algorithm for the Joint Replenishment of Inventory with Interdependent Ordering Costs. In: Cantú-Paz, E., Foster, J.A., Deb, K., Davis, L., Roy, R., O'Reilly, U.-M., Beyer, H.-G., Kendall, G., Wilson, S.W., Harman, M., Wegener, J., Dasgupta, D., Potter, M.A., Schultz, A., Dowsland, K.A., Jonoska, N., Miller, J., Standish, R.K. (eds.) GECCO 2003. LNCS, vol. 2724, pp. 2416–2417. Springer, Heidelberg (2003)
16. Pollack, J.B., Blair, A.D.: Co-Evolution in the Successful Learning of Backgammon Strategy. Machine Learning 32(3), 225–240 (1998)
17. Prestwich, S.D., Tarim, S.A., Rossi, R., Hnich, B.: A Steady-State Genetic Algorithm With Resampling for Noisy Inventory Control. In: Rudolph, G., Jansen, T., Lucas, S., Poloni, C., Beume, N. (eds.) PPSN 2008. LNCS, vol. 5199, pp. 559–568. Springer, Heidelberg (2008)
18. Prestwich, S.D., Tarim, S.A., Rossi, R., Hnich, B.: A Cultural Algorithm for POMDPs from Stochastic Inventory Control. In: Blesa, M.J., Blum, C., Cotta, C., Fernández, A.J., Gallardo, J.E., Roli, A., Sampels, M. (eds.) HM 2008. LNCS, vol. 5296, pp. 16–28. Springer, Heidelberg (2008)
19. Van Roy, B., Bertsekas, D.P., Lee, Y., Tsitsiklis, J.N.: A Neuro-Dynamic Programming Approach to Retailer Inventory Management. In: Proceedings of the IEEE Conference on Decision and Control (1997)

20. Schaffer, J., Whitley, D., Eshelman, L.: Combinations of Genetic Algorithms and Neural Networks: A Survey of the State of the Art. In: International Workshop on Combinations of Genetic Algorithms and Neural Networks, pp. 1–37 (1992)
21. Stanley, K.O., Miikkulainen, R.: Evolving Neural Networks Through Augmenting Topologies. Evolutionary Computation 10(2), 99–127 (2002)
22. Tarim, S.A., Manandhar, S., Walsh, T.: Stochastic Constraint Programming: A Scenario-Based Approach. Constraints 11, 53–80 (2006)
23. Thierens, D.: Non-Redundant Genetic Coding of Neural Networks. In: International Conference on Evolutionary Computation, Nagoya, Japan, pp. 571–575 (1996)
24. Yao, X., Liu, Y., Lin, G.: Evolutionary Programming Made Faster. IEEE Transactions on Evolutionary Computation 3(2), 82–102 (1999)

Determining a Minimum Spanning Tree
with Disjunctive Constraints

Andreas Darmann[1], Ulrich Pferschy[2], and Joachim Schauer[2]

[1] University of Graz, Institute of Public Economics, Universitaetsstr, 15, 8010 Graz,
Austria
andreas.darmann@uni-graz.at
[2] University of Graz, Department of Statistics and Operations Research
Universitaetsstr, 15, 8010 Graz, Austria
{joachim.schauer,pferschy}@uni-graz.at

Abstract. For the classical minimum spanning tree problem we introduce disjunctive constraints for pairs of edges which can not be both
included in the spanning tree at the same time. These constraints are
represented by a conflict graph whose vertices correspond to the edges of
the original graph. Edges in the conflict graph connect conflicting edges
of the original graph. It is shown that the problem becomes strongly
$\mathcal{NP}$-hard even if the connected components of the conflict graph consist
only of paths of length two. On the other hand, for conflict graphs consisting of disjoint edges (i.e. paths of length one) the problem remains
polynomially solvable.

Keywords: minimum spanning tree, conflict graph.

1 Introduction

The minimum spanning tree problem (MST) is a classical problem in combinatorial optimization. It has been extensively studied within its original framework
as well as in connection with fields as social choice theory, in particular with aspects of fair division. The issue of fairly assigning the costs of a minimum spanning
tree to individuals (represented by nodes) has first been raised by Bird [2], recent
works often consider characterizations of fair division rules (e.g., Bogomolnaia and
Moulin [4], Dutta and Kar [6] and Kar [11]). Another approach to fairness in connection with spanning trees, based on preference relations on the edges instead of
monetary costs, has been considered in Darmann, Klamler and Pferschy [5].

This paper however deals with a different aspect of the link to social choice
theory: the minimum spanning tree problem with conflict graph (MSTCG) addresses the question of finding a minimum spanning tree in a weighted, undirected graph, given there are incompatibilities for certain pairs of edges. The
incompatibilities mean that from each such conflicting pair of edges at most one
edge can occur in the spanning tree.

A practical example could be the installation of an oil pipeline system connecting various countries. Each country needs to be connected to the pipeline

F. Rossi and A. Tsoukis (Eds.): ADT 2009, LNAI 5783, pp. 414–423, 2009.
© Springer-Verlag Berlin Heidelberg 2009

system, but there are many ways to hook them up. Such a situation can be represented by means of a weighted undirected graph: the nodes are the countries, the edges are the possible links between the countries and the edge weights are the costs of the respective link between the countries. The cheapest way to install the system corresponds to a minimum spanning tree in the graph. However, for technical or political reasons many different firms can be required to install the system, and we assume that each link can be constructed by exactly one firm. Now certain conflicts may arise. A reason might be, that some firms are not willing to cooperate with each other, and thus certain links (edges) cannot both be contained in a solution. The MSTCG asks for the cheapest way to install the pipeline system subject to these conflicts.

It is natural to represent such symmetric conflict relations by means of an undirected *conflict graph*, where every vertex of the conflict graph corresponds uniquely to an edge in the original graph and an edge in the conflict graph implies that the two adjacent vertices, i.e. edges in the original graph, cannot occur together in an MST solution.

For a formal definition of the *minimum spanning tree problem with conflict graph*, let $G = (V, E)$ be an undirected connected graph with n vertices and m edges, where each edge e has associated a weight $w(e)$ (w is a weight function $w : E \rightarrow \mathbb{R}$). Furthermore, an undirected graph $\bar{G} = (E, \bar{E})$ represents a conflict graph where each of the m vertices corresponds uniquely to an edge $e \in E$ of G. An edge $\bar{e} = (i, j) \in \bar{E}$ implies that the two vertices incident to $\bar{e}$ – that is, the two edges $i, j \in E$ – cannot occur together in a spanning tree of G. In contrast to G, $\bar{G}$ is not necessarily connected and may contain isolated vertices (i.e. edges of G which can be combined with every other edge in the minimum spanning tree solution). MSTCG asks for a minimum spanning tree T in G, given that adjacent vertices in $\bar{G}$ are not both together included in T.

For a set of vertices $F \subseteq V$ in G let $E(F)$ be the set of edges in G that have both of its endpoints in F. Then MSTCG can be stated by the following integer linear programming (ILP) formulation:

$$(MSTCG) \quad \min \sum_{e \in E} w(e) * x_e \tag{1}$$

$$\text{s.t.} \sum_{e \in E} x_e = n - 1 \tag{2}$$

$$\sum_{e \in E(F)} x_e \leq |F| - 1 \quad \forall \emptyset \neq F \subseteq V \tag{3}$$

$$x_e + x_f \leq 1 \quad \forall (e, f) \in \bar{E} \tag{4}$$

$$x_e \in \{0, 1\} \quad \forall e \in E \tag{5}$$

Obviously, (1)–(3) and (5) is a classical ILP-model for MST (e.g. see [1]):

- the constraint (2) ensures that $n - 1$ edges are chosen,
- the constraint (3) implies that the chosen edges do not form a cycle and
- the constraint (5) states that x_e is a binary variable, indicating whether or not edge e is being contained in the spanning tree.

The remaining constraint (4) adds the conflict relations represented by the conflict graph. It is worth noting that, in contrast to the classical minimum spanning tree problem, due to this conflict constraint there might be no feasible solution for MSTCG even if the graph G is connected.

In this paper we will characterize the complexity of MSTCG and identify the graph classes for the conflict graph $\bar{G}$ where the problem changes from polynomially solvable to strongly $\mathcal{NP}$-hard.[1] These are graphs whose connected components are edges resp. paths of length two (we define the length of a path as the number of edges in the path). For obvious illustrative reasons we introduce the following terminology.

Definition 1. *A 2-ladder is an undirected graph whose components are paths of length one, i.e. edges connecting pairs of vertices.*

Definition 2. *A 3-ladder is an undirected graph whose components are paths of length two.*

It will be shown in Section 2 that MSTCG is already strongly $\mathcal{NP}$-hard if the underlying conflict graph is a 3-ladder. In particular we show that, given that the conflict graph is a 3-ladder, MSTCG is $\mathcal{NP}$-hard even when the edge weights are restricted to $\{0, 1\}$. On the other hand, it can be shown by a matroid intersection argument in Section 3 that the problem remains polynomially solvable for a 2-ladder as a conflict graph.

In contrast to the latter result, it should be noted that the shortest path problem with pairwise disjoint forbidden pairs of edges (i.e. with a 2-ladder conflict graph) is known to be strongly $\mathcal{NP}$-hard [8]. Results of the same flavour were recently derived for the classical 0-1 knapsack problem with conflict graphs. While this problem is strongly $\mathcal{NP}$-hard for arbitrary conflict graphs, it was shown in [12] that pseudopolynomial algorithms (and hence also fully polynomial approximation schemes) exist if the given conflict graph is a tree, a graph of bounded treewidth or a chordal graph. Bin packing problems with special classes of conflict graphs were considered from an approximation point of view by [10] and [9]. Complexity results for different classes of conflict graphs for a scheduling problem under makespan minimization are given in [3]. Further references on combinatorial optimization problems with conflict graphs can be found in [12].

2   A Strongly $\mathcal{NP}$-hardness Result for MSTCG

In this section we show that MSTCG is already strongly $\mathcal{NP}$-hard if the conflict graph $\bar{G}$ is a 3-ladder. E.g. for $e_1, e_2, e_3 \in E$ let a component of $\bar{G}$ be made up of the path (e_1, e_2, e_3). Then, in terms of the underlying graph G, a feasible spanning tree for MSTCG that contains e_2 must include neither edge e_1 nor edge e_3. However, a feasible tree that contains e_1 must not contain e_2, but it may contain e_3.

---

[1] A problem is said to be strongly $\mathcal{NP}$-hard if it is $\mathcal{NP}$-hard even when its numerical parameters are bounded by a polynomial in the length of the input, i.e. the number of input values.

2.1   The Idea of the Reduction

We reduce an $\mathcal{NP}$-complete subproblem of 3-SAT on the decision problem corresponding to MSTCG with a 3-ladder as conflict graph.

Let I be an arbitrary instance of 3-SAT with k clauses $C_1, \ldots, C_k$ and n variables $x_1, \ldots, x_n$, such that each literal x_i occurs in ℓ_i clauses and its negation $\bar{x}_i$ occurs in $\bar{\ell}_i$ clauses. We restrict ourselves to instances with $\ell_i + \bar{\ell}_i \leq 5$. The decision problem whether or not there exists a satisfying truth assignment for I is $\mathcal{NP}$-complete [8]. Considering this decision problem, we construct an instance I' of MSTCG by defining a graph G_{3-SAT} and a conflict graph $\bar{G}_{3-SAT}$ as described in Subsection 2.2.

It should be emphasized that in the graph G_{3-SAT} some nodes represent the clauses in I, whereas some edges represent the literals occuring in I.

However, the reduction of the above subproblem of 3-SAT to MSTCG is then performed in two steps (it is presented in full detail in Subsection 2.3). Given that I is a "YES"-instance of I, we construct an optimal solution T with weight k of the instance I' of MSTCG from a satisfying truth assignment for I. This is basically done by first choosing those edges to be in T that represent literals set "TRUE" under the truth assignment. The rest of the tree is being constructed in a quite straightforward way by regarding the conflict graph $\bar{G}_{3-SAT}$.

On the other hand, we show that an optimal solution T with weight k implies the existence of a satisfying truth assignment for I. Here the main idea is to show that for each clause of I at least one edge that represents one of the literals that make up the clause is contained in the tree. Thus, as a consequence of the conflict graph, the truth assignment that sets "TRUE" exactly those literals that are contained in the tree T constitutes a satisfying truth assignment for I.

2.2   The Graphs G_{3-SAT} and $\bar{G}_{3-SAT}$

Construction of G_{3-SAT}. Unless otherwise stated, each edge of G_{3-SAT} has zero weight. The graph G_{3-SAT} is being built as follows (see Figure 1): For each variable x_i, $1 \leq i \leq n$, we introduce

- the edges $x_i = (a_i, b_i)$ and $\bar{x}_i = (\bar{a}_i, \bar{b}_i)$ corresponding to the literals with the same label,
- a vertex i that is connected to b_i and $\bar{b}_i$ via the edges y_i and $\bar{y}_i$ respectively and
- one path of length ℓ_i starting in vertex i and ending in a_i. Let the edges of this path starting at i be called $w_{i0}, w_{i1}, w_{i2}, \ldots$. Each vertex of this path is connected to vertex b_i by the edges $z_{i1}, z_{i2}, \ldots$, where z_{ij} is adjacent to $w_{i(j-1)}$, $j \geq 1$. An analogous path is defined for i and $\bar{a}_i$ with the corresponding connections to $\bar{b}_i$.

For each clause C_j, $1 \leq j \leq k$,

- a vertex labelled C_j is introduced,
- for each x_i contained in clause C_j we insert a path of length 4 consisting of edges $(e_{ij}, f_{ij}, g_{ij}, h_{ij})$ starting in vertex a_i and ending in vertex C_j. Furthermore, a shortcut is constructed by joining vertex a_i to the vertex incident to f_{ij} and g_{ij} via an edge Δ_{ij}.

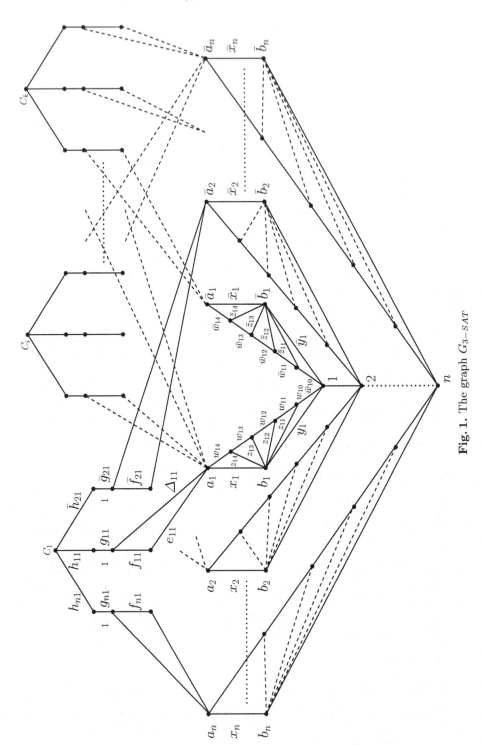

Fig. 1. The graph G_{3-SAT}

Analogously, the path $(\bar{e}_{ij}, \bar{f}_{ij}, \bar{g}_{ij}, \bar{h}_{ij})$ connects $\bar{a}_i$ with C_j if literal $\bar{x}_i$ is contained in C_j with an analogous shortcut $\bar{\Delta}_{ij}$.

Finally, edges connecting the parts of G_{3-SAT} described above are introduced. For $1 \le i \le n-1$ we introduce an edge connecting vertex i with vertex $i+1$.

A weight of 1 is associated with all edges g_{ij} and $\bar{g}_{ij}$. Note that these are the only edges of non-zero weight in G_{3-SAT}.

Construction of $\bar{G}_{3-SAT}$. The conflict graph $\bar{G}_{3-SAT}$ on the edges of G_{3-SAT} is defined in the following way. Denote the clauses containing literal x_i resp. $\bar{x}_i$ by $C_{x_{i0}}, \ldots, C_{x_{i(\ell_i-1)}}$ resp. $C_{\bar{x}_{i0}}, \ldots, C_{\bar{x}_{i(\bar{\ell}_i-1)}}$, where the order in which theses clauses are chosen is arbitrary but fixed. Then we introduce in $\bar{G}_{3-SAT}$ the edge $(w_{i0}, f_{ix_{i0}})$ and the paths $(z_{i1}, w_{i1}, f_{ix_{i1}}), \ldots, (z_{i\ell_i-1}, w_{i\ell_i-1}, f_{ix_{i\ell_i-1}})$. Again we construct equivalent components of $\bar{G}_{3-SAT}$ for the clauses $C_{\bar{x}_{i0}}, \ldots, C_{\bar{x}_{i(\bar{\ell}_i-1)}}$ containing literal $\bar{x}_i$.

Furthermore we add the edges $(x_i, \bar{x}_i)$, the edges $(\Delta_{ix_{i0}}, g_{ix_{i0}}), \ldots, (\Delta_{ix_{i(\ell_i-1)}}, g_{ix_{i(\ell_i-1)}})$ and the edges $(\bar{\Delta}_{i\bar{x}_{i0}}, \bar{g}_{i\bar{x}_{i0}}), \ldots, (\bar{\Delta}_{i\bar{x}_{i(\bar{\ell}_i-1)}}, \bar{g}_{i\bar{x}_{i(\bar{\ell}_i-1)}})$. This procedure is performed for all variables.

Remark 1. Note that $\bar{G}_{3-SAT}$ is not a 3-ladder. To be more precise, $\bar{G}_{3-SAT}$ consists of a subgraph being a 3-ladder, a subgraph which is made up of components consisting of a single edge, and of isolated vertices. However, by introducing "dummy edges" $\bar{G}_{3-SAT}$ can easily be transformed into a 3-ladder.

2.3   MSTCG with a 3-Ladder Conflict Graph Is Strongly $\mathcal{NP}$-hard

Theorem 1. *Let $G = (V, E)$ be an undirected graph and let the conflict graph $\bar{G} = (E, \bar{E})$ be a 3-ladder. Then MSTCG is strongly $\mathcal{NP}$-hard.*

Proof

Let I be an instance of 3-SAT where each variable occurs in at most 5 clauses. Let the instance I' of $MSTCG$ be defined by the graph G_{3-SAT} and the conflict graph $\bar{G}_{3-SAT}$ constructed as described in Section 2.2. Let τ be the set of feasible solutions of I'.

Let $T \in \tau$. It is easy to see that T must have a weight of at least k: The set $G_j := \{g_{qj} | 1 \le q \le n\} \cup \{\bar{g}_{qj} | 1 \le q \le n\}$ is a cut set, i.e. it separates the vertex C_j from the rest of the graph. Therefore, every spanning tree must contain at least one edge from G_j. Since each of the mentioned edges has a weight of 1 we have $w(T) \ge k$.

We proof the theorem by showing that the following equivalence holds:

$$\exists \text{ a satisfying truth assignment for } I \iff \exists\, T \in \tau : w(T) \le k$$

"$\implies$": Given a satisfying truth assignment t_I for instance I we construct a feasible solution T of I' with $w(T) = k$ in a quite straightforward way. Let $T := \emptyset$ and let $X := \{x_{i_1}, \ldots, x_{i_r}\}$ and $\bar{X} := \{\bar{x}_{k_1}, \ldots, \bar{x}_{k_s}\}$ be the sets of literals set "TRUE" under t_I (recall that setting literal $\bar{x}_{k_j}$ "TRUE" means to set variable x_{k_j} "FALSE", $1 \leq j \leq s$).

$$T := T \cup X \cup \bar{X}$$

$$T := T \cup \{y_i\} \cup \{\bar{y}_i\} \quad \forall i \in \{1, \ldots, n\}$$

$$T := T \cup \{(i, i+1)\} \quad \forall i \in \{1, \ldots, n-1\}$$

Label all clauses C_j "unmarked". Let $C(x_i)$ (resp. $C(\bar{x}_i)$) be the set of all clauses containing x_i (resp. $\bar{x}_i$). We complete T by performing the following algorithmic statements:

> for $i \in \{i_1, \ldots, i_r\}$:
> $T := T \cup \{\bar{w}_{i0}\}$
>   for $j \in \{1, \ldots, \ell_i\}$:
> $T := T \cup \{z_{ij}\}$
>   for $j' \in \{1, \ldots, \ell_i\}$:
> $T := T \cup \{\bar{w}_{ij'}\}$
>   for $C_j \in C(x_i)$:
> $T := T \cup \{e_{ij}, f_{ij}, h_{ij}\}$
>   if C_j is "unmarked": (a)
> $T := T \cup \{g_{ij}\}$
>   Label C_j "marked".
>   for $C_j \in C(\bar{x}_i)$:
> $T := T \cup \{\bar{\Delta}_{ij}, \bar{e}_{ij}, \bar{h}_{ij}\}$
>
> for $k \in \{k_1, \ldots, k_s\}$:
> $T := T \cup \{w_{k0}\}$
>   for $j \in \{1, \ldots, \bar{\ell}_k\}$:
> $T := T \cup \{\bar{z}_{kj}\}$
>   for $j' \in \{1, \ldots, \ell_k\}$:
> $T := T \cup \{w_{kj'}\}$
>   for $C_j \in C(\bar{x}_k)$:
> $T := T \cup \{\bar{e}_{kj}, \bar{f}_{kj}, \bar{h}_{kj}\}$
>   if C_j is "unmarked": (a)
> $T := T \cup \{\bar{g}_{kj}\}$
>   Label C_j "marked".
>   for $C_j \in C(x_k)$:
> $T := T \cup \{\Delta_{kj}, e_{kj}, h_{kj}\}$

It is easy to check that T is a feasible solution of I'. Since t_I is a satisfying truth assignment for I, each clause C_j contains a literal set "TRUE". As mentioned before, in order to reach a vertex C_j at least one edge with weight 1 has to be added to T. By (a) in the construction of T for each vertex C_j exactly one such edge is contained in T. Thus we get $w(T) = k$.

"$\Longleftarrow$": Let $T \in \tau$ with $w(T) = k$. Since each vertex C_j has to be reached, at least one edge $G_j = \{g_{qj} | 1 \leq q \leq n\} \cup \{\bar{g}_{qj} | 1 \leq q \leq n\}$ has to be contained in T,

$1 \leq j \leq k$. However, since each edge of G_j has weight 1, exactly one edge of G_j is contained in T, $1 \leq j \leq k$. Recall that each vertex C_j corresponds to a clause with the same label. Let for vertex (clause) C_j the edge g_{ij} be the unique edge in T that is an element of G_j. We will now show that this implies $x_i \in T$ and $\bar{x}_i \notin T$:

Let γ be the vertex incident to both g_{ij} and f_{ij}. As T is a tree there has to be a unique simple path p in T between vertex γ and vertex 1. Due to the fact that g_{ij} is the only edge in T that is contained in the set G_j the path p cannot pass through vertex C_j. Analogously, p cannot pass through any of the vertices $C_r \neq C_j$, $1 \leq r \leq k$. Furthermore, the edge Δ_{ij} cannot be contained in p because the edge (Δ_{ij}, g_{ij}) is a component of the conflict graph $\bar{G}_{3-SAT}$. Thus $f_{ij}, e_{ij} \in p$ (and hence $f_{ij}, e_{ij} \in T$) has to hold.

Now assume $x_i \notin T$. Then in order to reach vertex 1 the path p must contain $w_{i(\ell_i-1)}$. Note that $w_{i(\ell_i-1)} \in T$ implies $z_{i(\ell_i-1)} \notin T$ because of $\bar{G}_{3-SAT}$. Analogously, we have $w_{ik} \in T$ and $z_{ik} \notin T$ for $1 \leq k \leq \ell_i - 2$ and $w_{i0} \in T$ as well. Thus we have $w_{ik} \in T$ for all $0 \leq k \leq \ell_i - 1$ which contradicts $f_{ij} \in T$ due to the construction of $\bar{G}_{3-SAT}$. Hence $x_i \in T$, and again by $\bar{G}_{3-SAT}$ we get $\bar{x}_i \notin T$.

Summarizing the facts we have that for each vertex C_j exactly one edge x_i, resp. $\bar{x}_i$, is contained in T that is connected to C_j by the path $(e_{ij}, f_{ij}, g_{ij}, h_{ij})$, resp. $(\bar{e}_{ij}, \bar{f}_{ij}, \bar{g}_{ij}, \bar{h}_{ij})$. That is, for each clause at least one of the literals the clause is made up of is contained in the tree. Thus, the truth assignment that sets a variable x_i "TRUE" if $x_i \in T$ and x_i "FALSE" if $\bar{x}_i \in T$ constitutes a satisfying truth assignment for I. $\qquad\square$

Since MSTCG is strongly $\mathcal{NP}$-hard given the conflict graph is a 3-ladder, MSTCG is also strongly $\mathcal{NP}$-hard in case the conflict graph is a path. Finally both results obviously imply that MSTCG is strongly $\mathcal{NP}$-hard for general conflict graphs.

Corollary 1. *Given the conflict graph is a path, MSTCG is strongly $\mathcal{NP}$-hard.*

Corollary 2. *MSTCG is strongly $\mathcal{NP}$-hard.*

3 MSTCG with Disjoint Conflicting Pairs of Edges Is in $\mathcal{P}$

In this section we focus on the MSTCG where the conflict graph is a 2-ladder, i.e. the conflict graph represents pairwise disjoint forbidden pairs of edges of E. We will first give a representation of the set τ of feasible solutions of this problem by using matroid intersection. With the help of that representation, Edmonds' famous matroid-intersection theorem (Edmonds [7], cf. [13]) yields that an optimal solution of MSTCG with disjoint conflicting pairs of edges can be computed in polynomial time.

Let us consider the ILP-formulation of our problem. Condition (3) of this formulation induces the well-known *graphic-matroid*. To be more precise, the graphic matroid $\mathcal{M}_1 = (E, \mathcal{I}_1)$ is being formed by all subsets of E which do not form a cycle [13]. More formally, $\mathcal{I}_1$ is the set of all subsets of E such that (3) is satisfied. Obviously a base of $\mathcal{M}_1$ consists of $n - 1$ edges (a base corresponds to a spanning tree) and thus (2) is satisfied by all bases of $\mathcal{M}_1$. However, the fact that the conflict graph $\bar{G}$ is being made up of disjoint conflicting pairs of edges of G induces a matroid as well. As we will show below, defining $\mathcal{I}_2$ as the set of all subsets of E that do not contain a conflicting pair of edges (represented by $\bar{G}$) yields a matroid $\mathcal{M}_2 = (E, \mathcal{I}_2)$. This matroid will be called *conflict-free matroid*. More formally, the conflict-free matroid $\mathcal{M}_2 = (E, \mathcal{I}_2)$ is defined by

$$\mathcal{I}_2 := \{E' \subseteq E | \nexists (e, f) \in \bar{E} : \{e, f\} \subseteq E').$$

Lemma 1. $\mathcal{M}_2 = (E, \mathcal{I}_2)$ *is a matroid.*

Proof

Obviously, $\emptyset \in \mathcal{I}_2$ is satisfied.

Let J be an element of $\mathcal{I}_2$ and I any subset of J. Then I cannot include a conflicting pair as J does not and hence $I \in \mathcal{I}_2$ holds. Thus $J \in \mathcal{I}_2$ implies $I \in \mathcal{I}_2$ for all $I \subseteq J$.

Let $I, J \in \mathcal{I}_2$ and $|I| < |J|$, then we have to show that $(I \cup \{z\}) \in \mathcal{I}_2$ for some $z \in J \setminus I$:

- *Case 1: $I \subset J$.* This case is trivial.
- *Case 2: $I \cap J = \emptyset$.* Assume that no such edge z exists: Then for every edge $e \in J$ the set $(I \cup \{e\})$ is not in $\mathcal{I}_2$. This implies that there exists an edge $e^c \in I$ such that (e, e^c) is a conflicting pair in $(I \cup \{e\})$. But since $\bar{G}$ is a 2-ladder for each edge $e' \in E$ there is at most one edge that is in conflict with e'. Hence, every edge $e^c \in I$ can belong to only one conflicting pair (e, e^c) with $e \in J$. Thereby, we get a contradiction to $|I| < |J|$.
- *Case 3: $I \cap J \neq \emptyset$.* Let $I' := I \setminus (I \cap J)$ and $J' := J \setminus (I \cap J)$. Then we have $I', J' \in \mathcal{I}_2$, $|I'| < |J'|$ and $I' \cap J' = \emptyset$. From *Case 2* we get that there is a $z \in J'$ such that $(I' \cup \{z\}) \in \mathcal{I}_2$. It follows from $z \in J$ that z can not be in conflict with any edge in $I \cap J$ and hence $(I \cup \{z\}) \in \mathcal{I}_2$.   □

Clearly, any feasible solution of MSTCG with disjoint conflicting pairs of edges corresponds to a common base of the graphic matroid and the conflict-free matroid. Thus, an optimal solution of MSTCG is a common base of $\mathcal{M}_1$ and $\mathcal{M}_2$ with minimum weight. As a consequence, Edmonds' weighted matroid intersection algorithm gives the answer on the question of the computational complexity of MSTCG with disjoint conflicting pairs of edges.

Theorem 2. *(Edmonds [7], cf. [13])*
Let S be a set and let $c : S \to \mathbb{R}$. Given two matroids $M_1 = (S, I_1)$ and $M_2 = (S, I_2)$, a common base of M_1 and M_2 with minimum weight can be found in strongly polynomial time.

Since any optimal solution of MSTCG corresponds to a minimum-weight common base of the graphic matroid and the conflict-free matroid the above theorem yields the following result.

Theorem 3. *MSTCG with disjoint conflicting pairs of edges can be solved in strongly polynomial time.*

References

1. Ahuja, R.K., Magnanti, T.L., Orlin, J.B.: Network flows: theory, algorithms, and applications. Prentice Hall, Englewood Cliffs (1993)
2. Bird, C.G.: On cost allocation for a spanning tree: a game theoretic approach. Networks 6, 335–350 (1976)
3. Bodlaender, H.L., Jansen, K.: On the complexity of scheduling incompatible jobs with unit-times. In: Borzyszkowski, A.M., Sokolowski, S. (eds.) MFCS 1993. LNCS, vol. 711, pp. 291–300. Springer, Heidelberg (1993)
4. Bogomolnaia, A., Moulin, H.: Sharing the cost of a minimal cost spanning tree: beyond the folk solution. Rice University, Mimeo (2008)
5. Darmann, A., Klamler, C., Pferschy, U.: Maximizing the minimum voter satisfaction on spanning trees. Mathematical Social Sciences 58, 238–250 (2009)
6. Dutta, B., Kar, A.: Cost monotonicity, consistency and minimum cost spanning tree games. Games and Economic Behavior 48, 223–248 (2004)
7. Edmonds, J.: Matroid intersection. Annals of Discrete Mathematics 4, 39–49 (1979)
8. Garey, M.R., Johnson, D.S.: Computers and Intractability: A Guide to the Theory of NP-Completeness. W. H. Freeman & Co., New York (1979)
9. Jansen, K.: An approximation scheme for bin packing with conflicts. In: Arnborg, S. (ed.) SWAT 1998. LNCS, vol. 1432, pp. 35–46. Springer, Heidelberg (1998)
10. Jansen, K., Öhring, S.: Approximation algorithms for time constrained scheduling. Information and Computation 132(2), 85–108 (1997)
11. Kar, A.: Axiomatization of the shapley value on minimum cost spanning tree games. Games and Economic Behavior, 265–277 (2002)
12. Pferschy, U., Schauer, J.: The knapsack problem with conflict graphs. Optimization Online 2008-10-2128 (2008)
13. Schrijver, A.: Combinatorial Optimization, Polyhedra and efficiency, vol. B. Springer, Heidelberg (2003)

An Inductive Methodology for Data-Based Rules Building

J. Tinguaro Rodríguez[1,*], Javier Montero[1], Begoña Vitoriano[1], and Victoria López[2]

[1] Faculty of Mathematics, Complutense University of Madrid, Plaza de Ciencias 3
28040 Madrid Spain
{jtrodrig,javier_montero,bvitoriano}@mat.ucm.es
[2] Faculty of Informatics, Complutense University of Madrid,
Profesor José García Santesmases s/n 28040 Madrid Spain
vlopez@fdi.ucm.es

Abstract. Extraction of rules from databases for classification and decision tasks is an issue of growing importance as automated processes based on data are being required in these fields. An inductive methodology for data-based rules building and automated learning is presented in this paper. A fuzzy framework is used for knowledge representation and, through the introduction and the use of dual properties in the valuation space of response variables, reasons for and against the rules are evaluated from data. This make possible to use continuous DDT logic, which provides a more general and informative framework, in order to assess the validity of rules and build an appropriate knowledge base.

Keywords: rules induction, DDT logic, fuzzy inference systems, dual predicates.

1 Introduction

Fuzzy rules and algorithms are widely used in several real-life applications of computational intelligence, as targeted e-commerce marketing and advertising [14], natural disaster and emergency management (see for instance [7] and [13]) or control [8]. In many of these applications, decision and classification rules are obtained by means of experts and the subsequent knowledge engineering.

Nevertheless, in many cases, the necessary information to build these rules is contained in databases rather than in experts' heads. Moreover, as the underlying realities of these applications are evolving, it is also necessary to undergo a continuous learning process in order to adapt to these changing situations. Last but not least, in many fields this learning process is needed to be automated.

For all these reasons, procedures are needed to extract and build a set of fuzzy rules from raw data contained in databases. In this paper, we propose a general methodology to do so, based upon an inductive approach in which rules are conceived as successively experienced relations among variables. However, the notion of *dual property* or *dual class* will be introduced in order to be able to evaluate these successive relations

---

* Corresponding author.

F. Rossi and A. Tsoukis (Eds.): ADT 2009, LNAI 5783, pp. 424–433, 2009.

as *reasons for* or *reasons against* a certain rule. This will allow us to introduce DDT logic [8] in order to obtain more granularity for assessing the rules Another methodology for building interpretable fuzzy rules from data is the one described in [3].

This paper is organized as follows: first, we describe the model of knowledge representation, which takes a database as input and produces, by means of a set of classes (crisp or fuzzy), a matrix representing the data in a categorical way. Next, we describe how to introduce continuous DDT logic in order to evaluate and build a knowledge base. This will be achieved through the introduction of the notion of *dual property*.

2 Knowledge Representation and Notation

In order to build up rules from data and carry out a useful inference process, it is necessary to previously define the general framework and mathematical models that are used to represent the information and knowledge we are going to work with. In other words, a mathematical model of knowledge representation is needed to give the data an appropriate shape or structure, in agreement with those required for the input of the rules building process.

Basic raw data is intended to be a database, which could be viewed as a real-valued matrix $D = (d_{ki})_{m \times n}$, having m instances and n variables $X_1, ..., X_n$. Range $R(X_i) \subset \mathbb{R}$ of each variable X_i is then partitioned into a set C_i of $c(X_i) = c_i$ classes or categories $A_{i1}, ..., A_{ic_i}$, which can be fuzzy or crisp. In this paper, these classes are intended to be linearly ordered, i.e. $A_{ij} < A_{ij'}$ iff $j < j'$, but a different structure could be given as explained in [9].

We will use the words *property* and *predicate* as synonyms of *class* or *category*. Given a set C_i of classes used to evaluate a variable X_i, the *valuation space* of this variable is defined as $[0,1]^{C_i}$. One valuation state is assigned to every element in the variable's range through a membership function $\mu_{X_i} : R(X_i) \to [0,1]^{C_i}$. This one assigns a vector $(\mu_{A_{i1}}(x), ..., \mu_{A_{ic_i}}(x))$ to every element $x \in R(X_i)$, $\mu_{A_{ij}}$ being the membership function of the class A_{ij}.

We will use lower case letters to denote the values of variables in the database, this is, $x_i^k = d_{ki}$ for $k = 1, .., m$ and $i = 1, .., n$. In the crisp case, it is supposed that the value x_{ki} lies in exactly one class j', i.e., $\mu_{A_{ij'}}(x_i^k) = 1$ and $\mu_{A_{ij}}(x_i^k) = 0$ if $j \neq j'$. In the fuzzy case, $\mu_{A_{ij}}(x_i^k) \in [0,1]$ and the classes not necessarily form a fuzzy partition in the sense of Ruspini [11], i.e., $\sum_{j=1}^{c_i} \mu_{A_{ij}}(x_i^k)$ need not to sum exactly 1 (see [1]). In fact, missing values of any variable are modelized assigning the value 0 to every of its classes.

In this way, first level of knowledge representation is constituted by a matrix $H = (h_{kj})_{m \times l}$, $l = \sum_{i=1}^{n} c_i$ being the total number of categories or classes and such that $h_{kj} = \mu_{A_{ij}}(x_i^k) = \mu_{A_{ij}}(d_{ki})$, for all $k = 1,..,m$, $i = 1,..,n$, $j_i = 1,...,c_i$ and $j = 1,..,l$. Reference to the i-th variable is removed in the h_{kj}'s as it is intended that categories are sorted by the variables to which they correspond.

Rules need some variables to play the role of *premises* or independent variables, the remaining ones being called *consequences* or dependent variables. Thus, from the set of n variables $X_1,..,X_n$, a subset of p premises variables is extracted, which left us with another subset of $q = n - p$ consequence variables. Since in this approach the conclusion for each consequence variable is independent of the conclusion for the rest of them, for the sake of simplicity in the exposition we will suppose without loss of generality that $q = 1$, i.e., that there exist only one consequence or dependent variable. In the subsequent, this one will be denoted by Y, being $\{X_1,..,X_p\}$ the set of premises or independent variables. Formally, for a rule we shall understand an expression of the type

$$R: \; if \; X_1 \; is \; A_1 \; and \; ... \; and \; X_p \; is \; A_p \; then \; Y \; is \; B_j, \tag{1}$$

where A_i is a class of variable X_i, $i = 1..p$, and B_j is one of the classes defined for dependent variable Y, i.e., $j \in \{1,..,c(Y)\}$. As usual, the expression between *if* and *then* is referred to as the *premise* of the rule. The expression after the word then is referred to as the *conclusion* of the rule.

3   Rules Induction

Matrix H constitutes the first level of knowledge. However, in order to have some inference capability, a second level knowledge, or meta-knowledge, is needed. This second level knowledge, to which we will refer as *rules*, has to be extracted from the first one, and therefore this is the reason for we say that these rules are *data-based*.

Conceptually, the methodology for rule extraction and evaluation described in this paper is based upon the idea that a rule is built up through the successive repetition and experience of similar situations. It is usually accepted that whenever a relation is experienced or successively repeated, its rule condition is strengthened.

The approach presented in this paper follows these ideas. Each instance of the database in which the same classes of different variables appear together is considered as a case for the existence of a relationship between these categories. In this sense, what in a first approach is going to be measured and translated into the rules is the trend of a dependent variable as some combinations of premises variables appear.

However, some instances of data can be interpreted as not only favouring a certain rule R, but also as unfavourable to another rule R'. For example, let's suppose we have a dataset containing the heights of a significant number of professional basketball players and that a set of fuzzy classes or properties have been defined over the range of variable *height*. Now, consider the rule R: *professional basketball players are very tall*. If we scan the database in order to measure how much data supports R and find an instance with height 150 cm., then it is possible to interpret this case not only as a reason for the rule R': *professional basketball players are short* but also as a reason against R, this is, that instance constitutes an exception to R. Moreover, an instance with height 175 cm. is less unfavourable to R than the former with height 150 cm., thus a data point being an exception to a certain rule R is a matter of degrees.

Therefore, given a rule of the type shown in (1), a t-norm T and an instance $d_k \in D$ such that $A_1 \wedge ... \wedge A_p(d_k) = T(\mu_{A_1}(x_1^k), .., \mu_{A_p}(x_p^k)) > 0$, the idea is to look at the relation between the value Y_k of the dependent variable and category B_j for being able to measure to what extent this instance is either an exception to or a case for the rule under consideration. Our basic assumption is that a rule having only cases for has to be more true than another rule having also exceptions or cases against. It is in order to formalize these ideas that we introduce the notion of *dual properties* in the valuation space of dependent variable Y.

3.1 Dual Properties

It has been shown in the previous example how an instance satisfying the height-related property *being short* constitute an exception to a rule which has the class *being very tall* as conclusion. This way, such an instance could be seen as a reason against that rule coming from data.

The basic feature of the properties *being short* and *being tall* that allows to consider instances verifying the first as clear exceptions to a rule concluding the last is that these properties are *antonymous*. As they are opposite concepts or antonyms, that rule is somehow empirically contradicted by such instances.

Of course, this is only a linguistic feature. Classes or properties could be defined with these labels but having not such an opposite semantics. In practice, what really matters is not the name or label of these categories, but its opposite meaning, which has to be reflected in the formal definition of these classes. However, once properties have been given a specific semantics through membership functions, it is possible to conclude that they are acting as antonymous or *dual* properties.

For example, let's suppose S is a set of $c \geq 3$ fuzzy classes A_i, $i = 1..c$, defined over the range of a bounded continuous variable X, and assume S is linearly ordered as said in previous section. Moreover, assume also that these properties have a triangular shape and form a partition in the sense of Ruspini [12]. Therefore, classes situated at opposite extremes of the order have disjoint supports, i.e., if $\mu_{A_1}(x) > 0$ then $\mu_{A_c}(x) = 0$, for all x in the range of X, and vice versa. This way, semantics associated to both properties could be interpreted as if A_1 and A_c were antonymous or dual predicates. This example also shows how to introduce degrees when considering exceptions to rules.

Words *dual* and *antonym* are used here in a similar way. The role of opposite concepts or antonyms as the base for basic valuation structures will be extremely relevant in future developments (see [10], [12] and also [9]).In the forthcoming, the dual of a given predicate P will be denoted by ∂P, this is, symbol ∂ has to be read as "dual of".

Last example also suggests that *distance* or maybe *separation* are important features to be taken into account when searching for dual properties. Though some rather general ways of obtaining dual classes could be devised (as specular images or disjoint support complementation), in our opinion their precise definition has to depend on the context of applications. Therefore, we do not expose here a specific method to obtain the dual of any given class. Instead of that, two reasonable properties these dual classes should satisfy are given:

$$P1: \partial(\partial A) = A \text{ for every property } A \tag{2}$$

$$P2: \text{supp}(\mu_A) \cap \text{supp}(\mu_{\partial A}) = \varnothing \text{ for every property } A \tag{3}$$

P1 states that duality operator is involutive, as also happen with the linguistic antonym operator. Second property states that a class and its dual has to be separated in terms of supports of their membership functions. Notice that P2 implies $T(\mu_A(x), \mu_{\partial A}(x)) = 0$ for all x and every t-norm T and, in fact, both propositions are equivalent in the case T is the classical minimum t-norm, and also when T is the product t-norm. Remarkably, it is not the case when T is the Lukasiewicz t-norm.

It should also be noticed that the notion of duality is different from that of complementation. For example, *being short* is not the complementary predicate of *being tall*, i.e., the former is not the same as *not being tall*. In fact, given a non-crisp, fuzzy predicate A, P2 assures that negation and dual operators can not coincide and that the following proposition holds:

$$\mu_{\partial A}(x) < n(x) \ \forall x \in \text{supp}(\mu_A) \tag{4}$$

where $n(x) = 1 - x$ is the usual negation operator.

Moreover, *being short* is a positively defined property, which is not the case for the complementary predicate *not being tall*. Note that, very often, learning processes start with some direct estimations given on a family of initial classes, each one being associated to a concept, which should be therefore intuitively known. Those classes should be positively defined in order to allow direct valuation. Direct estimation requires direct intuition, and direct intuition usually refers to concepts positively defined. Classifying between *tall* and *short* is correct, but classifying between *tall* and *non-tall* implies an estimation on *being not tall*. Most people will find serious difficulties in assigning direct estimation to such an elaborated class. Most people will simply assign a degree of membership to *tall* and then deduce the degree of membership to *non-tall*. Most people will not assign a direct estimation for *non-tall*. We do not properly classify between these two classes *tall* and *non-tall*: we simply find this classification because it can be built just from the information about the degree of membership of *being tall*.

For these reasons, we can conclude that introduction of dual predicates in the valuation space of dependent variable leads us into a bipolar approach, in the sense that, given a data instance $d = (x_1, ..., x_p, y)$ and a predicate A, it is necessary to evaluate the membership degrees of y to both the class A and its dual class ∂A.

Therefore, knowledge representation described in Section 2 has to be completed by adding to matrix H the values $h_{kj}^{\partial} = \mu_{\partial B_j}(y_k)$ for all $k = 1...m$ and classes $B_j, j = 1...c(Y)$.

3.2 Computing Reasons for and against a Rule from Data

Given a combination of premises classes $A = A_1 \wedge ... \wedge A_p$ and a predicate B in the valuation space of Y, we will denote the rule *if A then B* by $A \rightarrow B$. It is well known

that $A \rightarrow B$ holds iff $\neg A \vee B$ holds, where the symbol $\neg$ stands for the logical negation "not" and $\vee$ for the logical conjunction operator "or". This leads to the classical
(crisp) truth values of $A \rightarrow B$, showed in Table 1.

Table 1. Binary truth values of $A \rightarrow B$

$A \rightarrow B$	A	$\neg A$
B	1	1
$\neg B$	0	1

Methodology presented here only takes into account the cases in which the premise
A is true, i.e., cases described in the first column of Table 1. This is because if $\neg A$
holds, then there exists another combination $A' = A_1' \wedge ... \wedge A_p'$ of premises classes such
that A' is true, and then we should focus on the rule $A' \rightarrow B$ rather than in the former
one. This left in principle $A \wedge B$ and $A \wedge \neg B$ as the only possibilities that can hold
once it is known that A occurs.

Thus, a data instance $d = (x_1, ..., x_p, y)$ such that $A \wedge B$ holds for d is interpreted as a
reasons for the rule $A \rightarrow B$. Recall that in a fuzzy framework as the stated in Section
2, $A \wedge B$ can hold to a certain degree, i.e., the truth of $A \wedge B$ is evaluated through a t-
norm T and is equal to $T(\mu_{A_1}(x_1), ..., \mu_{A_p}(x_p), \mu_B(y))$.

On the other hand, in order to measure reasons against $A \rightarrow B$, we should in principle focus on instances for which $A \wedge \neg B$ holds. However, as said in last section, it is
not always easy or even convenient to work with classes defined through negation.
Moreover, reasons against a rule should come from instances clearly constituting
exceptions to such a rule, which could not be the case for some instances having
$n(T(\mu_{A_1}(x_1), ..., \mu_{A_p}(x_p), \mu_B(y))) > 0$. Furthermore, in a fuzzy framework it is possible
$A \wedge B$ and $A \wedge \neg B$ to be simultaneously true (to a degree), which would be equivalent
to say that a data instance is a reason for and a reason against a rule at the same time.
This case is not allowed in the methodology presented here.

So then, we propose to measure reasons against a rule $A \rightarrow B$ by evaluating
the truth of $A \wedge \partial B$, which in turn is equal to $T(\mu_{A_1}(x_1), ..., \mu_{A_p}(x_p), \mu_{\partial B}(y))$, and therefore
is a positively evaluated quantity, i.e., not obtained through negation. Considering that
a rule $A \rightarrow B$ is *true* (resp. *false*) for an instance d such that $A \wedge B$ ($A \wedge \partial B$) holds to a
certain degree, this approach then differentiate between those cases in which the rule
is *true* for d, those in which the rule is *false* and those in which the rule is either *not
true* or *not false* for that instance d. Non-true case correspond to the complementary
of being true, this is, the case in which $A \wedge \neg B$ holds, and non-false case correspond
to the situation in which $A \wedge \neg \partial B$ holds. As these non-true and non-false predicates are
obtained respectively from true and false predicates through the use of the negation,
only these last cases have to be evaluated.

In order to evaluate how much data support a rule or does not support it, proportion
of instances belonging to the true case and to the false case is calculated. As it is already
said, the former is interpreted as reasons for the rule $A \rightarrow B$, and will be denoted by

$A \to^+ B$. Proportion of instances for which $A \wedge \partial B$ holds is assimilated to reasons against $A \to B$ and will be denoted by $A \to^- B$. Such proportions are computed through the following formulae:

$$A \to^+ B = \frac{\sum_{k=1}^{m} T(\mu_{A_1}(x_1^k),...,\mu_{A_p}(x_p^k),\mu_B(y^k))}{\sum_{k=1}^{m} T(\mu_{A_1}(x_1^k),..,\mu_{A_p}(x_p^k))} \tag{5}$$

$$A \to^- B = \frac{\sum_{k=1}^{m} T(\mu_{A_1}(x_1^k),...,\mu_{A_p}(x_p^k),\mu_{\partial B}(y^k))}{\sum_{k=1}^{m} T(\mu_{A_1}(x_1^k),..,\mu_{A_p}(x_p^k))} \tag{6}$$

where $A = A_1 \wedge ... \wedge A_p$ and $d_k = (x_1^k,...,x_p^k,y^k)$ is the k-th instance of the database.

These values give a description of the proportion of instances supporting truth or falsehood of the rule $A \to B$. Notice that missing values of premises variables does not affect to their computation, though those of the dependent variable does.

3.3 A Four–Valued Evaluation Framework Based on Continuous DDT Logic

Given a knowledge base R of q rules $R_1,...,R_q$, it is possible to associate to every $R_i \in R$ the ordered pair (r_i^+, r_i^-), where r_i^+, r_i^- are computed from data as described above. As for $p \geq 2$ $T(a_1,...,a_p) \geq T(a_1,...,a_p,a_{p+1})$ holds, it follows that $0 \leq r_i^+, r_i^- \leq 1$, and also it is obvious that $r_i^+ + r_i^- \leq 1$. Therefore, the set $N_R = \{(R_i, r_i^+, r_i^-)/R_i \in R \ \forall i = 1..q\}$ constitutes a model of evaluation of rules that can be seen as an Attanasov's intuitionistic fuzzy set (see [2]) over R.

Nevertheless, though intuitionistic fuzzy sets allow for an indeterminacy measure, which some authors (see [9]) point has to be seen as a measure of the ignorance attributed to the fact an object x satisfies a certain predicate P, they does not allow to modelize contradictory situations in which there are reasons for asserting $P(x)$ as also as reasons against such an assertion. These contradictory situations are better treated on the frame of continuous DDT logic (see for instance [11]).

Since a possible situation that can appear when evaluating a rule from data is that there exist some data instances supporting and some others contradicting such a rule, it seems convenient to use continuous DDT logic for the purposes described in this paper.

For example, suppose we have a rule R_i and from data we obtain $r_i^+ = r_i^- = 1/2 - \varepsilon$, ε being small in comparison with $1/2$. This means there are as many reasons for as against R_i, and also that almost all instances satisfying the premise of the rule are classified either as a reason for or as a reason against. This clearly indicates a contradictory situation. However, from an intuitionistic-fuzzy-sets point of view, we would obtain that indeterminacy associated to this situation is equal to $2\varepsilon > 0$, this is, that ignorance and contradiction should coexist. As this situation is nearly an absurd, we conclude that another logical model is needed. As said above, continuous DDT logic will be our choice.

In order to introduce continuous DDT logic, a linear transformation of the set $S = \{(x, y) / x \geq 0, y \geq 0, x + y \leq 1\}$ into the set $S' = \{(x, y) / 0 \leq x, y \leq 1\}$ is carried out. This transformation is given by the function

$$F(x, y) = \begin{cases} (x + y, 2y) & \text{if } x \geq y \\ (2x, x + y) & \text{if } x < y \end{cases} \tag{7}$$

By applying F to the second and third components of the ordered 3-tuple in N_R, the set $M_R = \{(R_i, R_i^+, R_i^-) / R_i \in R \ \forall i = 1..r\}$ is obtained, i.e., $(R_i^+, R_i^-) = F(r_i^+, r_i^-)$ for all $i = 1...q$. Now, the only constraint over R_i^+, R_i^- is $0 \leq R_i^+, R_i^- \leq 1$.

In this situation, DDT logic allows to evaluate the four possible epistemic states that can arise when faced with the evaluation of a rule $R_i \in R$ from data: truth t, falsehood f, contradiction k and unknown or ignorance u. Following [8], these four values can be obtained through the following formulas:

$$t(R_i) = \min\{R_i^+, 1 - R_i^-\} \tag{8}$$

$$k(R_i) = \max\{R_i^+ + R_i^- - 1, 0\} \tag{9}$$

$$u(R_i) = \max\{1 - R_i^+ - R_i^-, 0\} \tag{10}$$

$$f(R_i) = \min\{1 - R_i^+, R_i^-\} \tag{11}$$

Notice that, as explained in [9], two different t-norms have to be used in order to make possible to recover R_i^+, R_i^- from these four values. Other conditions have also to be imposed, as for example $t(R_i) + f(R_i) + k(R_i) + u(R_i) = 1$ or the already specified assumption of contradictory and unknown cases being exclusive.

4 Conclusions

Once these four values have been computed for every rule in the knowledge base R, it is possible to make further considerations about their validity (see for instance [4]). Though the way these considerations could be done lies outside the scope of this paper, some conclusions can be shed.

For example, given a rule $R_i \in R$, if $t(R_i) < f(R_i)$ then rule R_i should be discarded from R as it is more false than true. A truth threshold δ_t could be imposed so that only rules satisfying $t(R_i) > \delta_t$ are considered valid. Another possibility is to consider that a rule is valid only if data present a small number of exceptions. Thresholds δ_k and δ_u could also be defined in such a way that when $k(R_i) > \delta_k$ it could be concluded that more premise or independent variables are needed in order to differentiate between those cases in which the dependent variable lies in the conclusion class and those in which dependent variable lies in its dual. If $u(R_i) > \delta_u$, then it could be appropriated to reconsider whether rule R_i is a necessary one, or whether more data is needed. Also, firing strength of a valid rule R_i could be computed in terms of $t(R_i) - f(R_i)$.

In any case, this methodology allows for more granularity in order to evaluate rules than those based on a true-false approach only. Particularly, conflicting and hesitating situations could be detected and differentiated between, as those in which a rule is not valid due to lack of information (ignorance) and those in which a rule have data supporting it and contradicting it at the same time (contradiction). Moreover, it has been showed with last examples how the framework of DDT logic allows a vast amount of possibilities in order to evaluate and validate an existing knowledge base from data.

Furthermore, if the number of premise variables is not too large, it should be possible to use this methodology in order to build a knowledge base from data by evaluating *all* possible rules. This approach works well in decision support systems like the one described in [13], which is intended to work with a reduced number of variables.

Future research will concern the introduction of graphs in the valuation space of dependent variables in order to modelize relations between classes and to allow considering more complex structures than linear orders (see [9]). Some of the index developed in [1] to measure the efficiency of a classification will be tested upon this approach. Moreover, further efforts will be put in the study of the relations between this methodology, rough set semantics (following [15] and [5]), logical analysis of data (see [6]) and possibility theory (following [11]).

Acknowledgments. This research has been partially supported by grants TIN2006-06190, CCGO7-UCM/ESP-2576, and I-Math Consolider C3-0132. This paper started to be written while the first author was visiting LAMSADE in a STSM funded by COST Action IC0602, the support of which is also gratefully acknowledged.

References

[1] Amo, A., Montero, J., Biging, G., Cutello, V.: Fuzzy classification systems. European Journal of Operational Research 156(2), 495–507 (2004)

[2] Atanassov, K.T.: Intuitionistic Fuzzy Sets. Physica-Verlag, Heidelberg (1999)

[3] Destercke, S., Guillaume, S., Charnomordic, B.: Building an interpretable fuzzy rule base from data using Orthogonal Least Squares. Application to a depollution problem. Fuzzy Sets and Systems 158(18), 2078–2094 (2007)

[4] Fortemps, P., Slowinski, R.: A graded quadrivalent logic for ordinal preference modelling: Loyola-like approach. Fuzzy Optimization and Decision Making 1(1), 93–111 (2002)

[5] Fortemps, P., Greco, S., Slowinski, R.: Multicriteria decision support using rules that represent rough-graded preference relations. European J. Operational Research 188(1), 206–223 (2008)

[6] Hammer, P.L., Bonates, T.: Logical Analysis of Data - An overview: From Combinatorial Optimization to Medical Applications. Annals of Operations Research 148(1), 203–225 (2006)

[7] Iliadis, L.S.: A decision support system applying an integrated fuzzy model for long-term forest fire risk estimation. Environmental Modelling & Software 20, 613–621 (2005)

[8] Mamdani, E.H.: Application of Fuzzy Algorithms for the Control of a Dynamic Plant. Proc. IEE 121(12), 1585–1588 (1974)

[9] Montero, J., Gómez, D., Bustince, H.: On the relevance of some families of fuzzy sets. Fuzzy sets and systems 158(22), 2439–2442 (2007)

[10] Novak, V.: Antonyms and linguistic quantifiers in fuzzy logic. Fuzzy Sets and Systems 124, 335–351 (2001)

[11] Öztürk, M., Tsoukiàs, A.: Modelling uncertain positive and negative reasons in decision aiding. Decision Support Systems 43(4), 1512–1526 (2007)

[12] Paradis, C., Willners, C.: Antonymy and negation—the boundness hypothesis. J. Pragmatics 38, 1051–1080 (2006)

[13] Rodriguez, J.T., Vitoriano, B., Montero, J., Omaña, A.: A decision support tool for humanitarian organizations in natural disaster relief. In: Ruan, D., et al. (eds.) Computational Intelligence in Decision and Control, pp. 600–605. World Scientific, Singapore (2008)

[14] Ruspini, E.H.: A new approach to clustering. Inform. Control 15, 22–32 (1969)

[15] Tsoukiàs, A.: A first-order, four valued, weakly paraconsistent logic and its relation to rough sets semantics. Foundations of Computing and Decision Sciences 12, 85–108 (2002)

[16] Yager, R.R.: Targeted e-commerce marketing using fuzzy intelligent agents. Intelligent Systems and their Applications 15(6), 42–45 (2000)

A Framework for Designing a Fuzzy Rule-Based Classifier[*]

Jonas Guzaitis[1], Antanas Verikas[1,2,**], Adas Gelzinis[1],
and Marija Bacauskiene[1]

[1] Department Electrical & Control Instrumentation,
Kaunas University of Technology,
Studentu 50, LT-51368, Kaunas, Lithuania
[2] Intelligent Systems Laboratory, Halmstad University,
Box 823, S-30118 Halmstad, Sweden
antanas.verikas@hh.se, jonas.guzaitis@stud.ktu.lt,
{adas.gelzinis,marija.bacauskiene}@ktu.lt

Abstract. This paper is concerned with a general framework for designing a fuzzy rule-based classifier. Structure and parameters of the classifier are evolved through a two-stage genetic search. The classifier structure is constrained by a tree created using the evolving SOM tree algorithm. Salient input variables are specific for each fuzzy rule and are found during the genetic search process. It is shown through computer simulations of four real world problems that a large number of rules and input variables can be eliminated from the model without deteriorating the classification accuracy.

Keywords: Classifier, Fuzzy rule, Genetic algorithm, Knowledge extraction, Variable selection.

1 Introduction

Neural networks and support vector machines are probably the most popular data classification techniques. However, classifiers based on these techniques are not transparent enough and are often considered as "black boxes". The transparency is very important in some application areas, such as medical decision support or quality control. By contrast, fuzzy rule-based systems and fuzzy decision trees are known for their transparency and ability of accounting for uncertainty. ANFIS [1], fuzzy ARTMAP [2] are examples of the most prominent fuzzy logic-based systems. It is well known that designing of fuzzy rule-based systems in high dimensional spaces is rather problematic. However, there are many problems characterized by a small or moderate number of variables. Moreover, quite often high dimensional data vary in a much lower number of dimensions if

[*] We acknowledge the support from the agency for international science and technology development programmes in Lithuania (COST IC0602).

[**] Corresponding author.

F. Rossi and A. Tsoukis (Eds.): ADT 2009, LNAI 5783, pp. 434–445, 2009.

compared to the dimensionality of an input space. System structure identification and parameter optimization are two main issues to consider when designing a fuzzy rule-based system [1]. Fuzzy partitioning, variable selection, and fuzzy reasoning are the tasks to be solved for identifying the system structure.

Various approaches have been used for dealing with the two main fuzzy rule-based system design issues. The initial system structure, often termed as fuzzy partitioning, is usually identified through K-Means [3], Fuzzy C-Means [4], Learning Vector Quantization (LVQ) [5] or SOM-based clustering [6,7] as well as incremental clustering [8,9] or by constructing a decision tree [10,11].

Variable selection based on: the output sensitivity to the input change [12,13], the output sensitivity combined with the correlation between variables [6], Fisher's interclass separability measure [14], variable correlation with the output [15] are the most popular variable selection techniques applied. However, quite often, variable selection is not considered at all [7].

It seems that the simple gradient decent [6,12], error correction [16], and genetic search [7,10] are the most popular parameter optimization techniques utilized in various studies. The combined optimization of both structure and parameters has also been considered by applying genetic algorithms (GA) [17], unsupervised and reinforcement learning [18], or simple heuristics [15]. In [17] the genetic search process focusses on "hard" data points by assigning a higher weight to such points. Such an approach has also been adopted for learning weights z_j^q of fuzzy rules [19]. In [20], genetic search-based multi-objective optimization was applied to design a fuzzy rule-based system. The task was to maximize $f_1(S)$, minimize $f_2(S)$, and minimize $f_3(S)$, where S is a set of fuzzy rules, $f_1(S)$ stands for correctly classified training samples, $f_2(S)$ is the number of fuzzy rules in S, and $f_3(S)$ is the total number of antecedent conditions in S. Thus, $f_3(S)$ can be considered as the total rule length. The optimization starts with all possible rules in the search space defined by the training patterns.

Generalization ability is an important issue to consider when designing a fuzzy rule-based classifier. The most popular technique applied to improve the generalization ability is rule pruning based on similarity of fuzzy sets [14,6]. Other approaches utilized are: GA [21], simulated annealing [22], similarity of fuzzy sets combined with GA [7], through forgetting by decaying the grade of certainty of fuzzy rules [16], pruning of rarely used rules [15].

1.1 Fuzzy Rule-Based and Nearest Neighbour Techniques

The fuzzy rule based classification techniques are closely related to nearest neighbour (NN)-based classification approaches. NN-based classification has a sound basis, since there is a considerable body of evidence from the literature that classification and recognition of patterns by humans is best explained as a form of interpolation between similar patterns [23]. NN methods are frequently criticized as requiring much greater use of memory than, for example, neural network algorithms. However, NN learning algorithms can reduce their memory usage by only retaining the full density of exemplars near to classification boundaries and thinning them in other regions [24,25]. The location of fuzzy sets, reference patterns in the NN approach, can be further optimized by applying the LVQ techniques. LVQ has been widely used to learn reference patterns for classification based on the NN approach. Each class C_j is described by several reference patterns $\mathbf{m}_j^l$ (fuzzy sets in the rule-based approach), which are properly placed within each class region. An unknown $\mathbf{x}$ is then determined to belong to the class k, if:

$$k = \arg\min_j[\min_l d(\mathbf{x}, \mathbf{m}_j^l), \quad l = 1, ..., N_j, \quad j = 1, ..., Q] \tag{1}$$

where Q is the number of classes, N_j is the number of reference patterns representing the class j and $d(\mathbf{x}, \mathbf{m}_j^l)$ is the distance between $\mathbf{x}$ and $\mathbf{m}_j^l$.

One more drawback of classical NN and fuzzy rule-based methods is the often exhibited poor generalization performance as compared to neural networks, for example. In NN methods, the degradation in performance often accompanies the addition of new unimportant features. However, there are many problems where different features are important in different regions of the input space. Fig. 1 provides an example illustrating such a situation.

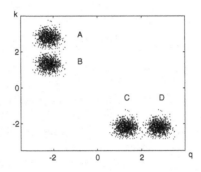

Fig. 1. Four decision classes in the two-dimensional space

The four data clusters illustrated in Fig. 1 represent four decision classes. It is obvious that the feature q is unimportant for discriminating the classes A and B, likewise the feature k is unimportant for discriminating the classes C and D. Thus, a subset of features used should be reference pattern or fuzzy rule dependent. However, in most of the known fuzzy rule-based classification algorithms, the feature selection problem is considered independently of the input space region or not considered at all. The objective of this work is to develop a fuzzy modeling framework capable of automatically generating a rule base for classification of numeric data, finding the optimal number of rules and input variables for each rule, and finding the optimal parameter values of fuzzy rules.

2   The Fuzzy Model

We use the Mamdani model [26], which is the most popular fuzzy model applied in various studies for fuzzy reasoning [7,12,6]. Concerning classification, the model is a collection of fuzzy rules R_j of the following form:

$$R_j : \text{IF } x_1 \text{ is } A_{j1} \text{ AND ... AND } x_n \text{ is } A_{jn} \text{ THEN class } C_q \text{ with } z_j^q \qquad (2)$$

where $A_{ji}(i = 1, ..., n)$ are fuzzy sets defined over the input variables x_i, C_q is a class label and z_j^q is a rule weight. Each fuzzy set is represented by a membership function. A triangular or a Gaussian function of the form

$$\mu_{ji} = \exp\left(-\frac{(x_i - c_{ji})^2}{\sigma_{ji}^2}\right) \qquad (3)$$

where c_{ji} and σ_{ji} are the center and the width of the Gaussian function, respectively, are common choices. We use Gaussian membership functions in this study. There are various ways to determine the rule weights z_j^q. In this work, the weight z_j^q is given by:

$$z_j^q = \frac{\sum_{\mathbf{x}_p \in C_q} \mu_{\mathbf{A}_j}(\mathbf{x}_p) - \sum_{\mathbf{x}_p \notin C_q} \mu_{\mathbf{A}_j}(\mathbf{x}_p)}{\sum_{p=1}^{N} \mu_{\mathbf{A}_j}(\mathbf{x}_p)} \qquad (4)$$

where N is the number of training patterns and the matching degree of the input pattern $\mathbf{x}_p$ with the antecedent part $\mathbf{A}_j = (A_{j1}, ..., A_{jn})$ is calculated using a T-norm

$$\mu_{\mathbf{A}_j}(\mathbf{x}_p) = T(\mu_{A_{j1}}(x_{p1}), ..., \mu_{A_{jn}}(x_{pn})) \qquad (5)$$

We use the min T-norm operator in this work. Weights z_j^q of this type were studied in [20].

A winning rule is used to make a decision. Thus, given a rule base S consisting of L rules, an input pattern $\mathbf{x}_p$ is assigned to the class q if

$$q = \arg\max_k \{T[\mu_{\mathbf{A}_j}(\mathbf{x}_p), z_j^k], \; j = 1, ..., L\} \qquad (6)$$

where T is the product T-norm operator, in this work.

Having defined the membership functions, we formulate the fuzzy modeling problem in the following way. Given N pairs of input-output patterns $(\mathbf{x}, y)$, create a minimal number of fuzzy rules r with the optimal number of features n_i for each rule and find the optimal values of parameters $(\mathbf{c}, \boldsymbol{\sigma}, \mathbf{z})$ of the fuzzy model $F(\mathbf{c}, \boldsymbol{\sigma}, \mathbf{z}, r, n_i, i = 1, ..., r)$.

3   The Approach

The procedure to construct the fuzzy rule-based classifier consists of the following steps.

1. Divide the data set into learning and test subsets.
2. Cluster the learning set data by applying the evolving SOM tree.
3. Based on the evolved tree, generate a population of sub-trees. Each sub-tree defines the initial structure of one fuzzy rule-based classifier. The generation is accomplished by randomly cutting branches of the tree grown in Step 2. The cutting occurs approximately between 25 and 75% of the tree depth.

4. Represent each node in the sub-tree population by a set of fuzzy sets with the Gaussian membership functions.
5. Encode the structure of each sub-tree of the population into a separate chromosome.
6. Take one sub-tree (classifier of a given structure) from the sub-tree population and encode features (used/not used) and parameters of the membership functions of the classifier into a chromosome. When encoding, enable feature selection independently for each fuzzy rule.
7. Generate a population of chromosomes encoding individual classifiers of the given structure. The individual classifiers differ in features and values of the parameters.
8. Apply the modified LVQ-3 algorithm to the individuals of the population and test the fitness of the individuals.
9. Apply genetic operations (to features and parameters) and generate a new population.
10. Repeat Steps 8–9 until convergence.
11. Take the best individual of the given structure.
12. Repeat Steps 6–11 for the whole population of sub-trees.
13. Apply genetic operations (to structure of sub-trees) and generate a new population of sub-trees.
14. Repeat Steps 6–13 for a given number of generations.

Next, we briefly describe the main topics of the technique.

3.1   The Evolving SOM Tree

Like SOM, the evolving SOM tree [27] exhibits the self-organization property. The evolving tree structure enables the SOM tree to efficiently handle large scale problems. Moreover, there is no need of choosing the map size beforehand. Like in ordinary SOM, each node of the SOM tree has a weight vector $\mathbf{w}_i$. When training the tree, for each training vector $\mathbf{x}$, the best matching unit (BMU) is found by a greedy tree search. BMU is always a leaf node. Weight vectors of the BMU and its neighbours are then updated using the SOM adaptation rule:

$$\mathbf{w}_i(t+1) = \mathbf{w}_i(t) - h_{ci}(t)[\mathbf{x}(t) - \mathbf{w}_i(t)] \qquad (7)$$

where $h_{ci}(t)$ is the neighbourhood function. We used the Gaussian neighbourhood function

$$h_{ci}(t) = \beta(t) \exp\left(\frac{\|\mathbf{r}_c - \mathbf{r}_i\|^2}{2s^2(t)}\right) \qquad (8)$$

where $s(t)$ is the width of the Gaussian function, $\mathbf{r}_c$ and $\mathbf{r}_i$ denote location of nodes c and i, and $\beta(t)$ is the learning rate. The meaning of $s(t)$ and $\beta(t)$ is the same as in SOM [28], while the meaning of the norm $\|\mathbf{r}_c - \mathbf{r}_i\|$ is quite different. The basic idea of calculating the distance is to count how many "hops" are needed to get from the BMU to the considered node along the shortest path [27]. The distance $\|\mathbf{r}_c - \mathbf{r}_i\|$ is then given by the number of hops minus one.

3.2   The Modified LVQ-3 Algorithm

Assume that d_i and d_j are the Euclidean distances from the pattern $\mathbf{x}$ to the reference patterns $\mathbf{m}_i$ and $\mathbf{m}_j$, respectively. Then $\mathbf{x}$ is defined to fall into a window of the relative width λ, if

$$\min\left(\frac{d_i}{d_j}, \frac{d_j}{d_i}\right) > \frac{1 - \lambda}{1 + \lambda} \tag{9}$$

For all $\mathbf{x}$ falling into the window adapt:

$$\mathbf{m}_i(t + 1) = \mathbf{m}_i(t) - \alpha(t)[\mathbf{x}(t) - \mathbf{m}_i(t)] \tag{10}$$

$$\mathbf{m}_j(t + 1) = \mathbf{m}_j(t) + \alpha(t)[\mathbf{x}(t) - \mathbf{m}_j(t)] \tag{11}$$

where $\alpha(t)$ decreases with time and $0 < \alpha(t) < 1$, $\mathbf{m}_i$ and $\mathbf{m}_j$ are two closest reference patterns to $\mathbf{x}$, whereby $\mathbf{x}$ belongs to the same class as $\mathbf{m}_j$, but not as $\mathbf{m}_i$. If $\mathbf{x}$, $\mathbf{m}_i$ and $\mathbf{m}_j$ belong to the same class:

$$\mathbf{m}_k(t + 1) = \mathbf{m}_k(t) + \gamma\alpha(t)[\mathbf{x}(t) - \mathbf{m}_k(t)] \tag{12}$$

for $k \in \{i, j\}$. If $\mathbf{x}$ belongs to a different class than $\mathbf{m}_i$ and $\mathbf{m}_j$:

$$\mathbf{m}_k(t + 1) = \mathbf{m}_k(t) - \gamma\alpha(t)[\mathbf{x}(t) - \mathbf{m}_k(t)] \tag{13}$$

for $k \in \{i, j\}$, where γ is a parameter.

   The algorithm performs fine tuning of the centers of membership functions. The last adaptation step is not used in the original version of the Lvq-3 algorithm. We have found that the use of the step quite noticeably improved the accuracy of the algorithm.

3.3   Encoding

There are two loops of genetic evolution: the outer loop concerning structure evolution and the inner loop concerning features and parameters of membership functions. The structure is determined by a sub-tree and is encoded as a connected graph. The chromosome encoding features and parameters of the membership functions can be split into *sections* and *sub-sections*. The number of sections is equal to the number of leaf nodes in the actual sub-tree (the number of fuzzy rules in the classifier). The actual sub-tree is a sub-tree governing the inner loop of genetic evolution. There are $2n + 1$ sub-sections in each section, where n is the dimensionality of the input space. One subsection (n bits) encodes features (used/not used–0/1) and the other $2n$ sub-sections encode the centers c and the widths σ of the n membership functions μ.

3.4   Genetic Operations

Crossover and mutation are the genetic operations applied in both loops of genetic evolution: the loop concerning structure evolution and the loop concerning features and parameters of membership functions. The crossover and mutation operations are executed with the probability of crossover p_c and mutation probability p_m, respectively. When performing crossover for structure evolution, parts of two sub-trees are exchanged. Mutation in structure evolution amounts to taking one step up or down (the direction is selected randomly) along a randomly selected branch of the tree. Regarding a chromosome encoding features and parameters of membership functions, crossover and mutation are performed separately in each sub-section of the chromosome. The crossover points are randomly selected in the "feature mask" sub-section and the parameter sub-sections and the corresponding parts of two chromosomes chosen for the crossover operation are exchanged at the selected points. The mutation is accomplished by reversing the value of a bit in the "feature mask" and by adding a random value from a given, symmetric around zero, interval to parameters in the parameters subsections selected for mutation.

3.5   Fitness Function

The fitness value of the ith chromosome f_i is given by

$$f_i = \chi_i + \eta \frac{\sum_{j=1}^{r_i} n_j}{r_0 \times n} \tag{14}$$

where χ_i is the classification accuracy obtained from the classifier encoded in the ith chromosome, η is a parameter, n is the total number of available features, n_j is the number of features used by the jth rule, r_0 is the number of rules in the initial tree, and r_i stands for the number of rules used by the ith classifier. The selection probability of the ith chromosome p_i for genetic operations is given by

$$p_i = \frac{f_i}{\sum_{j=1}^{M} f_j} \tag{15}$$

where M is the population size. The roulette selection principle was applied.

4   Experimental Investigations

4.1   Data Used

Four data sets have been used in the tests.

US congressional voting records problem. The United States Congressional voting records data set consists of the voting records of 435 congressman on 16 major issues in the 98th Congress. The votes are categorized into one of the three types of votes: (1) (*Yea*), (2) (*Nay*), and (3) (*Unknown*). The task is to predict the correct political party affiliation of each congressman. The 98th Congress consisted of 267 Democrats and 168 Republicans.

The breast cancer diagnosis problem. The University of Wisconsin Breast Cancer Data Set consists of 699 patterns. Amongst them there are 458 benign samples and 241 malignant ones. Each of these patterns consists of nine measurements taken from fine needle aspirates from a patient's breast.

The diabetes diagnosis problem. The Pima Indians diabetes data set contains 768 samples taken from patients who may show signs of diabetes. Each sample is described by eight features. There are 500 samples from patients who do not have diabetes and 268 samples from patients who are known to have diabetes. These three data sets are available at: http://archive.ics.uci.edu/ml/.

Pavement tiles surface inspection problem. A pavement tile surface is to be assigned into a *quality* or *defective* class. Features for the classification are extracted from a camera image. Five features characterizing the image texture and the grey level distribution [29] have been used to design a classifier. Fig. 2 presents four examples of pavement tile surfaces used in the study. In total, 200 quality and 200 defective surfaces were available.

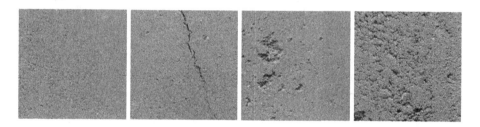

Fig. 2. Examples of pavement tile surfaces: a quality surface on the left and three defective surfaces

4.2   Experimental Setup

We randomly assign the available data points into the learning D_L and test D_T data sets. The data are normalized to have zero mean and unit variance. We run an experiment 30 times with different random splits of the database into the sets D_L and D_T. The results obtained are averaged over the 30 runs.

4.3   Optimization Parameters

The optimal size of the Lvq-3 window depends on the number of training samples. If a large number of samples is available, a narrow window would guarantee the most accurate location of the decision boundary. For good statistical accuracy, however, the number of samples falling into the widow must be sufficient [5]. The optimal value of γ depends on the size of the window, being smaller for narrower windows [5]. After some experiments the following values of the Lvq-3 parameters have been used: $\lambda = 0.05$, $\alpha = 0.02$, and $\gamma = 0.4$.

There are two loops of genetic evolution: the outer loop concerning structure evolution and the inner loop concerning features and parameters of the membership functions. The genetic search lasted for 100 generations (for both loops) with the following parameters: the population size was set to 50 and the number of offsprings produced for creating the next population was equal to 50. The number of generations was determined experimentally by monitoring changes of the fitness function value. The number chosen was such that no fitness function value increase was observed in the last 10 generations. The values of crossover and mutation probabilities were found experimentally. The following values worked well in the tests: $p_c = 0.95$ and $p_m = 0.01$. The appropriate value of the parameter η was found to be $\eta = 0.05$.

4.4   Results

In the first set of experiments, the feature selection has not been applied and the classification accuracy obtained from the fuzzy rule-based classifier was compared with the accuracy achieved by other techniques. The multi layer perceptron (MLP), k-NN, and LVQ-3 classifiers have been used for the comparison. The appropriate number of hidden nodes in the MLP and the k value of the k-NN classifier were found experimentally. The leaf nodes created by the evolving SOM tree were used as initial reference patterns for the LVQ-3 classifier. Table 1 presents the average test data set classification accuracy (%) obtained from the different classifiers using all available features. As can be seen from Table 1, the proposed fuzzy rule-based classifier provided the highest classification accuracy for all the problems studied.

Table 1. The average test data set classification accuracy (%) obtained from different classifiers using all available features

Data Set\Classifier	k-NN	MLP	LVQ-3	Proposed
Voting	91.24	93.78	77.41	94.68
Breast cancer	88.73	97.18	75.87	98.54
Diabetes	71.19	71.49	64.22	71.92
Surface inspection	77.30	81.63	78.13	84.13

In the next set of experiments, feature selection was activated and features, specific for each rule were found through the genetic search. Table 2 presents the average test data set classification accuracy obtained from the approach proposed using selected features. The classification accuracy obtained using all the available features is also presented for the sake of comparison. The obtained improvement in classification accuracy should be obvious from Table 2. Assuming that the classification errors are log-normally distributed and applying the t-test it was found that the difference between the classification accuracy obtained using the selected and all features is significant with 95% confidence, except for the Breast cancer data.

Table 2. The average test data set classification accuracy obtained from the approach proposed using all and selected features

Data Set\Features	All features	Selected features
Voting	94.68	98.68
Breast cancer	98.54	99.02
Diabetes	71.92	75.92
Surface inspection	84.13	99.63

Table 3 presents information on the number of rules and features used to classify the data. In the parentheses given are the number of initial rules and the number of available features. Ranges in the "Features" column indicate the minimum and the maximum number of features used by different rules. As can be seen from Table 3, the number of features used by different rules varies significantly. Observe that even if the number features used by two different rules is the same, the features used are often different. Thus, features used are rule specific, indeed.

Table 3. The number of rules and features used to classify data from the different data sets

Data Set	# Rules	# Features
Voting	15 (25)	05–09 (16)
Breast cancer	10 (14)	12–20 (30)
Diabetes	10 (16)	02–08 (8)
Surface inspection	09 (13)	03–04 (5)

Next, the influence of crossover and mutation probabilities, p_c and p_m, on classification accuracy was studied. The same p_c and p_m values were used for both structure and parameter evolution. To speed up the convergence, the studies were performed without employing feature selection. A very similar performance was observed for p_m values raging from 0 to 0.05. A value of $p_m = 0.01$ was selected. When studying the influence of p_c, the p_m was set to $p_m = 0$. Table 4 presents the average test set classification accuracy obtained for different p_c values.

Table 4. The average test set classification accuracy obtained for different p_c values

Data Set	$p_c = 0.25$	$p_c = 0.50$	$p_c = 0.75$	$p_c = 1.0$
Voting	92.89	93.01	93.21	94.68
Breast cancer	94.12	93.32	94.78	95.02
Diabetes	67.21	68.45	69.53	71.92
Surface inspection	82.01	82.10	83.02	84.13

5 Conclusions

Proposed is a general framework for designing a fuzzy rule-based classification system. The two-stage GA developed partitions the search space and enables evolving both structure and parameters of the classifier. Salient input variables, specific for each fuzzy rule, are also found during the search process.

Computer simulations of four real world problems have shown that the performance obtained from the classifier is comparable or even higher than the best performance obtained by other authors when using "black box" models. The proposed variable selection tool allowed to significantly increase the classification accuracy if compared to the case of using all the available input variables. It was shown through computer simulations that a large number of rules and input variables can be eliminated from the model without deteriorating the classification accuracy. Moreover, the classification accuracy of the test set data increased due to the reduction.

References

1. Jang, J.R.: ANFIS: adaptive-network-based fuzzy inference system. IEEE Trans. Systems Man Cybernetics 23, 665–685 (1993)
2. Carpenter, G.A., Grossberg, S., Markuzon, N., Reynolds, J.H., Rosen, D.B.: Fuzzy ARTMAP: A neural network architecture for incremental supervised learning of analog multidimensional maps. IEEE Trans. Neural Networks 3(5), 698–713 (1992)
3. Duda, R.O., Hart, P.E., Stork, D.G.: Pattern Classification, 2nd edn. John Wiley & Sons, New York (2001)
4. Bezdek, J.C.: Pattern Recognition with Fuzzy Objective Function Algorithms. Plenum Press, New York (1981)
5. Kohonen, T.: The self-organizing map. Proceedings of the IEEE 78(9), 1461–1480 (1990)
6. Chen, M.Y., Linkens, D.A.: A systematic neuro-fuzzy modeling framework with application to material property prediction. IEEE Trans. Systems Man, Cybernetics, Part B 31(5), 781–790 (2001)
7. Zhou, E., Khotanzad, A.: Fuzzy classifier design using genetic algorithms. Pattern Recognition 40(12), 3401–3414 (2007)
8. Kasabov, N.: Evolving fuzzy neural networks for supervised/unsupervised on-line, knowledge-based learning. IEEE Trans. Systems, Man and Cybernetics 31(6), 902–918 (2001)
9. Minku, F.L., Ludermir, T.B.: Clustering and co-evolution to construct network ensembles: An experimental study. Neural Networks 21, 1363–1379 (2008)
10. Abonyi, J., Roubos, J.A., Szeifert, F.: Data-driven generation of compact, accurate, and linguistically sound fuzzy classifiers based on a decision-tree initialization. International Journal of Approximate Reasoning 32(1), 1–21 (2003)
11. Pulkkinen, P., Koivisto, H.: Identification of interpretable and accurate fuzzy classifiers and function estimators with hybrid methods. Applied Sof Computing 7, 520–533 (2007)
12. Castellano, G., Castiello, C., Fanelli, A.M., Mencar, C.: Knowledge discovery by a neuro-fuzzy medeling framework. Fuzzy Sets and Systems 149, 187–207 (2005)

13. Verikas, A., Bacauskiene, M.: Feature selection with neural networks. Pattern Recognition Letters 23(11), 1323–1335 (2002)
14. Roubos, J.A., Setnes, M., Abonyi, J.: Learning fuzzy classification rules from labeled data. Information Sciences 150, 77–93 (2003)
15. Nauck, D., Kruse, R.: Obtaining interpretable fuzzy classification rules from medical data. Artificial Intelligence in Medicine 16(2), 149–169 (1999)
16. Nozaki, K., Ishibuchi, H., Tanaka, H.: Adaptive fuzzy rule-based classification systems. IEEE Trans. Fuzzy Systems 4(3), 238–250 (1996)
17. Ozyer, T., Alhajj, R., Barker, K.: Intrusion detection by integrating boosting genetic fuzzy classifier and data mining criteria for rule pre-screening. Journal of Network and Computer Applications 30(1), 99–113 (2007)
18. Er, M.J., Zhou, Y.: Automatic generation of fuzzy inference systems via unsupervised learning. Neural Networks 21(10), 1556–1566 (2008)
19. Nakashima, T., Schaefer, G., Yokota, Y., Ishibuchi, H.: A weighted fuzzy classifier and its application to image processing tasks. Fuzzy Sets and Systems 158, 284–294 (2007)
20. Ishibuchi, H., Nojima, Y.: Analysis of interpretability-accuracy tradeoff of fuzzy systems by multiobjective fuzzy genetics-based machine learning. International Journal of Approximate Reasoning 44(1), 4–31 (2007)
21. Chang, P.C., Liao, T.W.: Combining som and fuzzy rule base for flow time prediction in semiconductor manufacturing factory. Applied Soft Computing 6, 198–206 (2006)
22. Mohamadi, H., Habibi, J., Abadeh, M.S., Saadi, H.: Data mining with a simulated annealing based fuzzy classification system. Pattern Recognition 41(5), 1824–1833 (2008)
23. Lowe, D.G.: Similarity metric learning for a variable-kernel classifier. Neural Computation 7, 72–85 (1995)
24. Chang, C.L.: Finding prototypes for nearest neighbour classifiers. IEEE Trans. Computers 23, 1179–1184 (1974)
25. Verikas, A., Bacauskiene, M., Malmqvist, K.: Learning an adaptive dissimilarity measure for nearest neighbour classification. Neural Computing & Applications 11(3-4), 203–209 (2003)
26. Mamdani, E.H., Assilian, S.: An experiment in linguistic synthesis with a fuzzy logic controller. International Journal of Man-Machine Studies 7(1), 1–12 (1975)
27. Pakkanen, J., Iivarinen, J., Oja, E.: The evolving tree—A novel self-organizing network for data analysis. Neural Processing Letters 20, 199–211 (2004)
28. Kohonen, T.: Self-Organizing Maps, 3rd edn. Springer, Berlin (2001)
29. Guzaitis, J., Verikas, A.: An efficient technique to detect visual defects in particleboards. Informatica 19(3), 363–376 (2008)

Anytime Self-play Learning to Satisfy Functional Optimality Criteria

Andriy Burkov and Brahim Chaib-draa

Laval University, Quebec QC, Canada
burkov@damas.ift.ulaval.ca
http://www.damas.ift.ulaval.ca

Abstract. We present an anytime multiagent learning approach to satisfy any given optimality criterion in repeated game self-play. Our approach is opposed to classical learning approaches for repeated games: namely, learning of equilibrium, Pareto-efficient learning, and their variants. The comparison is given from a practical (or engineering) standpoint, i.e., from a point of view of a multiagent system designer whose goal is to maximize the system's overall performance according to a given optimality criterion. Extensive experiments in a wide variety of repeated games demonstrate the efficacy of our approach.

1 Introduction

Until now, the main body of the state-of-the-art multiagent learning (MAL) research [1] has been focused on finding a learning rule possessing specific properties. For example, when adopted by all agents of a multiagent system (MAS), such rule could bring to each agent an accumulated reward, "optimal" in a certain sense. I.e., a learning rule were considered to be good if the rewards accumulated by the agents (also called "players") were close to some values satisfying a certain criterion of optimality[1]. Two most widely used optimality criteria in the context of learning in repeated games are: closeness of the value accumulated by each player to the value of (a) a Nash equilibrium and (b) a Pareto-efficient joint strategy (which need not be an equilibrium).

The scenario where all agents use the same algorithm is called "self-play". Most of the existing multiagent learning rules assume self-play [2,1,3,4]. An important reason for that is because "self-play" multiagent systems (SPMAS) are of a great practical interest. Indeed, given an algorithm capable of achieving a utility satisfying a given optimality criterion in self-play, an engineer can create a number of identical agents (in the case of software agents, one can just make as many copies of one agent as required) put these agents into a given environment and let them converge.

However, in an arbitrary repeated game, the values corresponding to different optimality criteria can vary substantially from one criterion to another. So, when the game

---

[1] As a matter of fact, the terms "optimal" and "optimality" are not always appropriate in MAS. Indeed, there are often multiple entities (agents) having different interests in a MAS. In the classical game theoretical literature such terms as "Pareto efficiency" or "equilibrium" are used in place of "optimality". Nevertheless, we will use these terms to unify and simplify the presentation.

F. Rossi and A. Tsoukis (Eds.): ADT 2009, LNAI 5783, pp. 446–457, 2009.

being played is unknown, it is usually hard to chose the best learning rule. Another problem, when using algorithms satisfying such optimality criteria as (a) or (b) listed above, is that, from a practical point of view, neither of those criteria can be satisfactory for all SPMAS. Let us clarify this claim.

Let suppose we are an engineer that receives from a client a problem that needs to be solved by a number of identical agents. (We will call this problem "the environment" and this environment is supposed to be unknown in terms of players' rewards for different actions.) The agents (if embodied) are provided by the client, but we are free to decide about the algorithms used by the agents to solve the problem. The client expects the good solution to satisfy a certain quantitative criterion based on the values accumulated by the agents. For example, this criterion can require that the solution maximize a given algebraic function of player's accumulated rewards. In this case, which of the existing MAL algorithms and their corresponding optimality criteria will we choose?

One solution would be to run each algorithm on the given problem, observe the results, and pick the best. However, such an approach can be time and ressource expensive, and does not guarantee optimality. Another approach is to use a learning algorithm capable of solving problems in SPMAS in a way to directly satisfy functional criteria.

These functional criteria are opposed to such criteria as (a) and (b), which we call "relational", meaning that they are defined by taking into account relations between the values accumulated by each individual agent. In this case, the absolute values themselves *are secondary*. For example, a joint-strategy of multiple players is said to be a Nash equilibrium (criterion a) if the expected reward of *each player* is maximized given that *the other players* have their strategies fixed. In a similar manner, a joint-strategy is said to be Pareto-efficient (criterion b) if by changing this strategy so as to increase the expected value of *any subset of players*, there will necessarily be *a player out of this subset* whose value decreases. The same reasoning is applicable to a number of other relational optimality criteria (for example, correlated equilibrium [5]).

In this paper, we propose an approach to multiagent learning in repeated game self-play when the goal is to satisfy a given functional optimality criterion. We show that in such a setting our learning algorithm, called *Anytime Self-play Learner*, is (i) a better choice than a whole family of equilibrium and Pareto-efficient strategy learning algorithms, and (ii) anytime, i.e., the quality of solution increases gradually with the number of repeated game plays; thus the learning process terminated at an arbitrary moment still results in a reasonably good agents' behavior.

2   Formal Notions

For simplicity of exposition, our presentation will be given for two-player repeated game case. Extensions to an n-player setting (for an arbitrary $n > 2$) as well as to the multistate problems are also possible but omitted due to space limits.

2.1   Matrix Games and Their Solutions

A finite repeated two-player general-sum matrix game Γ (henceforth, a repeated game) consists of a set P of two players, p and q, with $(p, q) \in \{(1, 2), (2, 1)\}$. Player p has a finite number $M^p \in \mathbb{N}^+$ of actions it can choose from. The game is played iteratively. At iteration $i = 1, 2, \ldots$, each player p chooses an action $a_i^p \leq M^p$ and the vector $\mathbf{a}_i = (a_i^p, a_i^q) \in A$ gives a *joint action*. A is called the joint action space of players. For each player p there is a M^p-by-M^q matrix R^p defining the real-valued reward of that player after playing a joint action $\mathbf{a}_i$.

To choose an action from M^p at any iteration, each player uses its *strategy*. A strategy can be viewed as a function mapping the player p's current internal states into actions. A player's strategy can be *stationary* or *non-stationary*. Let π_i^p denote the rule, following to which player p chooses its action at iteration i. Then, p's strategy π^p is called stationary if $\pi_i^p = \pi_0^p$, $\forall i$. This means that p's strategy does not depend on current iteration, or, in other words, that it cannot change with time. Otherwise the strategy is called non-stationary.

A strategy profile $\pi = (\pi^p, \pi^q)$ is a joint strategy of players. To compare strategies and strategy profiles between them one can assign a metric to a strategy. We are using the expected limit of the means (ELM) metric. ELM assigns a unique value to an expected sequence of rewards that is obtained by a player following a given strategy profile π in an infinite sequence of iterations:

$$u^p(\pi) = E_\pi \left[\lim_{T \to \infty} \frac{1}{T} \sum_{i=1}^{T} R^p(\pi_i) \right] \tag{1}$$

In the above equation, $u^p(\pi)$ is the ELM value of strategy π. $R^p(\pi_i)$ denotes the expected immediate reward obtained by player p at iteration i if both players follow the strategy π at that iteration.

Nash equilibrium is a strategy profile $\hat{\pi} = (\hat{\pi}^p, \hat{\pi}^q)$ such that the following condition holds:

$$u^p(\pi^p, \hat{\pi}^q) \leq u^p(\pi) \text{ and } u^q(\hat{\pi}^p, \pi^q) \leq u^q(\pi), \forall \pi^q \neq \hat{\pi}^q, \pi^p \neq \hat{\pi}^p \tag{2}$$

Let $M(\Gamma)$ denote the set of all strategy profiles of game Γ. A Pareto-efficient solution of Γ is a strategy profile $\bar{\pi} \in M(\Gamma)$ such that the following condition holds:

$$u^p(\pi') < u^p(\bar{\pi}) \text{ or } u^q(\pi') < u^q(\bar{\pi}), \forall \pi' \neq \bar{\pi} \tag{3}$$

2.2 Optimality Criteria

The equations (2–3) define two relational optimality criteria discussed in the previous section. And as we claimed above, there are tasks where a use of relational criteria is not justified from a practical standpoint. In such environments, we would prefer agents to learn (and to use thereafter) strategies maximizing some mathematical function of their utility. This, functional, optimality criterion, depending on the task, can be based on such functions as *max*, *sum*, *product* or any other desirable function of players' individual utilities. If the utility is defined using the ELM metric then the functional optimality criterion $u(\pi)$ for a strategy profile π can be defined as $u(\pi) = \text{Op}_p(u^p(\pi))$, where $u^p(\pi)$ is the utility of player p defined using equation (1). In the latter equation, Op denotes a certain mathematical operator. For a given problem, it needs to be replaced by max, $\sum$, $\times$ or any required function.

2.3 Self-play

We *explicitly* focus on the self-play setting; and this is a *controlled* self-play, not an accidental coincidence of learning algorithms of agents. The latter property differs our approach from the main body of modern multiagent learning research proposing algorithms whose *behavior is justified for* (or *examined in*) self-play. Recall that by definition, self-play is a MAS setting in which *all* agents are *identical*. Until now, it has been typically assumed that agents' *algorithms*, or, in other words, rules of strategy update when learning, are identical. Other properties that can also be identical, such as (i) initial knowledge of agents, (ii) their utility metrics and (iii) optimality criteria, have escaped the attention of researchers. In this paper, we aim to fill this gap.

More precisely, in our controlled self-play scenario, which we call CSPMAS, we assume that both players, (1) *use the same learning algorithm*, (2) *have internal variables initialized with the values known to both players*, (3) *use the same utility metric* and (4) *optimize the same functional criterion*. We claim that in any *controlled* self-play scenario, Assumptions 2–4 are as well natural as Assumption 1, which is made in many previous multiagent learning papers [2,1,3,5,4]. In particular, this means that in any SPMAS,

Assumption (1) can intentionally be satisfied $\iff$ Assumptions (2–4) can intentionally be satisfied.

Also, we assume that players can observe each other's actions and their own rewards after a joint action is executed. This is as well a common assumption for many of MAL algorithms [2,3,4,6]. To stay as much general as possible, in our approach we assume that the reward observed by an agent p, $\forall (p, q) \in \{(1, 2), (2, 1)\}$, at iteration i after playing a joint action $\mathbf{a}_i$, is not necessarily deterministic. We only suppose that $\forall \mathbf{a} \in A$, the reward $R_i^p(\mathbf{a})$ observed after playing a joint action $\mathbf{a}$ is an instantiation of a random variable with certain mean and variance that do not change with time.

2.4 Information and Communication

Two important questions characterizing any MAS are (i) whether the agents know their own reward function and the reward function of the other agent, and (ii) whether communication between agents is available during learning. In this paper, we assume that the answer to both questions is *No*. Indeed, an affirmative answer to the first question makes the learning unnecessary, since the agents can compute an optimal joint strategy using the reward matrices. Accordingly, if the agents can communicate, the simplest scenario is to explore the reward structure of the game by executing joint actions one by one. Then, using communication, agents are able to share the acquired data. This, again, will make a further learning unnecessary.

3 Extended Strategies

To present our new algorithm, we first need to discuss one important implication of using the product criterion: emerging of strategies extended in time, or simply "extended strategies". These non-stationary strategies are known to be able to maximize the product of players' individual utilities to a greater extent than any stationary strategy [7]. As we will demonstrate later, our Anytime Self-play Learner algorithm is able to learn and execute extended strategies.

When two players p and q play a joint action $\mathbf{a} = (a^p, a^q)$ their rewards can be visualized as a point $\mathbf{x} = (x^p, x^q) = (R^p(a^p, a^q), R^q(a^p, a^q))$ in a two-dimensional space.[2] Let the set X contain all such points: $X = \{(R^p(a^p, a^q), R^q(a^p, a^q)) : a^p \leq M^p, a^q \leq M^q)\}$. Players can achieve any point in X as their ELM values by playing the corresponding joint action at every iteration. The convex hull of the set X contains all points that can be obtained as a linear combination of a subset of points of X. It is easily observable that the points laying on the boundary of the convex hull are always constructed as a linear combination of only two points of X. In terms of players' strategies, a point $\mathbf{z}$ on the boundary (recall that a point in X is a vector of players' ELM values) can be achieved by the players by playing a joint action, corresponding to a certain point $\mathbf{x}$, a w-fraction of all iterations, and by playing another joint action, corresponding to a point $\mathbf{y}$, the $(1 - w)$-fraction of all iterations (where w, $0 \leq w \leq 1$, defines the coefficient of linear combination).

Definition 1. *Given $l \in \mathbb{N}^+$, $0 \leq w \leq 1$, $\mathbf{a} \in A$ and $\mathbf{b} \in A$, an Extended Joint Action (EJA) is a joint strategy in which players play $\mathbf{a}$ during the first $k = \lceil l \cdot w \rceil$ iterations and $\mathbf{b}$ during the following $l - k$ iterations.*

Definition 2. *An Extended Joint Strategy (EJS) is an EJA repeated infinitely often.*

We call l the *length* of an EJA and k its *switch point*. Notice that for each $\mathbf{z}$ obtained as a combination of two points $\mathbf{x}$ and $\mathbf{y}$ from X, $\exists$ an EJS with certain $\mathbf{a}$, $\mathbf{b}$, l and w.

When the product criterion is used, the boundary of the convex hull is of a particular interest because the point $\mathbf{z}$ maximizing this criterion is always found on the boundary [8]. For two given points of X, $\mathbf{x} = (x^p, x^q)$ and $\mathbf{y} = (y^p, y^q)$, forming an edge of the boundary, the value of w maximizing the product criterion on this edge can be computed as follows [7],

$$w = \frac{-y^q(x^p - y^p) - y^p(x^q - y^q)}{2(x^q - y^q)(x^p - y^p)} \tag{4}$$

If $w < 0$ or $w > 1$, the maximum is achieved at respectively $\mathbf{x}$ or $\mathbf{y}$. To find w^* maximizing the product criterion, it is only required go over all pairs of points of X, compute w using Equation (4) and then pick a pair $\mathbf{x}^*$ and $\mathbf{y}^*$ of points for which w is maximized.

As one can note, an EJS will achieve the optimal ELM value defined by w^*, $\mathbf{x}^*$ and $\mathbf{y}^*$ only when $l \to \infty$. Let us show that as $l \to \infty$ the error induced by using a finite value of l rapidly decreases.

Proposition 1. *Let R denote the ELM value of an optimal point $\mathbf{z}$ on the boundary of X defined by the values w^*, $\mathbf{x}^*$ and $\mathbf{y}^*$ found as described above. Let l be the length*

---

[2] In this subsection, we use a simplified notation introduced by Littman [7]. According to it, $\mathbf{x} = (x^p, x^q)$ and $\mathbf{y} = (y^p, y^q)$ denote the vectors of players' rewards for two different joint actions viewed as points in a two-dimensional space.

of an EJA defined for two joint actions **a** *and* **b** *from A corresponding to the points* $\mathbf{x}^*$ *and* $\mathbf{y}^*$ *from X. Let* $\tilde{R}$ *denote the ELM of this EJA. Let* $\epsilon = \tilde{R} - R$ *define the error of using* $l < \infty$ *in this EJA. Then* $\epsilon \to 0$ *as* $l \to \infty$.

Proof. We have $l^2 R = (lwx^p + (l - lw)y^p)(lwx^q + (l - lw)y^q)$ and $l^2 \tilde{R} = (\lceil lw \rceil x^p + (l - \lceil lw \rceil)y^p)(\lceil lw \rceil x^q + (l - \lceil lw \rceil)y^q)$. We know that for any natural x, $\lceil x \rceil < x + 1$. Thus, we can write that $l^2 \tilde{R} < ((lw+1)x^p + (l - (lw+1))y^p)((lw+1)x^q + (l - (lw+1))y^q)$. The difference between $l^2 \tilde{R}$ and $l^2 R$ is then bounded as follows: $l^2 \tilde{R} - l^2 R < l(x^p - y^p)(x^q - y^q)(2w - 1) + 2x^p x^q - x^p y^q - y^p x^q$. Since $l > 1$, the error $\epsilon = \tilde{R} - R$ is bounded as follows: $\epsilon < \frac{(x^p - y^p)(x^q - y^q)(2w - 1)}{l} + \frac{2x^p x^q - x^p y^q - y^p x^q}{l^2}$.

Therefore, as l tends to ∞, ϵ tends to 0 with a rate inversely proportional to l.

When l and w^* are known to the players (i.e., defined by the designer of the MAS before to start learning) they are able to construct the EJS maximizing, to the extent of the error induced by using a finite value of l, the product criterion.

4   Anytime Self-play Learner

In this section, we present our new algorithm called Anytime Self-play Learner (ASPL). Its main steps are given in Algorithm 1.

Algorithm 1. Main steps of Anytime Self-play Learner

---

1. While exploring
   (a) Play an exploration action,
   (b) Observe the reward R,
   (c) Replay the same action proportionally to R,
   (d) Update counters of the other player's play.
2. While exploiting
   (a) Optimize according to the criterion and counters,
   (b) Play optimally.

4.1   Internal Variables

Our algorithm has one internal variable that needs to be initialized to all agents with the same value before the learning is started. This variable, called R_{max}, reflects the maximum utility that an agent can obtain in the game. In many practical tasks, this value can be set by the designer depending on how the utility is defined. E.g., for floor cleaning robots this value can be set based on the maximum possible surface one robot can clean given the initial volume of detergent in its tank. It is assumed that $R_{max} \geq R^p(a^p, a^q)$ for any player p and for all $a^p \leq M^p$ and $a^q \leq M^q$. I.e., R_{max} does not underestimate any of the rewards of players.

During learning, an ASPL agent p also maintains several other variables. The variables $K^p(a^p, a^q)$, $\forall a^p \leq M^p, a^q \leq M^q$, reflect the number of times a particular joint action (a^p, a^q) has been played. The variables $L^p(a^p, a^q)$, $\forall a^p \leq M^p, a^q \leq M^q$, reflect the number of times that the action a^q has been played by q at the iteration $i + 1$ following an iteration i at which players were playing (a^p, a^q).

4.2 Exploring

To exhibit a good joint behavior, players obviously need to explore the game they play. More specifically, an agent not only needs to learn its own rewards for joint actions, which, as we noticed, can be an instantiation of an unknown random variable. Also, in order to optimize the functional criterion, it needs to have "an idea" of the reward function of the other agent.

For this purpose, the algorithm proceeds as follows. Both ASPL players are explicitly synchronized (this is an advantage of self-play). At each odd iteration i, an ASPL player p randomly uniformly plays an action a_i^p, and observes its own reward R_i^p and the action a_i^q played by the other player. Then it updates its estimate $\tilde{R}^p(\mathbf{a}_i)$ of the reward $R^p(\mathbf{a}_i)$ for the joint action $\mathbf{a}_i = (a_i^p, a_i^q)$ as follows:

$$\tilde{R}^p(\mathbf{a}_i) = \tilde{R}^p(\mathbf{a}_i) + \frac{1}{k_{\mathbf{a}_i} + 1}\left(R_i^p - \tilde{R}^p(\mathbf{a}_i)\right) \tag{5}$$

where $k_{\mathbf{a}_i}$ represents the number of times the joint action $\mathbf{a}_i$ has been played so far. Finally, player p increments its counter $K^p(\mathbf{a}_i)$.

At the next iteration, $i + 1$, player p replays the action a_i^p with probability $\delta = R_i^p/R_{max}$. Otherwise, with probability $(1 - \delta)$ player p plays a random action different from a_i^p and picked uniformly from the remaining actions. Player p then observes the action played by player q, updates again its reward estimate for the played joint action using Equation (5), and, finally, updates its counter $L^p(\cdot)$ accordingly.

4.3 Exploiting

As soon as when exploring both players follow the same procedure, therefore, at any moment of time, player p can assume for any joint action $(a^p, a^q) \in A$ that $\frac{L^p(a^p,a^q)}{K^p(a^p,a^q)}$ is an unbiased estimator of the unknown reward function of player q. More precisely, it can believe that,

$$\frac{R^q(a^p, a^q)}{R_{max}} \approx \frac{L^p(a^p, a^q)}{K^p(a^p, a^q)}$$

Indeed, since the value R_{max} is the same and is known to both players, and as the players replayed their action a_i^p, played on the previous iteration i, according to the proportion $R^p(a_i^p, a_i^q)/R_{max}$, the values of counters $L^p(a^p, a^q)$ and $K^p(a^p, a^q)$ can give to player p a good estimate of the real value of $R^q(a^p, a^q)$. More precisely, player p can compute an estimate of the other player rewards as follows,

$$\tilde{R}^q(a^p, a^q) = \frac{L^p(a^p, a^q)R_{max}}{K^p(a^p, a^q)} \tag{6}$$

By so doing, it becomes possible to compute the strategy maximizing any given functional criterion and to execute this strategy thereafter. For example, if the functional criterion is sum, the optimal strategy can be computed by the players as,

$$\pi_i^p = \operatorname*{argmax}_{a^p:(a^p,a^q)\in A} \left(\tilde{R}^q(a^p, a^q) + \tilde{R}^p(a^p, a^q)\right) \forall i \tag{7}$$

Similarly, if the functional criterion is max, the optimal strategy can be computed as,

$$\pi_i^p = \underset{a^p:(a^p,a^q)\in A}{\operatorname{argmax}} \left(\max(\tilde{R}^q(a^p,a^q), \tilde{R}^p(a^p,a^q)) \right) \forall i \tag{8}$$

Finally, if the functional criterion is *product*, the optimal strategy can be computed as shown in Algorithm 2.

Algorithm 2. Procedure to find the optimal strategy for the *product* criterion

---

1. For all pairs of points $\mathbf{x}$, $\mathbf{y}$ from the set X, such that, $\mathbf{x} = (\tilde{R}^p(a^p,a^q), \tilde{R}^q(a^p,a^q))$ and $\mathbf{y} = (\tilde{R}^p(b^p,b^q), \tilde{R}^q(b^p,b^q))$ (where $a^p, b^p \leq M^p$ and $a^q, b^q \leq M^q$) compute w using Equation (4) and construct the corresponding EJS using Definitions 1–2.
2. Pick the EJS having the highest ELM value (according to the *product* functional optimality criterion).

---

5   Explore, Exploit and Coordinate

As one could remark, ASPL can be viewed as an anytime algorithm. This means that every two iterations the agents improve their estimates of the reward functions. If the learning (exploring) is stopped after a certain iteration, the players are able to choose the best strategy as yet and to play on it. However, in a general case two important issues need to be resolved in order to assure a good joint performance. These can be formulated as two questions: (1) when to stop exploring and start exploiting (known as the exploration-exploitation dilemma [9]) and (2) how to choose a coordinated joint strategy (i.e., the coordination problem [2]).

While the exploration-exploitation dilemma is a well-known problem in machine learning, several words need to be said about the coordination problem in MAS. Let suppose we have a game as follows,

$$R^{1,2} = \begin{pmatrix} 1,2 & 0,0 \\ 0,0 & 2,1 \end{pmatrix}$$

After a certain number of exploration iterations, players have certain estimates of their reward functions. Two points, $(1,2)$ and $(2,1)$, in X have the same ELM value. However, as the number of exploration iterations is always finite, the players have an error in estimates of their own and each other's rewards. Therefore, the strategies computed by the players independently can belong to different joint strategies. For example, player 1 can decide that the optimal strategy is to play the row 1, expecting to see the outcome $(1,2)$, but player 2 will play the column 2 foreseeing the outcome $(2,1)$. As the result, they will collect the suboptimal outcome $(0,0)$.

In practice, the designer of CSPMAS can sufficiently know the environment in order to be able to apply the "explore-then-exploit" principle [10]. More precisely, this means to separate the learning process into two phases (exploration and exploitation) and to fix the length of the exploration phase before the learning starts. For example, the room to clean by two robots can be the same, but the positions of chairs and tables can change. In this case, the designer can know that the agents need to continuously explore during the first T time steps, and then they can synchronously switch to exploit. The coordination problem can also be easily solved then. For instance, one agent can be assigned to play the role of *leader*, i.e., to choose the best joint strategy. The second agent, in turn, will act as *follower*, i.e., it will observe the strategy followed by the leader at the previous iteration and adapt to it. Such roles can be assigned at random since the ELM value of a joint strategy does not depend on a particular choice of leader and follower.

5.1 Mixed Exploring and Exploiting

On the other hand, if the designer is not able to fix in advance the length of the exploration phase, he can prefer to mix exploration and exploitation in a certain way, and let agents "decide" online. One technique to do this involves using a GLIE (for "greedy in the limit with infinite exploration") learning strategy. An example of such a strategy is Boltzmann exploration strategy:

$$\pi_i^p(a^p) = \frac{e^{\beta_i Q_i(a^p)}}{\sum_{b^p \leq M^p} e^{\beta_i Q_i(a^p)}}$$

where $\pi_i^p(a^p)$ denotes the probability with which player p chooses to play the action a^p at each odd iteration i, β_i is an exploration factor slowly increasing with time (Singh et al. [11] show how it can be defined in order to assure convergence in the limit). Q_i can be viewed as a certain current preference of agent p over its actions. In our experiments, to solve the coordination problem, we compute $Q_i(a^p)$ at each odd iteration using a heuristic called Combined Optimistic Boltzmann (COB) proposed by Claus and Boutilier [2]. It consists of assigning higher values to the "promising" strategies weighed by the likelihood to be simultaneously chosen by the other player. As the exploration decreases (with the increasing factor β) this heuristics permits the agents to effectively coordinate on the same joint strategy and to play on it most of the time. Notice that there exist more efficient methods (in terms of exploration time and cost) to mix exploration and exploitation, such as a Bayesian approach [12].

6 Experimental Results

It is only fair to compare a new algorithm with the existing ones if it uses the same or relaxed assumptions and is searching for the same kind of solution. In our case, there is no other algorithm capable of learning strategies optimizing functional criteria in MAS (two exceptions and their limitations are discussed in Section 7). On the other hand, there exist a number of MAL algorithms, as those cited above, which, while using different assumptions, converge to the same kinds of relational solutions like Pareto-efficient or Nash equilibrium. So, in our case we will indirectly compare our algorithm with all these algorithms by comparing the ELM value of the solution found by ASPL with the corresponding values (according to the same criterion) of different relational solutions. The goal of this comparison is to demonstrate that when the goal of the designer is to satisfy a given functional optimality criterion, ASPL is the best choice.

We empirically tested ASPL on two different testbeds. The first series of testbeds, called "Random Games M" (or, RGs M, for short), contains randomly generated two-player repeated games with the number of player actions, $M = M^p = M^q$, equal respectively to 2, 3, 5 and 10. In each game from Random Games M, the rewards of players are integer values uniformly distributed between 0 and 100, and new values are generated each time a game is started.

The second testbed, called "Conflict Games" (CGs), contains 57 games listed by Brams [13]. These are two-player two-action repeated games whose rewards are integer values between 1 and 4. These games were called "conflict" because contain no outcome that simultaneously maximizes the ELM value of both players. Conflict Games are especially suitable to make a comparison of solutions computed by ASPL for different functional criteria with other possible solutions usually found by other MAL algorithms in self-play (e.g., Nash equilibrium and Pareto-efficient solution).

We conducted our experiments in the following way. From each testbed, a game was randomly picked and played during 100,000 iterations. This process (called an *experiment*) was repeated 100 times and then the obtained data were averaged. Table 1 presents the ELM values of the strategies to which ASPL players converge in different games. The alternative values are respectively (APE column) the average utility (in terms of the corresponding ELM value) of all pure stationary Pareto-efficient solutions and (ANE column) the average utility of all stationary Nash equilibria found by the Lemke-Howson algorithm. We did not compare the ASPL's solution with non-stationary Pareto-efficient solutions because there are no algorithms whose convergence to such kind of solution was proved in a non-special case (one exception is discussed in Section 7). As one can see, in both testbeds the solution found by ASPL outperforms all other solutions of those games. The advantage of ASPL is especially pronounced if the functional optimality criterion is *product*. In this case, ASPL often converges to an extended strategy, which is typically more effective in optimizing this criterion.

Table 1. Utility of ASPL for different function optimality criteria compared to the utilities of other solutions

max				*sum*				*product*		
	ASPL	APE	ANE		ASPL	APE	ANE		ASPL	APE	ANE
CGs	4.00	3.62	3.44	CGs	6.41	6.29	6.02	CGs	10.36	8.29	9.09
RGs 2	88.71	82.11	80.02	RGs 2	140.73	126.88	128.10	CGs 2	5179.85	4512.07	4537.66
RGs 3	94.68	88.85	84.19	RGs 3	161.14	151.02	150.34	CGs 3	6555.84	4190.74	4672.07
RGs 5	98.18	94.29	87.83	RGs 5	174.05	162.14	157.98	CGs 5	7715.53	5432.45	5652.87
RGs 10	99.46	96.60	92.68	RGs 10	187.15	182.09	175.93	CGs 10	8745.05	6466.64	6305.67

The curves of Figures 1 (a and b) reflect the evolution of the ELM value during learning in games from Random Games M. For each learning iteration, the curves present the current ELM value[3] according to the *sum* and *product* functional criteria (the curves for *max* look similar and were omitted due to the space limits). We can observe that for each functional optimality criterion, the ELM value of ASPL becomes close to the optimal one after a reasonably small number of learning iterations.

---

[3] These values were averaged over 100 experiments.

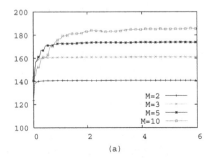 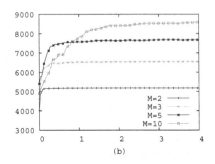

(a)   (b)

Fig. 1. The evolution of ELM value of the learned policy in Random Games M according to different functional criteria: (a) *sum* and (b) *product*. The X axis represents learning iterations ($\times 10^3$); the Y axis represents ELM value.

Both "explore-then-exploit" and "mixed exploring-exploiting" strategies of exploration/coordination performed well in our experiments. The only concern about the latter strategy is that GLIE principle guarantees the convergence to an optimal behavior only in the limit. Further, since COB is a heuristic, in a relatively small number of runs agents could coordinate on a suboptimal strategy. However, we observed that even in such rare cases, agents still chose a joint strategy "close" to an optimal one and never the worst one.

7 Related Work

Here, we would emphasize two related works. In the first one, by Greenwald [5], the desired solution of the learning problem is correlated equilibrium. When several equilibria are possible, the author proposes to choose a unique one by using an "objective function", an analog of our functional criterion. However, this implies that agents have two opposite goals: (1) to be selfish (inclination to equilibrium solution implies the agents to be selfish) and (2) to want to sacrifice, by selecting, using the objective function, an equilibrium, which is probably sub-optimal to itself. Generally, if the agents are supposed to want to sacrifice, there is no need in seeking after an equilibrium solution.

In the second work, Crandall and Goodrich [6] propose an approach to the learning of multi-step strategies, analogous to our extended strategies. When the length l of an extended joint action is fixed to 1 (the only value used in their experiments), this yields in a relatively small number of learning iterations. However, by increasing l (to allow more complex extended joint actions and thereby obtain other values inside the convex hull) the number of joint strategies to explore becomes exponentially large, but only a (very) small number of them is really interesting. In our approach, we find the best extended joint strategies directly, i.e., without enumeration of all pairs of action sequences of length l. Besides, the approach of Crandall and Goodrich does not permit satisfying a given functional criterion.

8   Conclusions

In this paper, we presented a novel approach to learning in self-play. We argued that when the learning problem is a controlled self-play, a good learning algorithm should get additional benefit from this. E.g., the agents can have internal variables initialized with the same values, use the same utility metric and optimize the same function. We then presented the notion of functional optimality criterion, as opposed to relational optimality criteria such as Nash equilibrium. We pointed out that the solution of a problem found by an algorithm seeking to satisfy a relational criterion can be suboptimal if a functional optimality criterion needs to be satisfied. We demonstrated that in such problems, our algorithm is a better choice than the classical algorithms for self-play.

References

1. Bowling, M., Veloso, M.: Multiagent learning using a variable learning rate. Artificial Intelligence 136(2), 215–250 (2002)
2. Claus, C., Boutilier, C.: The dynamics of reinforcement learning in cooperative multiagent systems. In: Proceedings of AAAI 1998 (1998)
3. Hu, J., Wellman, M.: Nash Q-learning for general-sum stochastic games. Journal of ML Research 4, 1039–1069 (2003)
4. Banerjee, B., Peng, J.: Performance bounded reinforcement learning in strategic interactions. In: Proceedings of AAAI 2004 (2004)
5. Greenwald, A.: Correlated-Q learning. In: AAAI Spring Symposium (2003)
6. Crandall, J., Goodrich, M.: Learning to compete, compromise, and cooperate in repeated general-sum games. In: Proceedings ICML 2005 (2005)
7. Littman, M., Stone, P.: A polynomial-time Nash equilibrium algorithm for repeated games. Decision Support Systems 39(1), 55–66 (2005)
8. Nash, J.: The Bargaining Problem. Econometrica 18(2), 155–162 (1950)
9. Sutton, R.S., Barto, A.G.: Reinforcement Learning: An Introduction. MIT Press, Cambridge (1998)
10. de Farias, D., Megiddo, N., Cambridge, M., San Jose, C.: Exploration-Exploitation Trade-offs for Experts Algorithms in Reactive Environments. In: Advances in Neural Information Processing Systems 17: Proceedings of The 2004 Conference. MIT Press, Cambridge (2005)
11. Singh, S., Jaakkola, T., Littman, M., Szepesvári, C.: Convergence Results for Single-Step On-Policy Reinforcement-Learning Algorithms. Machine Learning 38(3), 287–308 (2000)
12. Chalkiadakis, G., Boutilier, C.: Coordination in multiagent reinforcement learning: A bayesian approach. In: Proceedings of the Second International Joint Conference on Autonomous Agents and Multiagent Systems (AAMAS 2003), Melbourne, Australia (2003)
13. Brams, S.: Theory of Moves. American Scientist 81(6), 562–570 (1993)

Author Index